Systems Engineering & Management

Volker Ahrens

Systems Engineering & Management

A generic approach to the design and control of socio-technical systems

Springer Vieweg

Volker Ahrens
Elmshorn, Germany

ISBN 978-3-658-46874-3 ISBN 978-3-658-46875-0 (eBook)
https://doi.org/10.1007/978-3-658-46875-0

Translation from the German language edition: "Systems Engineering & Management" by Volker Ahrens, © Der/die Herausgeber bzw. der/die Autor(en), exklusiv lizenziert an Springer Fachmedien Wiesbaden GmbH, ein Teil von Springer Nature 2024. Published by Springer Fachmedien Wiesbaden. All Rights Reserved.

This Springer Vieweg imprint is published by the registered company Springer Fachmedien Wiesbaden GmbH, part of Springer Nature.
The registered company address is: Abraham-Lincoln-Str. 46, 65189 Wiesbaden, Germany

- for Uta -

Acknowledgements

The genesis of this book follows the approach described in it. As is so often the case, systematic thinking and action have proven their worth in organizing the insights and experiences that have flowed in. Above all, however, institutions and people in my personal and professional environment have contributed significantly to the success of this work.

First and foremost, I would therefore like to thank Springer Verlag for taking on the printing and using its good reputation to disseminate the results of my work. My special thanks go to Dipl.-Ing. Eric Blaschke, who made a significant contribution to the success of this work as an editor. I would also like to thank the readers for their interest in a prominent position.

The beginnings of my work on this book go back to my studies of mechanical engineering at the Leibniz University of Hanover with a focus on production technology and a special interest in ergonomics. Even back then, I got to know people from a wide range of disciplines and made friends for life, who still provide me with access to their respective professional perspectives today. These include, in particular, the pastor Angelika Breymann and her husband and theologian Dr. Hans-Hermann Tiemann, as well as the pastor Hans-Jürgen Pabst. The next generation now includes my godchild Paul Hendrik Tiemann, M.Sc., who is currently studying for a doctorate in electrical engineering and is not only a brilliant professional, but also socially and ecologically committed. And my son Benedict, who is studying political science and history, has already left his mark in terms of content. Finally, Dipl.-Psych. Rainer Meiser, a psychologist and successful management consultant who combines his approach to people with a keen eye for business and society, has also joined us. I would like to take this opportunity to express my sincere thanks to all of them for their support, both in terms of content and morale.

Special thanks also go to my esteemed colleague Prof. Dr.-Ing. Burkhard Schmager, Head of the Engineering & Management program at the Ernst Abbe University of Applied Sciences in Jena and a member of the German Accreditation Council, who reviewed the manuscript and provided very valuable suggestions. And I would also like to take this opportunity to express my particular gratitude to my now unfortunately former Research Assistant, Dipl.-Inform. Martin Hieronymus, because without his tireless support over many years, this book would never have been written.

I would like to dedicate this book to my partner, Prof. Dr. Uta Gaidys, Head of the Department of Nursing and Management at the Hamburg University of Applied Sciences and member of the German Council of Science and Humanities.

Declarations

Competing Interests The author has no conflicts of interest to declare that are relevant to the content of this book.

The author is a professor at a private, state-recognized university of applied sciences in Germany. He teaches Systems Engineering, Production and Quality Management, Logistics and Process Management, and Project Management. In addition, he heads the degree programs in Engineering & Management (B.Sc.) and International Engineering & Management (B.Eng.), which is conducted in cooperation with a college in the United States. Finally, he coordinates the engineering sciences department.

The author is a board member of the Regional Group North of the REFA Federal Association for Work Design, Business Organization and Corporate Development, the oldest German organization in this field (founded in 1924). He is also a member of the F 204 "Modeling and Simulation" technical committee of the Society for Production and Logistics of the Association of German Engineers (VDI) and he is a member of the Lower Saxony Chamber of Engineers.

Ethics Approval This study was performed in line with the principles of the Declaration of Helsinki.

Prologue

In view of numerous fundamental works on Systems Engineering (SE) such as those by Reinhard Haberfellner et al. (2018), Joe Jenney et al. (2010), Alexander Kossiakoff et al. (2011) or Petra Winzer (2016), the handbook of the International Council on Systems Engineering (INCOSE) (Walden et al. 2023) and a whole series of other treatises on special areas such as Model-based Systems Engineering (MbSE) (e.g. Weilkiens 2014) or Requirements Engineering (e.g. Ebert 2019), the need for books in this field seems to be covered. Therefore, the publication of another fundamental book requires justification. Five reasons are presented below.

(1) The *division of labour* and *specialization* associated with the *industrial revolution* has led to a significant increase in *efficiency* and *effectiveness*, not only in practice but also in science. In this respect, it seems prudent to continue along this path. However, like everything else, this development does not only have advantages. One of the disadvantages is that connections between the specialized domains can fall by the wayside. This can and should be avoided through interdisciplinarity. In this respect, everyone is probably in agreement. However, there is not always agreement when it comes to striking a balance between specialization and generalization. For example, the claim of protagonists of *interdisciplinarity* such as Günter Ropohl (2009, p. 72) to counteract what he calls the atomization of scientific knowledge by striving for a new unity of science in the sense of a mathesis universalis goes too far[1]. And the conviction that Norbert Wiener (1961, p. 26 f.) shares with his close companion Arturo Rosenblueth, namely that the most fruitful areas for the flourishing of science are those that are neglected as no man's land between the specialized disciplines, but that it is precisely these border areas of established sciences that are the most unruly, but would offer the qualified researcher the richest opportunities, probably overshoots the mark a little.

On the other hand, Günter Ropohl (ibid., p. 73) is of course right when he states that problems in practice do not follow the division of subjects at universities and that their solution requires not only the contribution of many different specialist areas, but also the connections between them. However, it is not enough to occasionally think outside the box,

[1] Friedrich Dürrenmatt (1985) has already anticipated what the discovery of the world formula would entail with his physicists. In this respect, such an aspiration does not seem worth striving for anyway.

as they say, from the perspective of a specialist discipline. Interdisciplinarity is, even if this appears to be a contradiction at first glance, a special discipline in its own right[2].

Combined disciplines such as *Engineering & Management* are already picking up on this insight and moving in this direction (FFBT/VWI 2019, p. 21). However, they have so far largely limited themselves to training and practice - and have been very successful at that. Specific research funding programs, on the other hand, are usually sought in vain. Industrial engineers can look to engineers for this, but then have to reckon with not being recognized as quasi half engineers, and they usually have the same fate in the economic sciences. Against this background, this volume is intended to contribute to providing specific foundations for interdisciplinary training and practice as well as for research aimed at this.

The ability to enable and promote interdisciplinarity is primarily attributed to *systems theory* (Ahrens 2017, p. 10 ff.). In reality, however, attempts to include a range of other relevant disciplines, such as psychology and the social sciences in particular, if they are undertaken at all, often remain in their infancy. Günter Ropohl (2009, p. 71 ff.), for example, illustrates how great the difficulties can be in the struggle for an engineering position in relation to other systems theory perspectives across all three editions of his highly recommended work on a general systems theory of technology, many parts of which are taken as a basis here[3]. Other authors, including those mentioned at the beginning, almost completely dispense with such discussions.

There is no doubt that it is risky to enter unfamiliar territory. The reception is not always warm. Nevertheless, if the present volume attempts to incorporate findings such as those of socio-logical systems theory (e.g. Luhmann 1996a; Parsons 1951) and psychology based on systems theory (e.g. Watzlawick et al. 2013) to a greater extent than is the case in other works, this is already combined with a request to the representatives of these theories to be indulgent, as this will inevitably not do justice to their discipline-internal state of research in every respect. Above all, no contributions to basic research are to be expected here. Rather, the results of *application-oriented research* are presented, which are aimed at bringing basic theoretical knowledge to practical application.

In this respect, the relevant explanations are also intended as a guide to further and, in many cases, theoretically more in-depth works, the supplementary reading of which is expressly recommended for experienced readers. However, the explanations are deliberately not limited to purely practical instructions, as Günter Ropohl (2009, p. 73) rightly classifies the *scientification of practical problem solving* as one of four roots of contemporary systems thinking. The apparent complexity of the problems at hand cannot be mastered with anecdotal trivializations. In this respect, the considerations and insights presented here are also intended as a bridge between the theory of science and Systems Engineering & Management, which is primarily geared towards practical application.

---

[2] In this sense, there are many interdisciplinarities that depend on the disciplines involved. The focus here is on the disciplines of technology, sociology and psychology, which are occasionally supplemented by other disciplines such as mathematics or biology.

[3] Günter Ropohl (1979, p. 49) expressed his opinion as an engineer on Niklas Luhmann's social science systems theory in a downright scathing manner: Since this system-theoretical dilettantism annoyingly burdens the reputation of even the most serious systems theory in the scientific public, we must explain our position in the following with the greatest thoroughness in order not to fall prey to such critics who, informed about systems theory only in the aforementioned socio-philosophical distortion, beat the sack, although they can only mean the donkey.

(2) Furthermore, Systems Engineering sometimes finds it difficult to distinguish between its own field of activity of *design* (Engineering) and that of *control* (Management). For example, what Reinhard Haberfellner et al. (2018, p. 61 ff.) call macrology, which describes the design phase of the system life cycle[4]. (Sect. 3.1), ends with the end of the design phase. And even the widely known V model (e.g. Walden et al. 2023, p. 35 ff.) does not go beyond this. Conversely, those who are primarily concerned with the control of systems base the entire system life cycle on the control loop model, although this is only suitable for ongoing operation, and are surprised that the control is highly unstable in phases of change (design) (e.g. Küppers 2019, p. 62). This in turn results in the difficulty of getting to grips with the transition from design to control. No matter how hard *Change Management* tries at this point, the introduction of something new is still too rarely successful without often significant losses (e.g. Lauer 2014).

The *entire* system life cycle will therefore be considered here. The focus will be on the first phases of design through to commissioning, as this is and remains the preferred field of activity of Systems Engineering. However, the inclusion of the subsequent phase of controlling what was previously designed and introduced and what should not change significantly in the future, i.e. Systems Management (SM), has two advantages:

- On the one hand, design and steering are based on a *polarity* in the sense of Georg Wilhelm Friedrich Hegel (2013, p. 14), i.e. they are in a *complementary relationship* to each other. This in turn means that, on the one hand, they are mutually exclusive and cannot be reduced to each other *dichotomy*, but at the same time they also complement each other and can only be understood in their context. In addition, commissioning, i.e. the connection between the two poles, can be understood in a narrower sense as a *singularity* and, in a somewhat less pronounced form, as a transient phase, so that in this transitional phase from shaping to steering, laws apply that cannot be adequately described either by those of the preceding phase or those of the subsequent phase. Here, completely separate laws apply.
- On the other hand, the design objects are intended for ongoing operation once they have been commissioned. However, as already indicated, ongoing operation follows different laws than the design phase. However, since the design objects are intended to function according to these other laws, the designers must know these laws just as well and take them into account during the design phase as the laws according to which they themselves design. The often impressive state of science and technology can be taken as evidence that there are actually no recognizable deficits in this respect. However, this can be countered by the fact that numerous systems are not (or no longer) stable, including the current environmental conditions of the Holocene, which are particularly threatened by anthropogenic climate change, or the economic system, which is in danger of slipping away, including the international financial system. Against this backdrop, despite great respect for what has been achieved so far, one can certainly come to the conclusion that there is still a considerable need for action in terms of shaping and maintaining equilibria: As we all know, the better is the enemy of the good.

[4] This life cycle is often referred to as the product life cycle (e.g. Kaldasch 2013). With regard to the generalization intended here, which also includes natural systems, the narrowing down to products, i.e. artificial systems (artefacts), appears too narrow

(3) A third reason for adding the work presented here to the already existing and thoroughly recommendable works is the point of view that perhaps most strongly expresses the problem experiences of recent systems research (Luhmann 1996a, p. 45): *Complexity*. For the most part, this term is not even defined in more detail and thus fulfills the function of catalyzing common experiences (ibid.). The persistent difficulties in dealing with complexity can be explained, for example, on the basis of evolutionary epistemology, according to which human thinking is not phylogenetically designed to recognize complex relationships, but simply to survive in its own ecological niche (Riedl 1987, p. 79 ff., 294 ff.). This correlates with the *constructivism* on which this study is based, according to which humans have no access in principle to real reality[5], and certainly not to one that is complex[6].

Nevertheless, Systems Engineering & Management aims to make a contribution to mastering complexity (e.g. Haberfellner et al. 2018, p. 23; Walden et al. 2023, p. 25; Winzer 2016, p. 20). As a strategy, Reinhard Haberfellner et al. (2018, p. 35) suggests simply avoiding complexity and reducing it to simple systems as far as possible[7]. Petra Winzer (2016, p. 20) expressly agrees with this. If we follow evolutionary epistemology and constructivism, it is indeed impossible to avoid or reduce complexity. However, we should not make a virtue of this necessity for the time being. On the contrary, it follows from the basic cybernetic *law of requisite variety* (Ashby 1956, p. 206 ff.) that those who want to master complexity must have at least the same degree of complexity in the sense of a stock of variation as the system to be mastered, so that they have a reaction option ready for every situation in which the system may find itself (Heylighen and Joslyn 2001, p. 14).

There are now indications in biology that complexity can also be controlled in a very simple, even trivial way. For example, Physarum polycephalum, a slime mold (Myxogastria) from the Physarida order, is able to find the shortest path in a labyrinth (Nakagaki et al. 2000, p. 470), even though it consists of only a single cell. Such intelligent behavior is possible because although the organism does not have an internal brain, it does have one outside itself, namely in the form of a self-produced mucus that represents a kind of memory. After all, such findings have already been applied technically by using a slime mold for the central control of a robot (Tsuda et al. 2005, p. 202 ff.). Nevertheless, research in this direction is still in its infancy, meaning that it is not yet possible to derive strategies for mastering complexity that are based on simplification. Until such strategies are developed, we will have to make do with the basic cybernetic law of requisite variety (Ashby 1956, p. 206 ff.). This volume will therefore devote more attention to complexity and ways of mastering it than other works on Systems Engineering do.

(4) Furthermore, there are repeated efforts to standardize Systems Engineering (e.g. Walden et al. 2023; Winzer 2016, p. xi). Standardization is undoubtedly at the heart of the

---

[5] In everyday language, the term reality is used for this. Here, however, this play on words is deliberately chosen to draw attention to the fact that, in the light of the underlying constructivism, reality is not what people think it is when they speak of reality or, for example, objectivity.

[6] This insight can be found, for example, in the german national standard DIN 69901-1:2009-01 on Project Management (p. 6).

[7] That would be just like someone coming up with the idea of creating a fair tax system in a modern society in which a tax return fits on a beer mat. Such a Such a tax system would undoubtedly be simple, but would not do justice to a modern society's understanding of fairness.

design and use of technology: standards and guidelines are an integral part of engineers' everyday lives. And just as an aside, it should be mentioned that the exclusive offer of certified and, in particular, time-limited qualifications can also be used to develop very lucrative business models. Whoever succeeds in this should be granted it. Here, however, things are moving in the opposite direction, because standards also have their pitfalls. Particularly when they are linked to business models, they merely focus on the perspective of individual organizations, which is primarily intended to serve one purpose: to cement unique selling points. Standards and guidelines, on the other hand, are ideally developed by representatives of all those involved and affected (stakeholders). In this case, however, they represent the *lowest common denominator*: anything that cannot be agreed upon is left out or formulated so nebulously that everyone can find themselves in it. Neither the perspective of individual organizations nor the lowest common denominator of standards and guidelines promotes *innovation*. But innovation is the intention here. This is explicitly not about standardization, but about making a contribution to the progress of the SEM.

(5) In view of developments that attempt to distance themselves from the SEM, this seems entirely appropriate. This applies in particular to so-called *agile approaches* such as Scrum. They claim to be better and, in particular, faster and more flexible than established methodologies. The aim of the work presented here is to recapture the agile approaches, so to speak, as otherwise the *generic claim of systems theory* could not be upheld. For this to succeed, however, the social science and psychological findings already mentioned above under point (1) must be included, because in this respect agile approaches are actually a step ahead of the established SEM (e.g. Gloger 2016, p. 27 ff.). Only the consistent recourse to the general systems theory on which SEM is based makes it possible to make up for the partial lead and, if possible, even promote SEM back into a leading position. Agile approaches are by no means completely refuted by this, but they are further underpinned and harmonized with established methodologies such as SEM. Both SEM and agile approaches should benefit equally from this.

With these intentions, the following explanations go far beyond what is standardized in standards and guidelines such as ISO/IEC/IEEE 15.288:2023. Anyone looking for a kind of instruction manual for the *state of the art* in Systems Engineering is better served by the standards and guidelines themselves, by explanatory works such as that of the International Council on Systems Engineering (e.g. Walden et al. 2023) or by those of the other authors mentioned above. Here, on the other hand, the state of the art is also presented, but primarily with the aim of *further developing* the SEM, especially with regard to the further systematization of terms, process models, methods, procedures and tools,

- with the aim of merging previous approaches into an integrative SEM,
- with the aim of strengthening interdisciplinarity,
- with the aim of understanding complexity in the context of SEM even better than before,
- with the aim of overcoming paradigms such as process orientation and
- with the aim of reconciling agile approaches with SEM.

Contents

Acronyms

Acatech	German national academy of science and engineering
AHP	Analytic Hierarchy Process
AI	Artificial Intelligence
AIAG	Automotive Industry Action Group
ANSI	American National Standards Institute
ANT	Actor-Network Theory
APQP	Advanced Product Quality Planning
AOA	Activity-on-arc network
AON	Activity-on-node network
AS-i	Actuator Sensor Interface
B2B	Business to Business
B2C	Business to Consumer
BA model	Barabási-Albert model
bdd	Block Definition Diagram (SysML)
BBSR	German Federal Institute for Research on Building, Urban Affairs and Spatial Development
BCG	Boston Consulting Group
BCM	Business Continuity Management
BDI	Federation of German Industries
BGB	German civil code
BI	Business Intelligence
BimSchG	German immission control act
BMBF	German Federal ministry of education and research
BMI	German Federal Ministry oft he Interior
BoP	Bottom of the socioeconomic pyramid
BPMN	Business Process Model and Notification
BRL	Basis reference lottery
BUS	Back (mounted) panel unit
CAD	Computer-aided Design
CAN	Canada
CC	Corporate Citizenship

CD	Compact Disk
CDIs	Controller-device Interfaces
CE	Certainty equivalent
CERN	Controller-device Interfaces
ChemG	German chemicals act
CIP	Continuous Improvement
CMM	Capability Maturity Model
CORBA	Common Object Request Broker Architecture
CRISP-DM	Cross-industry standard process for data mining
CRM	European organization for nuclear research
CSA	Canadian Standards Association
CSR	Corporate Social Responsibility
DES	Double Exponential Smoothing
DesignG	German trademark and design law
DFG	German research foundation
DIN	German institute for standardization
DM	Data Mining
DMS	Document Management System
DOE	Design of Experiments
DRIFT	Do it right the first time
DRK	German red cross
DV	Dependend variable
DW	Data Warehouse
EDD	Earliest Due Date
EDP	Electronic Data Processing
EF	Elementary function
EIA	Electronic industries association
EIS	European Interconnected System
EMAS	Eco Management and Audit Scheme
EN	European Standard
EON	Event-on-node network
EPC	Event-driven Process Chains
ERM	Entity-Relationship Model
ERP	Enterprise Ressource Planning
ePVO	ePrivacy Regulation of the European Union (EU)
ETL	Extract - Transform - Load
EU	European Union
FAST	Functional Analysis System Technique
FCFS	First Come, First Serve
FGG	Basic command areas of the German armed forces
FIFO	First In, First Out
FIRE	Financial independence, retire early
FMEA	Failure Mode and Effects Analysis
FOR-DEC	Facts – Options – Risks – brief pause for thought – Decision – Execution – Check (the procedure for an aviation task force)

FTA	Fault Tree Analysis
FTP	File Transfer Protocol
GDSS	Group Decision Support Systems
GF	Basic Function
GG	Basic law for the Federal Republic of Germany
GfSE	German Association for Systems Engineering
GmbH	Limited Liability Company
GNU	Free operating system
GRR	Gauge Repeatability and Reproducibility
GTIN	Global Trade Item Number
HDv	Army service regulation of the German armed forces
HF	Main Function
HGB	German commercial code
HalblSchG	German semiconductor law
HMI	Human-Machine Interface
HoQ	House of Quality
HQR	Qualifications Framework for German Higher Education Qualifications
HTML	Hypertext Markup Language
HTTP	Hypertext Transfer Protocol
iCT	Industrial Computer Tomography
IEC	International electronical commission
IDEF	Integrated Definition
IEEE	Institute of electrical and electronics engineers
IfSG	German infection protection act
insO	German insolvency code
IO-Link	SDCI according to DIN EN 61131-9:2015-02
IoT	Internet of things
IP	Internet Protocol
IPA	International phonetic alphabet
IPO	Input — Processing — Output
IRR	Internal rate of return
ISI	Fraunhofer Institute for systems and innovation research
INCOSE	International Council on Systems Engineering
INUS	Insufficient but non-redundant part of an unnecessary but sufficient condition
ISO	International Standardization Organization
IV	Independend variable
KDD	Knowledge discovery in databases
KMK	Standing conference of the ministers of education and cultural affairs
KonTraG	German corporate control and transparency act
KPI	Key Performance Indicator
KrW-/AbfG	German closed substance cycle and waste management act
LAN	Local Area Network
LC	Load cycles
LCM	Life cycle management

LkSG	German supply chain obligations act
MA	Moving averages
MbSE	Model-based Systems Engineering
MDDBS	Multidimensional database
MDA	Machine data acquisition
MES	Manufacturing Execution System
MMR	Multi-moment recording
MPG	Max Planck Society for the Advancement of Science
MSA	Measurement system analysis
MSE	Mean square error
MTBF	Mean Time Between Failures
MTM	Methods Time Measurement
MTTR	Mean Time To Repair
MuSCoW	Must have, should have, could have, won't have
NASA	National Aeronautics and Space Administration
NP	A class of problems that cannot be solved deterministically in polynomial time (NP-complete)
NF	Secondary function
NIS	Network Information Service
NPV	Net present value
NTM	Non-trivial machine
ODE	Ordinary differential equation
OCEAN	Openness, conscientiousness, extraversion, agreeableness, neuroticism
OEM	Original equipment manufacturer
OLAP	Online analytical processing
OMG	Object Management Group
ORGALIME	Umbrella organization for the engineering industries of mechanical and plant engineering, the electrical industry and metal processing
OR	Operations research
PatG	German patent act
PDA	Production data acquisition
PDCA	Plan — Do — Check — Act
PDE	Partial Differential Equation
PERT	Program Evaluation and Review Technique
PPAP	Production Part Approval Process
PLC	Programmable Logic Controller
PM	Project Management
PPC	Production Planning and Control
QFD	Quality Function Deployment
QM	Quality Management
QTC circle	Quality (performance), adherence to deadlines and costs
RAMSS	Reliability, availability, modifiability, safety, security
R&D	Research & Development
RDB	Relational database

REFA	German association for work design, business organization and business development
RKI	Robert Koch Institute
req	Requirement diagram (SysML)
ROI	Return on investment
RPA	Robotic process automation
SADT	Structured analysis and design technique
SARS	Severe acute respiratory syndrom
SCADA	Supervisory control and data acquisition
SCCI	Source code control interface
SCM	Supply Chain Management
sd	Sequence diagram (SysML)
SDCI	Single-drop digital communication Interface for small sensors and actuators (DIN EN 61131-9:2015-02)
SE	Safety equivalent or Systems Engineering (contextual)
SES	Single exponential smoothing
SECI	Socialization, externalization, combination, internalization
SEM	Systems Engineering & Management
SGB	German social security code
SIPOC	Supplier — Input — Processing — Output — Customer
SM	Systems Management
SMART	Specific, measurable, accepted, realistic, timely
SoI	System of interest (focal system)
StGB	German penal code
SOP	Start of production
SortSchG	German plant variety protection law
SoS	System of systems
SPICE	Software process improvement and capability determination
SPT	Shortest processing time
SQL	Structured query language
stm	State machine (SysML)
SVM	Stakeholder Value Management
SWOT	Strengths, weaknesses, opportunities, threats
SysML	Systems modeling language
TA	Technology Assessment
TCP	Transmission Control Protocol
TDD	Test driven development
TES	Triple exponential smoothing
TF	Partial function
THE MESS	Testing, history, instrument change, statistical regression toward the mean, experimental mortality, selection, selection interaction
THW	German federal agency for technical relief
TM	Trivial machine
TMU	Time measurement unit (MTM)
TQM	Total Quality Management

TRIZ	Theory of inventive problem solving
UA	Utility analysis
UDP	User Datagram Protocol
UrhG	German copyright law
USA	United States of America
USP	Unique Selling Point
UVgO	German sub-threshold procurement directive
UVV	German accident prevention regulations
UWG	German law against unfair competition
VDA	Association of the German automotive industry
VHS	Video home system
VNM	Von Neumann-Morgenstern expected utility function
VSM	Viable systems model
WHG	German water resources act
WLAN	Wireless local area network
YP	Yellow pages
ZDF	Second German television channel

Chapter 1
Introduction

This introduction to Systems Engineering & Management aims to clarify what it is and what it is used for. The National Academy of Science and Engineering (Acatech) in Germany defines technological science as the creation of cognitive prerequisites for innovation in technology and the application of technical knowledge, as well as laying the foundations for reflecting on their implications and consequences. It defines technology as artificial, purpose-driven objects and processes with material and immaterial elements. However, technology is always and without exception linked to people[1]: it is produced and used by people (Ropohl 2009, p. 44). Conversely, humans are generally dependent on technology, as they cannot survive in large parts of nature without clothing, housing or tools (Ropohl 1987, p. 36 with reference to Gehlen 2016). Therefore, SEM is not only about pure technology, but also and above all about what is referred to as socio-technical systems[2] (e.g. Mikl-Horke 2000, p. 164 ff.) and psycho-technical systems (Ahrens 1998, p. 54). And the environment (synonym: surroundings) in which such systems are embedded, in that they are subject to their conditions and have consequences for them, is also included (Fig. 1.1). This extension of the view beyond system boundaries is an important prerequisite for the now widespread concern to protect the natural environment more than in the past.

The fact that the term system has already been used without having been defined beforehand shows that there is no getting around this term in general linguistic usage. In fact, it is omnipresent: in road traffic, for example, we speak of a zipper system, in society of a healthcare system and in the financial world of a payment system. This universality of the system concept is entirely intentional. However, it is used more abstractly in the general

[1] This explicitly contradicts statements according to which artifacts, i.e. purely technical systems (material systems), can be autonomous or intelligent and make decisions independently (e.g. Bendel 2017, p. 168).

[2] Other aspects, such as ecology, could easily be added to this list of relevant aspects. Since social systems cannot exist without such aspects, it would also be possible to dispense with the explicit addition of further aspects altogether. The fact that the technical aspect is nevertheless mentioned in the following is due to the fact that this is primarily about the design and control of systems that are primarily characterized by technology.

V. Ahrens, *Systems Engineering & Management*, https://doi.org/10.1007/978-3-658-46875-0_1

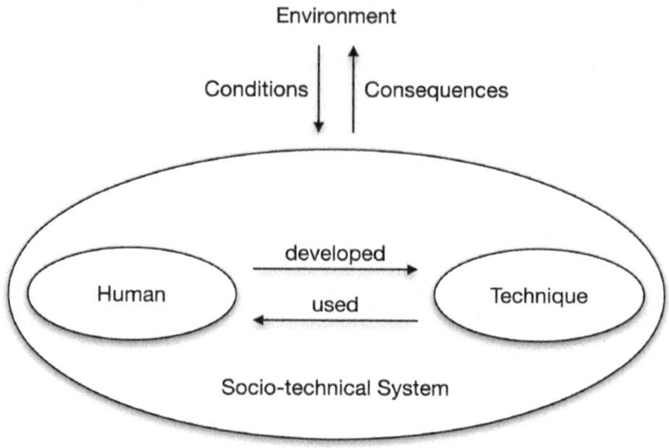

Fig. 1.1 Socio-technical system (Ropohl 2009, p. 44)

systems theory on which the SEM is based than in everyday life. It will become clear later on that this allows us to gain remarkable insights.

What initially remains here is the *universally valid claim* of systems thinking. Linked to this is the possibility of being able to apply the insights gained on this basis in other fields of knowledge in which systems play a role (Fig. 1.2), be it in biology, sociology, psychology or technology (Luhmann 1996, p. 16). Ludwig von Bertalanffy (1951, p. 306) derives a major advantage of systems theory from this *interdisciplinarity*: General Systems Theory will be an important means to facilitate and to control the application of model conceptions and the transfer of principles from one realm to another. It will no longer be necessary to duplicate or triplicate the discovery of the same principles in different fields isolated from each other. In other words, thinking in systems contributes to the fact that, as they say, you don't have to keep reinventing the wheel. Once findings have been made, they can rarely be transferred directly, but often analogously to other things.

In this respect, Systems Engineering & Management does not invent anything fundamentally new. For example, it is well known how something is developed. There are countless specific instructions on this in all relevant fields of knowledge (e.g. Bender and Gericke 2021, p. 27 ff., Ehrlenspiel and Meerkamm 2013, Lewin 1958, p. 210 f., Naefe 2018, p. 40 ff., Nyamsi 2019). The specific feature of systems theory in general and SEM in particular is to make *generic statements* about it. These in turn can contribute to simplification and thus to economic efficiency, support interdisciplinary work and *promote innovation*.

In SEM, for example, it is customary to select a few examples from the variety of process models, present them in more or less detail and discuss them (e.g. Böhm 2005, p. 1 ff., Haberfellner et al. 2018, p. 82 ff., Walden et al. 2023, p. 25 ff., Winzer 2016, p. 23 ff.). However, this will be largely dispensed with here with reference to sources such as those mentioned as examples. Instead, the advantage of systems thinking cited by Ludwig von Bertalanffy (1951, p. 306) (see above) will be used to explain the generic process model of Systems Engineering in the required detail and to justify its general validity (Chapter 4). This also implicitly explains all alternative process models that make sense. One exception, due to its increasing prevalence, is the Scrum approach, which is included

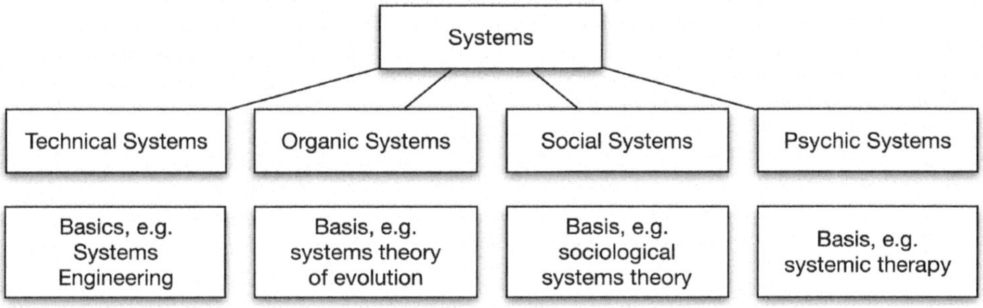

Fig. 1.2 Characteristics of systems theory (Luhmann 1996, p. 16)

as a representative of all agile approaches in order to make it clear that it does not differ significantly from established Systems Engineering in the form described here.

More important than a comparison of process models for the development of systems appears to be a comparison of the development on the one hand and the application of the development results on the other (Chapter 3). Although the differences between these two fundamental phases in the *life cycle* of systems are immense, too little attention is usually paid to them. It could be argued that Systems Engineering is only concerned with development, so that the application phase is of secondary importance, and to a certain extent this is true. However, the results of developments are made for application, so the developers also need an understanding of the laws that apply in the application phase. In addition, knowledge is improved when it can work through differences. For example, you can get a good idea of the color black if you also have the color white in front of you, and good can be recognized particularly clearly if you have an idea of evil (Watzlawick et al. 2013, p. 21). The same applies here: if you know which laws apply to the application of development results, you can better understand the laws according to which development must take place in order for it to be successful.

The Systems Engineering process model to be developed from these comparisons depicts the chronological course of a problem solution. This process is divided into different phases. In each of the phases, specific methods are used that help to complete the respective work efficiently and effectively (economically). They are therefore explained in as much detail as is necessary for an understanding of Systems Engineering. In this context, however, it is neither sensible nor possible to cover all methods exhaustively. Works that do this, such as those on *empirical social research*, often run to hundreds of pages (e.g. Andler 2015, Atteslander 2010). Their supplementary reading is recommended.

It follows from the previous considerations that Systems Engineering is a methodology based on systems thinking for solving socio-technical problems, i.e. for developing solutions to problems. However, Reinhard Haberfellner et al. (2018, p. 9 f.) rightly point out that a methodology alone does not solve a problem. However, they leave their list of what is also necessary, for example specialist knowledge, situational knowledge, experience, behavior/psychology, ethics of action and so on, open and do not provide a scientifically based foundation. This can be found in the *competency model*[3] of the Qualifications Framework for German Higher Education Qualifications (HQR 2017, p. 6 ff.) of the Standing Conference of the Ministers of Education and Cultural Affairs of the federal states in the Federal Republic of Germany (KMK) and will be used as a basis here.

In general, competence refers to the cognitive abilities and skills that can be learned by individuals and that are required to *solve problems*, as well as the associated motivational, volitional and social willingness and ability to use problem solutions successfully and responsibly in variable situations (Weinert 2001, p. 27 f.). Competence is therefore a multidimensional concept that needs to be broken down further with regard to its practical application.

Heinrich Roth's (1971, p. 389 f.) differentiation between *self-competence, professional competence and social competence* is widely recognized, i.e. the ability to act responsibly for oneself, the ability to act and make judgements within certain specialist areas, and the ability to contribute to the economy, politics, society and culture (Fig. 2.4). The German Qualifications Framework for Lifelong Learning (DQR) supplements these competences with *methodological competence*, which, as a cross-sectional competence, serves in particular to develop the other competences (AK DQR 2011, p. 4). This ultimately results in the competency model of the Qualifications Framework for German Higher Education Qualifications (HQR) of the Standing Conference of the Ministers of Education and Cultural Affairs (KMK) shown in Fig. 1.3.

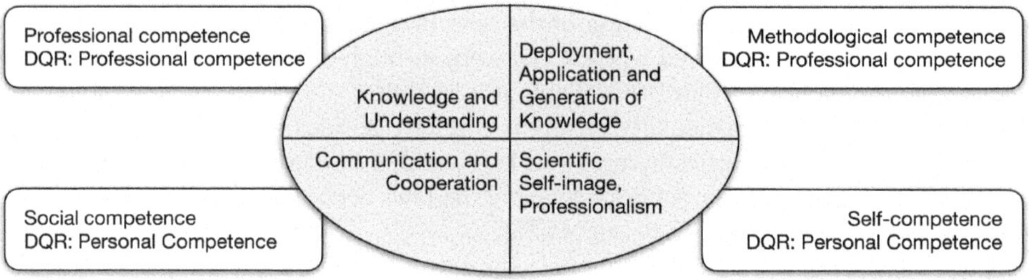

Fig. 1.3 Competence model of the Qualifications Framework for German Higher Education Qualifications (HQR) of the Standing Conference of the Ministers of Education and Cultural Affairs (KMK) (DQR 2011, p. 4)

The methodological competence that is the focus here is therefore only one of several prerequisites for successfully solving problems, but it establishes important connections to and between the other required competencies. Although we will not go into this in detail here, an example may clarify what is meant.

Systems-based thinking helps to recognize *patterns*, for example. This can be based on basic principles such as *recursion* and *self-similarity*, which can be applied to route planning problems, for example. For example, similar problems are repeatedly encountered at

[3] In organizational theory, the terms competence and authority are often used synonymously. Here, authority is understood to mean, in contrast to competence, the authorization granted by law or order to intervene in other areas of responsibility.

different spatial scales: minimizing the distances to be covered in the event that a scarce resource starts from a starting point, successively travels to several stations at different locations and finally returns to the starting point. This is also known as the traveling salesman problem, which is dealt with in *Operations Research* (OR) and *theoretical computer science* (e.g. Domschke et al. 2015, p. 151 ff.) and represents an NP-hard problem, i.e. one that cannot be solved efficiently for typical applications, due to the associated combinatorics[4].

- This problem occurs in logistics, for example, on a scale that is usually measured in kilometers, when a truck has to deliver goods to several customers on a milk run tour.
- On a scale usually measured in meters, this problem is found within factory halls (intralogistics) when industrial trucks transport goods to several workstations on a tour.
- On a scale of centimetres and fractions thereof, a comparable problem can be found on a workbench, for example, when the worker performs different tasks one after the other with his hand-arm system.
- In the order of magnitude of millimetres and fractions thereof, for example, it is necessary to program a robot for assembling printed circuit boards in such a way that the distances covered are minimized.

In the light of the principle of self-similarity, recursive thinking reveals that the problems are comparable at all scales - but not identical problems, because in logistics vehicles move in two dimensions (earth's surface, water surface, prescribed flight altitude), while the hand-arm system of a worker or a robot arm can cover three dimensions. What is important, however, is that systematic thinking can help to recognize and understand basic principles so that different knowledge domains such as logistics, ergonomics and automation technology can be transcended. This in turn helps with application, for example on the basis of the *Theory of Inventive Problem Solving* (TRIZ). This methodology lists recursion and self-similarity as the 7th principle in the list of innovative principles and uses the metaphor of the Russian matryoshka to illustrate this (e.g. Koltze and Souchkov 2017). Methodological competence, in this case the application of basic principles, therefore helps to acquire and apply professional competence.

With this in mind, this book is structured in such a way that the first part (Chapter 2) lays the foundations for thinking in systems. This begins immediately with an embedding of technology in its social context, which is likely to demand a lot from engineers in particular, as it leads them into unfamiliar territory. However, *providing food for thought* is one of the objectives stated above, and those who engage with it will have to reckon with *irritations*, especially since this is precisely what irritations (synonymously: perturbations, disturbances) are essentially about in the psychological and sociological context, so that these will be discussed in detail anyway.

In the second part (Chapter 3), the five generic process models for the design (Engineering) and control (Management) of systems are first introduced in an overview in order to provide orientation for the subsequent explanations.

- The overview begins with the *system life cycle* (Section 3.1), which forms the connecting bracket for the four other process models, i.e. a meta-model.

[4] This means that under the as yet unproven assumption that the complexity classes P and NP can be distinguished from each other, there is no algorithm that can calculate a shortest round trip in polynomial computer runtime.

- After the introduction of the meta-model, a section on the cognitive science-based *distinction between problems and tasks* is added to the theoretically founded distinction between projects (design) and processes (control), which appears to be particularly relevant in view of the process orientation paradigm (Section 2.5) that has prevailed since the early 1990s (especially Hammer and Champy 2003). This is followed, with regard to the practical application of this distinction
- the *problem-solving cycle* for the design phase (Section 3.3) and
- the *control circle model* for the application or control phase (Section 3.4).
- The phase transition (Section 3.5) between development and application is often particularly critical in practice.
- The phase transition between development and application is regularly particularly critical in practice: *Change Management* experts state that around 75% of all change projects do not lead to the desired success (Lauer 2019, p. 49). As the approaches used to date are therefore obviously unsatisfactory, suggestions for further development are provided here.
- Furthermore, recent experiences such as those in connection with the coronavirus pandemic in particular have made it clear that the *management of exceptional situations* (synonymous: emergency, disaster) must be better prepared in future, especially as they are now expected more frequently than in the past. For this reason, approaches are also presented in Section 3.7
- This third Chapter concludes with a critical examination of the *PDCA cycle*[5]. This discussion is actually at odds with the intention, which has otherwise been largely maintained, to refrain from comparing different process models, as is common in other works of this kind (e.g. Haberfellner et al. 2018, p. 82 ff.), which can be looked up there if necessary and could distract from the core of the concern here. The reason for the exception to the rule is the experience that the PDCA cycle is very widespread, but is almost always misunderstood. To paraphrase Albert Einstein, it can be said in advance that everything should be made as simple as possible - but not simpler. The apparent triviality of the PDCA cycle may be tempting, but it is not helpful. This is explained further in Section 3.8

The third part (Chapter 4) then goes into the nitty-gritty, as they say, and explains the Engineering in detail. To this end, the phases of the problem-solving cycle are virtually explained in detail by describing relevant methods and tools for each phase, as well as models and the theories on which they are based. This can be read as a roadmap or instruction manual if you wish, but the claim goes beyond this in that the respective aspects are always explained and justified. This is based on the insight that it is not enough to simply apply standards, especially in complex design projects, which is what we are talking about here. As important as standards are, in practice they must always be interpreted, adapted or further developed (tailoring), especially when it comes to innovation. But then the only thing that helps is to think for yourself, get creative, question and improve. In turn, it helps immensely if you have understood the basic principles. Everything else arises from this in the specific application. However, such confidence in success, which is particularly

[5] As you know, the four letters stand for P = Plan, D = Do, C = Check and A = Act.

important with regard to innovation, presupposes that you have actually understood the principle beforehand, and the explanations in Chapter 4 are intended to contribute to this.

The fourth part (Chapter 5) on the phase transition is comparatively short, but important, because so many things go wrong in practice when introducing something new, especially during the transition from development to application (see above). After a process model as well as methods and tools have already been described in detail in Section 3.5, the fifth chapter is primarily concerned with explaining the special and, above all, particularly difficult and still not fully understood laws that are assumed to be the causes of the frequent and often far-reaching losses that phase transitions entail.

The fifth and final part (Chapter 6) deals with the control of systems (Management). However, this cannot and should not be a complete outline of Business Administration, Production Management and so on. Each of the relevant aspects in this context fills volumes in its own right. The explanations are therefore limited to a systems theory perspective, which has two advantages: on the one hand, the connections with the design phase, which is also based on systems theory, can be made particularly clear on this basis and, on the other hand, innovations can be derived from the generic considerations.

As much as the author may wish for this, it is unlikely that the book will be worked through chronologically from the first to the last page. In order to support cross-reading and make it easier to find essential content, corresponding keywords are printed in italics in addition to the deeply structured table of contents.

Chapter 2
Systems Thinking

Systems thinking is *abstract thinking*. However, abstract thinking has a reputation for being difficult, at least more difficult than its opposite: concrete thinking. One stands for *theory*, the other for *practice*. After all, Kurt Lewin is said to have claimed that nothing is more practical than a good theory (Wegener et al. 2011, p. 47). This insight will be examined in more detail below because, as will become clear, thinking is always abstract by nature. As a result, access to the real world, i.e. to what is commonly associated with practice, is generally largely blocked. Nevertheless, the solutions to problems should prove themselves in the real world, i.e. in practice. The following considerations are intended to help close this gap between human thinking on the one hand and reality on the other, at least to some extent, in order to increase the probability of success of problem solutions beyond the level that can be achieved by chance or amateurish action.

The fact that there is a significant difference between reality and the image people have of it has been known since ancient times. Plato (2014), for example, was aware of this in his allegory of the cave. To this day, this insight has not changed significantly. Immanuel Kant, for example, also distinguished between the *phenomenon* as a sensually perceived representation of reality and the *thing itself* as an actual reality. Referring to both ancient philosophers and Kant, Karl Raimund Popper (1989, p. 41 ff.) ultimately founded today's widely recognized scientific *theory of critical rationalism*.

Through abstraction, those aspects of phenomena are singled out that appear essential (relevant, decisive, significant) for what one wants to know or achieve (e.g. Bjørke 1995, p. 10 f.). Initially, this applies in general and in this respect also to every subconscious, unconscious and lay conscious perception. Systems thinking contributes to making abstractions professionally. However, this requires at least a basic understanding of *thinking*, as methods and procedures for conscious problem solving have proven to be particularly successful when they are compatible with natural thought structures and processes. Professional problem solving is therefore essentially based on findings from cognitive science (e.g. Dewey 1997; Dörner 1979; Duncker 1974; Lüer/Spada 1990; Noë 2010; Sell and

V. Ahrens, *Systems Engineering & Management*, https://doi.org/10.1007/978-3-658-46875-0_2

Schimweg 2002), but also includes other domains such as anthropology (e.g. Tomasello 2022).

Abstract thinking allows for *generalizations*. What is described as a system must be comparable, at least in some essential respects, to something else that is also described as a system. Otherwise, the two could not be subsumed under a common term. And the term system is used in very different disciplines: in Electronic Data Processing (EDP), for example, we speak of a system into which data is entered or which delivers the results of data processing; in psychology, we speak of a therapy system when a family dispute needs to be resolved (e.g. Watzlawick et al. 2013, p. 24), and in technology, for example, we speak of an energy supply system.

The aim of systems thinking is to disregard (abstract) domain-specific characteristics and identify those aspects that are common to all systems. These commonalities can be *isomorphisms* (patterns) and/or *regularities*. One technological example is *bionics* (Nachtigall 2010, p. 98), which transfers biological structures and processes to technical applications. One well-known result is the winglets at the end of the wings of modern aircraft, which reduce the induced air resistance and are now being further developed into spiroids (split-wing loops). These innovations originated from the analogy with bird wings. The technical reproduction of the nodular protrusions known as tubercles, which can be found on the leading edges of the flippers of humpback whales, on the wings of airplanes or on the rotor blades of wind turbines, for example, is also expected to lead to energy and acoustic advances. Analogy formation is a very important method for pattern recognition (e.g. Watzlawick et al. 2013, p. 23).

The example of bionics also shows that systems thinking supports *interdisciplinary work*[1], whereby the disciplines can be much further apart than is the case between the natural sciences (here: biology) and technology. Since the 16th century, the humanities have been part of that part of science that deals with *explaining phenomena* (in the sense of René Descartes: res extensa). The humanities, for example, which attempt to *understand noumena* (in the sense of René Descartes: res cogitans), are far removed from this (Dilthey 1924, p. 143 f.; Watzlawick et al. 2013, p. 127 ff.). And even if Max Weber (1972, p. 4) distances the social sciences somewhat from the humanities by describing their mission as explaining understanding, they are still relatively far removed from natural science and technology. But even between the two, isomorphisms can be identified.

One example of this is the reaction of physical and social systems to observations. It is often assumed that an observer and the observed system are independent of each other, so that the observer only believes to observe the properties and behavior of the observed system (Laurikainen 1985, p. 273 ff.). In fact, however, there are many situations in which the observer's influence on the observed system is so great that it cannot be neglected; an observer always observes himself as well. In physics, this is regularly the case when measuring quantum mechanical phenomena (ibid.), but empirical social research is also familiar with this problem in relation to social systems (Atteslander 2010, p. 80 ff.). Furthermore, such considerations lead to *second-order cybernetics*, which explains that and why observation is an even more far-reaching problem (von Foerster 1995).

[1] Sometimes the concept of interdisciplinarity is replaced by that of transdisciplinarity (e.g. Mittelstraß 2002), but this is only an unnecessary exaggeration and without any additional gain in expression or knowledge. Engineering thinking and action is not dependent on such rhetorical stylistic devices.

Recognizing and understanding this fundamental problem lays the foundation for all further considerations and opens up new perspectives for innovation.

2.1 Constructivism and Second-Order Cybernetics

With second-order cybernetics, Heinz von Foerster (1995) is regarded as one of the founders of radical constructivism, according to which all *perception is subjective* (Watzlawick 2002), as it is not carried out by sensory organs (doubts about phenomenological evidence), but by specific sensory regions of the brain (Roth 1986). The task of the sensory organs is merely to convert the multitude of different physico-chemical events such as light quanta, sound pressure waves, scent molecules or vibrations into a single class of events, namely electrical nerve potentials. As a result, the environmental signals lose their specificity. It is no longer possible to see what is encoded in the potentials. And their origin can no longer be reconstructed. If we consider against this background that the electrical nerve potentials can also be generated by internal processes such as thinking, we come to the conclusion that reality is essentially constructed by the observer himself. Perception is [...] the assignment of meaning to neuronal processes that are in themselves meaningless, is construction and interpretation (ibid., p. 170).

Engineers, most of whom are well versed in mathematics, may find this particularly illuminating if they quote the small calculation made by Heinz von Foerster (1985, p. 51): Since we only have around 100 million sensory cells, but our nervous system contains around 10,000 billion synapses, we are 100,000 times more receptive to changes in our internal environment than to changes in our external environment. Against this background, Ernst von Glaserfeld (1996), another founder of radical constructivism, explains at a higher level of aggregation that all knowledge, however one may define it, exists only in the minds of people and that the thinking subject can only construct its knowledge on the basis of its own experience (ibid., p. 22). A more or less conscious focus of the gaze on an object illustrates this. What lies in the visual field has a significantly greater chance of being perceived and processed than anything that lies in the visual field or even in the peripheral visual field (Bokranz and Landau 2012, vol. 2, p. 176 f.; Schlick et al. 2018, p. 486).

Based on this example, it is easy to understand that the majority of the environment does not even penetrate a person's brain, so that even this first filter makes it clear how little the image that a person forms of their environment can be congruent with the real environment (Roth 1987, p. 247). However, evolution has meant that the image fits at least as far as it is sufficient for survival in one's own ecological niche under normal conditions. As will be explained in Section 2.2 it must be viable (von Glaserfeld 1996, p. 43) – nothing more. In this respect, radical constructivism must be weakened somewhat. As an alternative, the methodological constructivism (also: Erlangen constructivism) used as a basis in the following suggests itself, which assumes that methodological support of cognitive processes, as also offered by Systems Engineering & Management, improves the viability of the constructs.

For the SEM, this results in three essential questions: (1) How can people communicate with each other if everyone only understands the world with their own perception, and

(2) how can people find access to at least that part of real reality that is relevant to their lives and survival, in and for which they therefore also develop their technical solutions. The latter aspect in particular is of great importance insofar as technical solutions are realized and then have to prove themselves in reality, the real reality. With its realization, however, technology leaves people behind in their illusory world, so to speak, which is one reason why people are often no longer able to control what they create and bring into the world in every respect. Johann Wolfgang von Goethe was already aware of this in his Sorcerer's Apprentice. But not much has been learned from it to this day. Therefore, and in view of the increasingly far-reaching effects of technology, the fundamental limits of its controllability must be considered more than ever (Jonas 1984). Among other things, *technology assessment* (TA) contributes to this (e.g. Simonis 2013). In addition, all those involved and affected are called upon to participate.

2.1.1 Emergence

The answer to the first question (1) is based on the phenomenon of *observational observation* (Ahrens 1998, p. 17): Two systems, of which system A intends to direct system B, face each other. Both observe each other in that each observer forms an image of the other observer as well as an image of the image that the other presumably forms of him and in turn an image of the image of the image and so on – here you can already see how convoluted such a recursion can become. In detail, system A first forms an image of system B and, on this basis, draws up an intervention plan according to which it attempts to persuade system B to behave in a way that system B would not do on its own. However, system B does not perceive this as an intervention, but merely as a new boundary condition (disturbance, perturbation). As a result, B forms an image of A's intentions and comes to terms with them in its own way. As a result, it reacts to A, but not in the way A expected. At least A can now observe how B reacts to his interventions and take this into account in further interventions. However, B then observes how A adapts its interventions and develops reactions that are now no longer only attuned to the intervention, but also to its change. In turn, A then observes how B reacts to his intervention and to his adaptation of his intervention, and now also adapts the way he adapts his interventions, which B observes once again and also adapts to.

So rules arise from rules from rules... Systems theorists refer to this as *rule emergence*. Both systems control themselves more and more from the perspective of the other system, but design this perspective themselves. As a result, actions relate to the expectation of expectations, to the expectation of an expectation of expectations and so on. In this way, actions very soon become indeterminate and susceptible to disruption. There is a coupling between the systems, but it is unstable and can get out of control at any time. For Wolfgang Krohn and Günter Küppers (1990, p. 121), the problem arises in summary from a planning of the control of systems by self-control in systems on the basis of external descriptions in self-descriptions, and in their estimation there is no universally valid solution for such action situations, as this gets caught up in recursive loops and self-referential paradoxes.

They therefore consider the failure of interventions to be normal. Successful understanding is only possible in exceptional cases.

At first glance, the second question (2) appears to be of a scientific-technical nature – but only at first glance. What is then missing is (3) the connection between humans and technology in the sense of socio-technical and psycho-technical systems (Fig. 1.2), and establishing this connection is not so easy. Nevertheless, an attempt will be made to answer this third question in the following. But first things first.

The neurobiological insight that the human brain only processes electrical nerve signals and chemical messengers, some of which are stimulated by internal processes such as thinking and only triggered to a limited extent by the environment, and that this is also filtered and greatly reduced, is referred to as *operational closeness* (Maturana and Varela 2018). This is associated with a whole series of other aspects that cannot be explained in detail here, but which will be introduced in Systems Engineering & Management and will be taken up again and again and further elaborated on later, as they are of great importance for the design of socio-technical and psycho-technical systems, but have so far received too little attention.

It follows from the operational closeness of psychological and social systems that these systems – in the case of the human being under consideration here, the brain – are actually only concerned with themselves (Maturana and Varela 2018). This is referred to as *self-reference* or autopoiesis: Thoughts that a person has as a mental system can only be perceived – i.e. observed – by other thoughts of this person. There is no way to introduce thoughts into a mental system from outside or to observe them outside a mental system. At this elementary level of processing thoughts, there is therefore neither input nor output for a mental system, but only circular closure in the strict autopoietic sense (Willke 2006, p. 64). There is no direct contact between different systems of consciousness (Luhmann 1985, p. 404).

Paul Anton Watzlawick et al. (2013, p. 24 f.) show that this insight is also supported by group theory (Glosauer 2016) (Sect. 2.4.4). The first of the four axioms of this theory, which is referred to as the axiom of closure, states that every connection of every element $a \in G$ of a group $(G, *)$ (almost synonymous with system) with itself or with every other element of the group produces another element of the group. To illustrate this, Paul Anton Watzlawick et al. (2013, p. 24) use the example of the face of an analog clock, whose elements are the integers between 1 and 12: $a = \{1, 2, ..., 12\}$. In this case, the link is made by addition or subtraction and each addition of two elements results in a new element of the group. For example, if you add 6 h at 8 o'clock in the morning, the result is 2 o'clock in the afternoon. The fact that and how group theory as a mathematical discipline is applied isomorphologically to mental systems will be clarified later on.

However, operational closure is not to be understood as the opposite of openness, but as a prerequisite for it, albeit for a special kind of openness. In the case of mental systems, this is *cognitive openness*. The brain forms the infrastructure, so to speak, that makes thinking possible. However, there is no direct connection between the brain and thinking (Fig. 2.1). It is not the brain that thinks.

In the context of SEM, an analogy to road traffic is suitable for understanding this. Here, too, the availability of roads and vehicles merely forms the infrastructure, i.e. the prerequisite for the development of road traffic. It is true that activities can also be observed

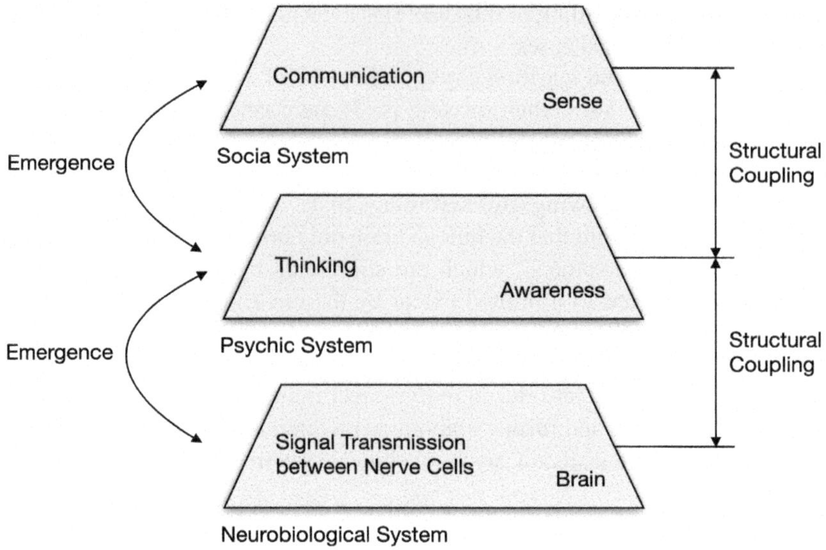

Fig. 2.1 Structural coupling

without traffic, such as the construction or maintenance of parts of the infrastructure, just as the brain shows its own activities on a neurobiological level, which would also take place without thinking. However, road traffic as such, just like thinking, develops according to its own logic.

At a quasi microscopic level, i.e. at the level of individual road users, this logic does not arise by chance or as an end in itself, but from the purposeful actions of the sum of all road users. This individual action is dependent on the infrastructure, as road users usually drive on the existing roads and only rarely off them (although more frequent cross-country driving in similar places may well lead to a new road being built there at some point). In terms of systems theory, the traffic that forms is therefore *structurally determined* and in group theory the structure corresponds to the given combination possibilities of the elements (Glosauer 2016). Nevertheless, the infrastructure gives road users so many degrees of freedom that they can use them according to their own logic. And so, although thinking is also dependent on the structure of the brain and is therefore structurally linked to it, as Fig. 2.1 shows (e.g. Maturana 1985, quoted from Willke 2006, p. 62 ff.), it nevertheless follows its own logic.

On a macroscopic level, however, the behavior of the overall system can no longer be explained by the sum of the individual activities if it is sufficiently *complex*[2] (Ratter 2006, p. 115). In systems theory, this phenomenon is referred to as *emergence*. For a simplified explanation of this phenomenon, reference is often made to a sentence attributed to Aristotle, according to which the whole (here: road traffic) is more than the sum of its parts (here: the individual road users). However, Wolfgang Metzger's (1975, p. 6) modification

[2] A detailed explanation of what is meant by complexity here can be found in Section 6.1

of this theorem, derived from Gestalt theory, comes closer to today's insights, according to which the whole is not more than its parts and their interaction in a quantitative sense, but is different.

In complex systems, in which many individuals interact in a network, design and control concepts are required at a macroscopic level that start with the whole, as the targeted influencing of individuals at a quasi-microscopic level cannot lead to the desired results[3]. In road traffic, for example, the probability of congestion can be reduced from a certain traffic density onwards by standardizing the speed of all vehicles with the help of a suitable traffic guidance system, but not by each road user taking individual measures.

Phenomena such as these are investigated in *synergetics*, for example to physically explain the emergence of laser light (e.g. Haken 2012). In psychology, Paul Anton Watzlawick et al. (2013, p. 118 ff.) describe a change triggered by a super-system in its sub-systems as a *second-order solution* and explain this with the help of the *logical type theory* founded by Alfred North Whitehead and Bertrand Russell (1919-1913). This describes systems as classes or sets (hence the well-known term set theory in mathematics) and is based, among other things, on the fundamental theorem that whatever concerns the totality of a class (set) must not itself be part of that class (ibid., vol. 1, p. 37).

Against this background, the preferred *bottom-up approach* in agile approaches (Goll and Hommel 2015, p. 79) appears problematic, as it is based on the assumption that the sum of all initially developed elements (fragments) ultimately results in the whole. Nevertheless, Joachim Goll and Daniel Hommel (ibid.) explain that there are examples of this approach not always working, but without specifying the examples. Although this does not lead to a departure from the bottom-up approach, it does lead to the addition that the associated lack of understanding of the desired overall system should be supplemented by the initial development of a shared vision of the whole. In this respect, agile approaches do not manage entirely without a *top-down approach*. However, in view of the insight developed above that there is no 1:1 relationship between the whole and the sum of its parts, a top-down approach is not reliable either. The traffic system mentioned above is an example of this: although the infrastructure (especially traffic routes) is predetermined, processes (e.g. traffic light circuits such as the so-called green wave) are designed and technical aids (e.g. bicycles, cars) are provided for individual use of the structures and processes, it is not possible to reliably plan and control how the individual road users ultimately use all of this. The mass panic that occurred after the Love Parade in Duisburg on June 24, 2010, in which 19 people lost their lives, is a shocking example of this. Although it can be assumed that such phenomena can be prevented through better planning and control, and that this is always possible is not to be denied, the insights from systems theory on which this is based suggest that the assumption of a linear and seamless transition neither from the elements

---

[3] This contradicts the hope formulated by INCOSE (2015, p. 16) of being able to explain and design emergent properties and behaviors of an overall system through an understanding of the links between the various system elements and their relationship to the whole (system). Rather, when designing technology, the aim must always be to avoid emergence by building trivial machines in the sense of Heinz von Foerster (1993, p. 247 ff.). Building non-trivial machines would be irresponsible, as such machines are uncontrollable. This is currently being discussed, for example, with regard to the further spread of artificial intelligence (AI) applications, as the underlying neural networks do not allow any causal conclusions to be drawn about the occurrence of results.

to the system (bottom up) nor vice versa (top down) leads reliably to the result. However, more on how emergent behavior can be taken into account later.

2.1.2 Linking Perception and Action

Now that it has been clarified how thinking is structurally linked but emergent and thus follows its own logic, virtually emancipated from the brain, the next step is to explain how thinking as such takes place. To do this, the unconscious must be transformed into *language*: We only know what exactly we think, desire and feel when we succeed in putting the contents of the mind into words, explains Peter Bieri (2007, p. 45). This applies above all to verbal language, but in Systems Engineering & Management there are other forms of expression that contribute significantly to the formulation of thoughts: mathematical expressions, technical drawings, modeling notations such as the Systems Modeling Language (SysML), programming languages such as Java or Python and even physical models. The more and the better an engineer masters such languages, i.e. formalized means of expression, the better he is able to design and control. The terms discussed here also serve this purpose, defining the space and expanding it in comparison to other works on Systems Engineering & Management, in which knowledge, understanding, ability and evaluation and thus also innovation are possible.

Equipped with such means of expression (languages), a mental system can now begin to think. Due to this fundamentally self-referential (autopoietic) processing (Roth 1987, p. 246), contributions from outside do not enter the mental world of the focal system (system of interest) either in the form of thoughts or as copies of relevant parts of the environment. The sensory stimuli, which are already greatly reduced during perception and which, when converted into nerve impulses and in the course of their further processing in the brain, can no longer be seen where they come from, merely represent *irritations* or *perturbations*: they disturb.

New information, wherever it originates, is now compared with what is stored in *memory* in a linguistically retrievable form. As the memory contains much more information than the newly received information, it is easy to imagine that the stored information has a much greater influence on thinking than new information. This is why memory is classified in neurobiology as the central organ of perception (Roth 1987, p. 247). This is also considered plausible because the processing capacity of a mental system would quickly be exhausted if it were confronted with the full complexity of its environment. *Selective perception*, which is also popularly known and often portrayed as a deficit, is in fact a vital protection against excessive demands.

In addition, when new information is incorporated into the already stored information base, checks take place (Roth 1997, p. 21) that can distort what is coming in from outside. Criteria are partly innate, partly developed in early childhood and partly derived from later experiences (ibid., p. 125). In addition, only information that is capable of connecting to existing information is incorporated (ibid., p. 107). The *theory of cognitive dissonance* (Festinger 1978) explains that mental systems strive for harmony, consistency and congruence in their mental world and change anything that disrupts this harmony so that it fits seamlessly into the remaining thoughts and memories (ibid., p. 253). For example, a person who fundamentally strives to protect the natural environment may try to justify the frequent preference for private cars over bicycles or public transportation by pointing to the large number of jobs created by automobile production. If and to the extent that such a line of thought brings the inner structure of conviction into harmony, it is incorporated as a solution.

To support the connectivity of the SEM to other disciplines, a model should be presented here that is particularly widespread in the human sciences and represents what has been explained so far on the same basis, but with slightly different intentions. This is the *situation circle* developed by Thure von Uexküll together with Wolfgang Weisack (2003, p. 213 ff.) and shown in Fig. 2.2, which is based on the function circle introduced by his father Jakob Johann von Uexküll[4] .

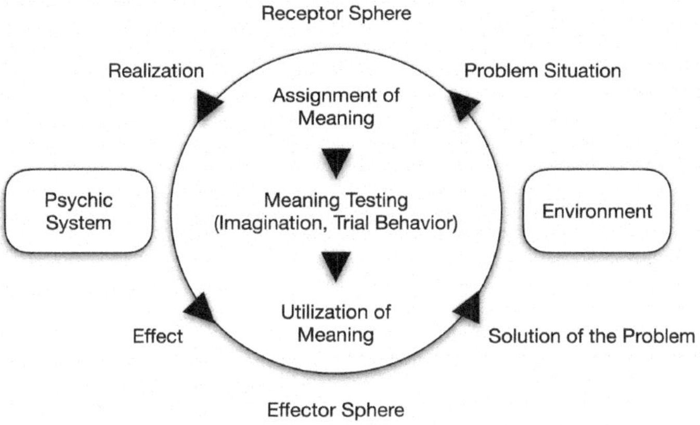

Fig. 2.2 Situation circle (Uexküll and Weisack 2003, p. 213 ff.)

The situation circle is a sensorimotor circle in which perception (sensory) and action (motor) are cyclically linked. The emergence with which actions arise from perceptions is represented by the interposition of the imagination. Initially, a meaning is attributed to what is perceived (hypothesis). On this basis, alternative meanings or actions are mentally played through. Siegmund Freud (1911, p. 233) coined the term *reality testing* for this. Only after a promising strategy for action has been developed in this way is it put into practice. The problem solution must then prove itself in the environment. If it fails to do so, this is perceived as a persistent or newly arising problem situation and triggers the cycle once again. It is also crucial in this model that the degrees of freedom in the perception, imagination and development of an action strategy are so great that, in the constructivist sense, no 1:1 relationship between a mental system and its environment is possible.

[4] Humberto Romesín Maturana also developed his contributions to the systems theory of cognition developed together with Francisco Javier Varela García on the basis of Jakob Johann von Uexküll's work. The results of their work in turn had a significant influence on the systems theory contributions of Heinz von Foerster and Niklas Luhmann. These diverse interrelationships can only be traced in part here. In-depth reading of the original literature is therefore recommended.

In practical terms, the previous considerations mean that an understanding between people on the one hand and an understanding of what happens in a person's environment on the other is rather unlikely (Gadamer 2010). Therefore, attempts to increase this probability are of great importance, and in this respect, SEM can learn from *agile approaches* such as *Scrum*.

Epics[5] and *user stories* offer a starting point for this (Pichler 2008, p. 34 ff.; Sutherland 2015, p. 124 ff.; Wirdemann 2017, p. 50 ff.). At first glance, this is a simple renaming (Wirdemann 2017, p. 10), because at its core, a user story is nothing more than what is usually referred to as a requirement (Sect. 4.2.4.9) in SE (Walden et al. 2023, p. 101; Winzer 2016, p. 29), in Quality Management (DIN EN ISO 9001) and especially in Requirements Engineering (Ebert 2014) (Wirdemann 2017, p. 10). In view of the previous considerations, however, a second look reveals a remarkable detail: the *conversation* (ibid., p. 50 ff.).

The insight that the interpretation of requirements depends above all on communication between those who set requirements and those who are supposed to fulfill them is based on the experience that requirements are generally incomplete and ambiguously formulated and that they are also constantly changing (ibid., p. 26). The latter is primarily due to the fact that every problem solution involves a learning process (Sect. 3.2). In this respect, statements that approaches such as Systems Engineering are preferable when everything is known, while agile approaches are predestined when not all requirements are known at the beginning of a design project and when these can still change in the course of the project (Goll and Hommel 2015, p. 79), seem completely unsuitable as an attempt to differentiate between the approaches.

Every problem solution requires potential misunderstandings between different stakeholders to be minimized. In agile methodologies, a user story noted on a story map (Fig. 2.3) forms the starting point for this. The probability that the sum of all user stories does not include all relevant requirements and that each individual user story can be misunderstood is initially high. It is therefore not sufficient on its own to ensure that all requirements are fulfilled as far as possible. In addition, agile methodologies such as Scrum ensure that solution developers have the opportunity to ask questions at any time if a requirement seems incomplete or unclear. To this end, the technical project manager assumes the role of *product owner* as it is known in Scrum (ibid., p. 9). Their task is to put themselves in the shoes of the stakeholder groups (Sect. 4.2.4.5), understand their requirements and communicate them within a corresponding design project.

In view of the limits that self-reference and operational unity place on mutual understanding, the possibility of communication between participants with different roles and competencies at any time appears to be a viable option. In this respect, this approach is certainly recommended. However, this cannot be done without any objections, because in fact the problem of mutual understanding is merely shifted to the product owner. The product owner needs a good deal of empathy and a pronounced ability to empathize with other people's problems and to adopt the expectations of others, not only beyond the boundaries of psychological systems, but also beyond the boundaries of social systems, which will be discussed below. The fact that this approach can nevertheless be successful within the limits to be expected from a systems theory perspective, at least more successful than ap-

---

[5] These correspond to the user stories, but are more general and must be converted into user stories at a later date. For the sake of simplicity, only user stories are discussed below.

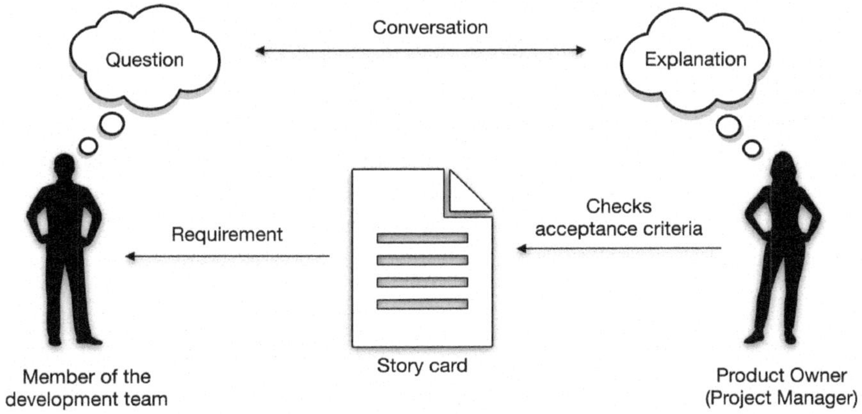

Fig. 2.3 Story cards as an agile tool to avoid misunderstandings

proaches that include the conversational aspect less explicitly, should not be doubted here. On the contrary: a recommendation for application has already been made. Nevertheless, the participation of stakeholder groups explained in Section 3.5 can enable even greater success because it dispenses with the intermediary of an authority (product owner) and brings the participants together directly.

Of the three system levels shown in Fig. 2.1, the two lower levels of neurophysiological and psychological systems have now been explained. In the following, the top level, i.e. the social level, will be supplemented, whereby many of the aspects already described, such as operational closure, self-reference, structural coupling and emergence, can be applied analogously and therefore do not need to be discussed again here. However, they will play a central role again and again in the further course, so that they will be further elaborated in the respective contexts.

First of all, it is important to realize that, according to systems theory, social systems do not consist of people or persons, but rather of symbolically and linguistically mediated *communication* between people (Luhmann 1996, p. 191 ff.; Tomasello 2022; Willke 2006, p. 65).

In his empirical studies on evolutionary anthropology, Michael Tomasello (2022, p. 13) identifies pointing and gesture play as basic symbolic forms as a psychological platform for *linguistic communication*, more precisely as a socio-cognitive and socio-motivational infrastructure. For such a simple form of communication as pointing to an object to be understood, it requires shared attention, shared experience and shared cultural knowledge (ibid., p. 15 ff.). In addition, a pro-social motivation of helping and sharing is required, a *shared intentionality* that distinguishes humans from animals. Pointing is based on the natural tendency of humans to follow the line of sight of others to external objects, and gesturing is based on the natural tendency to interpret the actions of others as intentional (ibid., p. 20). Finally, the skill of shared cultural learning and imitation is needed to create and pass on commonly understood communicative conventions and constructions (ibid., p.

23)[6]. On this basis, language finally replaces the naturalness of pointing and signing with a shared history of social learning (which all know mutually to be common) (ibid., p. 20) (ibid., p. 20) and thus increases the efficiency of communication.

In functionally differentiated social systems, part-systems such as the economy, politics, the community and culture (Parsons 1986, p. 50; 1996, p. 20) each form specific semantic codes or *guiding differences* that determine what can and cannot be communicated within them. What is expressed in the respective semantic codes makes *sense* (ibid., p. 43 ff.). Fig. 2.4 shows the AGIL scheme introduced by Talcott Parsons (also Münch 1992).

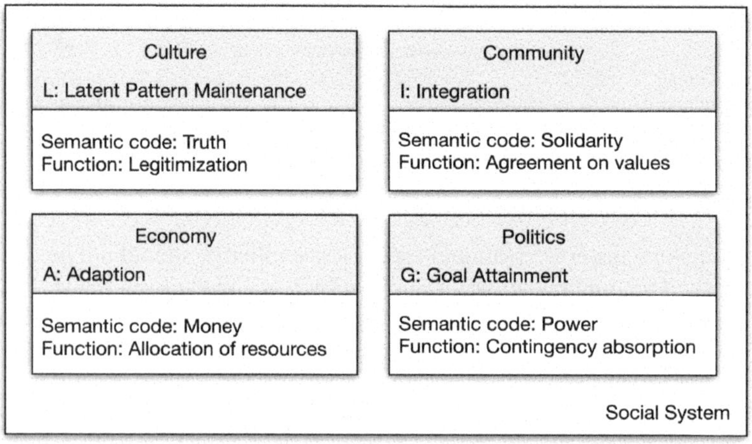

Fig. 2.4 AGIL scheme (Münch 1992, p. 35)

- Within a social system, the economic part-system is responsible for adapting to the environment (adaptation, therefore abbreviated as A). To achieve this, scarce resources must be used in the best possible way (efficiently and effectively, i.e. economically) (allocation). In doing so, the economy concentrates on what is expressed in means of payment (Willke 2006, p. 66 f.). It is largely blind to everything else. This can be seen, for example, in the treatment of common goods (Ostrom 1999) such as air, whose pollution is currently to be reduced by putting a price on it through the allocation of tradable pollution rights. Only then does climate protection become an object of economic action.
- The political part-system is responsible for the goal-oriented action of the respective social system (Goal Attainment, therefore abbreviated to G). Contingency is absorbed through decisions (March and Simon 1993, p. 165; Simon 2015, p. 66 ff.), or as Luhmann (1994, p. 170) puts it more vividly: something that is possible (contingent) in this way, but also differently becomes in this way and not differently through a decision. The basis for this is not primarily factual knowledge, but power calculations. Why this is the case

[6] The skill of imitation as a prerequisite for social learning is taken up again in connection with the SECI model (Fig. 3.33).

and what it means for system design and control is so important and at the same time so much in need of explanation that this aspect is discussed in more detail in Section 2.6.2.3

- In the community part-system, a value system develops and establishes itself that unites the members of the community in solidarity (integration, therefore abbreviated to I).
- Finally, the cultural part-system is responsible for legitimizing (justifying) the social system (latent pattern maintenance, therefore abbreviated to L). This is associated with a claim to the truth of statements, which in science, for example, must be logical (provable) and intersubjectively comprehensible, or in religion must be derivable from the respective underlying original works such as the Bible, the Koran or the Talmud.

These introductory considerations should initially suffice to make clear for all three system levels shown in Fig. 2.1 the context in which systems theory places central aspects such as operational closure, self-reference or autopoiesis, structural coupling, emergence, irritation or perturbation, language, memory, communication, meaning and semantic codes or guiding distinctions. The fact that this does not by any means explain everything that is needed for understanding becomes clear when you consider that the relevant works in the respective specialist disciplines generally comprise hundreds of pages, and some, like Luhmann's complete works, even thousands. A comprehensive understanding can therefore only be gained by studying these works. Nevertheless, the aspects mentioned will be taken up again and again in the further course and then deepened at least to some extent in order to make them accessible to the engineering-oriented SEM at least to the extent that it is useful for the design of socio-technical systems.

2.2 Models

On the basis of constructivism, the statement that models are simplified images of reality (Czichos 2019, p. 27; DIN 69901-1:2009-01, p. 6; Stachowiak 1973, p. 131 ff.) does not seem very informative, because humans cannot perceive anything other than images of reality (phenomena). It therefore makes little sense to define models in terms of their relationship to reality. To do this, one would have to be able to define what actual reality (the thing in itself) is. Another approach can be found by first defining what people usually recognize as reality. For example, the concept of image can be defined for this purpose. The actual reality would then be the *archetype*. In contrast to the image, a *model* can then be defined as something that is specifically created in order to fulfill a certain purpose for a certain subject in a certain situation. These criteria correspond to what Herbert Stachowiak (ibid.) calls *pragmatism*.

This pragmatism also corresponds to the modest claim that Ernst von Glaserfeld (1996) describes with the concept of *viability*: Actions, concepts and conceptual operations are viable if they fit the purposes or descriptions for which we use them (ibid., p. 43). For example, Isaac Newton's physics is viable for small speeds, but not on the scale of the speed of light; Albert Einstein's physics should be taken as a basis here. Moreover, pragmatism is of such fundamental importance for Systems Engineering & Management that a separate Section 2.3 is dedicated to it.

Karl Raimund Popper (1984) developed *critical rationalism* based on this world view, which forms the basis of contemporary *scientific work*. It is assumed that people are fundamentally capable of recognizing truths about actual reality, namely to the extent that these prove themselves in practice, but that they cannot be certain whether the insights held to be true correspond to objective reality. It is therefore important to constantly and critically scrutinize the insights gained and to use identified errors to gain even more valid insights. Even knowledge that appears to be certain always remains hypothetical. An assertion of objectivity is merely an attempt to evade responsibility for the subjectivity of what is believed to be true.

The model-based cognition cycle outlined in Fig. 2.5 follows from these considerations. The section of reality considered to be real (archetype, thing in itself) that appears relevant with regard to an intended problem solution is modeled through *abstraction*. This means that everything that does not promise to contribute to the immediate problem solution is dispensed with (e.g. Bjørke 1995, p. 10 f.). In addition, what remains is simplified as far as possible. This is followed by the *verification and validation* of the model (Rabe et al. 2008). In the course of verification, the aim is to check whether the model is internally consistent, i.e. whether, to put it simply, 1 + 1 really add up to 2. Validation should clarify whether the model correlates sufficiently with the part of reality it is supposed to represent. Subsequently, insights are gained from the model. These findings initially only apply in relation to the model. Whether and, if so, to what extent they prove themselves in reality only becomes apparent when they are applied in practice. This may result in modified or additional questions that trigger the model-based knowledge cycle once again. Here too, it is important to realize that human knowledge is always based on images or models of real reality, even if these are consciously, purposefully or even professionally modeled.

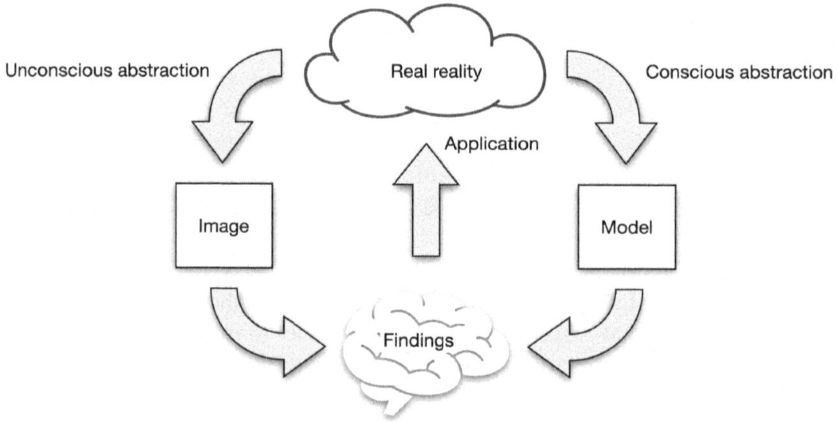

Fig. 2.5 Model-based cognitive process (Reinhardt 2003, p. 8)

A large repertoire of *methods, procedures and tools* is available for modeling and working with models. Of these, this volume focuses in particular on those that can be attributed to Systems Engineering & Management[7]. Since all further explanations deal with this, the relevant explanations here will initially only address very fundamental aspects.

This includes the fact that colloquial language is one of the most important and most frequently used tools. The struggle for words, the attempt at clear and easily understandable definitions, the constant reassurance of having been understood correctly, this and much more makes it clear how omnipresent modeling is and how important *natural language* is in this context. This is why error-free spelling (orthography) and punctuation, differentiated linguistic expression (Ogden and Richards 2013) and even rhetoric are fundamental prerequisites for professional work. In addition, there are *specialist languages* that offer more precise means of expression in specific domains. Take Latin, for example, which is still widely used in the humanities. In addition to the spoken word, however, a whole range of other, very general forms of expression are also widespread: (technical) *drawings* (e.g. Hoischen 2020), *mathematical formulas* (e.g. Papula 2018), *notations* such as the Systems Modeling Language (SysML) (e.g. Weilkiens 2014), *programming languages* and, in this context, *simulation languages* for formulating experimentable models (e.g. Gutenschwager et al. 2017) through to physical models made from a wide variety of *materials* such as cardboard, clay or matches (e.g. Ahrens 2019).

Several criteria must be applied when using methods, procedures and tools. Initially, it seems obvious to use the most suitable means for the respective application. In practice, however, such an ideal would not always be helpful, and this refers not only to real-life practice but also to scientific practice. In fact, it is also permissible to qualify this undoubtedly desirable criterion with further criteria. These include the following, whereby the order is expressly not intended to express a ranking. Moreover, the list is not exhaustive; further criteria may be added, and some may be omitted:

- Availability: If resources are already available, their use is justified if and to the extent that the objective of the modeling can be achieved.
- Cost-effectiveness: This criterion follows from availability, among other things, because the procurement of resources is usually associated with costs that can be avoided or at least reduced if available resources are used.
- Safety: This primarily refers to occupational safety, i.e. the avoidance of injuries and illnesses for all those involved and potentially affected. This applies in particular to the performance of experiments, but also to the construction of material models. Data protection and data security also serve to prevent the violation of personal rights, legitimate interests (e.g. confidentiality of trade secrets), etc.
- Environmental protection: Here too, the focus is primarily on experiments and material models. Among other things, the principle of proportionality applies here, as environmental pollution is generally unavoidable. It should therefore be proportionate to the benefits of modeling. In addition, certain experiments may be prohibited altogether if the damage to the environment is unacceptable and/or irreversible (worst-case scenario, catastrophic potential).

---

[7] For example, music and dance are also forms of expression, albeit ones that have not yet been harnessed for Systems Engineering & Management.

- Law and morality: Any modeling must be in line with the law and prevailing morality. In doing so, interests and values that are distant in terms of space (e.g. Supply Chain Obligations Act (LkSG)) and time (intergenerational justice) must be taken into account.

2.3 Pragmatism

Since Systems Engineering & Management serves the concrete design and control of real systems, the measures aimed at this should be practical, i.e. effective in reality. In the light of constructivism, however, humans have no direct access to this actual reality. As already explained in the previous Section using Fig. 2.5, we only have an image (model) of the original. If we follow the founder of pragmatism, Charles Sanders Peirce (Pape 2002), the usefulness[8] of a model – to put it simply – can only be measured by whether and, if so, to what extent the action based on the model proves itself in practice, i.e. in reality (pragmatic maxim). Therefore, the *experiment* as a scientific method is of particular importance. This allows a bridge to be built, so to speak, between a model in the sense of an idea according to which an action could be successful in reality and the direct confrontation with reality, which can punish a wrong idea with serious consequences under certain circumstances. Interposing an experiment can and should mitigate such consequences.

However, pragmatism should not be misunderstood to mean that problems and tasks can largely be solved or completed with *common sense*, as the saying goes. Natural judgment based on a broad social consensus is also necessary for professional Systems Engineering & Management – but not sufficient (Watzlawick et al. 2013, p. 13). Pragmatism also does not mean gaining knowledge solely through intuition or self-observation (introspection), although this also provides Systems Engineering & Management with a very good starting point.

According to the ideas of Charles Sanders Peirce (ibid.), the principle-based fallibility of intuition and intellect can only be alleviated through continuous attempts at falsification in the sense of critical rationalism (fallibism), i.e. through *scientific work*. Pragmatism therefore does not mean purely or predominantly practical work, but rather theoretically founded work that must prove itself in practice. However, pragmatism calls to mind the limits of such work and is temporarily content with the fact that not every approach, not every method and, in particular, not every result of human action does optimal justice to reality. However, this acceptance of limits is temporary in that every encounter with limits provides an impetus for further optimization. Charles Sanders Peirce hopes that such an ongoing process of research will bring us closer and closer to reality, but without expecting that a final correspondence between the image (model) and the archetype (reality) can ever be achieved.

Charles Sanders Peirce's work on pragmatism is also of considerable importance for Systems Engineering & Management insofar as it was continued by John Dewey (1997), among others, to whom the problem-solving cycle on which this article is based can be traced back (Haberfellner et al. 2018, p. 70). At the same time, John Dewey, together with George Herbert Mead, gave significant impetus to the Chicago School of Sociology, which

---

[8] The aspect of usefulness is emphasized above all by William James (Pape 2002).

is still influential today. In turn, Herbert Blumer emerged from this school as the founder of symbolic interactionism, a theory that provides a basis for the approach to mediating between implicit and explicit knowledge in the course of Change Management explained in Section 3.5.4. This makes it clear that the deeper roots of Systems Engineering can be found in the USA. The Anglo-Saxon culture prevalent there meets continental European culture in this volume, for example the purely deductively derived sociology of Niklas Luhmann (1996), which in turn is confronted with the sociology of the US American Talcott Parsons[9].

2.4 Modeling of Technical Systems

The previous considerations on constructivism have far-reaching consequences for the models widely used in Systems Engineering & Management. In order to be able to work this out, modeling approaches that are widespread in the technological context are first introduced below. This is followed by a discussion of these approaches in the light of constructivism in order to finally arrive at a model conception that brings together all the aspects explained so far.

2.4.1 Functional Model of Technical Systems

Fig. 2.6 shows a very frequently used type of system modeling in Systems Engineering & Management. According to this model, a system processes (P) inputs (I) into outputs (O) according to a function. The model is therefore also referred to as the IPO model. The Six Sigma methodology extends this model further to the SIPOC model, in which the S stands for Suppliers and the C for Customers (e.g. Niemann et al. 2021, p. 20). When modeling the dynamic behavior of systems in particular, the interfaces between the system and the environment are also referred to as *sources* on the input side and *sinks* on the output side (Wenzel 2018, p. 7).

A distinction is usually made between mass, energy and information (Czichos 2019, p. 10), whereby information in this context is subdivided into data and commands, while mass can be organic or inorganic. Depending on the field of application, some of these generic variables can be neglected, mapped to each other or concretized. In computer science, for example, mass and energy often only serve as information carriers. In physics, mass and energy can be converted into each other using the squared speed of light as a proportionality factor. In production and logistics, mass is treated as a material and in economics, information is often concretized into monetary units.

Such flow variables are modeled, for example, in discrete form using so-called *tokens*, which pass through the system after certain time intervals (intermediate arrival times). In other domains such as electrical engineering or communications engineering, *signals* are used when information is assigned to a measurable physical quantity.

[9] This can be seen as an attempt to build bridges between influential occidental scientific cultures.

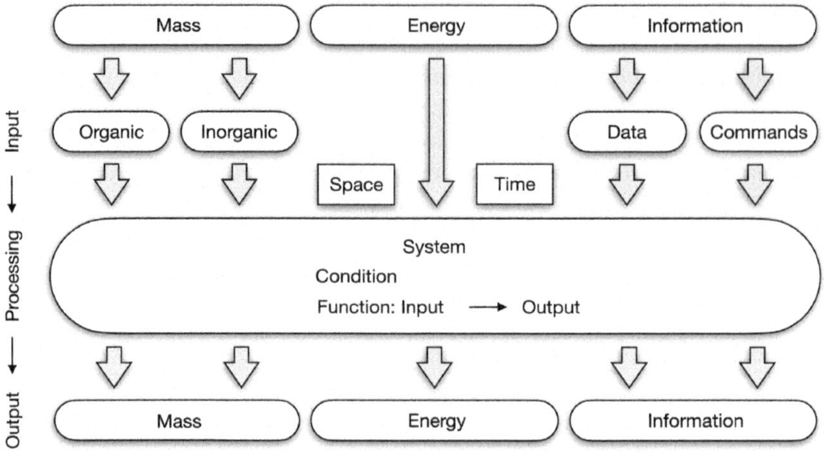

Fig. 2.6 Functional model of technical material systems (Ropohl 2009, p. 97)

In addition to the intended state and flow variables, unintended variables must also be taken into account and included in the model if required. On the input side, these are in particular *disturbance variables* such as mechanical vibrations or impulsive shock and impact processes, electromagnetic fields and voltage or current peaks, thermal influences and temperature fluctuations as well as atmospheric and climatic influences such as humidity and gas contamination (Czichos 2019, p. 11). *Dissipation effects* must be taken into account on both the input and output side as well as in the course of the processing process, i.e. friction that leads to the conversion of mechanical energy into heat, wear in the form of material or shape changes or information losses, for example during the conversion of analog to digital signals (ibid., p. 12). Finally, *losses* must be taken into account on the output side, for example by specifying a degree of efficiency.

In view of the increasing importance of data processing and communication (so-called digitalization), it may be of interest that mass and energy cause scarcity in equal measure: For example, only one of several people can own an object at any one time. If there are fewer objects than interested parties, there will always be some who go away empty-handed. If one person gives their object to another, they will no longer have it. The situation is different with information: if someone gives information to someone else, they still have it afterwards: the information has multiplied. And what's more: when information is passed on, new information is generated, in this case, for example, the information that one person has given information to another. Information therefore does not create scarcity, but abundance. In addition, unlike mass, information cannot be easily demanded back, with the result that the person giving it back no longer has it. Once given out, information remains in the world, so to speak.

Basic equations are available to formulate the relationship between input and output variables, i.e. the so-called transition function (Czichos 2019, p. 27). For example, stored masses, energies and momentum are represented in the form of balance equations; physico-chemical equations of state take the form of constitutive equations; processes such as heat conduction are represented with phenomenological equations; several irreversible processes are modeled using entropy balance equations and linking equations are used to represent the interconnection of processes.

In detail, the relationship between input and output variables is described by the so-called *transition function*. Simple test functions, such as the step function shown in Fig. 2.7, are usually used as input variables to investigate the signal transmission behavior. In the *stationary*, i.e. time-independent case, the reaction of the output variable v to the input variable u can be described as the function $v = f(u)$ and represented graphically by a $u - v$ characteristic curve (here and in the following Czichos 2019, p. 15). The *dynamic behavior* is described on the one hand by the frequency response and the *root locus plot* and on the other hand by the *Bode diagram*:

- If the input signal $u(t) = u_0 \cos \omega t$ is applied with the initial amplitude u_0 and the angular frequency ω, this results in the amplitude v and the phase shift φ at the output in the steady state: $v(t) = v_0 \cos(\omega t + \varphi)$.
- The locus curve is displayed in the complex plane either in Cartesian form with the real part on the abscissa and the imaginary part on the ordinate or in polar coordinates with the pointer length $A = v(\omega)/u$ and the angle φ.
- In the Bode diagram, the amplitude gain (magnitude) and the phase shift (argument), referred to as the phase response, are displayed logarithmically one above the other as a function of the frequency for the steady state. Logarithmizing facilitates the creation of complex diagrams by replacing the multiplication of partial functions by their addition. In addition, the frequency-dependent behavior can be clearly displayed over a very wide range.

Many representations do not show the model outlined in Fig. 2.6 from top to bottom, but from left to right, and declare the inputs and outputs in an application-specific manner (Fig. 2.8, top left). Examples can be found in Reinhard Haberfellner et al. (2018, p. 32) and Petra Winzer (2016, p. 90), but also in completely different fields of knowledge such as the historical sciences (e.g. Goetz 2014, p. 22). Further differentiations and additions are occasionally added. For example, Christof Ebert (2019, p. 30) adds boundary conditions in connection with *Requirements Engineering* (Fig. 2.8, top right) and interprets the model in this form as a functional requirement. Joe Jenney (2010, p. 96) extends the IPO model to a context model (Fig. 2.8, bottom left) by adding controlling influencing variables, i.e. instructions and constraints (boundary conditions), as well as supporting influencing variables (resources such as people, machines and infrastructure) to the input and output. This corresponds to the IDEF0 model of the *IDEF model family* developed by the US Air Force, which in turn originated from the *Struktured Analysis and Design Technique* (SADT) model. A final example of modified forms of the IPO model can be found in the statistical *Design of Experiments* (DOE) explained in Paragraph 4.4.5.1.2.4 4.4.5.1 (Schneider 2020, p. 255 ff.; Siebertz et al. 2010, p. 3), which distinguishes between targeted and non-targeted changeable inputs (Fig. 2.8, bottom right). Occasionally, the IPO model is also interpreted as a process model (e.g. Winzer 2016, p. 90) and the flow it contains does indeed indicate dynamic behavior. In processes, however, the IPO model only forms one process element (DIN EN ISO 9001:2015-11, p. 12); only the combination of at least two such process elements results in what is commonly understood as a process (Sect. 2.4.5).

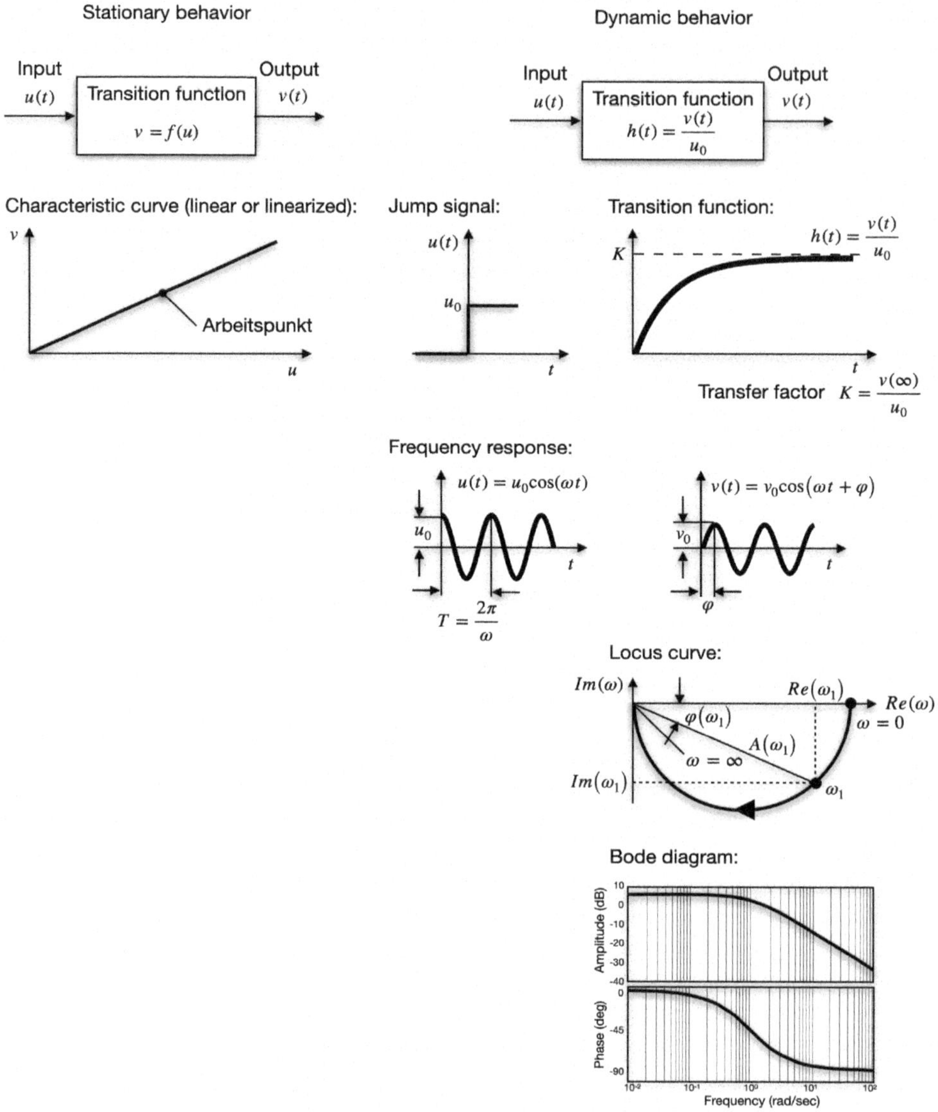

Fig. 2.7 Characterization of stationary and dynamic properties of technical material systems (Czichos 2019, p. 16)

Systems modeled in this way are open in every respect: an observer knows the function or process according to which an input is transformed into an output. The flow variables flow through the system virtually loss-free (unless losses are explicitly modeled), so that the system also follows the commands that are issued from outside in the course of the information flow without resistance. This corresponds to the idea of what Heinz von Foerster (1993, p. 245) calls a *trivial machine* (TM). As an example, he cites a cash register, such as

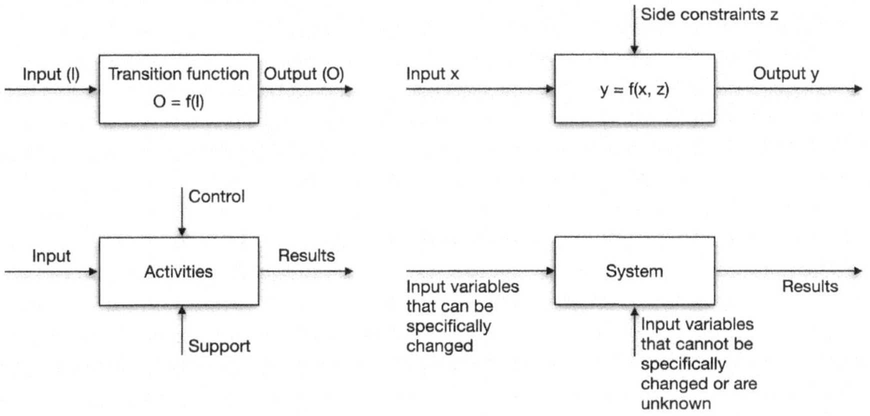

Fig. 2.8 Variants of the functional model of technical material systems

the one in a supermarket, with Noodles ... $3.50 on the receipt when the item's barcode is scanned. However, everyday experience shows that it is sometimes impossible to scan the barcode, even after several attempts, so that cashiers sometimes feel compelled to enter the Global Trade Item Number (GTIN) manually. Even such simple technical systems appear to be more complex than is generally thought, but more on that later. First, other forms of modeling commonly used in Systems Engineering & Management will be presented to show that they all involve almost the same problems.

Horst Czichos (2019, p. 13 ff.) demonstrates the potential of interdisciplinary thinking using the function model by replacing the usual flow variables of material, energy and information with the four-pole representation (also: two-port representation), which is particularly common in electrical engineering (Fig. 2.9), in which two terminals are attached to the input and output side. On the input side, an effort must be made, which in the model world of *bond graphs* (Roddeck 2019, p. 9) is generalized as efford $e(t)$ (cause) and from which an effect, also generalized as flow $f(t)$, results on the output side. The product of these two variables results in the power output P. The efford-flow relation is formed by a domain-specific resistance variable, in electrical engineering for example by the ohmic resistance R with the resulting efford-flow relation $v = R \cdot i$ (Ohm's law).

After the previous considerations have increased the degree of abstraction, the function carriers can also be concretized in the opposite direction. In mechatronics, for example, it is common to distinguish between (1) sources and (2) sinks, (3) transformers, (4) converters and (5) storage units (Czichos 2019, p. 28). (1) *Sources* release material, energy or information from a supply. In mechanics, they can take the form of springs or weights (potential energy); in electrical engineering, they can be capacitors (electric field), coils (magnetic field) or accumulators (electrochemistry). In fluid mechanics, tanks in particular perform this function. (2) In mechanics, *transmitters* are, for example, gears, in electrical engineering transformers and in thermodynamics heat exchangers. (3) In mechanics, *transducers* are manufacturing processes in accordance with DIN 8580:2022-12, i.e. forming, reshaping, cutting, joining, coating, assembling and changing material properties. In electrical

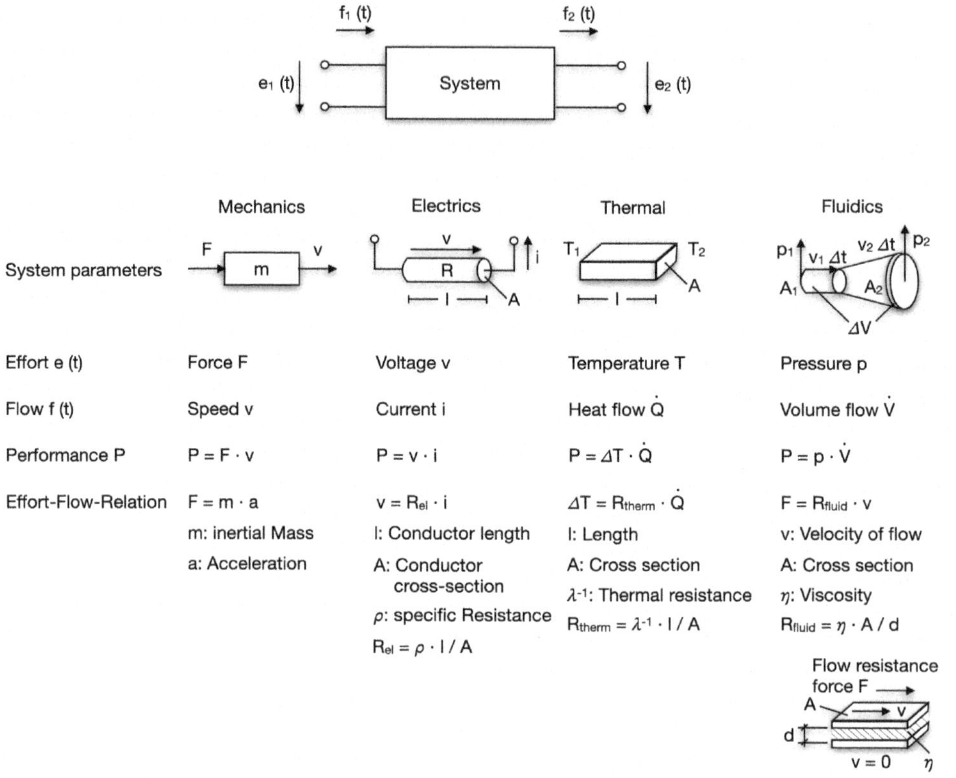

Fig. 2.9 Analogies between mechanics, electrics, thermics and fluidics using the generalized four-pole model (Czichos 2019, p. 14)

engineering, motors or generators and electromagnets are used. The combustion engine is a thermal example and pumps and fans are used in fluid mechanics. (4) *Accumulators* temporarily absorb flow variables and (5) *sinks* lead the flow variables out of the system. They can also be used to model dissipations, for example through dampers, hydraulic throttles, electrical resistances or friction.

2.4.2 Structural Models

If you combine several functional models, you get the structural model shown in Fig. 2.10 (Haberfellner et al. 2018, p. 37; Ropohl 2009, p. 76). From this perspective, the focus is primarily on sub- and part-systems as well as *elements* and their *relations* to each other. Fig. 2.10 only shows relations between elements. However, relations between other definable units such as systems, sub-systems or part-systems are also possible. A structure therefore describes the arrangement pattern of elements and their relations to each other. In concrete application, this approach is widely used as an *Entity-Relationship Model* (ERM). This often additionally characterizes elements through attributes. For example, the element employee is given a first name, a surname and a personnel number. There are numerous variants of the notation, such as IDEF1X from the IDEF model family developed by the US Air Force or the Chen notation, which goes back to the originator Peter Chen. ERMs are particularly widespread in computer science, for example in *semantic data modeling*.

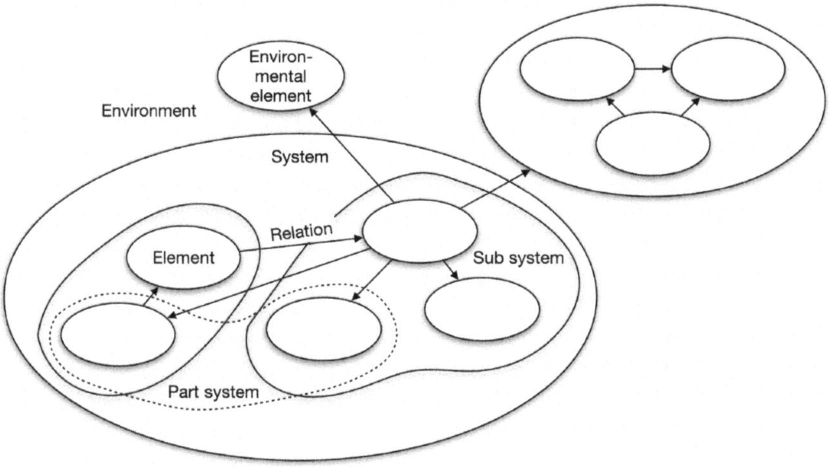

Fig. 2.10 Structural model (Haberfellner et al. 2018, p. 28)

This can be, for example, the arrangement of workstations in a factory hall that are related to each other through material flows. A possible structure would then be that of flow production (flow shop), according to which the stations are always arranged in the sequence of the same work steps, or that of workshop production (job shop), in which similar workstations are grouped together in so-called workshops (setup principle), for example all lathes in the turning shop (e.g. Wiendahl 2014, p. 42). Fig. 2.11 shows typical *topologies* that are specifically designated in different application areas. For example, the topology already referred to as flow production in the context of production corresponds to the line topology or the BUS topology (BUS: Back (mounted) Panel Unit), while the line topology in logistics, for example, is referred to as the supply chain. In logistics, the ring topology is referred to as a milk run and the star topology as a hub-and-spoke system. Names that have evolved historically and domain-specifically can therefore be traced back to basic patterns that are transferable across disciplines.

In the social sciences, structures of social systems and networks are analyzed using *sociometry* (Diekmann 2017, p. 495 ff.). The results are, for example, sociomatrices (relationship matrices) and sociograms (Fig. 2.10). As a rule, they depict communication behavior and thus take into account the insight explained in Section 2.6.2.2.3 that communication acts form the elements of social systems and not, as one might commonly think, people or persons. To put it simply, typical results of sociometric analyses can be described

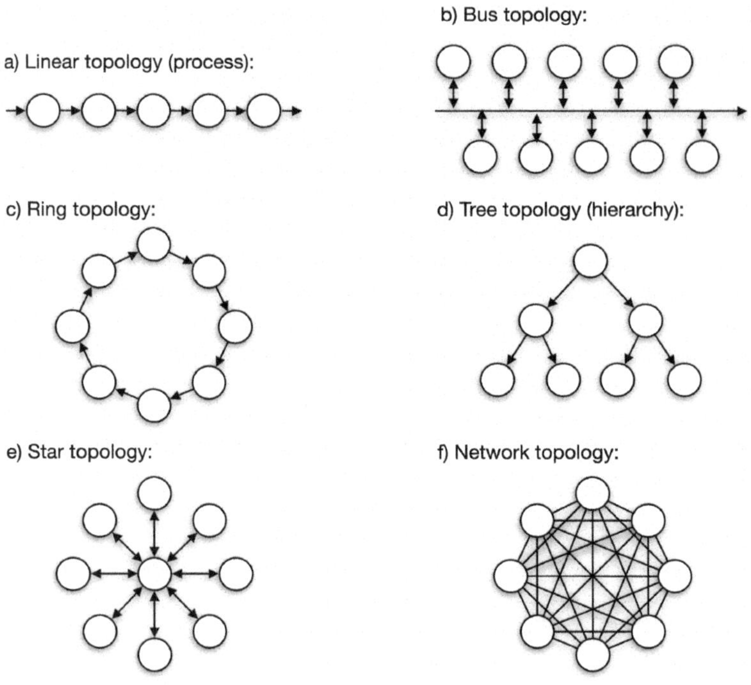

Fig. 2.11 Ideal (trivial) topologies

as answers to questions such as: who talks to whom about what and why[10]? Methods such as observation (Sect. 4.2.4.3), interviews (Sect. 4.2.4.2) and document studies (Sect. 4.2.4.1) are used for this purpose. The latter deals with process descriptions in organizations, for example, and primarily reveals the formal structures, while the methods mentioned above also and especially make informal structures transparent – more on this later. The results of such analyses are an essential basis for *organizational development*, among other things (e.g. Schiersmann and Thiel 2018).

Fig. 2.10 also shows two *sub-systems* and one *part-system*. While the designation as a sub-system refers to a hierarchically subordinate relationship to the super-system, part-systems arise independently of a hierarchical integration. For example, it would be conceivable to designate departments of a company's parent organization, which is generally set up to carry out routine activities, as sub-systems. Project groups that are set up on a temporary basis to solve problems and to which personnel are sent who come from different departments and remain anchored there could then be referred to as part-systems that are delimited quasi transversely to the core organization. However, it should also be noted that social systems created in this way do not consist of people or individuals, but of the acts of communication that they contribute.

[10] Other questions such as where or for how long can easily be added.

Based on *graph theory*, which provides mathematics with an important basis for discrete modeling (e.g. Bungartz et al. 2009, p. 21), further characteristics can be named for the structures of systems:

- They therefore consist of a finite *number* V of elements (optionally also sub- and/or part-systems) and an equally finite number E of relations. If their order is important, mathematicians combine these two numbers into so-called tuples (V, E). An infinite number of elements is sometimes mathematically acceptable, but makes no sense from a systems theory perspective, as it would then no longer be possible to distinguish a system from an environment and/or other systems, and an infinite number of relations would also make no sense from a systems theory perspective, as structures could then no longer be formed.
- The relations can each point in a certain *direction* (directed relations). This is indicated by arrows. Alternatively, two elements can only be connected to each other (undirected relations). This is represented by simple lines.
- The *neighborhood* v of an element can also be specified. This refers to the number of other elements connected to the respective element. The term valence is also used synonymously for this, which describes the number of bonding electrons and therefore the number of possible bonds of an atom in Bohr's atomic model, for example.
- In a number of elements connected by directed relations, any two elements are connected to each other by a *path*, i.e. a sequence of relations. If not all elements of a system are connected to each other, there are several connection components. An *Eulerian tour* is a path that includes each element exactly once.
- A path that starts and ends at the same element is called a *cycle*. In Fig. 2.11 this is shown as a ring topology. A *Hamiltonian circle* is a cycle that includes each element exactly once.

The term *architecture* is used synonymously with the term structure in Systems Engineering (e.g. ISO/IEC/IEEE 42010:2011). However, Reinhard Haberfellner et al. (2018, p. 189) note that, unlike the term structure, the term architecture is rarely defined and used precisely. However, because it is seemingly impossible to get rid of, they provide a definition in such a way that a system architecture not only describes the arrangement pattern of the elements and their relationships, like the structure, but also the assignment of functions to the elements. However, they fail to explain why a separate term is required for this and why the concept of architecture should be predestined for this purpose. Joachim Goll and Daniel Hommel (2015, p. 19) subsume under the architecture of a system not only the structure, but also the sequence of functions, by which they presumably mean what is usually referred to as a process. Rüdiger Kaffenberger et al. (2012, p. 92), on the other hand, define architecture as the selection of elements, their properties and their arrangement. Even this brief selection of attempted definitions shows that Reinhard Haberfellner et al. (2018, p. 189) are correct in their observation of an imprecise definition of the term architecture.

In fact, since ancient times, architecture has meant the art of building, i.e. an artistic or at least aesthetic approach to built space (e.g. Glancey 2022), while the technology of building is the responsibility of civil engineering, and many a civil engineer will have despaired of the artistic ideas of architects if they have strayed too far from functional or economic possibilities, so that a solution only seems possible with the help of the well-known Siemens air hook. Architects are allowed to do this, but it has so little to do with engineering that the term architecture should remain where it belongs. In any case, it is no longer used here.

In view of the fact that further considerations will focus on how the concept of *complexity*, which is also often only used in a vague way, can be defined more concretely, we will first briefly explain why systems form structures in the first place. The cybernetic *law of requisite variety* (Ashby 1956, p. 206 ff.) is chosen as the starting point, according to which a system that wants to assert itself in a complex environment (external complexity) must have at least the same degree of complexity – here in the sense of a stock of variation. Added to this is the resulting own (internal) complexity, which also requires a corresponding level of complexity to be mastered (Ahrens 1998, p. 14 f.). With increasing complexity, this would quickly lead to the systems being overwhelmed. In order to avoid this, systems first form a *boundary* between themselves and their environment. This is constitutive for systems and distinguishes them from their respective environment in that the latter has no boundary from the perspective of the focal system.

In the second step, systems use their elements and their relationships (relations) to each other to form structures and processes that reproduce (reconstruct) the complexity of their respective environment. However, the number of elements and their respective linking capacity (valence) are limited. As a result, no system can produce a point-to-point mapping of the external complexity. Instead, every system is forced to reduce complexity, i.e. to create a simplified image of its environment (abstraction). This corresponds to the constructivist understanding explained in Section 2.1. At the same time, it becomes clear at this point that the reduction of complexity in itself is not a virtue that should be recommended without further ado. In fact, it is unavoidable out of necessity, while it would actually be necessary to reconstruct the external complexity within the system as precisely and comprehensively as possible in the sense of the law of requisite variety (ibid.).

Against this background, Niklas Luhmann (1996, p. 46) defines systems as complex if, due to immanent limitations of the linking capacity of the elements, each element can no longer be linked to every other element. While this approach is discussed in more detail and then quantified in Section 6.1.1 on the basis of its practical application to business organization, only the reverse conclusion will be drawn here, namely that systems that are capable of linking each of their elements with every other element (regardless of whether they actually do so) can be described as trivial. Fig. 2.11f shows such a structure as a model (abstract) and companies with very few employees are a practical example of this, but only in this respect; they can also be complex in other respects.

Furthermore, the need to develop the system-internal structures and processes in such a way that they allow the system to assert itself in its environment as well as to cope with its internal complexity means that this cannot be left to chance. For social systems, it can be said with reference to Niklas Luhmann (ibid., p. 92 ff.) that the structures must make *sense*. This in turn leads to the insight that systems whose structures arise by chance can also be described as trivial.

The distinction to trivial structures now allows the conclusion that meaningful structures must result from corresponding *formation laws*. Such laws are examined as models by *network theory*, among others. Insights that can be gained from this will be explained below using the example of the BA model by Albert-László Barabási and Réka Albert (2016, p. 164 ff.). For this purpose, the term network is defined more generally than usual, in that all connections of elements, regardless of topology, are understood as networks. Line-like, star-shaped or tree-like structures (Fig. 2.11) are then special forms of networks.

The formation law of the BA model makes the probability of linking two elements dependent on their attractiveness, which is expressed in the number of valences v already present before a new link is added. This assumption is also referred to as *preferential attachment*. If a new element always attached itself to the existing element with the highest number of valencies, a centralized structure, i.e. a star topology (Fig. 2.11e), would result. If every new bond were left to pure chance, a network topology would emerge. However, the topology sketched in Fig. 2.11f is only a basic example of this, as all elements are linked to all others, whereas random networks with a smaller number of valences have to make do with fewer links.

The star structure on the one hand and the network structure on the other are two trivial extreme forms of networks that are represented by the BA model. In between, the formation law stipulates that new elements continue to link preferentially with elements that already have a large number of links, but that other elements also have a chance (Dittes 2021, p. 112). The probability p_i of the element i attaching is therefore proportional to its valence v, where n is the number of all existing elements:

$$p_i = \frac{v_i}{\sum_j^n v_j} \tag{2.1}$$

The development of a network begins with a finite number m_0 of elements. These can be connected to each other in any way by any number of relations as long as each element is related to at least one other element. A new element is then added at each time step, which is linked to $m \leq m_0$ existing elements according to probability p_i. Fig. 2.12 shows that a few particularly strongly networked elements are established in this way, as well as a number of other elements with a significantly lower degree of networking and a very large number of elements that are hardly networked at all. In logistics, such networks are referred to as hub-and-spoke systems: Hubs form the hubs (depots) of a wheel, so to speak, from which the spokes (depots) branch out into the surface. This practical example demonstrates how well models like this can reconstruct reality.

However, the chosen optics can be deceptive: The interpretation just outlined is more or less tacitly based on the assumption that such networks spread out quasi-horizontally. This in turn suggests a lack of hierarchy. This is also often the case, for example in the linking of Internet pages or in the frequency with which publications are cited in other publications. Nevertheless, it is quite possible to interpret structures such as the one shown in Fig. 2.12 as a hierarchy according to Fig. 2.11d. These are usually thought of vertically. These are usually thought of vertically, and so Fig. 2.13 shows the same structure as Fig. 2.12, but in the form of a chain of instructions from top to bottom. In this case, the diameter of the positions can be interpreted as the line margin.

Regardless of how structures that arise according to educational laws such as that of the BA model are interpreted, what they have in common is that they can be shown to be complex and therefore, in the sense of the law of requisite variety (Ashby 1956, p. 256

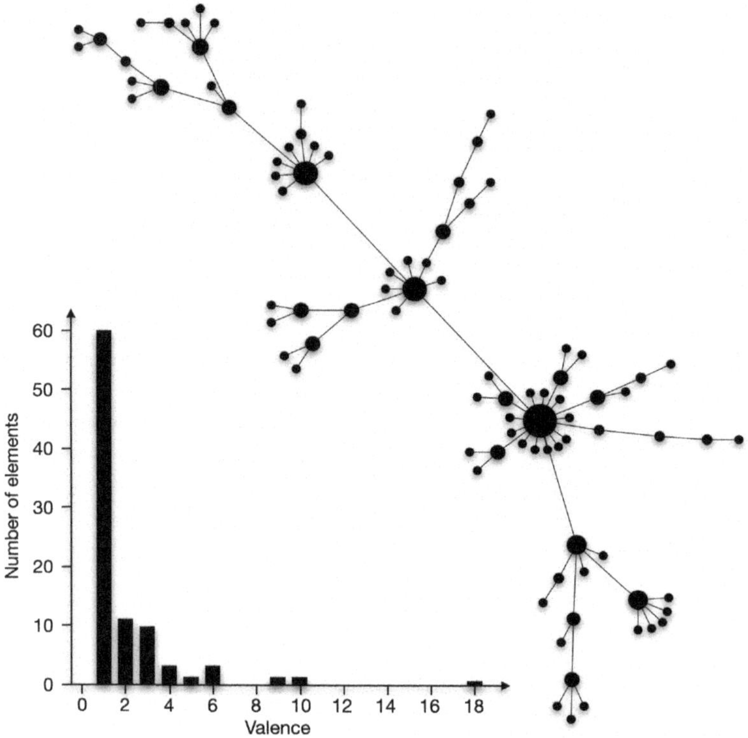

Fig. 2.12 Topology of a complex network based on the BA model with valence number v = 1

ff.) are suitable for contributing to the design and control of complex systems, whether in business organization (see Section 6.1 for more details) or in a technical context, for example in the form of the automation pyramid outlined in Fig. 4.60 (Heinrich et al. 2020, p. 4; Seitz 2015, p. 24) or energy supply networks (Dittes 2021, p. 118).

The latter were initially structured hierarchically within countries in accordance with Fig. 2.14 after the so-called electricity war between Thomas Alvar Edison, who initially built decentralized power plants with his preferred direct current technology, and Nikola Tesla and George Westinghouse, who prevailed with their preferred alternating current and the energy transmission over long distances that this made possible:

- Large power plants at the top of the hierarchy feed into the extra-high voltage grid at 220 kV or even 380 kV,
- Medium-sized power plants and large consumers such as the railroads or industrial companies with high energy requirements are connected to the high-voltage grid below them, which is usually 110 kV,
- Urban power plants and smaller industrial companies are connected to each other via the medium-voltage grid with up to 35 kV, and
- At the lowest level of the hierarchy, there are low-voltage grids (city grids, local grids) with 400 V and 230 V.

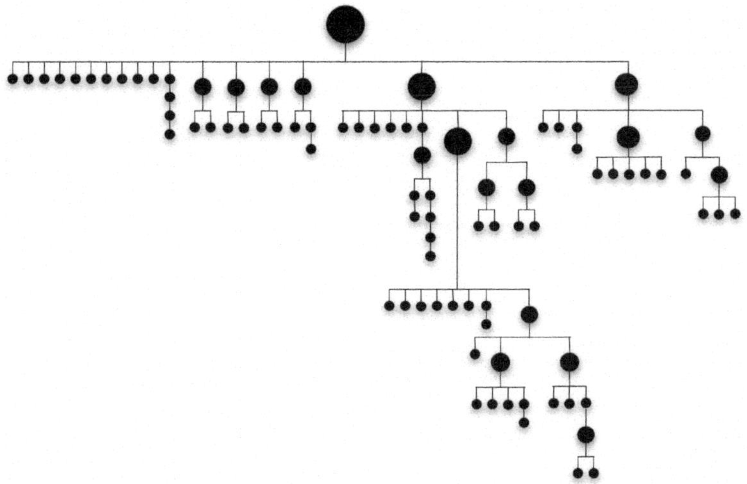

Fig. 2.13 Hierarchical interpretation of the network in Fig. 2.12 as a chain of command

Such grids are controlled by central switching lines (synonymously: control centers, control rooms), which provide so-called system services such as keeping the grid frequency and voltage constant or restoring the energy supply after an outage. In this way, the energy supply grid is operated as an interconnected grid. In Europe, this includes the European Interconnected System (EIS). At this level, the grid already has the characteristics outlined in Fig. 2.12. Such structures are also increasingly expanding at the lower levels of the grid, as renewable energies such as wind turbines and solar collectors are primarily being installed on a decentralized basis. At the same time, the role of former energy consumers is changing, as they are increasingly generating energy at the same time, for example by installing solar systems on their own roofs. As a result, the complexity of the energy grid is increasing significantly, making centralized operation ever more challenging and disruptions ever more likely – but also because the infrastructure is increasingly becoming the target of attacks by hackers, for example. It therefore remains to be seen to what extent the energy supply grid will move closer to the structure outlined in Fig. 2.12 and thus possibly lose at least part of its interconnected structure and promote island grids that can continue to be operated regionally if other parts are disrupted.

To demonstrate the complexity of such networks, let us first consider the temporal development on which the BA model is based. Although the number of valences, their so-called degree, is always a natural number ($v_i \in \mathbb{N}$), the dynamics can be expressed by the following differential if, for better computability, it is approximated that the degree can assume all real numbers ($v_i \in \mathbb{R}$):

$$\frac{dv_i}{dt} = m \cdot p_i \tag{2.2}$$

For large numbers, the solution is integration:

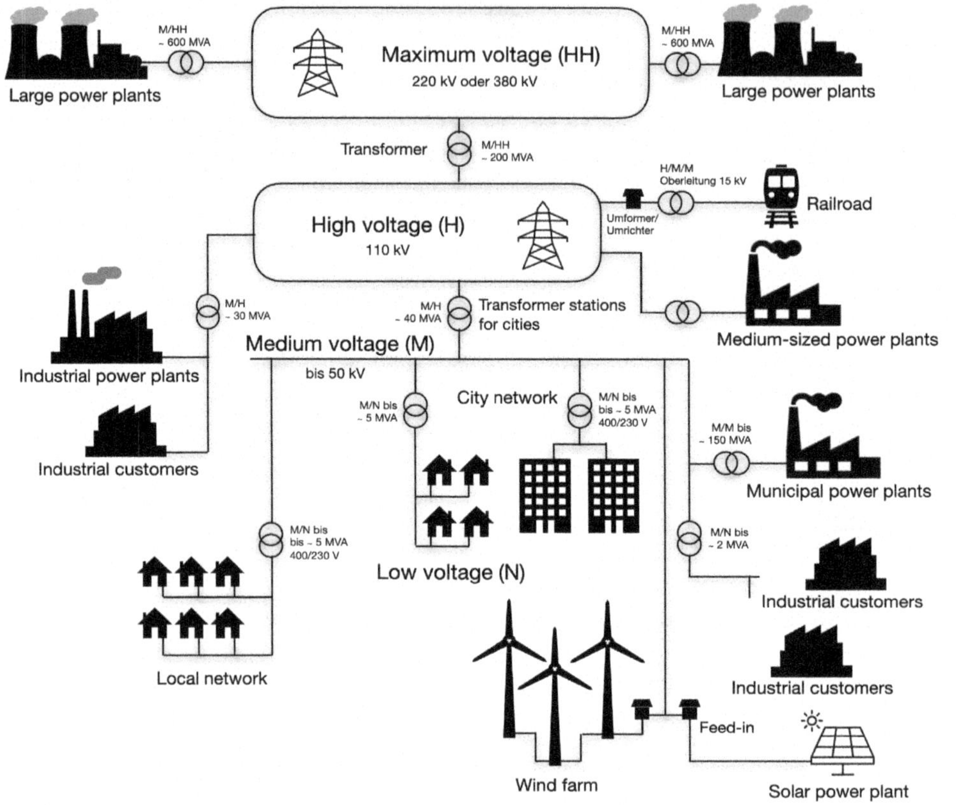

Fig. 2.14 Practical example of a hierarchically structured system: interconnected energy supply network

$$v_i\,(t) = m\left(\frac{t}{t_i}\right)^{\beta} \quad \text{with } \beta = \frac{1}{2} \tag{2.3}$$

In this, t_i is the time at which the i-th element is added to the network. The more elements the network contains, the greater the competition for new relations. As a result, the rate of accumulation of relations decreases. Furthermore, older elements have an advantage over younger ones, as they have had more time to accumulate a larger number of relations.

The distribution of the degree of valence among the elements follows a *power law*:

$$p_v = \frac{2m\,(m+1)}{v\,(v+1)\,(v+2)} \tag{2.4}$$

If the frequencies are plotted over the valence numbers in double logarithmic representation, a straight line is obtained (Fig. 2.15), since it is generally true that $\log(y) =$

$-n \cdot \log(x)$, if $y = 1/x^n$. In this specific case, the exponent is -2.5 and is independent of the average number of valencies of an element. This is a *scale-free network*. Scale-invariant structures show similar properties (self-similarity) in different orders of magnitude (dimensions). The advantage of such quasi-natural structures is their *robustness* (fault or failure safety), as a large network can break down into smaller individual networks in the event of a failure, even of nodes with a high degree of interconnection, which continue to operate at least in their catchment area. If, on the other hand, the central node in a centralized network (star structure) were to fail, the entire network would cease to function. An overly interconnected system would again be similar to a randomly formed network and would be complex (expensive) due to *redundancies* and slow (sluggish) due to long distances. Therefore, when designing networks, it is important to choose a medium degree of networking that allows a compromise between robustness on the one hand and effort and inertia on the other (Dittes 2021, p. 115). This may require different solutions in different areas, depending on the functions performed by elements in a network (e.g. clients or servers).

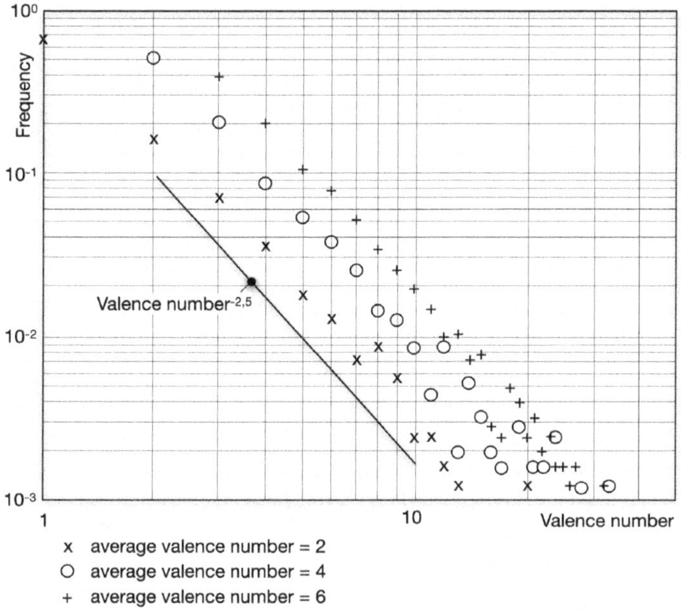

Fig. 2.15 Distribution function of the valences of a network with preferential attachment (Dittes 2021, p. 113)

To generalize, one can say that complex systems form the greatest possible variety of structures with the help of power laws (ibid., p. 42). The power laws express the fact that the elements of the corresponding systems or systems are coupled to each other and thus correlate (interact). Complex systems are always pushed to the limits of their capacity in order to achieve the optimum exploitation of what needs to be exploited in each specific case. In this context, the limit is not the system limit, but the state at which exceeding the limit jeopardizes the existence of the system. Systems always go to the extreme, so to speak, and are always in a critical situation (ibid., p. 22 ff.). The latter is explained further in Section 3.1. Here, therefore, only a simple, illustrative example is given. The development of a system is generally determined by the product of the number of offspring and the reproduction rate, whether these are atoms in the course of a chain reaction or children of parents. If the product is greater than one, i.e. more offspring survive from generation to generation, the system explodes. If it is smaller than one, the development is dampened and eventually comes to a standstill. A system can only maintain a permanent equilibrium at the boundary between explosion and damping. This equilibrium is critical in that even slight deviations from it can lead to one extreme or the other, thereby jeopardizing the existence of the system. The sensitive reaction of systems to minor changes is a characteristic of complex systems. The strong causality principle, according to which similar causes have similar effects, is called into question by the so-called butterfly effect (Argyris et al. 2017, p. 303 f.), according to which small changes can have large effects and vice versa.

Having previously discussed two-dimensional structures with networks, these are to be supplemented with *three-dimensional structures* below. Such structures can be found in solid state physics and chemistry as well as in mineralogy on the order of 1 nanometer in the form of seven crystal systems categorized according to their symmetry. If the coordinate origin is placed at the center of a crystalline body, triclinic, monoclinic, orthorhombic, tetragonal, trigonal, hexagonal and cubic geometries (point groups) can be distinguished. Crystalline structured bodies are contrasted with amorphous bodies whose elementary structure shows no structures. In the order of 100 picometers, the same bodies show an atomic structure, which is modeled quantum mechanically.

Fig. 2.16 shows the 14 known *Bravais lattices* (space groups), which are used in Kirstal lography to differentiate between possible lattice systems. The cubic lattice shows the highest form of symmetry: all three axes are of equal length and have angles of 90 degrees to each other. The body-centered cubic and face-centered cubic lattice, as well as the hexagonal lattice, are found in metals (Arnold 2013, p. 53 ff.). In materials technology, abbreviations such as krz lattice for body-centered cubic lattices, kfz lattice for face-centered cubic lattices or hdP lattice for hexagonal lattices with the densest packing can be found. In crystallography, the Hermann-Mauguin symbols or Pearson symbols are also widely used. The lattice with the least symmetry is the triclinic lattice: all edge lengths differ from each other, as do all angles. It is found, for example, in the mineral anorthite, which can be used for the extraction of aluminum as well as for the production of glass fibers and refractory materials.

2.4.3 Hierarchical Models

The possibility of bringing system structures into the form of a hierarchy (Fig. 2.17) is often highlighted in systems engineering as an independent option for modeling (e.g. Haberfellner 2018, p. 32; Walden et al. 2023, p. 12 ff.). In agreement with Petra Winzer (2016, p. 90), however, it should be noted that the hierarchy is actually only one of several

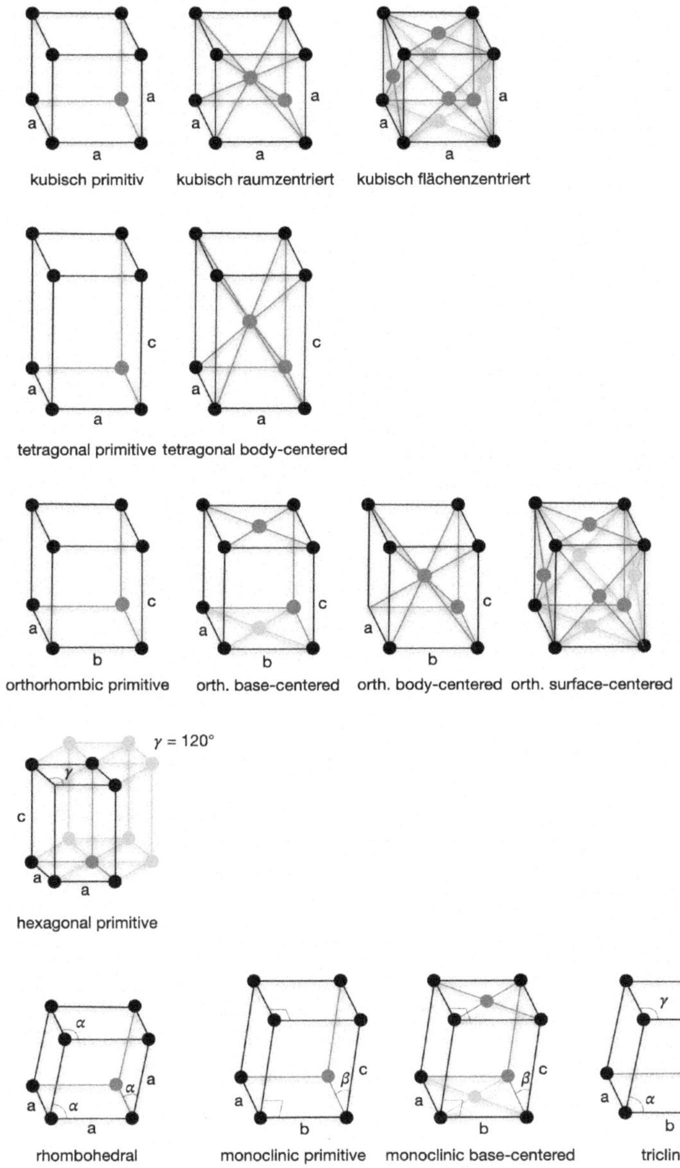

Fig. 2.16 Three-dimensional structures

structural forms. It is shown as a tree structure in Fig. 2.11. The fact that it is nevertheless given its own Section here is due to its widespread use.

It is used, for example, in *structural parts lists* (Wiendahl 2014, p. 162 ff.). The system that is the focus of consideration is referred to as the *focal system* or the *System of Interest* (SoI). This system defines the *reference level*. From there, a system that is located on a

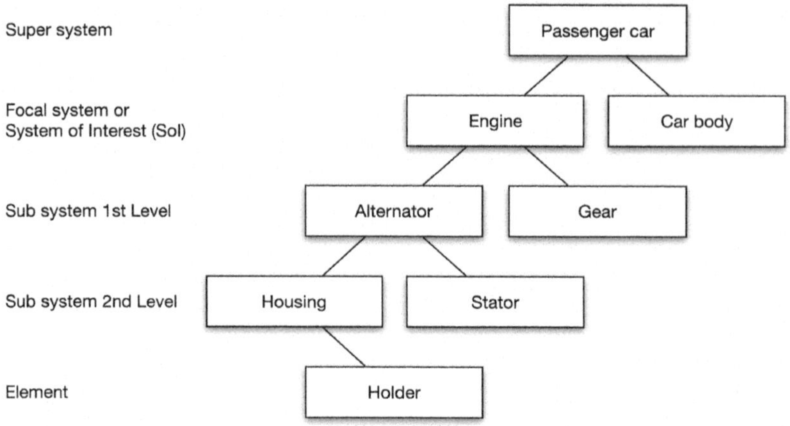

Fig. 2.17 Hierarchical model (Ropohl 2009, p. 122)

higher hierarchical level is referred to as a *super-system*, and one that is located below it as a *sub-system*. The distances can be specified as *orders*. The smallest unit that cannot be broken down further for the purpose of modeling is referred to as an *element*. This does not mean that it cannot be broken down further for other purposes, for example into atoms or protons, neutrons and electrons. The decisive factor is that further decomposition appears irrelevant for the respective purpose.

A useless box is used below as a further application example. Fig. 2.18 shows how it works. In the initial situation (1), the microswitch disconnects the motor from the battery. When a user presses the toggle switch (2), this starts the motor, which moves a lever. This opens the lid of the box and releases the microswitch (3). The lever then moves the toggle switch back to its original position (4), which reverses the polarity of the motor and thus its direction of rotation, so that the lever moves back towards the microswitch (5). Finally, the lever touches the microswitch so that it de-energizes the motor (5). This restores the box to its initial position (1).

Fig. 2.19 visualizes the hierarchical structure of the box, whereby small items such as fastening elements and cables are not shown. The structure is modeled using the *Systems Modeling Language* (SysML) used in *model-based systems engineering* (MBSE). This methodology aims to base design projects not only on documents, but also and above all on a central model on which all participants work together (Alt 2012, p. 38 f.). The model is stored in a database for this purpose. Work on the model is carried out using appropriate computer programs. Models of this type are also referred to as *digital twins* (Bracht et al. 2018, p. 13). In the course of digitalization, the idea is to implement a so-called *metaverse* parallel to the real world, in which a digital object (digital twin) modeled as realistically as possible exists for each real object, which, for example, enables experimentation (simulation) without disrupting ongoing operations in order to be able to transfer solutions found in the virtual world to reality as quickly as possible (in real time) and with as little loss as possible (Mozumder et al. 2022, p. 256 ff.). To this end, the twins should be in a permanent data exchange. Data that flows from the real object to the virtual

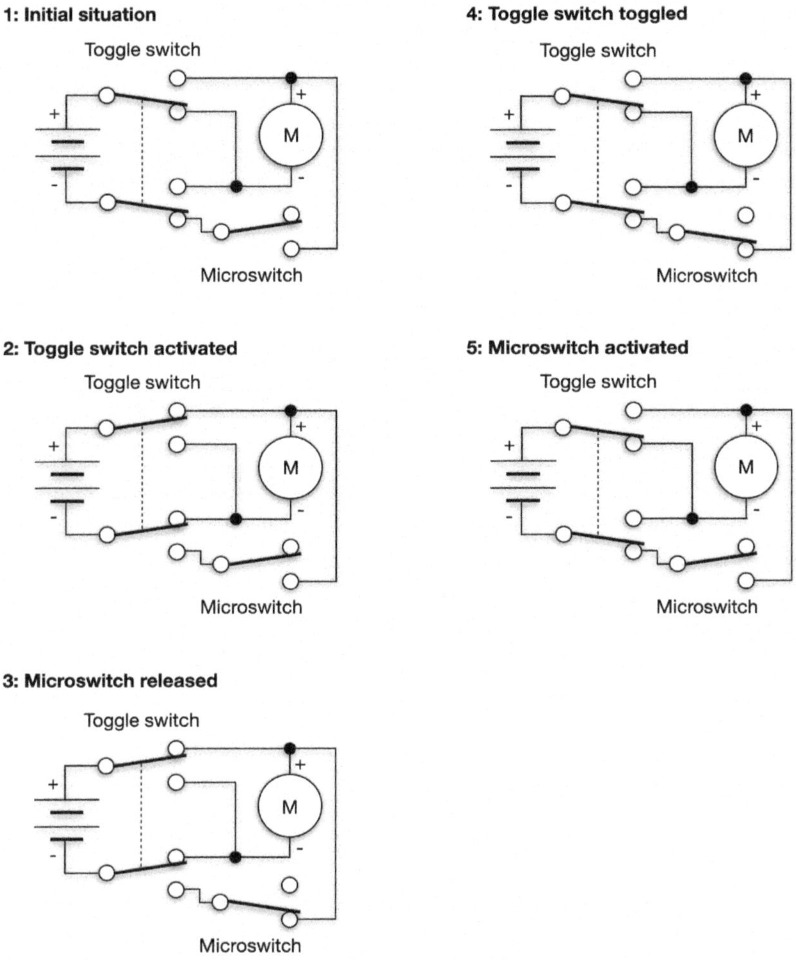

Fig. 2.18 How a Useless Box works

object is referred to as a *digital shadow*, while data that flows in the opposite direction is referred to as a *digital trigger*.

SysML offers different views of the central model, which are represented graphically as so-called diagrams. Fig. 2.19 shows the block definition diagram (bdd). Blocks correspond to what is referred to as a class in object-oriented programming. This makes it clear that SysML, which is standardized by the Object Management Group®(OMG), is based on the fundamental principles of object orientation, such as the principle of inheritance.

The Figure shows the technical-physical structure (ibid., p. 95) of the Useless Box outlined in Fig. 2.18 in abstract form. The focal system Useless Box, which is arranged at the top level of the hierarchy, consists of the sub-system Box with its elements housing and lid, the sub-system Battery holder with four batteries, the sub-system Gear motor with

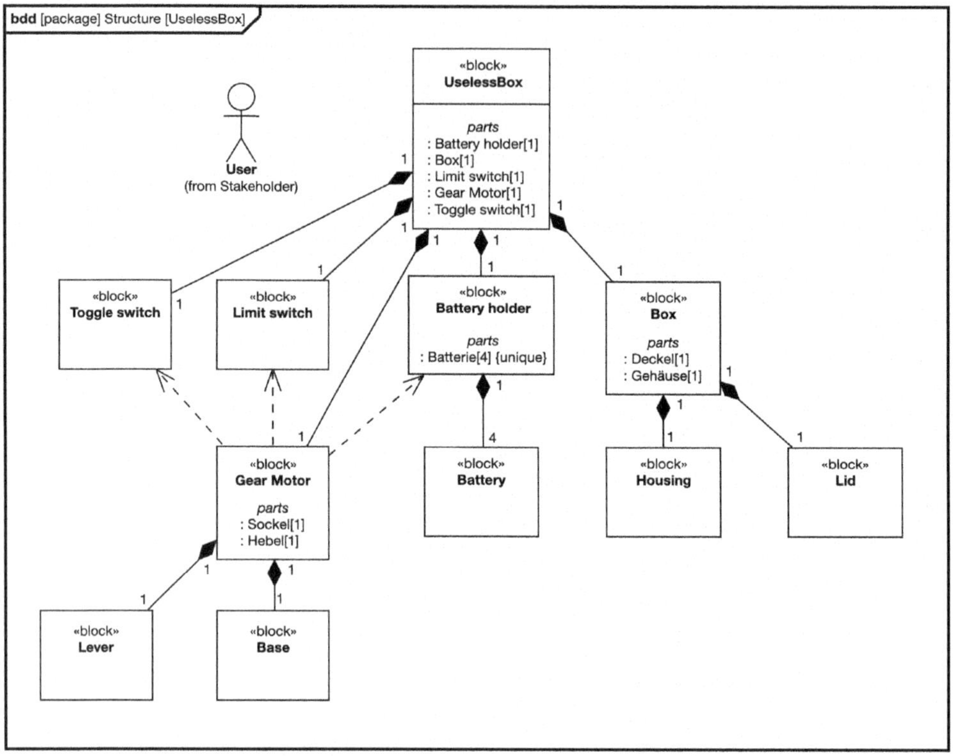

Fig. 2.19 Example of a Useless Box structure in SysML notation, created with Enterprise Architect software

a base for attachment to the housing and with the lever that opens the lid and switches off the toggle switch. Further elements are the toggle switch and the limit switch.

The focal system and its sub-systems and elements are connected by two types of relations: The *composition relationship*, which is shown as a solid line with a diamond at the end, describes the affiliation of sub-systems and elements to super-systems. The diamond is always located at the respective super-system. The quantities are indicated by the composition relationships: for example, a battery holder belongs to the Useless Box, while the battery holder holds four identical batteries. *Dependencies* are shown as further relations using dashed lines with an arrow at the end. For example, the voltage supplied by the batteries is required to operate the motor. The arrow always points from the dependent to the independent block.

2.4.4 Group Theory and Logical Type Theory

The group-theoretical interpretation of systems is of considerable importance for further considerations. Particularly for *Change Management*, this results in key insights in conjunction with the logical theory of types (Watzlawick et al. 2013, p. 22 ff.), which will be explained in detail later on. At the same time, the application of mathematical findings in psychology and sociology once again demonstrates the strength of systems theory with regard to interdisciplinary work and, in particular, pattern recognition.

In mathematics, group theory forms an essential basis for geometry and algebra, but also has numerous other applications, for example in crystallography, the theory of forms in art, linguistics, quantum theory and the theory of relativity (ibid., p. 23). In addition to the previous statements on systems, it specifies their properties (Glosauer 2016, p. 21 ff.). Therefore, although it is not applicable to all systems, it is applicable to those to which the group theory axioms apply, and such systems are particularly relevant for Systems Engineering & Management.

What is described as a group depends on the application. Numbers can be treated as a group (algebra), as can objects, organisms or events (Watzlawick et al. 2013, p. 24). A group is generally described as a pair $(G, *)$ of the set G of the elements $(a, b, c) \in G$ grouped together in connection with the type of link between two elements (relations). A two-digit (binary) link is usually shown with an asterisk ($*$). If the sequence of the link is important (consecutive execution), this is symbolized with a circle ($\circ$) and referred to as a composition. For example, the link $a \circ b$ (read: b after a) means that b is applied first and then a (Glosauer 2016, p. 4).

Elements of a system form a group if the following conditions are met (ibid., p. 21 f.):

- The set must be *self-contained*. This means that the combination of two elements of the set always produces an element of the same set: $* \; G \times G \longrightarrow G, (a, b) \longrightarrow a * b$. For example, if you add 6 h to the time 8 o'clock in the morning on an analog clock with $G = (1, 2, ..., 12)$, the result is 2 o'clock in the afternoon (Watzlawick et al. 2013, p. 24).
- The *associative law* applies to the link, so that it is not necessary to use parentheses. The following therefore applies for all $a, b, c \in G : (a * b) * c = a * (b * c)$. For example, starting from a certain point on a surface, you will always reach the same end point after any number of moves of individual length and direction, regardless of the order of the moves, if the number of moves and their length and direction are maintained. For example, if you choose the four cardinal points as the direction and one kilometer as the length, you will reach the starting point again after four moves (e.g. first north, then west, etc. or first east, then north, etc.) regardless of the chosen order (Watzlawick et al. 2013, p. 25 f.).
- A group is also characterized by the fact that it contains a *neutral element e* (often abbreviated to *id*), which has no effect when linked to another element: $a * e = e * a = a$ for all $a \in G$ (Glosauer 2016, p. 21). With an additive link, $e = 0$, with a multiplicative link, $e = 1$. In the sum of all tones, the neutral element is formed by silence (Watzlawick et al. 2013, p. 26).

- Furthermore, each element of a group $a \in G$ has an *inverse element* $a^{-1} \in G$, which results in the neutral element e when combined with $a : a * a^{-1} = e$. For example, the addition of $5 + (-5) = 0$.
- Finally, there are groups in which the *commutative law* applies, so that the same result is obtained despite swapping the operators: $a * b = b * a$ for all $a, b \in G$ (Glosauer 2016, p. 21). In honor of the late Niels Henrik Abel, one of the founders of group theory, such groups are also referred to as Abelian groups.
- One characteristic of a group is its power $|G|$, which indicates the number of elements of G. This is also referred to as the order of the group. This is also referred to as the order of the group (ibid., p. 22).

While Paul Anton Watzlawick et al. (2013, p. 21 ff.) use group theory to explain why and how systems stabilize and consequently resist change, they use *logical type theory* to provide starting points for how systems can nevertheless be persuaded to change in the sense of Change Management (ibid., p. 27 ff.). The starting point is the insight already cited in Section 2.1 that the whole (system) is different from the quantitative sum of its parts (elements). Alfred North Whitehead and Bertrand Russell (1910-1913, vol. 1, p. 37) express this on a mathematical basis to the effect that whatever concerns the totality of a class (set) must not itself be part of this class. Technically, this can be explained using the example of the control loop, in which the reference variable w must be specified from the outside in order to change the state of equilibrium in a targeted and sustainable manner; a control loop can only maintain an equilibrium based on its internal logic. However, socio-technical systems do not follow a change request as easily as purely technical systems, as the psychological and social influences are far more complex (more stubborn) (Nassehi 2021, p. 25; Watzlawick et al. 2013, p. 28). This makes it all the more impressive that Paul Anton Watzlawick et al. (ibid.) nevertheless succeed in deriving their findings on the possibilities of Change Management in socio-technical systems from the strict logic of mathematics. However, they do not do this by mathematical means, but by using analogies, which are common in interdisciplinary work (ibid., p. 23).

Possible applications of this insight can be found, for example, in the context of so-called *swarm intelligence* (Hamann 2019). While the intelligence of individuals such as birds, fish or ants appears relatively limited, the corresponding collectives exhibit intelligent behaviour on the basis of decentralized, non-hierarchical but correlated decision-making processes that significantly exceeds the sum of individual contributions. Conversely, people as individuals are generally comparatively intelligent, while the collective does not appear to be able to ensure its own survival, particularly through peace, justice and the preservation of creation (Seifert 1999, p. 7) (Watzlawick et al. 2013, p. 28). Armin Nassehi (2021, p. 30 ff.) formulates the unease resulting from this observation in the form of the *sociodicy* originally introduced by Pierre Bourdieu (1993, p. 252 f.), which in turn refers to the theodicy formulated in Gottfried Wilhelm Leibniz's Monadology and asks the question of how society (social system) can allow so much suffering and problematic behavior while it seems to have the means to counteract it. Why would the agents (psychological systems) not strive for the best, even though they have the means to do so?

The following answers can be given at this point: (1) Psychical systems (more precisely, their acts of communication, see Section 2.6.2.2.3) are elements of social systems and, in the light of group-theoretical laws, are not in a position to influence the superordinate system precisely (deterministically). At best, they can disturb (irritate) it, but whether and, if so, how the super-system reacts to this is up to the super-system. (2) And even society as a social system cannot change itself precisely (deterministically) if one takes the laws of group theory as a basis. Armin Nassehi (2021, p. 18) describes this as the *excessive demands* of modern societies. The obvious inability of societies (not to mention the global community) to stop anthropogenic climate change is an example of this. At the level of organizations, this correlates with the observation that organizations do not strive for the best possible (optimal) organizational form (even if this is usually continually claimed), but are satisfied with a viable form (Weick 1995). (3) What remains is the possibility of an impulse for change from the environment that can be derived from the logical theory of types. In relation to the example of climate change, this could possibly be significantly noticeable consequences of climate change. Just as children are said to avoid touching a hot hob until they have burnt themselves on it, society may also only learn from harm. Given the apocalyptic scale of the damage that is feared, this may be an unfortunate insight, but based on the insights into the functioning of social systems cited here, there seems to be no better idea to date. However, this should not prevent us from continuing to search for a better idea (Martin Luther's apple tree). Further system-theoretical foundations for this follow in Section 3.5.2.

The realization that a super-system follows a different logic than its sub-systems or elements leads to the insight that the hierarchy discussed in Section 2.4.3 is an oversimplification, especially in large socio-technical and psycho-technical systems. Consequently, the widespread view in agile approaches that a functioning whole emerges almost automatically from the sum of fragments (Goll and Hommel 2015, p. 79) is only permissible for very simple, purely technical systems, for example in the form of parts lists (Wiendahl 2014, p. 162 ff.). However, since systems in practice generally act as socio-technical and/or psycho-technical systems, this approach regularly reaches its limits. In fact, there is always a hierarchy of logical types, i.e. different levels of logical abstraction (Watzlawick et al. 2013, p. 28). Neither the bottom-up approach common in agile approaches nor the top-down approach found in systems engineering do justice to this. Systems, sub-systems and elements follow their own logic at each hierarchy level, which cannot be adequately derived from the logic of the other hierarchy levels. What follows from this insight for Change Management is explained in more detail in Section 3.5.

2.4.5 Process Models

As a first approximation, it can be said that structures are static (so-called infrastructure). The *dynamics*, i.e. the behavior of a system that changes over time (time-variant), unfolds on them. An example of an infrastructure is the road network including all technical (e.g. traffic lights) and organizational (e.g. traffic regulations, registration offices, road traffic police, courts) facilities. This traffic infrastructure remains unchanged over a long period of time, while the road traffic on it is dynamic, i.e. it changes over time. On the basis of such an idea, models will be introduced below that make it possible to model the dynamics of systems. However, it should be noted in advance that structures are not quite as static as first glance might suggest. In fact, they are also subject to a dynamic, albeit one that is much slower than that of processes. For example, new roads are built and traffic regulations are changed over time. This should not be ignored insofar as the design of structures is one of the main problems of systems engineering, while the control of processes is the subject of Systems Management.

2.4.5.1 Modeling of the Random Time-Variant Behavior

Times and time-dependent variables such as the speed in kilometers per hour or the arrival rate of customers per hour at a post office counter can be modeled in the form of *scalar variables* or with the help of *probability distributions*. Distribution functions can in turn be described by *shape parameters* (e.g. number of degrees of freedom of the chi-square distribution), *position parameters* (e.g. arithmetic mean, mode, median) and *scale parameters* (e.g. standard deviation of the normal distribution, lifetime of the Weibull distribution). Some of the distributions commonly used in practice are introduced below as examples. A distinction is made between continuous distributions and discrete distributions.

One of the best-known continuous distributions is the *normal distribution* shown in Fig. 2.20 (Bungartz et al. 2009, p. 40). It is used for variables that fluctuate around a certain value. For example, *physical phenomena* such as the duration of sunshine [hours/day] are normally distributed. A normally distributed random variable X with the value range $Wx = R$ with the parameters $\mu \in \mathbb{R}$ (expected value) and $\sigma \in \mathbb{R}$ (standard deviation) has the density

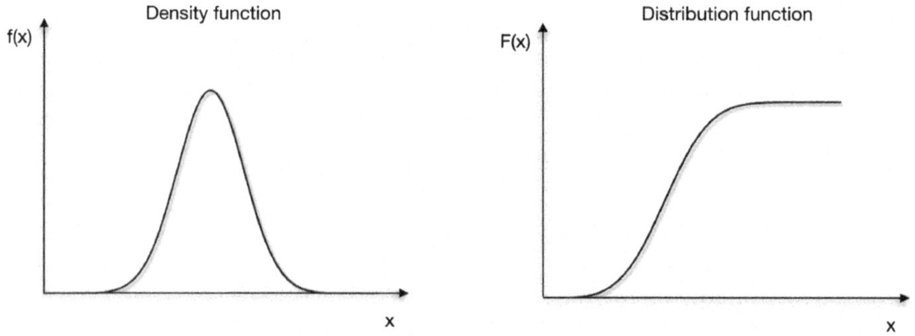

Fig. 2.20 Density and distribution function of a normal distribution

$$fx\,(x) = \frac{1}{\sqrt{2\pi\sigma}} \cdot e^{\frac{(x-\mu)^2}{2\sigma^2}} \tag{2.5}$$

Using the master function, this results in the distribution function

$$F_X(x) = \frac{1}{\sqrt{2\pi\sigma}} \cdot \int_{-\infty}^{x} e^{\frac{(t-\mu)^2}{2\sigma^2}} dt \tag{2.6}$$

However, as the density function cannot be integrated in a closed form, the values for $F_X(x)$ can only be calculated approximately by numerical integration and can be looked up in relevant tables.

Of particular importance is the standard normal distribution with the parameters 0 and 1. Its distribution function is known as the Gaussian function, the corresponding graph as the *Gaussian bell curve*. The function is

$$\Phi(x) = \frac{1}{\sqrt{2\pi}} \cdot \int_{-\infty}^{x} e^{-\frac{t^2}{2}} dt \tag{2.7}$$

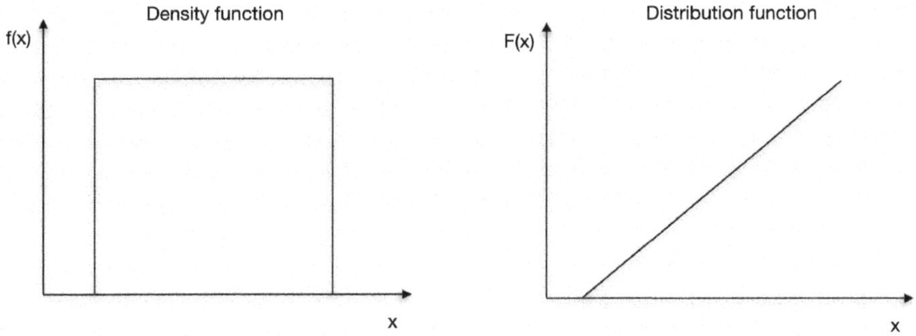

Fig. 2.21 Density and distribution function of a uniform distribution

Another frequently used distribution is the uniform distribution. Fig. 2.21 shows the continuous form. In the interval $\Omega = [a, b]$, the density of the random variable X is calculated as

$$f_X(x) = \begin{cases} \frac{1}{b-a} & \text{for } x \in [a, b], \\ 0 & \text{otherwise} \end{cases} \tag{2.8}$$

Using the master function, this results in the distribution function

$$F_X(x) = \int_{-\infty}^{x} f(t)\mathrm{d}t = \begin{cases} 0 & \text{for } x < a, \\ \frac{x-a}{b-a} & \text{for } a \le x \le b, \\ 1 & \text{for } x > b \end{cases} \tag{2.9}$$

The expected value $E(X)$ and the variance $V(X)$ are calculated as follows:

$$E(X) = \frac{a+b}{2} \qquad V(X) = \frac{(b-a)^2}{12} \tag{2.10}$$

The standard deviation $\sigma(X)$ and the coefficient of variation $\varrho(X)$ are calculated as follows:

$$\sigma(X) = \frac{b-a}{2\sqrt{3}} \qquad \varrho(X) = \frac{b-a}{\sqrt{3}(a+b)} \tag{2.11}$$

The uniform distribution is used when there is no preference for any value, i.e. when all values are equally likely or must be assumed to be equally likely due to a lack of corresponding information. The latter is based on the so-called *Laplace principle*, which is applied in decision-making situations. If the probabilities of occurrence of situations are not known, the *theorem of insufficient reason* is applied. The decision is then made on the basis of the expected value rule. The decision alternative with the highest expected value is then selected.

In addition to continuous distribution, discrete uniform distribution is also common. A simple example is the roll of the dice, in which only the six integer values 1 to 6 can occur. The probability of rolling each number is known to be the same, namely one-sixth. It is interesting to note, however, that this only applies when a single die is used. If several dice are used, the probability of the total number of dice is normally distributed.

Another continuous distribution is the *negative exponential distribution* (Fig. 2.22). It is used, for example, to model intermediate arrival times between two events. This can be the time that a machine runs fault-free between two failures in the life phase in which only random failures are to be expected (Sect. 4.3.4.5) (Mean Time Between Failure: MTBF). This distribution is the only continuous distribution that is memoryless (Sect. 2.5). A subsequent event therefore does not depend on the previous event.

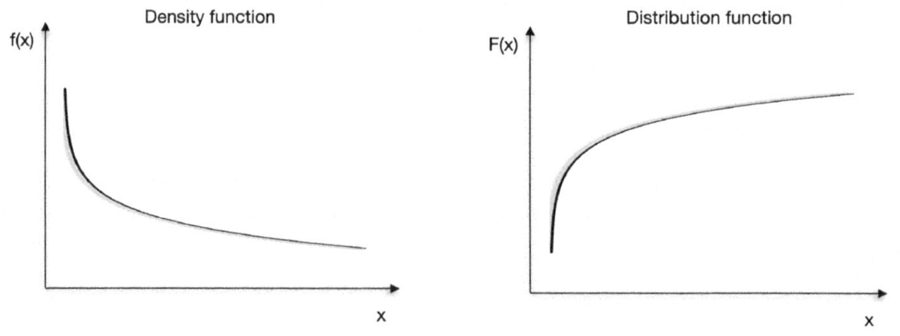

Fig. 2.22 Density and distribution function of a negative exponential distribution

With the parameter $\lambda > 0$, the density of the negative exponential distribution is calculated as follows

$$f_X(x) = \begin{cases} \lambda \cdot e^{-\lambda x} & \text{if } x > 0, \\ 0 & \text{otherwise} \end{cases} \tag{2.12}$$

Their integration results in the distribution function

$$F_X(x) = \int_0^x e^{-\lambda t} \mathrm{d}t \tag{2.13}$$

By partial integration, the expected value $E(X)$ and the variance $V(X)$ are calculated as follows

$$E(X) = \frac{1}{\lambda} \qquad V(X) = \frac{1}{\lambda^2} \tag{2.14}$$

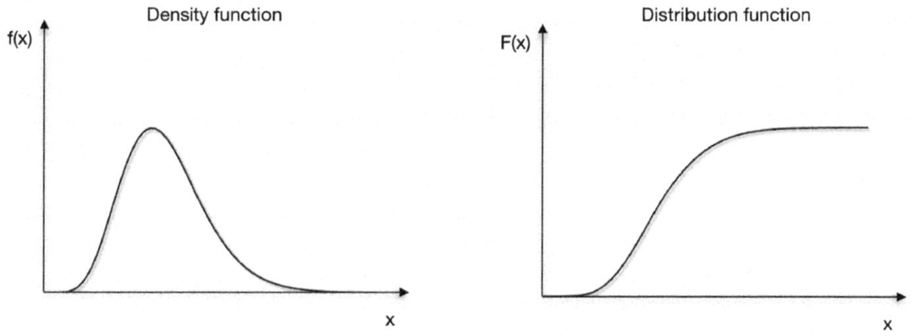

Fig. 2.23 Density and distribution function of an Erlang distribution

The *Erlang distribution* will now be introduced as a further continuous distribution, as it is used in addition to the negative exponential distribution, for example to model the duration of a machine failure (Mean Time To Repair: MTTR). It is a generalization of the exponential distribution as well as a special case of the gamma distribution and takes into account the fact that very short downtimes and very long downtimes are less likely than downtimes of medium duration, whereby the distribution is left-skewed (Fig. 2.23). The density is calculated with the two parameters $\lambda > 0$ and $n > 0$ as follows

$$f_X(x) = \begin{cases} \frac{\lambda^n \cdot x^{n-1}}{(n-1)!} e^{-\lambda x} & \text{if } x \geq 0, \\ 0 & \text{if } x < 0 \end{cases} \tag{2.15}$$

The distribution function is

$$F_X(x) = \begin{cases} 1 - e^{-\lambda x \sum_{i=0}^{n-1} \frac{(\lambda x)^i}{i!}} & \text{for } x \geq 0, \\ 0 & \text{for } x < a \end{cases} \tag{2.16}$$

The expected value $E(X)$ and the variance $V(X)$ of the Erlang distribution are calculated as follows:

$$E(X) = \frac{n}{\lambda} \qquad V(X) = \frac{n}{\lambda^2} \tag{2.17}$$

Of the discrete distributions, the *Poisson distribution* (Fig. 2.24) is of particular interest with regard to modeling the time behavior of systems. It represents a limiting case of the binomial distribution and is suitable for describing the number of events within certain time intervals, for example the number of customers who visit a post office per hour or the number of Prussian cavalry soldiers who are killed by horse kicks every year. The latter led Ladislaus Bortkewitsch in 1898 to the law of small numbers, also known as the two-thirds law, which will be explained here using the example of the game of roulette: With $\lambda = 1$, after a series of rotations with 37 coups each, it can be determined that only about two-thirds of the numbers are hit, about half of them multiple times. However, according to the law of large numbers, the probability of hitting each number is the same, namely one thirty-seventh[11].

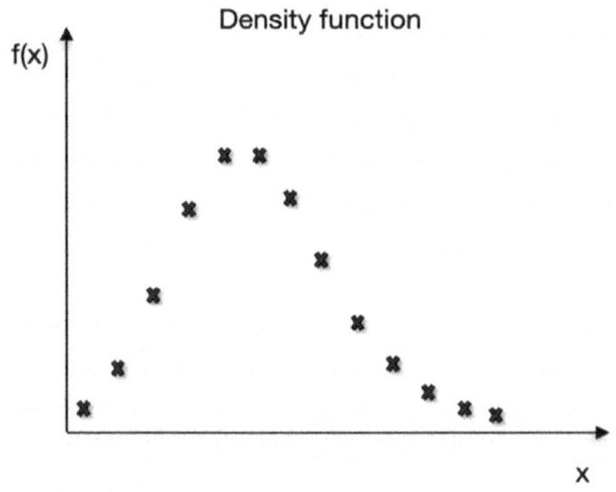

Fig. 2.24 Distribution function of a Poisson distribution

With a parameter $\lambda > 0$, the density function in the value range $\mathbb{N}_0$ is

$$x(x) = \frac{e^{-\lambda}\lambda^i}{i!}, \quad i \in \mathbb{N}_0 \tag{2.18}$$

Expected value $E(X)$ and variance $V(X)$ are

[11] No winning strategy can be derived from this insight.

$$E(X) = \lambda \qquad V(X) = \lambda \tag{2.19}$$

2.4.5.2 Static Models

Now that the temporal behavior of the elements or systems involved in a process has been described, the next step is to model the interaction of the elements or systems (relations). A distinction can be made between static and dynamic models. Static models depict the dynamic behavior of systems by describing characteristic states.

The assumption that processes are predominantly linear sequences of activities is particularly widespread in so-called *Process Management* (e.g. DIN ISO 21500:2016-02, p. 16; Gaitanides 2012; Haberfellner et al. 2018, p. 46). An equally simple and widespread type of notation is the arrow shape (Gaitanides 2012, p. 168). A popular concretization of this notation can be found in the *value chain model* by Michael E. Porter (2000, p. 66), which is shown in Fig. 2.25.

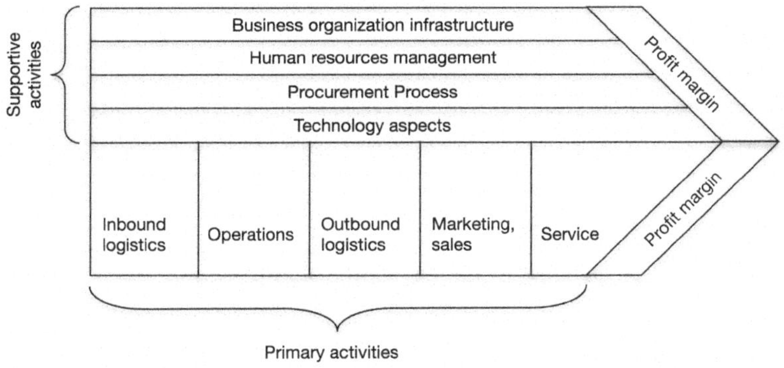

Fig. 2.25 Model of the value chain by Michael E. Porter (2000, p. 66)

The linear sequence of activities is represented particularly consistently by the sequence of arrows in the form of a so-called *spaghetti diagram* (Thonemann 2015, p. 148 ff.). Fig. 2.26 shows a simple example of this.

The spaghetti diagram is extended by the so-called *swimlane diagram* (Fig. 2.27), which is also known as a *makigami* and assigns the process steps to the organizational units (swimlanes) shown line by line (Fischermanns 2013, p. 120; Gaitanides 2012, p. 172; Weilkiens 2014, p. 256 f.) in order to include parts of the organizational structure in addition to the process organization. Although the process in this notation traverses the organizational units involved, this also generally represents a largely linear process flow.

Fig. 2.26 Spaghetti diagram (Thonemann 2015, p. 148 ff.)

Pure linearity is extended in more differentiated process models such as *event-driven process chains* (EPC) (Allweyer 2005, p. 133 ff.; Thonemann 2015, p. 154 ff.), or the *Business Process Model and Notification* (BPMN) (Freund and Rücker 2014, p. 22 ff.). Notations such as these also provide for branching (concurrency) and synchronization (Fig. 2.28). However, the dominance of the largely linear sequence of activities remains intact. Although the authors mention the possibility of loops (Freund and Rücker 2014, p. 76 ff., p. 171) or jumps back (Allweyer 2005, p. 134), both Guido Fischermanns (2013, p. 241 ff.) and Jakob Freund and Bernd Rücker (2014, p. 134 f.) unanimously declare that this should be minimized. Network-like structures are largely excluded.

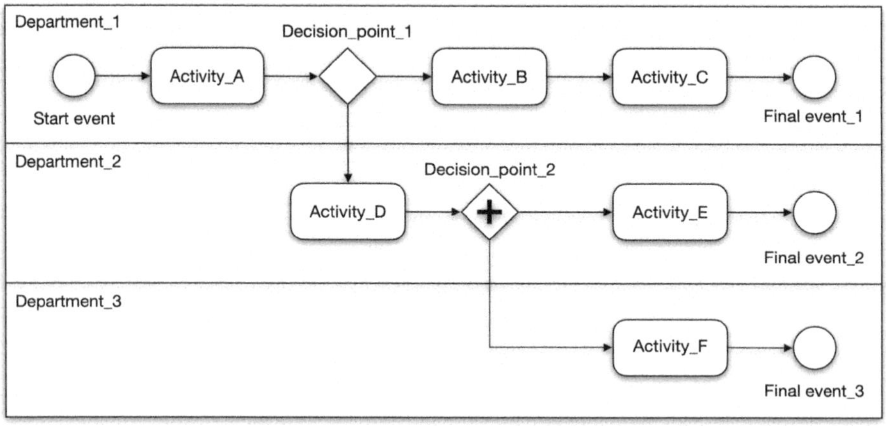

Fig. 2.27 Example of a swimlane diagram (e.g. Fischersmanns 2013, p. 120)

Another widely used option for statically mapping the dynamic behavior of systems is to model networks on the basis of *graph theory* (e.g. Nickel et al. 2022, p. 151 ff.). In this theory, systems and elements are referred to as *nodes*, as their functions and properties are not relevant for the investigations in detail. The relationships between the systems and elements are called *edges*. All nodes and edges together form a *network*. At this very abstract level, graph theory provides a whole series of generally valid insights (ibid.), which, however, will not be explored further here in their abstract form.

Instead, the *network planning* technique known from *Project Management* will be used to illustrate the temporal behavior of networks using static models (e.g. Bjørke 1995, p. 7; Jakoby 2021, p. 181 ff.; Nickel et al. 2022, p. 151 ff.). This method in turn uses its own terminology. Systems and elements are referred to as activities, as only the times required to perform their respective functions are of interest. The transition from one state of the network to the subsequent state at a certain point in time, which is assumed to be timeless, is referred to as an *event*.

The network diagram shown in Fig. 2.29, which was created according to the Program Evaluation and Review Technique (PERT), is used as a concrete example (Böhm 2005, p. 300 ff.; Jakoby 2021, p. 189 ff.; Nickel 2022, p. 168 ff.). This is an *Acitivty-on-node network* (AON), as the events are assigned to the nodes and the relations (relationships) to the arcs,

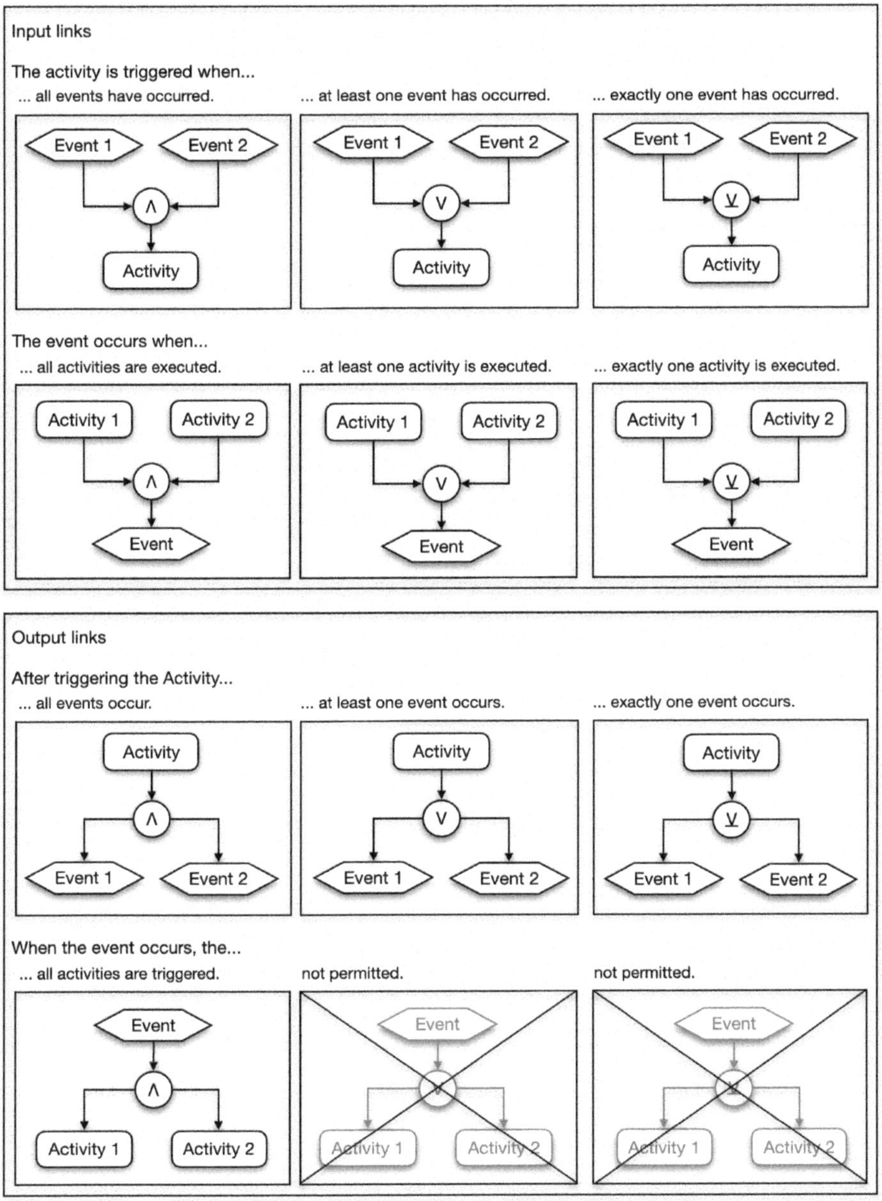

Fig. 2.28 Possibilities of linking in an event-driven process chain (Thonemann 2015, p. 156)

which places the events at the center of attention. Alternatively, activity-oriented networks are used, including *activity-on-arc networks* (AOA), in which the events are modeled as nodes and the activities as arcs between two nodes, as well as activity-on-arc networks, in which the activities are represented as nodes and the events as arcs.

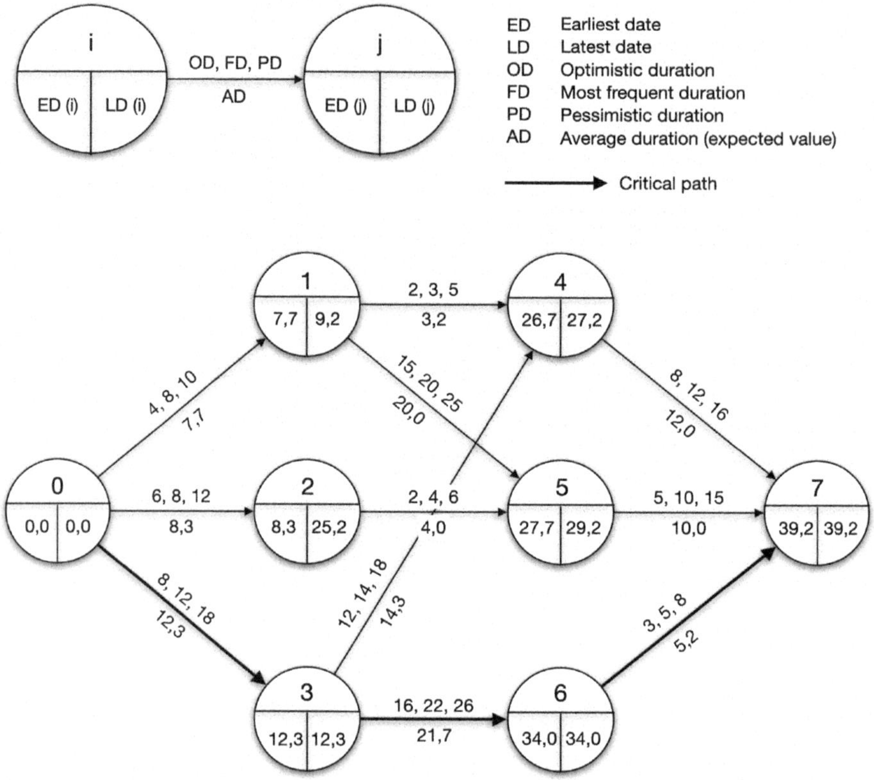

Fig. 2.29 Event node representation of a network plan for scheduling according to the PERT method, based on Böhm (2005, p. 305)

In the Program Evaluation and Review Technique (PERT), the optimistic, probable and pessimistic durations are noted ob the edges (events) according to the so-called three-time estimation, and the mean value and variance calculated from this on the basis of a beta distribution are noted under the edges (Böhm 2005, p. 300 ff.; Nickel et al. 2022, p. 168 ff.).

$$AD = \frac{OD + 4 \cdot FD + PD}{6} \tag{2.20}$$

By adding these times, the earliest start date can be calculated for each activity using *forward scheduling*. This ultimately results in the earliest completion date for the entire network. Based on this, the latest finish dates for all activities are calculated by subtracting the times along the paths through the network in the course of *backward scheduling*. The difference between the earliest and latest dates is the slack, which corresponds to the buffer time in activity-oriented networks and indicates whether and, if so, how far an event can be postponed without affecting the end date of the entire project. The path on which there is no slack is referred to as the *critical path*, as a postponement of an activity on this path always has an impact on the overall project.

In addition to discrete methods such as graph theory, numerical methods can also be used to model the dynamic system behavior. One example of this is *differential equations* (Bungartz et al. 2009, p. 68 ff.; Şanal 2020, p. 487 ff.). These are mathematical equations for functions that depend on one (ordinary differential equation: ODE) or more (partial differential equation: PDE) independent variables and also contain derivatives of the function. The ODE describes how these variables change in relation to each other. As a simple example, the so-called *logistic differential equation* and its solution (Fig. 2.30) will be presented below, as it can be used to model the growth and saturation phase of the *system life cycle* shown in Fig. 2.47, among other things. The differential equation has the form of a Bernoulli's differential equation with $\alpha = 2$ and reads

$$y' = ky(M - y), \qquad 0 < y < M \tag{2.21}$$

M is interpreted as the size of the population that limits growth. The proportionality factor k summarizes the combined effect of reproduction and reduction (starvation). y indicates the respective population; the difference $M - y$ thus quantifies the remaining capacity.

To solve the differential equation, the variables are separated from each other in the first step:

$$\frac{\mathrm{d}y}{\mathrm{d}x} = ky(M - y) \qquad | \cdot \mathrm{d}x \cdot \frac{1}{y(M - y)} \tag{2.22}$$

$$\frac{1}{y(M - y)}\mathrm{d}y = k\mathrm{d}x \tag{2.23}$$

Now comes the integration:

$$\int \frac{1}{y(M - y)}\mathrm{d}y = \int k\mathrm{d}x \tag{2.24}$$

The left-hand side of the equation is simplified by partial fraction decomposition. The approach for this is

$$\frac{1}{y(M - y)} = \frac{A}{y} + \frac{B}{M - y} \qquad | \cdot y(M - y) \tag{2.25}$$

A and B are now determined for this:

$$1 = A(M - y) + By \tag{2.26}$$

For the sake of simplicity, the two zeros of the old denominator are used:

$$y = 0 : 1 = A \cdot M \Rightarrow A = \frac{1}{M} \tag{2.27}$$

$$y = M : 1 = B \cdot M \Rightarrow B = \frac{1}{M} \tag{2.28}$$

The integral is now written as follows:

$$\int \left(\frac{\frac{1}{M}}{y} + \frac{\frac{1}{M}}{M - y} \right) \mathrm{d}y = \int k \mathrm{d}x \tag{2.29}$$

The constants are drawn in front of the integrals:

$$\frac{1}{M} \int \left(\frac{1}{y} + \frac{1}{M - y} \right) \mathrm{d}y = k \int \mathrm{d}x \tag{2.30}$$

The solution of the left integral is now obtained from the relevant tables:

$$\frac{1}{M}(\ln |y| - \ln |M - y|) = kx + c \qquad \text{with } c = const \tag{2.31}$$

This expression now only needs to be solved for y:

$$\frac{1}{M} \ln |\frac{y}{M - y}| = kx + c \qquad | \cdot M \tag{2.32}$$

$$\ln |\frac{y}{M - y}| = kMx + c \qquad | \cdot e^{\cdots} \tag{2.33}$$

$$|\frac{y}{M - y}| = e^{kMx} \cdot c, \qquad c > 0 \tag{2.34}$$

$$\frac{y}{M - y} = ce^{kMx} \cdot c, \qquad c \neq 0 \tag{2.35}$$

$$\frac{M - y}{y} = ce^{-kMx} \tag{2.36}$$

$$\frac{M}{y} - 1 = ce^{-kMx} \tag{2.37}$$

$$\frac{1}{y} = \frac{ce^{-kMx} + 1}{M} \tag{2.38}$$

$$y = \frac{M}{ce^{-kMx} + 1} \tag{2.39}$$

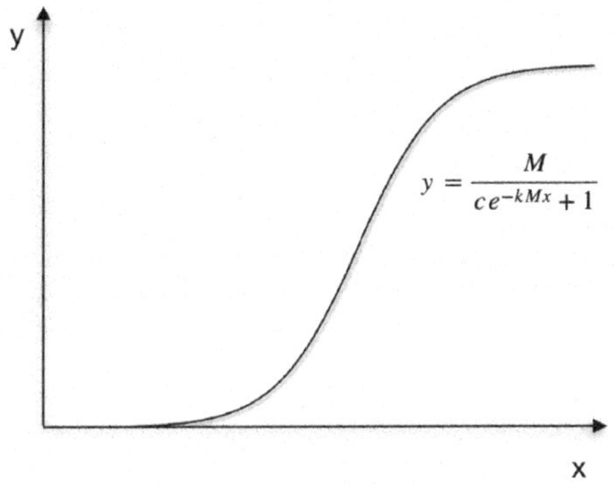

Fig. 2.30 Logistical function

Non-linear dynamics such as these can be modeled in various ways. System dynamics, which are closely associated with systems engineering, are presented below as an example (Forrester 2013). Networks of interacting elements and/or systems are represented using flowcharts. Fig. 2.31 shows this using a simple example (Haberfellner et al. 2018, p. 46). This shows so-called loops with positive (*reinforcing loops*) and negative (*balancing loops*) polarities: Frogs lay eggs and eggs hatch into frogs, which in turn lay eggs, and so on. The more frogs lay eggs, the more frogs hatch from them. Cause and effect point in the same direction. In this way, the frog population can quickly take on very large proportions (reinforcing loop). However, this is countered by the fact that frogs sometimes cross roads that are used by cars that run over frogs. The more frogs cross roads, the more of them die (*same direction*). This can cost many frogs their lives (*opposite direction*). This results in a balancing loop. In this way, the dynamic interactions in networks can be investigated. One conclusion that can be drawn from this is, for example, that too many frogs die when crossing roads, so that frog tunnels are built under roads to help stabilize the frog population (homeostasis).

Following on from the previously explained logistic equation, the following application example shows the modeling of the diffusion model by Frank M. Bass (1967). This model

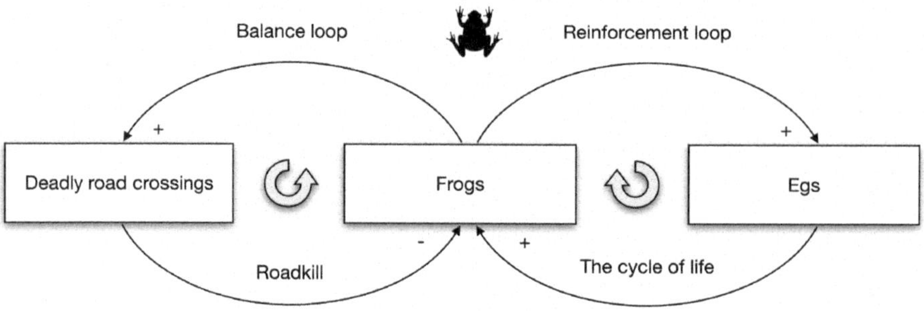

Fig. 2.31 Example of circular reasoning (Haberfellner et al. 2018, p. 46)

(Fig. 2.32) describes the market launch of innovative products under the assumption that, on the one hand, customers must be addressed who are interested in benefiting from innovations as early as possible (innovators, early adopters, lighthouse customers) and, later on, customers who primarily value the reliability of products (majority). This is relevant for marketing, among other things, as innovation-oriented customers are primarily reached through active product advertising, while a product that has been accepted by these customers will continue to spread through word-of-mouth advertising. The *path dependency* explained in Section 2.5 comes into play here, which is initially explained in a simplified way by the fact that success results from success (but also failure from previous failure).

It should be expressly noted that this *evolutionary economics approach* (e.g. Lehmann-Waffenschmidt and Peneder 2022) differs significantly from the more widespread *neoclassical microeconomics*. While the latter assumes that economic equilibria are formed on markets through supply and demand, i.e. that the whole is inferred from the many decisions made by individuals (bottom-up), evolutionary economics assumes that the decisions of individuals also depend to a large extent on the whole (top-down).

In the diffusion model, p describes the proportion of innovation-oriented customers and q the proportion of imitators. t marks the periods and x_t represents the sales in the respective period.

$$x_t = pM + (q - p)y_{t-1} - \frac{q}{M}y_{t-1}^2 \tag{2.40}$$

Fig. 2.32 shows how this model is formulated using system dynamics. In addition to Eq. 2.40, the adaptation rate through word-of-mouth advertising is multiplied by a contact rate. Otherwise, the dependencies shown correspond to Frank M. Bass' diffusion model. As expected, the progression of the total number of customers acquired corresponds to the progression shown in Fig. 2.30 according to the basic logistic function. In addition, the progression of customers acquired through active product advertising[12] and word-of-

[12] As the number of customers acquired through direct product advertising is much smaller than the number of customers acquired through word-of-mouth advertising, this curve is shown on a different scale, as it would otherwise not be recognizable.

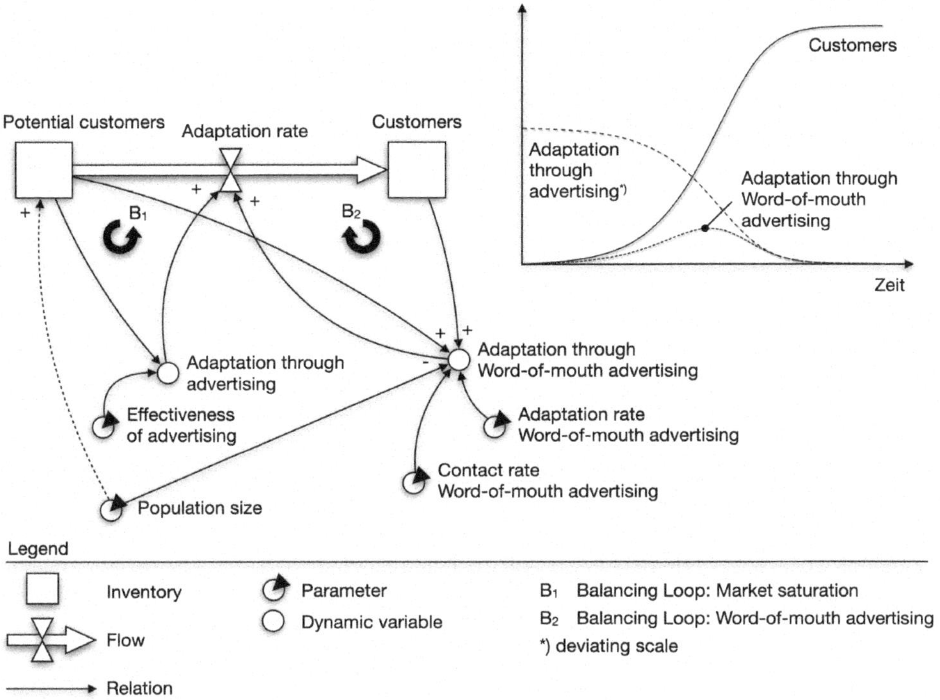

Fig. 2.32 Modeling of the bass diffusion model using system dynamics with the help of AnyLogic software

mouth advertising is shown. One advantage of this type of modeling is its clarity. However, specific software is required for modeling. Alternatively, the calculation and its graphical representation can also be created using a spreadsheet program. Although the dependencies are less transparent in this program, this is offset by the advantage of the widespread use of such programs.

The universal applicability of system dynamics can also be seen in the *system archetypes* modeled on this basis by Peter Michael Senge (2008, p. 455 ff.), which represent frequently observable patterns of human action.

If actors in an equilibrium process do not perceive a *delay in the effect of their actions*, this can lead to them reinforcing their actions (Fig. 2.33). As a result, they overshoot the target, as the saying goes. If this is perceived in the further course of the process, it can also lead to an exaggeration when the intensity of action is reduced. Peter Michael Senge therefore recommends either patience or reducing the delay of effects.

As a second archetype of human action, the author mentions *limits to growth* (Fig. 2.34). A reinforcing loop is coupled to a balancing loop with a time delay. The delay is caused by limiting resources which, in line with the logistical function explained above (Fig. 2.30), result in a constantly decreasing growth rate. In Fig. 2.29, the time-delayed coupling of the limiting resource is indicated by a double minus sign. The self-reinforcing cycle (Fig. 2.35, left) remains intact until the balancing cycle (Fig. 2.35, right) takes full effect. If the

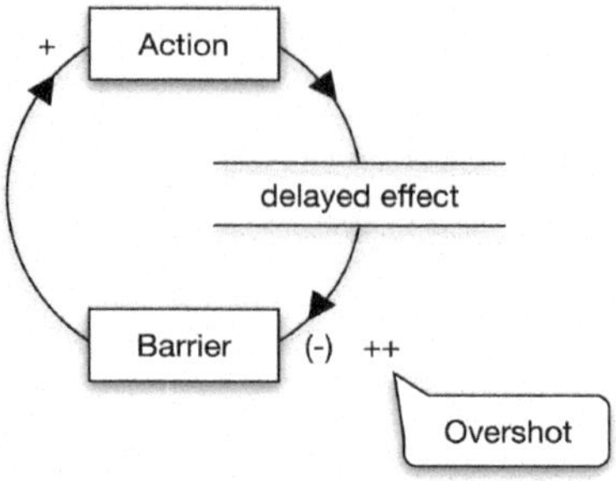

Fig. 2.33 Systemic archetype of delayed effect (Senge 2008, p. 455)

resources are not continuously regenerated (e.g. regenerative energies), the entire system eventually comes to a standstill. This leads to the realization that permanent (unlimited) growth is inherently impossible – an insight that should give pause for thought, especially in view of the goal of continuous economic growth, which is particularly widespread in industrialized nations, and points to the need for a *post-growth economy* (Paech 2012) and *sustainable economic activity* (Ekardt 2021; Meadows et al. 2020).

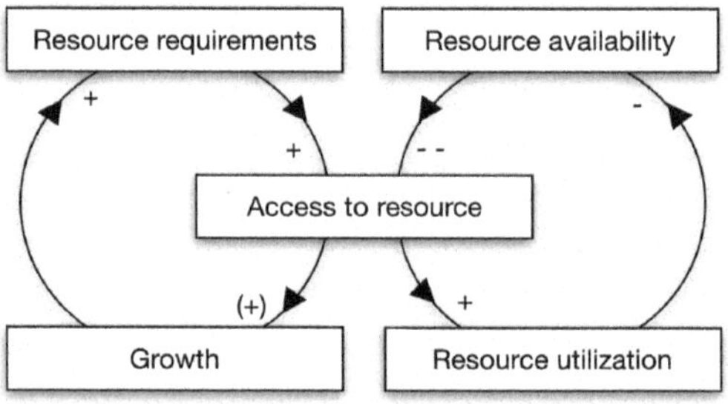

Fig. 2.34 System archetype of the limits to growth (Senge 2008, p. 456)

If a short-term solution (quick wins) is chosen to solve a problem, this can have an immediate effect but negative consequences in the long term. Fig. 2.36 describes this by linking a short-term balancing cycle to a long-term balancing cycle (Senge 2008, p. 458 ff.). Over time, this leads to a change in the overall structure by bridging the balancing cycle to develop long-term effective solutions, resulting in a combination of a short-term balancing cycle and a self-reinforcing cycle. As a result of cumulative side effects on the new path, the original problem is increasingly lost from view, causing the *problem to shift*. This corresponds to the development and maintenance of an addiction: the original problem is continually treated *symptomatically* until it ultimately loses its significance, while the new problem becomes dominant due to the self-reinforcing feedback loop. To avoid such a problem-solving strategy, the method of asking why five times (*5-W method*), which is part of the concept of *Lean Production* (Andler 2015, p. 57 ff.; Dombrowski and Mielke 2015, p. 84), can be used, for example, to get to the bottom of a problem by repeatedly

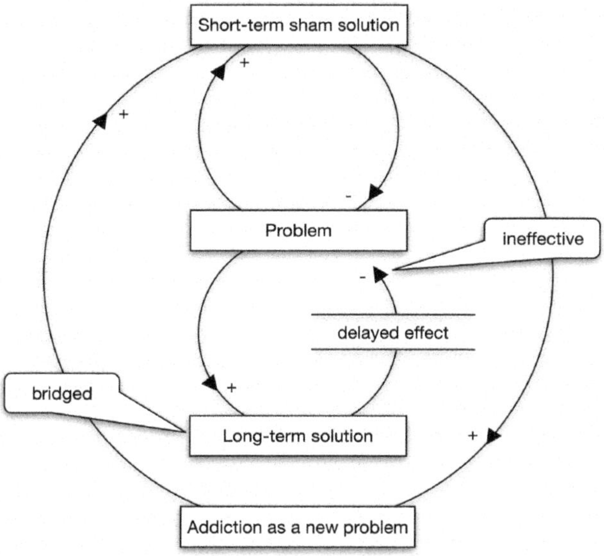

Fig. 2.35 Systemic archetype of problem shifting (Senge 2008, p. 458)

asking about the actual, often underlying causes. This method corresponds to the *Ishikawa diagram* shown in Fig. 4.11 (synonym: fishbone diagram).

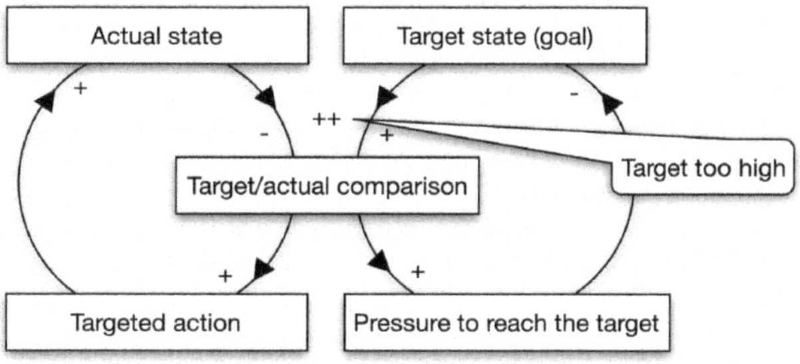

Fig. 2.36 System archetype of goal dilution (Senge 2008, p. 462)

A system archetype similar to problem shifting is the *softening of goals* (Senge 2008, p. 462 f.). The model of this archetype shown in Fig. 2.36 consists of two balancing cycles that are linked by a target-actual comparison. While the cycle shown on the left is oriented towards concrete actions, the cycle on the right is determined by expectations (goals). Initially, goals lead to corresponding actions. However, if the actions do not lead to success within a period of time that is considered appropriate, the pressure on the actors is increased. As a result, targets that are perceived as too high are lowered. To avoid this, realistic targets must be defined from the outset. This can also be done by gradually expanding target levels in order to divide a large step into several smaller and therefore more feasible steps.

Maturity models such as the Capability Maturity Model (CMM) (Hertneck and Kneuper 2011) or Software Process Improvement and Capability Determination (SPICE) (ISO/IEC 15504-5; Müller et al. 2016) are suitable for this purpose. Such models define the degree of target achievement in stages such as (1) starting, (2) repeatable, (3) defined, (4) managed and (5) optimized.

A competitive situation can result in *escalation* (Senge 2008, p. 463 f.). The model shown in Fig. 2.37 depicts this from the perspective of A by means of two balanced cycles, which, however, build each other up through their coupling. A's success reinforces B's efforts, which rebalances its cycle. However, this affects the relative success of A and now drives it to increase its efforts, and so on. If there are no limiting resources, this ultimately leads to the violent destruction of both cycles. The *arms race* is a practical example of this. Both sides consider their own actions to be a defensive reaction to the aggression of the other (security dilemma). Transparency offers a way out (arms control). Another possibility is that one side gives in (in the vernacular: the wiser gives in) and unilaterally disarms.

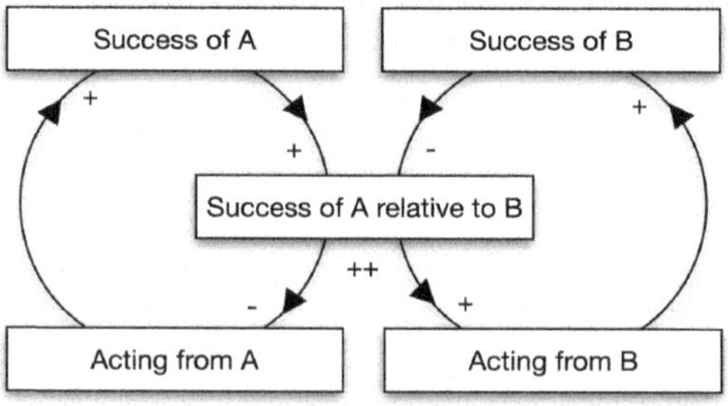

Fig. 2.37 System archetype of escalation (Senge 2008, p. 463)

The *success of the successful* party also results in positive feedback through the coupling of two balanced cycles (Fig. 2.38). In this case, however, the respective build-up works in opposite directions. This is due to different starting conditions. The connection point is a limiting factor that is controlled unilaterally from the outside. For example, A becomes more and more successful due to the fact that it is favored by the initial state, while the success of B continues to decline at the same time. In terms of game theory, this is a so-called zero-sum game (Nickel et al. 2022, p. 110 ff.). A practical example is research funding. As a result of the funding, a funded institute has a better chance of success than an institute that was unsuccessful in its application for funding and can build on this in the next round of applications for further funding. In order to avoid such a situation, the overarching system that distributes the resources is particularly important. For example, some of the resources can be reserved for those who have not yet received any resources (e.g. newcomers) or are structurally inferior. Furthermore, *empowerment*, i.e. measures to strengthen the underdogs, can be further-reaching, as being inferior in competition with the strong does not necessarily mean being weak. Finally, the *David versus Goliath* strategy can help by the previously inferior participant transforming supposed weaknesses such as a manageable size into strengths such as speed, flexibility and creativity.

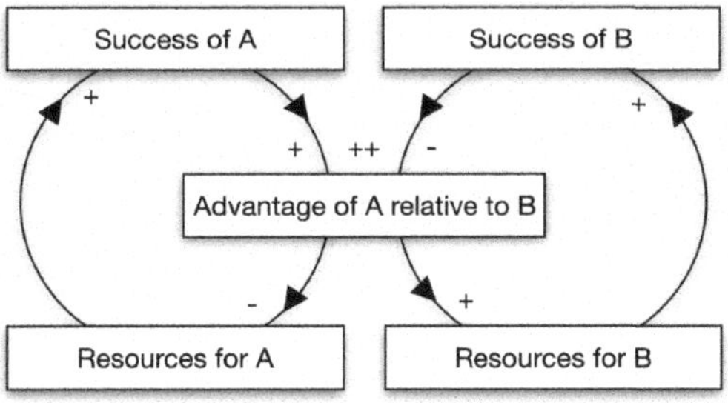

Fig. 2.38 Systemic archetype of the success of the successful (Senge 2008, p. 465)

One of the most consequential system archetypes, especially on a global scale, is the *tragedy of the commons* (Senge 2008, p. 467 f.). Common goods, services and rights are characterized by the fact that those who use them compete with each other, while their use by the provider is not intended or possible. As a result, no price can be enforced for the use of a private offer or no fee can be enforced for a public offer. It is completely impossible to exclude consumption in the case of natural environmental services, such as breathing air. In the case of public goods, such as the road network, users cannot be excluded, but will enter into partial rivalry if bottlenecks (traffic jams) occur. This is exacerbated by the so-called *free-rider problem*. For example, the mail order business contributes to *overuse* of the road network due to the associated delivery traffic, particularly on the so-called last mile (Wellbrock et al. 2022).

The model (Fig. 2.39) consists of two cycles, each of which is self-reinforcing. They are only balanced by the coupling that results from the shared use of limited resources. Although both users are free to choose the quantities they use, they always remain below the critical quantity. In contrast to the system archetype of limited growth (Fig. 2.34), however, resources are not exploited until they are completely used up. Instead, exploitation takes place rhythmically so that resources are only ever available for a short time. The second essential difference to limited growth is that the tragedy of the commons only comes to light in the case of overcritical exploitation. While the benefit that leads to overexploitation accrues in full to those who make use of it, the damage is distributed among all those involved, so that the great benefit is only offset by minor damage.

This represents a false incentive that can lead to market failure in economic systems and to far-reaching and possibly irreversible damage in ecological systems.

Avoiding the tragedy of the commons is only possible if the corresponding good is transformed into a type of good that enables pricing or the enforcement of collectively binding rules (e.g. exclusion from use). This insight is often used to justify the *privatization* of public goods, such as energy and water supply or the provision of transport infrastructure. In reality, however, this is merely one of several options, which often proves to be unsuitable because, among other things, it does not solve the problem of provision, i.e. it cannot ensure basic provision for the collective, particularly in a socially and ecologically responsible community (neoliberalism). The limits of advanced privatization of public goods and an economy largely left to its own devices have become apparent, at the latest with the onset of the state of emergency as a result of the coronavirus pandemic and Russia's war of

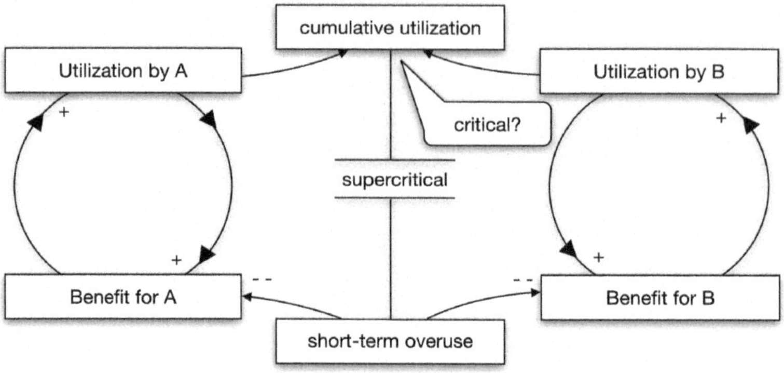

Fig. 2.39 System archetype of the tragedy of the commons (Senge 2008, p. 467)

aggression against Ukraine. At the same time, the now noticeable climate change reveals the urgent need for serious, far-reaching and sustainable environmental protection.

Emissions trading is considered a successful measure to avoid the tragedy of the commons (Lucht and Spangardt 2005). The issue of a limited number of emission certificates, each of which entitles the holder to emit a limited amount of carbon dioxide and other greenhouse gases, is in the hands of the state. Political (collective) targets can be incorporated into the allocation. In addition, the certificates can be traded on a corresponding market, resulting in a price. This price can be offset against investments in climate protection so that such investments can be profitable. In this way, external effects are internalized into the economic system. In sociological terms, this can be explained by the fact that an issue that could not previously be expressed using the semantic code of this sub-system of social systems (Sect. 2.1), i.e. in money, now becomes relevant in this sub-system. The measure conforms to the market, i.e. it makes use of the inherent logic of the economic system, although it serves purposes whose origin lies outside this system.

Whether and, if so, to what extent it is true that common goods, services and rights regularly cause tragedies has not yet been conclusively clarified. For example, non-profit works such as the collective contribution to the creation and further development of a free Internet encyclopaedia (Wikipedia) or the collective programming of freely accessible software such as GNU/Linux have so far been viewed positively in wide circles.

Error corrections that initially prove successful can lead to unexpected and possibly negative consequences as a result of a time delay (Senge 2008, p. 469 f.). In the model, such a *failure of corrections* is represented by an initially balanced cycle that transforms into a reinforcing cycle with a time delay (Fig. 2.40). This dynamic is similar to problem shifting (Fig. 2.33), but in this case the focus remains on the original problem. A classic example is an occasional electrical short circuit that is prevented by inserting a stronger fuse. This exacerbates the actual problem of a damaged power line (represented by a double plus sign in the model) and can lead to a cable fire. This problem can be avoided in the same way as the problem of problem shifting (see above), i.e. by avoiding (pre-)quick solutions (quick wins) and by a thorough root cause analysis (5-W method).

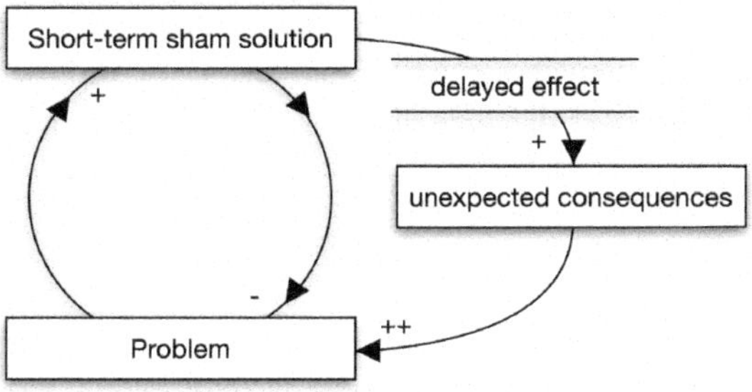

Fig. 2.40 Systemic archetype of the failure of corrections (Senge 2008, p. 469)

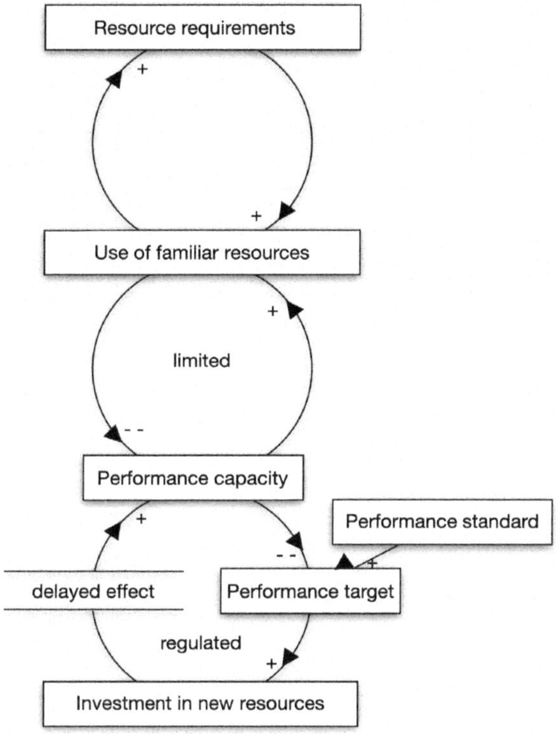

Fig. 2.41 Systemic archetype of growth in connection with underinvestment (Senge 2008, p. 470)

Finally, Peter Michael Senge (2008, p. 470 f.) also recognizes a system archetype in *growth in connection with underinvestment*. This results from linking the limits of growth (Fig. 2.34) with a softening of objectives (Fig. 2.37). The model shown in Fig. 2.41 shows the feedback of two balanced cycles with a self-reinforcing cycle. The solution to the resulting problem is already included in the model. The interaction of the two cycles shown in the Figure above shows that the limited use of familiar resources limits performance in the long term. This is illustrated by the double minus signs. In the lowest cycle, this leads to a reduction in possible performance targets, which is again indicated by the double minus signs. This can only be counteracted if new resources are developed in good time, which then become habitual resources once they are established, which usually takes some time.

2.4.5.3 Dynamic Models

The dynamic modeling of systems is also referred to as *simulation*. In contrast to static methods, the state sequences of a system that occur over time are developed endogenously on the basis of the interactions and time mechanisms depicted in the underlying model. This often allows significantly more interactions to be taken into account. The results are not calculated analytically, but determined experimentally (VDI 3633, Sheet 1:2014-12; Wenzel 2018, p. 1 ff.). The results are presented both with the help of key figures and in diagrams, which in particular also show the temporal behavior (VDI 3633 Sheet 11:2020-10).

A simple and illustrative example is *simulation games* that are carried out by people. For example, the Beer Distribution Game is used to provoke the so-called *bullwhip effect* (Forrester 2013; Schönsleben 2016, p. 94 f.), which describes how small fluctuations in order quantities at the end of a supply chain (customers) build up to large fluctuations at the beginning of the supply chain. To this end, groups of people take on different roles, such as that of a wholesaler or retailer, and carry out activities such as ordering and delivering beer according to predefined rules. Recording certain data such as stock levels and delivery quantities allows the system's behavior to be analyzed over time. A similar approach can be found in *experimental archaeology*, which looks for answers to historical questions by means of an afterlife under defined conditions (Coles 1979).

In industrial practice, however, more formal forms of simulation are common, which are usually carried out using *software* on computers. For this purpose, the sub-systems and part-systems as well as the elements13 of the real systems are modeled on the basis of the functional model of technical material systems (Fig. 2.6). In detail, the states contained therein are formulated as 9-tuples $\sum = (T, X, Y, Z, X^*, Y^*, F, g, \leq)$, whereby the following applies in detail (Wenzel 2018, p. 7):

- The sub-elements of the time set T are real numbers.
- The input alphabet X and the output alphabet Y are arbitrary sets with the elements x and y respectively.
- Z is the state alphabet and an arbitrary set with the elements z, which describe the individual states of the system under investigation.
- X is the signal space and a subset of X^T, i.e. the set of all time functions (signals) with $x : T \longrightarrow X, x(t) = x$ with $t \in T$ and $x \in X$.
- Y is the signal space and a subset of Y^T, i.e. the set of all time functions (signals) with $y : T \longrightarrow Y, y(t) = y$ with $t \in T$ and $y \in Y$.
- The transition function F is defined by $F : (T \times T \times Z \times X^*)' \longrightarrow Z$ and $F(t', t, z, x^*) = z'$, where z and z' denote the system states at time t and t' respectively. The input x arriving at time t triggers a transition to the state z' at time t'. $(T \times T \times Z \times X^*)'$ describes a certain subset $(T \times T \times Z \times X^*)$, as F does not have to be defined on the entire set.
- The result function g is defined by $(T \times Z \times X^*) \longrightarrow Y$ and $g(t, z, x) = y$.

- $\leq$ stands for the linear order relation on the time set T.

The model elements can describe physical or logical elements of the modeled system. A *physical* element would be a machine, for example, and a *logical* element would be the associated machine controller. Furthermore, model elements can behave *statically* or *dynamically*, whereby this distinction depends on the aspect under consideration (Sect. 2.4.6). For example, from the perspective of internal material flow, a pallet behaves dynamically, i.e. it can change its position, whereas a machine is to be regarded as static in this respect, while it behaves quite dynamically internally when processing workpieces. It is also possible to differentiate between *permanent* and *temporary* elements, as a workpiece enters the model unprocessed and leaves it in processed form, while a machine is a permanent part of the factory.

With regard to modeling the dynamic behavior of the elements, Sigrid Wenzel (2018, p. 10 ff.) distinguishes between continuous and discrete simulation at the top level (Fig. 2.42). *Continuous simulation* updates the state variables of the model elements over time in a continuous process. For this purpose, the dynamic behavior is described by a series of coupled differential equations. In contrast, *discrete simulation* only considers the state changes at certain points in time.

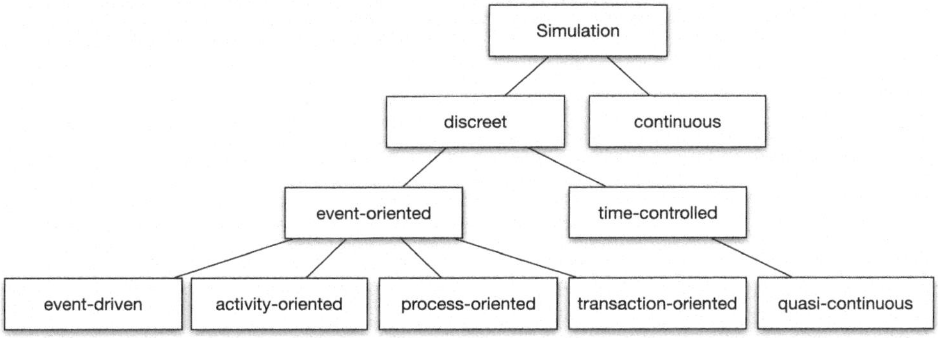

Fig. 2.42 Possibilities of dynamic modeling and simulation of systems (Wenzel 2018, p. 10 ff.)

This can be either event-driven or time-driven (synonym: synchronous) (Fig. 2.42). The latter is particularly common in its quasi-continuous variant. Here, progress is made over time at constant intervals t. At the end of each interval or at the beginning of the following interval, all state changes of all model elements that have occurred within the interval are taken into account. The smaller the intervals become, the closer you get to a continuous simulation, but the greater the computational effort. This method is particularly relevant for mapping processes that also run continuously in reality, such as fluids or pastes.

In contrast, event-oriented methods are suitable for processes that lead to state changes at irregular intervals in practice, as is the case in logistics when storing, handling and transporting unit loads, for example. An *event* is understood to be a change in the state of a model element that occurs at a specific point in time. The change of state is timeless, but then remains unchanged until a new change of state is triggered by an event. An event can be triggered both within the model (endogenous) and externally (exogenous), for example by manual intervention by the person carrying out a corresponding experiment. In addition to an event, an *activity* is understood to be a time-consuming operation that changes the state of a specific element. This can be, for example, the processing of a workpiece on a work system.

In an event-driven simulation, all variables are initialized after the start of a simulation run (Fig. 2.43) (März et al. 2011, p. 16). Furthermore, an event list is opened in which all events are entered and then processed in sequence. At the beginning, the start is always noted as an event and if a fixed end is planned for the simulation run, this is also entered now. The next event is then read from the event list. Immediately after initialization, however, such an event should not yet be entered. The simulation program then updates the states of all sub-systems, part-systems and elements (from now on only specified as elements). As all elements are usually still idle at the beginning, there will not be much to do. However, the source should be configured in such a way that it provides a token to run through the model at mostly random time intervals based on a probability distribution. To illustrate this, let us assume that this is a product that is to be processed at various processing stations according to a predefined work plan. The status of the source is therefore set to Token provided and a request to the means of transport to collect the product (token) from the source is entered in the event list as a subsequent event. If there is nothing else to do, the simulation time jumps to the next event and triggers the transportation vehicle to travel to the source. The addition of the travel time, which in the case of scale modeling is calculated from the distance and speed, depending on the level of detail also from the accelerations (starting, braking) and taking into account randomly determined disturbances on the basis of corresponding probability distributions (Sect. 2.4.5.1), results in the event at which the means of transport will arrive at the source. This event is entered in the event list and the status of the means of transport is set to busy. In addition, the journey of the means of transport is logged with all the usual data and, if necessary, with any additional data required so that it can be taken into account in subsequent evaluations. The additional data to be collected is set as part of the modeling. Standard simulation programs usually offer numerous parameterization and programming options for this.

Finally, all logged data is saved and, depending on the simulation program used, a whole series of prepared and therefore clear evaluations can be generated. In addition, it is usually possible to observe the simulation run during its execution (monitoring). Graphical representations can also be selected for this purpose, as shown in Fig. 2.43 using the example of the temporal progression of the occupancy of a workstation.

A generic simulation concept has just been presented in connection with event-oriented simulation (Wenzel 2018, p. 13 f.). Modeling is often based on object-oriented modeling. In addition, corresponding computer programs offer libraries that already contain prefabricated elements (model modules). These are domain-specific in that, for example, libraries that are intended for the simulation of logistical systems contain building blocks such as means of transport and warehouses, while libraries that are intended for the simulation of crowds in sports stadiums, for example to be able to investigate the collective behavior of crowds in panic situations, primarily contain building blocks that represent human behavior (VDI 3633, Sheet 6:2022-12).

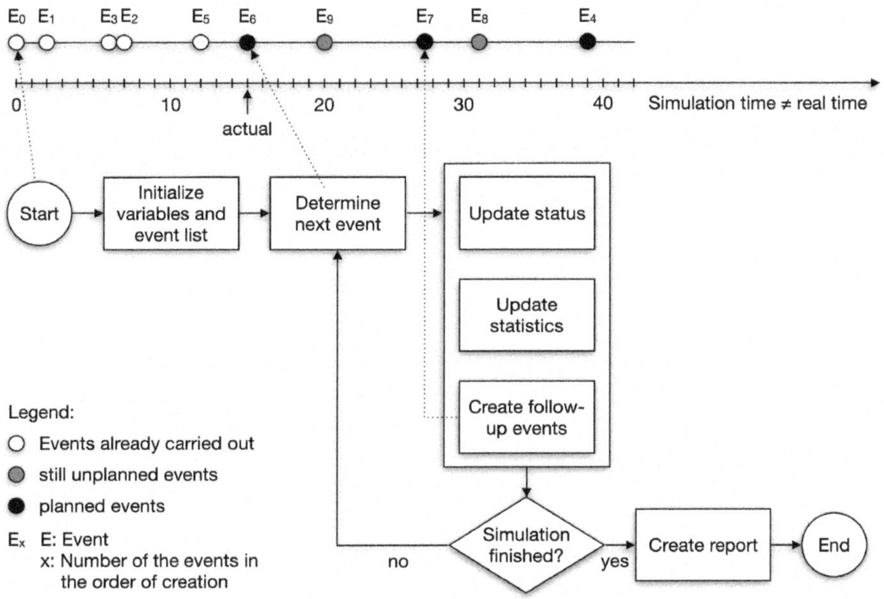

Fig. 2.43 Sequence of an event-driven simulation (März et al. 2011, p. 16)

The building blocks shown in Fig. 2.44, for example, are part of a production logistics application in which a line production was modeled in order to experimentally determine how the interlinking losses can be reduced by isolating the bottleneck from the upstream and downstream workstations using buffers and to what extent the throughput can be increased if, alternatively or additionally, the disruptions at the stations are reduced in the course of continuous improvement (CIP). Buffered production is thus opposed to the concept of Lean Production, which is distinct from it (Womack et al. 1992, p. 19), so that such a model can be used to investigate whether and, if so, to what extent one approach is actually superior to the other with regard to the aspect under investigation.

In addition to generic simulation concepts, language concepts are also widespread (Wenzel 2018, p. 13), which are either based on just one method and offer corresponding event and time control options or contain additional language constructs and modeling mechanisms, which are usually implemented in a block or function-oriented manner. One example is the open source software Xcos, which complements the freely available software Scilab for solving mathematical problems and graphically displaying the results. Commercial products with a comparable range of functions are also available.

Finally, *mathematical simulation concepts* are also widespread (ibid.). One example can be found in SysML. This can be shown using the Useless Box introduced in Section 2.4.2 Fig. 2.45 first shows the sequence diagram (sd). This shows the interactions between the subsystems and elements (interaction partners) actively involved in the processes, which were defined in the block definition diagram (Fig. 2.17), using the SysML notation (Alt 2012, p. 53 f.; Weilkiens 2014, p. 283 ff.). Corresponding messages specifically take the form of operation calls or signals and are represented by horizontal arrows pointing from the calling

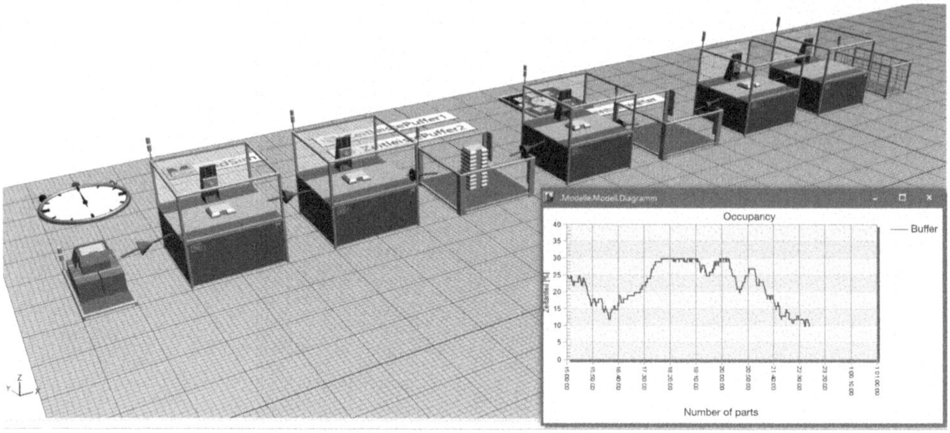

Fig. 2.44 Example of a simulation model created with the Plant Simulation software

to the called interaction partner. The time axis runs downwards, while the interacting sub-systems or elements are arranged next to each other. A dashed lifeline follows the time axis from each interaction partner. The bars shown on the lifelines (execution focus) indicate that the respective interaction partner is active in the corresponding time interval. One advantage of SysML is that it is largely self-explanatory to those who are familiar with its notation. And so the sequence diagram shown in Fig. 2.45 appears so self-explanatory that no further explanation seems necessary, especially in view of the functionality of the Useless Box explained above.

Another option for modeling the time behavior of systems using SysML is the *state machine* (stm) shown in Fig. 2.46. In it, the states of the active sub-systems and elements of the focal system are triggered by transitions, which are represented by arrows (Alt 2012, p. 59 f.; Weilkiens 2014, p. 269 ff.). The process is simulated by tokens, which are shown as filled circles. Oliver Alt (2012, p. 56) compares these tokens very vividly with the baton that relay runners pass to each other so that it is always clear which runner is currently taking part in the competition. The analogy chosen by Tim Weilkiens (2014, p. 246 f.) to the marble run, on which a marble rolls along the track (transitions) and triggers activities along the way, is equally illustrative. The state diagram is also largely self-explanatory in light of the already explained functionality of the Useless Box. Only the concurrency of control flows should be briefly mentioned (Weilkiens 2014, p. 262 ff.). If a token hits a splitting bar (ForkJoinH) shown as a black bar in the state diagram, as many tokens are generated as there are further transitions from it. In the further course, splitting bars are then used to synchronize concurrent control flows. If several transitions meet at a splitting beam, the control flow only continues there when tokens have arrived at all transitions. With the exception of one continuing token, all others are destroyed.

One advantage of behavioral diagrams such as the state diagram chosen here as an example is that computer programs that allow modeling with SysML also offer the option of simulating modeled processes. The documentation of a simulation run for the state diagram

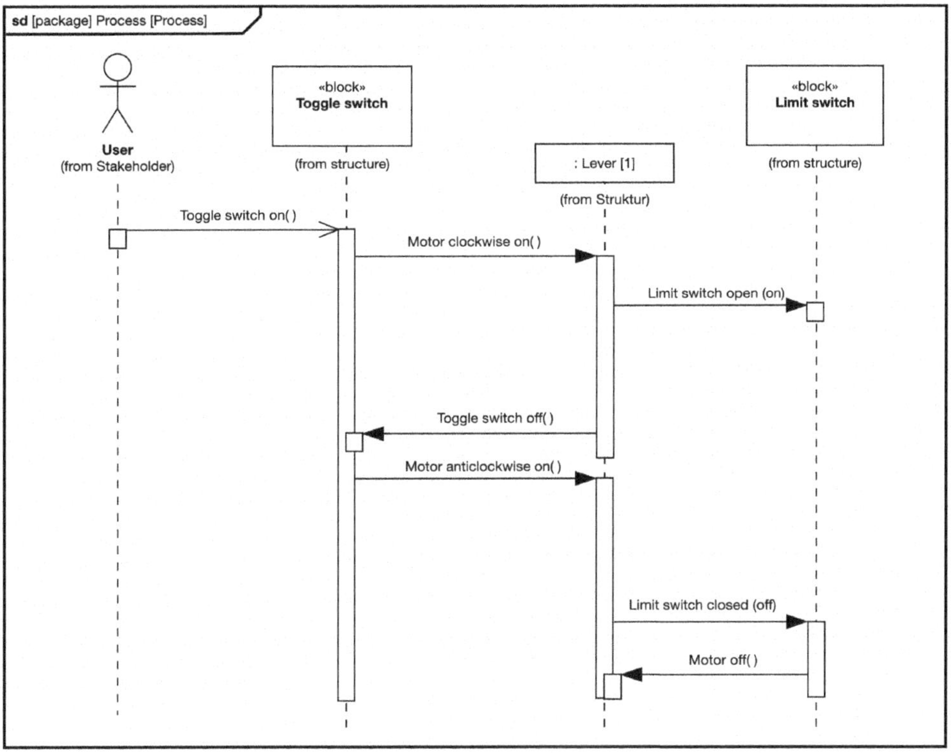

Fig. 2.45 Sequence diagram in SysML notation, created with Enterprise Architect software

shown in Fig. 2.45 is listed below. This shows that the modeled process corresponds to expectations and is error-free.

- [96466923] Configuration: mode: Interpreted simulation, platform: UML Basic
- [96466923] Preparing Simulation Data
- [96467518] Loading Machine
- [96467548] Simulation Started
- [96467549] Start
- [96468073] Toggle switch on
- [96468590] Motor clockwise on
- [96469105] ForkJoinH
- [96469664] Cover open
- [96469672] Limit switch open
- [96470201] Toggle switch off
- [96470203] ForkJoinH
- [96470729] ForkJoinH
- [96471275] Motor counterclockwise on
- [96471795] ForkJoinH

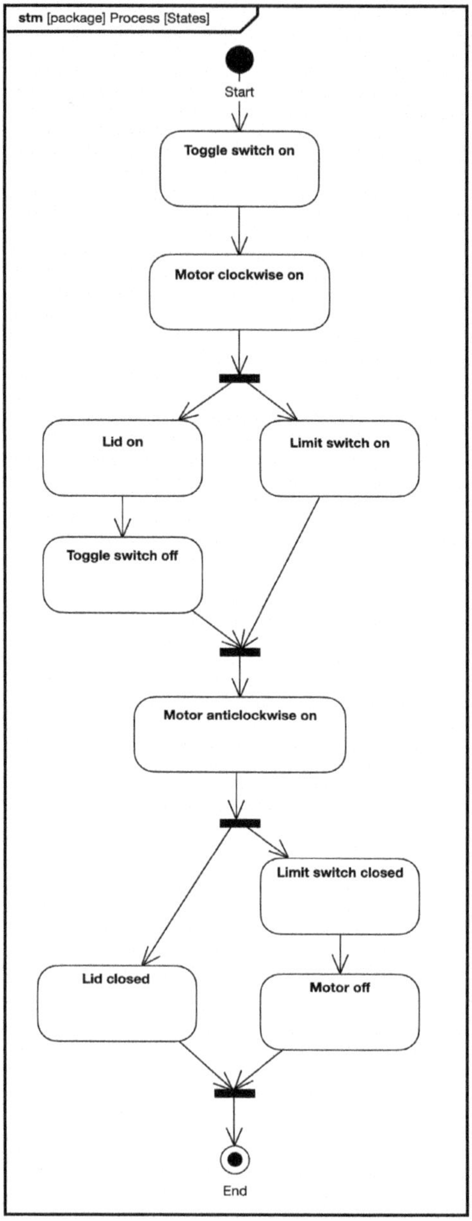

Fig. 2.46 State diagram in SysML notation, created with Enterprise Architect software

- [96472358] Cover closed
- [96472360] Limit switch closed
- [96472876] ForkJoinH

- [96472877] Motor off
- [96473404] ForkJoinH
- [96473953] End
- [96474010] Simulation Ended

2.4.6 Aspects of Systems

In addition to structures and processes, various aspects of a system are often brought to the center of attention. For this purpose, a kind of *filter* or *template* is placed over the system to highlight certain characteristics or properties (Haberfellner et al. 2018, p. 33). Examples include the flow of money, the flow of energy, the flow of information or the flow of materials (Fig. 2.47), although this selection deliberately makes it clear that there may be overlaps with other approaches, such as the IPO model in this case (Fig. 2.6). In a specific application, one of the approaches should be chosen to avoid such redundancies.

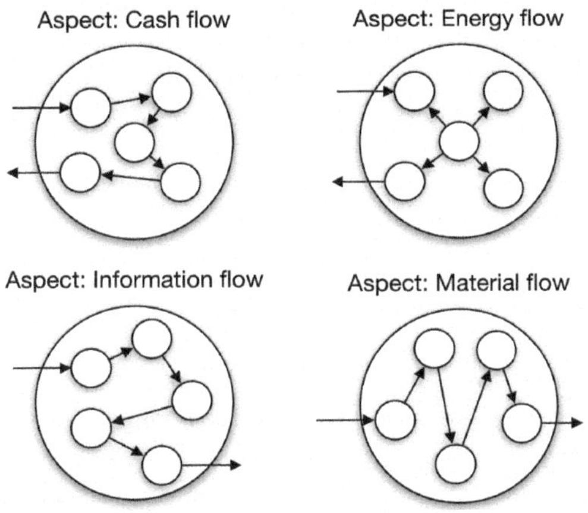

Fig. 2.47 Aspects of a system (Haberfellner et al. 2018, p. 33)

Model-based systems engineering (MBSE) offers a concrete application in which the underlying model is stored centrally, for example in a database. Fig. 2.48 makes it clear that different participants in a system design (viewers) only work with corresponding parts of this model under certain aspects. The *Systems Modeling Language* (SysML) pulls these parts together in a diagram and thus offers a specific view of the central model (Alt 2012, p. 39).

Attributes are used to characterize systems and elements. *Classifications* can be made based on these characteristics or properties of the objects under consideration. This means that the objects can be ordered (sorted) according to groups, quantities or categories. This allows patterns to be recognized, for example. This in turn is an important prerequisite for *standardization*, which is particularly important and very widespread in organization and

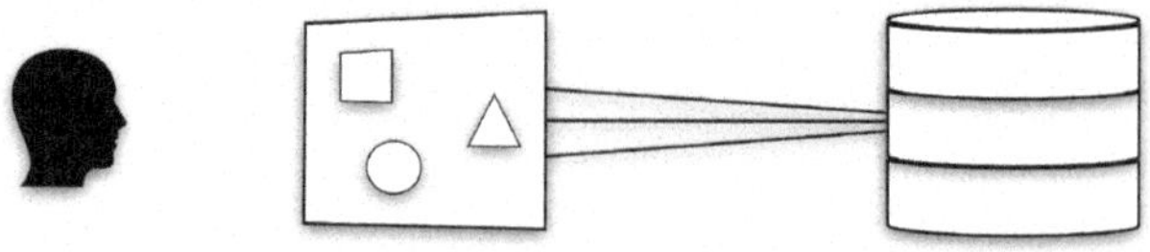

Fig. 2.48 Views of a model (Alt 2012, p. 39)

technology. A deliberately simple example here is the ordering of objects based on their geometric shape, according to which some objects can be described as spheres and others as cubes. For a concrete description, *variables* (placeholders, variables) are used that can be assigned values. For example, a sphere can be described with the two variables center and radius. A special form of variable is the *parameter*. This differs from a *constant* in that it is only constant in a currently observed situation, but can be varied in another situation. This is why a parameter is said to be arbitrary but fixed. For example, in a quadratic function $f(x) = b - x^2$, the opening b can be used as a parameter for a certain series of different function arguments x, whereby the graph is known to result in a parabola. If, on the other hand, x is used as a parameter and b is varied in the form $f(b) = x^2 - b$, the graph results in a straight line with the gradient $m = x^2$ through the origin of the $y - b$ plane.

2.5 Context and Path Dependency

What all the types of modeling presented so far have in common is that they only represent the technical part of the socio-technical system outlined in Fig. 1.1 However, as already explained in the introduction, technology alone cannot be meaningfully delimited. This even applies to the bracket shown as an example in Fig. 2.17, which is used to attach an alternator in the engine compartment of a vehicle, because a developer working for a plastics processing company will tend to make it out of plastic, whereas a representative of a metalworking company will tend to design it as a metal part. For all the advantages and disadvantages that the respective materials bring with them, it will not only depend on purely technical aspects as to which properties are placed in the foreground, but also on the social *context* in which the respective representatives are integrated and to which they feel committed. Despite all the objectifiability that technology brings with it due to its predictability, it itself leaves sufficient scope for social instrumentalization.

This is also shown by the fact that technical solutions to problems generally depend on their social history, i.e. in the constructivist sense they are shaped more by social memory than by the perception of external phenomena. However, this cannot be explained in scientific or technological terms, but rather in social scientific terms.

- One example of this is the passenger car, which is still modeled on a horse-drawn carriage in which only the horses have been replaced by an engine. Even today, the combustion engine is usually located where the horses used to run: in the front, and even today the coachman sits in the front and the driver in the back seat.

- Another frequently cited example is the QWERTZ keyboard (David 1985, p. 332 ff.). Originally, the arrangement of the characters on a keyboard was chosen so that the mechanics of a typewriter could not jam during fast keystrokes. With the introduction of computers, this was no longer necessary, so that an ergonomic arrangement would have been possible. In actual fact, however, the already widespread keyboard layout was retained.

In systems theory, such effects are referred to as *path dependency*, which, as the two examples have just shown, can be traced above all in technological developments, but is also generally valid. Douglass North, for example, was awarded the Nobel Prize in Economics in 1993 for demonstrating path dependency in economics. This phenomenon is still too little known in systems engineering, but is nevertheless of great importance. It is therefore discussed in more detail below.

For good reasons, technologists tend to think in terms of control loops in connection with the maintenance of *flow equilibrium* (homeostasis), which is often technically necessary. Deviations from the target state, for example from the position of a turning tool in relation to the workpiece surface to be machined, must be minimized by negative feedback via the feedback loop of the position control loop (Seitz 2015, p. 184 ff.). Economists also recognize advantages in the efficient and effective distribution of scarce resources, for example, in the market equilibrium that results from the dynamic interplay of supply and demand.

Path-dependent developments, on the other hand, are influenced by *positive feedbacks*. Fig. 2.49 shows that many possible solutions are pursued in the first phase. In a second phase, these increasingly converge on a specific development path. However, this path is formed emergently. This development can be modeled, for example, using the principle of preferential attachment according to Albert-László Barabási and Réka Albert (2016, p. 164 ff.) explained in Section 2.4.2. This means that the path ultimately chosen does not necessarily have to be an optimal path. Path dependencies can also lead to *suboptimal solutions*. Finally, there is a lock-in effect. In phase 3, only contributions that correspond to the path once chosen are classified as relevant. Only when this path reaches obvious limits does a radical break occur. From now on, a new first phase begins in which anything is possible again.

An example of the situation in phase 2 is the development of the standard for analog video magnetic tape recordings in the 1970s. Initially, the video home system (VHS) developed by JVC competed with the Betamax system launched by Sony and with the VCR/Video 2000 introduced by Grundig and Philips. VHS eventually prevailed, although it had a lower picture quality than the other two technologies. The decisive factors were more generous licensing, a longer running time, at least initially, more convenient handling of the cassettes and simple and reliable equipment. As plausible as these advantages may seem in retrospect, it was difficult to foresee at the time that they would be superior to the aspect of better picture quality.

An example of a situation in phase 3 is rail transport, which prevented a new, possibly superior technology such as the Transrapid from being developed where it was already in place. The fact that an attempt was nevertheless made to establish a new type of transport technology may be due to ignorance of or a lack of consideration for the phenomenon of path dependency. In this case, the associated investments would have to be classified as bad

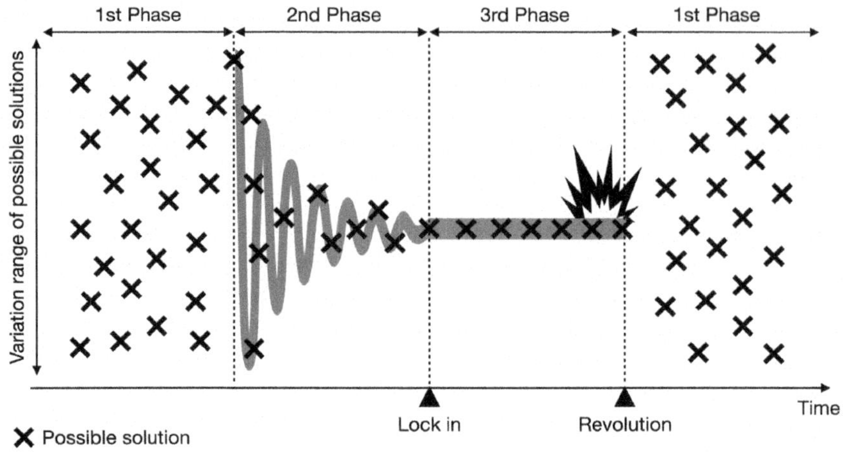

Fig. 2.49 Paradigmatic development

investments and would be regrettable. However, the attempt may also have been made with the courage and ambition to overcome the dependency on a path that is considered outdated and to replace it with a technology that is considered more promising for the future. Such an attempt would deserve respect.

The contributions to positive feedback can be manifold and are therefore not always easy to identify. These include investments that have already been made and need to be amortized, standards such as norms and guidelines that document the state of the art, patents, know-how among developers and users, interfaces to other technologies that are linked to the focal technology, and even habits, cultural anchors and traditions. When the model train stands in the cellar or attic and unites grandparents with their grandchildren in playing together and when Michael Ende's Jim Button and Luke the Engine Driver is part of the international repertoire of popular bedtime stories, it is not a purely rational, i.e. technical or economic, question as to whether the railroad will be replaced by a new technology. But economic consequences can be reason enough to stay on a path once it has been taken. Just imagine the costs that would be incurred by countries such as the United Kingdom or India if they were to switch to right-hand road traffic or, conversely, if continental Europe or the USA were to switch to left-hand traffic.

Thomas Samuel Kuhn (2003) has shown that path dependencies along the lines outlined in Fig. 2.49 also exist in science. He described the situation in the third phase as a paradigm. In such phases, it is primarily public funding that is concentrated on the predominant field of research. Specialist conferences align their main topics accordingly and specialist journals only accept corresponding contributions, as this increases the number of participants and circulations in view of the growing number of institutions and individuals turning to the paradigm. Above all, peer review, the mutual control among scientists, also contributes significantly to the fact that work outside the paradigm is hardly given a chance. One resulting advantage is that, thanks to the pooling of capacities in the predominant field of research, great progress can be made relatively quickly. The disadvantage, however, is that alternative solutions and other issues remain unresolved, regardless of their quality and urgency.

Problems like these can also be observed in organizations. The decline of the once very successful company Kodak is an example of this. The company held on to analogue technology for too long and missed the boat on digitalization as a result. To this day, little has been learned from this: standardized selection procedures when hiring new employees, KPI-induced targets that limit the scope for action, flexibility and creativity of employees managed in this way, or the preference for opportunistic behavior when it comes to promotions are just some of the mechanisms that encourage people to stick to a path once they have embarked on it.

There is also an effect that Markus Göbel and Johannes Schmidt (1998) refer to as *hyper-inclusion*. Management is particularly affected by this when it meets the expectation of constant availability, combined with excessive working hours and frequent business trips, and thus loses touch with the world outside the company, sometimes even outside the close circle of management (Goeudevert 1996, p. 180 ff.). This is now also referred to as *bubble formation* and favors positive feedback loops on suboptimal paths, from which an escape is ultimately only possible through a radical break, which, however, often results in insolvency for companies. One way of mitigating the consequences of hyper-inclusion would be to return the court jester so that there is at least one person in management circles who is allowed to point the way out of dead ends without having to reckon with punishment (exclusion) (McMaster et al. 2004, p. 129 ff.).

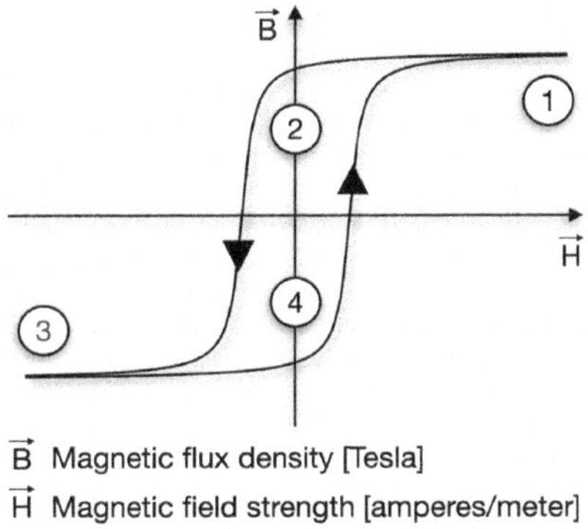

Fig. 2.50 Hysteresis as an example of a natural path dependency

Another model of path dependency is the *hysteresis* shown in Fig. 2.50. Here, the output variable B depends not only on the independently variable input variable H, but also on the previous state of the output variable. The magnetization of a ferromagnet is an example of this. If, starting from point 1, the magnetic field strength H is reduced to zero, the magnetic flux density B at point 2 does not go back to zero. A remanence flux density remains, as the previously fully aligned elementary magnets retain some of their orientation. If one goes from a reversed polarity of the magnet in point 3 back to the value zero, a remanence flux density remains again in point 4, but this time with a different value than before in point 2. Despite the same magnetic field strength $H = 0A \cdot m^{-1}$, the magnetic flux density B therefore assumes different values depending on the previous state. This type of path dependency can also be found in other areas, for example in physiology with the pressure-volume ratio of the lungs or in hydrology with the tide. Psychologically, the willingness to take risks can be explained with the help of hysteresis: as the saying goes, one learns from harm, and a longer phase without harm can lead to recklessness. Technically, hysteresis is used in two-point controllers, for example, to avoid oscillations in thermostats by installing a bimetal (Seitz 2015, p. 131).

Finally, Heinz von Foerster (1993, p. 247 ff.) used his *non-trivial machine* (NTM) to make it clear that path-dependent behavior can also result from the interaction of trivial machines (TM). The machine shown in Fig. 2.51 looks like a TM from the outside. On the inside, it consists of two TMs, the drive function F and the state function Z. In this respect, an NTM, like a TM, is also *synthetically deterministic*, i.e. uniquely defined. Unlike a TM, however, an NTM is not *analytically determinable* because, unlike a TM, the same stimuli x show different reactions y at different times.

The drive and state functions are linked in such a way that the transformation of the input x into the output y is determined not only by the drive function F, but also by the internal states z and z'. Z calculates the state z' recursively by expressing z' through itself at an earlier stage z. Z is also influenced by the input x.

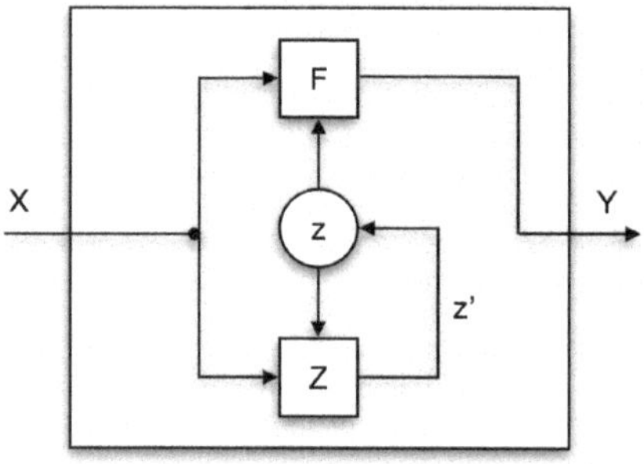

	I			II	
x	y	z'	x	y	z'
A	0	I	A	1	I
U	1	I	U	0	II
S	1	II	S	0	I
T	0	II	T	1	II

y = F (x, z), Drive function

z' = F (x, z), State function

Fig. 2.51 Non-trivial machine (von Foerster 1993, p. 247 ff.)

Assuming that the NTM can assume exactly two internal states z' = I and z' = II, the drive and state functions can be represented as shown in the table in Fig. 2.51.

- If a sequence of $x = A$ is applied to the NTM at the beginning, it delivers the result $y = 0$ for each of these stimuli.
- If a sequence of $x = U$ is applied to the NTM, it responds with a sequence of results $y = 1$.
- If the input $x = S$ is applied to the NTM, the result is $y = 1$. What cannot be observed from the outside, however, is that the internal state $z' = I$ is switched to the state $z' = II$ at the same time. If the first input $x = S$ is followed by a second input $x = S$, the NTM

does not return $y = 1$ again, but $y = 0$. This resets the internal state back to $z' = I$, so that the result $y = 1$ is output for a third input of $x = S$, and so on. The results therefore oscillate between 1 and 0.

- Heinz von Foerster (1993, p. 250) concludes his observations with the sequence USA, to which the NTM reacts either with a sequence of three ones or with a sequence of three zeros, depending on its internal initial state.

The NTM therefore not only exhibits path-dependent behavior. It is also analytically impenetrable, because with two possible states $n = 2$ and four possible input states A, U, S and T, an outside observer who wanted to try to recognize patterns from the machine's reactions to all possible combinations of the input states in order to deduce the internal structure would have to carry out $6 \cdot 10^{76}$ investigations (ibid.). This is not possible efficiently: the search for a solution would simply take too long.

For Systems Engineering & Management, the considerations on context and path dependency result in both opportunities and limitations. It should be noted in advance that in the third phase of the development outlined in Fig. 2.49, it can be equally useful to further develop the existing in the sense of continuous improvement (Kaizen) (Liker 2016, p. 349 ff.) or, as in the first phase, to be innovative in the sense of Joseph Schumpeter (1911, p. 158 ff.). It is crucial to be aware that the probability of failure is very high if you try to do something that does not fit the respective phase.

- Experience has shown that swimming against the tide in phase 3 or looking for new solutions outside of an intact paradigm does not seem very promising. Trying to do so would most likely mean wasting resources, as was the case with the development of the Transrapid, for example. Nevertheless, Chapter 3 attempts to overcome the paradigm of process orientation that has been widespread in organizational theory worldwide for around thirty years.
- At the same time, as the example of Kodak has shown, it can be fatal to overlook or even ignore a revolution and stick to old solutions in phase 1. In this phase, it is important to be particularly innovative, just as Apple, for example, helped the smartphone achieve its breakthrough with the introduction of the iPhone in such a phase.
- Finally, phase 2 harbors the risk of believing that you as an individual can have a significant influence on what leads to a new development path. This is the result of emergence. This does not mean that you should not make any active contributions yourself, but rather that you should recognize as early as possible where the journey is heading and not hold on to your own ideas for too long in this second phase. In this respect, Sony, Grundig and Philips obviously did everything right in the 1970s, both with the development of their own contributions and with their timely switch to VHS. Only Sony may have regretted selling the patent for VHS originally developed in-house to JVC.

2.6 Modeling of Socio- and Psycho-Technical Systems

On the basis of a constructivist understanding, the models of technical systems presented in Section 2.1 have another problem in addition to a lack of context and path dependency: they present systems as fundamentally open. This goes hand in hand with the idea of a more or less strictly deterministic behavior in the sense of a trivial machine in the sense of Heinz von Foerster (1993, p. 245). Although this does not rule out unintentional behavior, it assumes that in such a case there is also an underlying *causality* that can be identified and eliminated in the course of troubleshooting, for example by asking why five times (*5-W method*). (Andler 2015, p. 57 ff.; Dombrowski and Mielke 2015, p. 84). However, this is incompatible with a basic constructivist understanding. This does not fundamentally speak against the use of methods such as the 5-W method, but extends the horizon of observation to non-strictly causal relationships, i.e. to the actions of psychological and social systems.

This basic understanding takes as its starting point the relationship between system and environment. The *boundary* necessary for this distinction is always drawn by the system. Although this does not lead to the interruption of connections between system and environment, it does mean, and this is of great significance, that cross-border processes are subject to other conditions of continuation (Luhmann 1996, p. 34 ff.). These conditions are subject to a very extensive *autonomy* of the system and can only be recognized or even influenced from the environment to a very limited extent. At best, as the saying goes, one can try to put oneself in the system's shoes, for example with the help of *empathy*, in order to gain a certain understanding of the system's internal structures and processes and, in the sense of *hermeneutics* (e.g. Gadamer 2010), to understand why a system acts in a certain way, and one can of course also use methods, procedures and tools of SEM to increase the probability of mutual understanding.

It is important to note that from the perspective of a system, the environment is not a system. It only maintains its unity through the system that differentiates itself from it. And since every system only delimits itself from its environment, no two systems ever have an identical environment (Luhmann 1996, p. 34 ff.). This also applies to sub- and part-systems as well as to elements (the latter can also present themselves as systems from a different perspective, often even as systems whose complexity exceeds that of the super-system). For them, too, the super-system presents itself as an environment that is excluded as such by themselves. As a result, the super-system is perceived differently by each sub- and part-system as well as by each element. The differentiation of systems is therefore a process for increasing complexity (ibid., p. 38). This also applies to a hierarchical form of differentiation. Therefore, the following Section will essentially have to focus on giving greater weight to the autonomy of systems.

Before this, however, the boundary between system and environment must be examined in more detail. The starting point for the following considerations is the insight that the boundary is always attributable to the system, while the environment differs from systems primarily in that it has no boundary (ibid.). This raises the question of why a super-system is shown in Fig. 2.17 from the perspective of the focal system. From this perspective, this super-system should not appear to the focal system as a system, but as an environment, due to the relationship between system and environment that has just been introduced. And this is exactly what it actually does – but only from the perspective of the focal system.

Against this background, it can be stated that Fig. 2.17 represents a system hierarchy from the perspective of an observer, i.e. from an external point of view, so to speak. For certain purposes, this representation is very helpful, for example for building a structural parts list for a technical artifact (Wiendahl 2014, p. 164).

In practice, however, such a purely technical view can easily lead to false assumptions and expectations, which experience has shown often result in conflicts and disappointments, because *artifacts* usually behave differently than a system hierarchy in the sense of a structural BOM would suggest. This observation is based on the insight explained in Fig. 1.1 that artifacts are to be understood as *socio-technical systems*.

The consequences of this approach can be illustrated by almost all development processes in which there are regularly profound communication difficulties between those responsible for the development of different super- and sub-systems or elements. For example, a team designing an engine for a car often has its own ideas about the body in which its engine is to be installed and expresses these in a list of requirements (specifications) for the body development department (Ebert 2019). Conversely, this also applies to those involved in designing the body. As a result, intensive coordination processes are necessary in order to reach a sufficiently common understanding of what should ultimately emerge as a super-system (car).

Disappointments and conflicts can probably never be completely avoided in the course of such coordination processes, but the understanding that, for example, one's own system, to which one ascribes oneself, is not perceived as a system by another system with which one has to deal, but as an environment, can help to limit expectations of each other, misunderstandings from the outset and not to attribute them to the inability or unwillingness of others to cooperate, but to attribute them to the natural laws according to which cross-border processes are always subject to different conditions of continuation (Luhmann 1996, p. 34 ff. 34 ff.) and according to which these conditions are always subject to the autonomy of the respective system and are very difficult to recognize, understand or even influence from the outside. Understanding is, as Hans-Georg Gadamer (2010) states, highly improbable and can at best be made a little more probable through some effort.

A very significant advantage of Systems Engineering & Management over the hermeneutics proposed by Hans-Georg Gadamer is that it provides concrete methods, procedures and tools to support understanding between those involved in design and management projects. One example of this is *Requirements Engineering* (Ebert 2019). And *Process Management* (Gaitanides 2012) also set out in the early 1990s to improve the overcoming of boundaries between the departments of organizations with the help of organizational and technical measures. Everyday experience shows that success can certainly be achieved in this way, but at the same time experience also repeatedly shows the limits of such approaches. However, these are rarely due to the incompetence of those responsible for introducing and maintaining such measures, or the unwillingness or inability of those who are supposed to work accordingly. The differentiation of systems is, as already mentioned above, a procedure for increasing complexity (Luhmann 1996, p. 38) and in this respect is of decisive importance for organizations that have to assert themselves in a complex environment, in view of the basic *law of requisite variety* (Ashby 1956, p. 206 ff.). A certain degree of demarcation between the differentiated sub- and part-systems as well as the elements and the associated difficulties are virtually the price that organizations have to pay for this.

A complete elimination of autonomy and operational unity, if it were possible at all[13], would have a far greater impact on the survival of systems than accepting autonomy and its consequences. For this reason, autonomy will be discussed in more detail below.

2.6.1 Systems of Systems

Systems engineering has taken a first, albeit still inadequate, step in this direction with its considerations on systems of systems (SoS) (Haberfellner et al. 2018, p. 31 f.; Walden et al. 2023, p. 235 ff.). However, it is often difficult to make a precise distinction from other system descriptions, especially hierarchical structuring (Fig. 2.17) (Haberfellner et al. 2018, p. 31). To differentiate, Haberfellner et al. (ibid.) suggest that systems that are parts of systems should be granted greater *autonomy* compared to sub-systems. They illustrate this with the example of a holding company: if the individual companies of the group were subject to an overall corporate management that determined the strategic direction, the individual companies would be regarded as sub-systems of the holding company. If it were merely a financial holding company that managed the subsidiaries solely on the basis of monetary targets, it would be a system of systems.

Another example would be an airport that is managed by an operating company. In addition to this core business, the infrastructure of large airports also consists of other parts such as hotels and connections to other transportation systems. In principle, a hotel can also be operated independently of an airport and is often part of a chain that operates other hotels away from airports. Nevertheless, the integration of a hotel into an airport operation often results in specific offers. For example, in order to predict the arrival of a hotel guest, the hotel will automatically make a comparison with current forecasts of the landing times of the respective aircraft and, in the event of a delay, will not wait for a call from the guest, which is difficult to make from the aircraft. Such a service is rarely offered by hotels in other locations. And other transportation companies are generally run according to the same rules that apply to them everywhere else. Nevertheless, announcements are often also made in English on trains that stop at an airport, for example. Examples such as these make it clear that the degree of autonomy in systems of systems can indeed be assumed to be very high, while at the same time there are specific connections between the systems in certain respects that can be used as justification for speaking of a system at the higher level as well. For example, a passenger may not recognize any relevant difference between the airport and a hotel that is spatially and organizationally integrated into it.

On the basis of a constructivist understanding, autonomy is in fact quite suitable as a collective term for a whole series of characteristics of socio-technical systems such as operational coherence and self-reference or autopoiesis. What is missing, however, are models that depict this type of autonomy.

Psychological and sociological systems theory already has a model that takes account of the autonomy of such systems and which has gained a certain amount of currency. For example, Fig. 2.52 shows a notation proposed by George Spencer Brown (1969, p. 69 ff.) in connection with his *law of form*. The markers, known as crosses, model observations

[13] Strictly speaking, a system that is not autonomous would not be a system.

that consist of two individual operations: (1) *distinguishing* and (2) *labeling* (also Simon 2015, p. 52 ff.). What is perceived by the conscious mind is what makes a difference in contrast to something else, and in order for this difference to be consciously processed, it must be named.

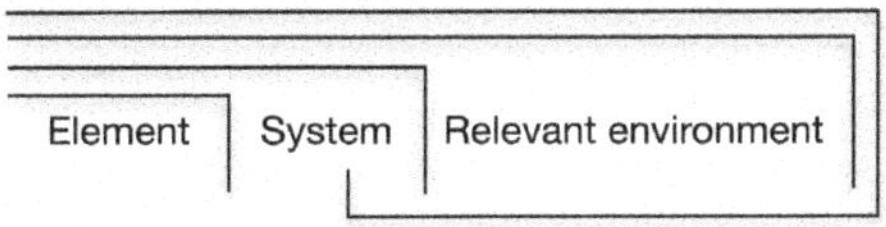

Fig. 2.52 Crosses for modeling observations (Simon 2019, p. 57)

(1) The crosses shown in Fig. 2.52 are to be interpreted as follows: What is to the left of the vertical line lies within the boundary drawn by a distinction, what is to the right lies outside it. What is below the horizontal line is an element of the system, i.e. is encompassed by the respective distinction. Since there are no input-output relationships between autopoietic systems, but the collision with these environments only causes irritations, an image of the environment(s) must be constructed *within* the organization on the basis of these irritations (Simon 2015, p. 58 f.). George Spencer Brown refers to this system-internal reconstruction of the difference between system and environment as *reentry* and notes it as a bracket around the system and the environment it considers relevant, as shown in Fig. 2.52.

(2) The designation can now take place, for example, in such a way that the system company (the organization) hires Mr. Meier as an employee and thus makes him an element[14] of the system, and identifies new sales opportunities in its own industry, which it designates with the name of the product with which it intends to exploit these opportunities.

The IPO model in Fig. 2.6 is a model that corresponds to the ideas of engineers in that it represents openness and transparency in the ways in which energy, mass and information (here meant as data and commands) are processed by and with technical systems, while George Spencer Brown's model in Fig. 2.52 reflects the ideas that psychologists and sociologists have of operationally closed, autopoietic systems. What is missing so far is a model that combines both perspectives.

2.6.2 Extended Functional Models

In the following, a proposal is made that goes back to the work of Günter Ropohl (1979, p. 105 ff.). These were revised for the first time in the 1990s (e.g. Ahrens 1998, p. 173 ff.). This in turn was reflected at least to some extent in further editions of the original work (Ropohl 2009, p. 89 ff.), albeit without citing the relevant sources[15]. The model is now

[14] The term element is not entirely correct here in relation to social systems, as such systems do not consist of people or persons, but of communication. More on this later, however. For now, the chosen formulation should be sufficiently understandable.

[15] As we learn in the preface to the third edition (2009), the volume of the work would have been disproportionately swollen by acknowledging the allegedly large number of contributions to the discussion.

undergoing a further, albeit less extensive, revision as a result of new insights that have been gained since the beginnings of this type of modeling.

Fig. 2.53 shows the *functional model of psycho-technical and socio-technical systems* in abstract form. In its external view, it corresponds to the functional model of technical material systems (IPO model) according to Fig. 2.6, in that it processes (P) the input variables (I) mass, energy and information into corresponding output variables (O). In this respect, it is an open system. Inside, based on the hierarchical model shown in Fig. 2.17, three sub-systems are distinguished in a hierarchical relationship to each other: the execution system, the information system and the decision system.

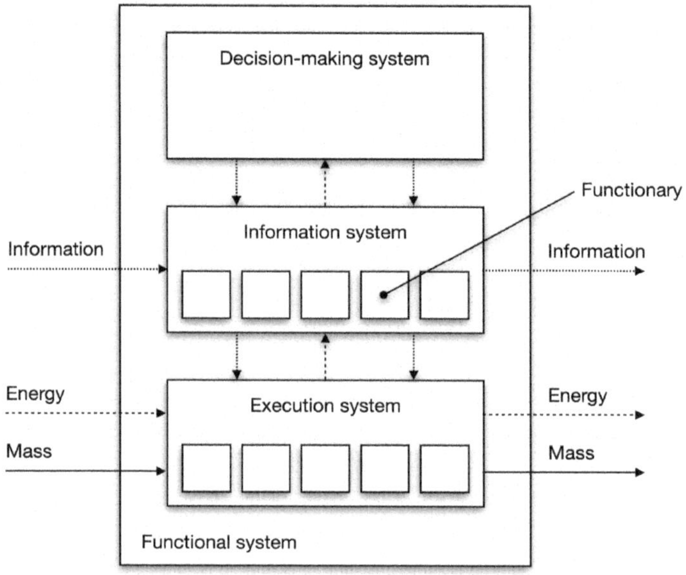

Fig. 2.53 Abstract model of socio-technical systems (Ahrens 1998, p. 53)

- The core function of the *execution system* is to transform all material-energetic variables. In automation technology (Seitz 2015, p. 24), this corresponds to the *field level*. The reference to the core of this function is necessary insofar as material and energetic processes also take place in the two other sub-systems, but without being part of the core of the respective functions.
- The core function of the *information system* is the processing of data and commands. The work processes in the execution system are controlled from here. Among other things, this creates control loops, which will be discussed in detail later. Another function of the information system is the coordination with other socio-technical systems through a corresponding exchange of information across system boundaries18. However, the model is by no means limited to technical aspects, but also includes all psychological and social aspects. Therefore, the concept of information will have to be differentiated later.

- The *decision-making system* represents a significant part of the autonomy of a socio-technical system. This is modeled by various properties. For example, this sub-system itself has no direct connection to the environment. In this respect, the system is self-referential (autopoietic) and operationally closed (e.g. Willke 2006, p. 61 ff.). The decision-making system only communicates with the environment in a roundabout way via the information system. However, this acts as a filter. A concrete example of this is an organization in which middle management, which can be assigned to the information system, prepares decision papers for the top decision-making bodies. Here, middle management partly selects itself what it considers relevant for decision-making and is partly free in the preparation of information. As a result, it can significantly influence the decisions of the highest decision-making bodies (Goeudevert 1996, p. 134). Furthermore, the decision-making system is not divided into different functionaries, which should make it clear that decisions cannot be delegated to technology. More on this in a moment.

2.6.2.1 Functional Model of Psycho-Technical Systems

These characteristics of the functional model of socio-technical systems, which were initially introduced very briefly, will be explained further below when it is used to depict psycho-technical and socio-technical systems. The first step is to apply the model to psycho-technical systems (Fig. 2.54).

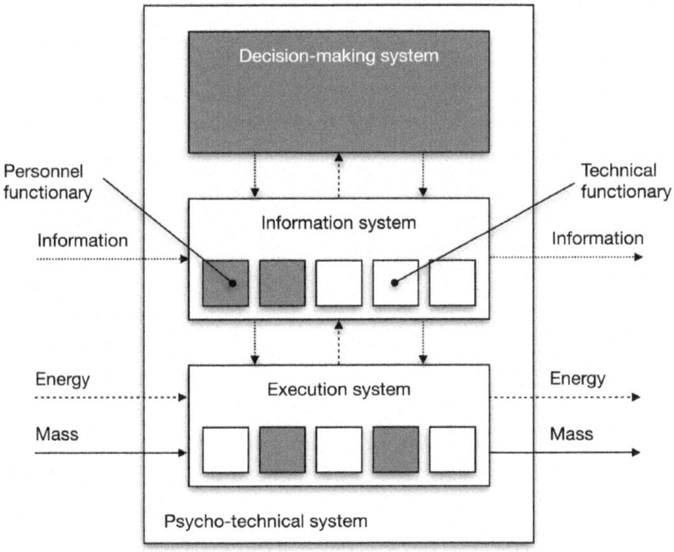

Fig. 2.54 Functional model of psycho-technical systems (Ahrens 1998, p. 54)

This distinguishes those functions that are carried out by the person from those for which the person uses technical function carriers. In the information system, for example, this could be the use of an AI system to analyze large amounts of data (big data); in the execution system, it could be the use of a milling machine to process metal parts.

What is important with regard to compatibility with a constructivist basic understanding is that such a delegation of the execution of functions to technical aids is not possible in the decision-making system[16]. Decisions are always tied to people. The reason for this is that decisions must be taken responsibly. This means that someone has to take responsibility for the consequences of a decision. However, putting a machine on trial and possibly in prison in the event of negative consequences would be pointless.

This is particularly remarkable in light of current developments in digitalization, as it is certainly associated with the idea that human decisions can be delegated to technology (e.g. Mertens et al. 2017, p. 8). One example of this is the so-called autonomous driving of vehicles in road traffic. There is no doubt that a high degree of automation is possible, which largely relieves people of the task of driving their vehicles. Nevertheless, they still make the decisions, even if it is only the decision to leave the driving of their vehicle to the relevant automated systems. However, if a situation or event relevant to responsibility arises, the human being will ultimately have to be held accountable. In this respect, they would do well to be aware of their decision-making authority and possible consequences and to act responsibly. This will not change even if technologies are used that are referred to as artificially intelligent, because no matter how one distinguishes such a form of intelligence from human intelligence: assuming maturity, taking responsibility is not a question of intelligence: ignorantia legis non excutat.

Furthermore, with the only indirect connection of the decision-making system to the outside world and with the effect of the information system as a filter, all aspects of constructivism explained in Section 2.1 can be depicted. A mental system modeled in this way is operationally closed and self-referential (autopoietic). It can only be irritated from the outside by perturbations in order to process them internally according to its own logic. In this respect, it corresponds to George Spencer Brown's model in Fig. 2.52. At the same time, however, the system modeled in this way is open in those respects in which it processes energy and material through the use of technology, among other things, and exchanges and processes the relevant data and commands. In this respect, the model of the socio-technical system retains the characteristics of the model of technical systems (IPO model) as shown in Fig. 2.6. This corresponds, for example, to the design of work systems in *ergonomics* (Schlick et al. 2018, p. 418). It structures the relevant design fields into the *energetic design*, which corresponds to the flow variable energy in the IPO model, the *informational design*, which corresponds to the flow variable information in the model, and the *anthropometric design*, which corresponds to the flow variable mass. This example shows that systems theory is well suited to structuring historically evolved areas of knowledge in a standardized way, thereby facilitating the connection possibilities of different domains.

---

[16] This is still the subject of controversial debate. In this respect, this statement is a well-founded contribution to it, but in principle also allows for other points of view, if and insofar as these are also comprehensibly justified. So far, however, no other point of view has been convincing.

The functional model of socio-technical factual systems is also comparable to Heinz von Foerster's non-trivial machine as shown in Fig. 2.51. It differs in that Heinz von Foerster constructed a machine that is as simple as possible. This is because it consists internally of two trivial machines. However, this already gave the NTM a complex behavior that cannot be deduced from the outside. The internal structure of the functional model of socio-technical systems is much more complex than the interaction of two TMs. This makes it far less transparent and directly influenceable from the outside than an NTM.

2.6.2.2 Functional Model of Socio-Technical Systems

The functional model of socio-technical systems is *self-similar*, i.e. it follows the *principle of recursion*. This is demonstrated by the fact that its basic structure can also be used on the social level, i.e. where at least two mental systems interact. This is expressed in the model (Fig. 2.55) by the fact that the function carriers in the three sub-systems are in turn socio-technical or psycho-technical systems.

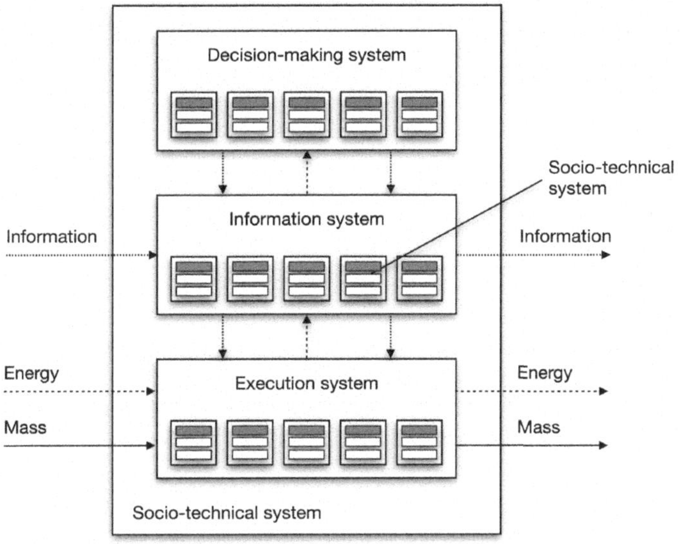

Fig. 2.55 Functional model of socio-technical systems, applied to an organization (Ahrens 1998, p. 55)

Here and in the following, the model will be applied to organizations. In principle, it can also be applied to other areas such as the institution of marriage. However, the relationships that apply there, some of which differ considerably from organizations, are not relevant in the context of SEM and will therefore not be discussed in depth. Organizations, on the other hand, are relevant because most professional problem solving takes place in their context and is therefore subject to the possibilities and limitations that prevail within them.

In organizations, middle management can be located in the information system, whose function is to direct the function holders in the execution system. These can be individual departments or specific individuals, for example. However, since these are themselves socio-technical or psycho-technical systems and therefore have all the characteristics explained above, i.e. they are operationally closed and self-referential, they do not react to instructions from hierarchically superior authorities in the mechanistic sense of command and obedience, but rather according to their own laws (Goeudevert 1996, p. 245). Instructions are merely perceived as perturbations; they interfere. It is true that the systems are structurally linked so that communication is possible in principle, and everyday experience shows that this is the case. At the same time, however, everyday experience is also proof of the limits of such understanding. Even sanctions, which can extend to the exclusion of an organizational member from the organization, can only prevent the self-will (autonomy) of organizational members to a limited extent.

This also applies to the decision-making system. Unlike in a psycho-technical system, it is perfectly possible in a socio-technical system to delegate decisions to functionaries. In an organization, this corresponds to the top management. According to the model explained here, however, these do not necessarily pursue the same goals. In fact, they may well be in competition with each other and serve individual goals, such as their own profile. Everyday experience shows that such behaviors do occur (ibid., p. 133 f.). And with regard to instructions to hierarchically subordinate instances, the same applies as in the relationship between middle management and functionaries in the execution system.

Hierarchy

However, the limitations of hierarchical coordination just mentioned should not obscure the fact that it makes a very significant contribution to the organization of complex systems. For a better understanding, however, it is helpful to distinguish between hierarchy and domination (Willke 2006, p. 70), because objections to hierarchy often lead those in power to worry that their power is being called into question at the same time.

The origin of such concerns may lie in the social development from monarchies to democracies. The abolition of a strictly hierarchical organization and the loss of power actually coincided. However, at the social level, it was also about the liberation of human beings from dependence and immaturity (ibid., p. 55), about the realization of general human rights and duties (A/RES/217, UN Doc. 217/A-(III)). At the level of organizations, on the other hand, power results from the generally recognized right to the power of disposal over property, which is not called into question here. Employees of an organization are not its citizens, but dependent employees. Their civil liberty is guaranteed by the fact that they voluntarily enter into a dependency limited to the employment relationship under their employment contract and can end this dependency at any time in a self-determined manner. And within the organization, they submit to the instructions of the owners or their representatives, but within the limits of law, contract and morality. That is and remains fine, and this is not changed here.

If one separates domination and hierarchy[17] in this way, the latter simply remains as a certain form of organization of specialization based on the division of labour, which can also be organized in a completely different way, for example by coordinating everyone with everyone else (Herbst 1976, p. 31). However, hierarchy is more successful than alternative organizational structures.

- Max Weber (1972, p. 56 f.) explains in relation to social systems that the decisive reason for the advance of the bureaucratic (i.e. hierarchical) organization has always been its purely technical superiority over any other form: A fully developed bureaucratic mechanism relates to these exactly as a machine relates to the non-mechanical ways of producing goods. Precision, speed, clarity, strict subordination, savings in friction, material and personal costs are increased to the optimum in strictly bureaucratic, especially monocratic administration by trained individual officials compared to all collegial or honorary and part-time forms.
- Helmut Willke (2006, p. 55) quotes Howard Pattee (1973, p. 73), a biologist who emphasizes the superiority of hierarchical organization for biological systems: Hierarchical organization is so universal in the biological world that we usually pass it off as the natural way to achieve simplicity of efficiency in a large collection of interacting elements. If asked what the fundamental reason is for hierarchical organization, I suspect most people would simply say, 'how else would you do it?'. Evolutionary developments therefore obviously also preferably produce hierarchical forms of organization of specialization based on the division of labour.

Helmut Willke, referring to Max Weber (1972), sees hierarchy as particularly suitable when it comes to working through simple (binary) conditional sequences that can be broken down into individual steps and whose partial solutions can just as easily be put together to form an overall solution. These are collectively relevant, comprehensive but clearly defined, often complicated but well-ordered problems that cannot be solved by individuals and whose solution requires the orderly, organized cooperation of professionally competent role holders (Willke 2006, p. 58)21. The hierarchy in the organization thus corresponds to the parts list of a technical object (Fig. 2.14): each individual part and each assembly can be assigned more or less to an organizational unit (position, department, etc.), and the further up the parts list you move towards the overall system, the higher you move up the hierarchical chain of command. And this is exactly how it actually works in practice: The plant manager responsible for vehicle assembly reports to the head of the powertrain production area, who in turn reports to the department head for transmission construction, etc.

Max Weber (1972) already had such a mechanistic understanding of organization: His concept of complexity is shaped by the 19th century model of technically complicated mechanical machines. His idea of rationality is that of a hierarchical division of labor along a logical, graduated order of ends and means (Willke 2006, p. 59). To this day, the so-called

[17] This distinction goes back to Helmut Willke (2006, p. 70), although he himself does not adhere to it consistently by contrasting democracy and hierarchy dichotomously in other places (ibid., e.g. p. 57 f.). The latter is to be avoided here in order to establish democracy as a form of rule, which at the same time allows hierarchy as a mere form of organization wherever this division of labour coordinates specialization most efficiently and effectively.

machine metaphor, according to which a company has to function like a well-oiled machine, underlies large parts of business administration (Kieser and Ebers 2014, p. 97). Although this is repeatedly lamented, this understanding of engineering science is sometimes not denied a certain degree of success (e.g. Gaitanides 2012, p. 2 f.). Incidentally, this also justifies the engineering in the title of this work. But perhaps it is precisely the engineering sciences' responsibility to reflect on their own limitations and, for their part, to build a bridge to the social sciences. The interdisciplinary field of Engineering & Management is particularly predestined for this, although it is not yet sufficiently aware of this itself (Ahrens 2017, p. 10 ff.).

This correspondence between the structure of a task or a problem, here using the example of the parts list of a vehicle (Fig. 2.10), and the structure of the organization, which in such a case best takes the form of a hierarchy, will be retained for a while in order to show next how hierarchy enables complexity to be managed. Herbert Alexander Simon (1978, quoted here after Willke 2006, p. 60 f.) has worked out how hierarchical systems are able to develop and stabilize partial solutions through the formation of sub-systems in almost any depth, which can also be used repeatedly to solve other overall tasks or problems, as is the case with vehicle batteries, for example. If necessary, such a partial solution can even be used to solve tasks or problems that did not (yet) exist or were not even considered when it was developed. For example, the car battery can also be used to supply power to an electric pasture fence. In this respect, hierarchy also enables innovation.

However, only those tasks or problems that cannot be completely disassembled and reassembled in every respect can be described as *complex* (for more details, see Section 6.1). These would merely be extensive and possibly *complicated* in this respect (Haberfellner et al. 2018, p. 34 f.). Tasks or problems are understood to be complex if they cannot be completely broken down without losing properties of the overall system. They are also not linked in a one-to-one relationship in all respects and are interdependent in their parts – this would be the case of a fully networked system (sic) in which no element can change without this change having a direct effect on the entire system (Willke 2006, p. 61). Rather, complex hierarchical systems are *modular systems* in the sense that the sub-systems are connected in certain moments that are strategically important for the system and communicate with each other via selective interfaces, but are independent of each other in many other moments and respects, so that not all events in the sub-systems are also relevant for the overall system because they are processed in the internal mode of operation of the sub-systems and prevented from influencing the overall system by threshold values, relevance criteria or similar (ibid.).

Talcott Parsons (1960) recognized a further advantage of hierarchy in the fact that it allows the separation of levels with the aim of setting areas, competencies, work contexts and social constellations autonomously against each other and yet achieving a conditioning of autonomy through clearly defined possibilities for intervention and intervention: Hierarchy protects against the unauthorized intervention of others in one's own work and precisely and precisely designates the few places from which intervention must be expected and which can be expected to intervene (Baecker 1994, p. 24) (Willke 2006, p. 62). 24). For example, the head of one department cannot give instructions to an employee of another department without further ado, even if he is hierarchically superior to the latter (Goeudevert 1996, p. 160). To do this, the chain of command must be followed, which runs at least through the head of the other department and possibly also through higher authorities, so that the employee in question can only expect instructions from one authority, namely the head of his own department. In a strict bureaucracy, not even direct instructions from superiors of the head of one's own department are permitted; these must normally also comply with the chain of command.

Finally, the right of superior bodies to issue instructions is limited to equitable discretion (fairness, appropriateness, reasonableness) and to the leeway left by agreements in the employment contract, provisions of a works agreement or an applicable collective agreement and statutory provisions (also BAG - 10 AZR 330/16) in accordance with Section 106 of the German Industrial Code (GewO). Therefore, if not every instruction has to be expected, but is limited to a repertoire that may not be clearly defined, but is nevertheless foreseeable, this reduces the complexity for all those involved and affected.

Hierarchy, as outlined above, allows the creation of extremely complex structures. The development, production and ongoing operation of aircraft, power plants or satellites would not be possible without a hierarchical organization. However, it is precisely this success that is ultimately the cause of the limits of this organizational model (Willke 2006, p. 67), because the larger hierarchies are, the greater the effort required to design and operate them, and at some point a point is reached where this effort exceeds the benefits.

- In a small craft business, the owner still looks after his staff himself, while doing the same work as his employees. From a certain size, he will have to hire a specialized employee to take care of personnel management. If the company continues to grow, an entire department will be needed for this, until finally, with further growth, a department will be needed to coordinate the work of all HR departments, then a department that coordinates all departments that coordinate the HR departments, and so on. In large companies, this can take on considerable proportions and, incidentally, this applies not only to HR, but to all areas[18].
- A similar situation can be observed in company regulations. In a small company, the owner passes on verbally the rules according to which certain tasks are to be carried out. As the company grows, such rules are written down, and in even larger companies, regulations are created that document how regulations are to be created, published, enforced, changed, withdrawn from circulation and archived (DIN EN ISO 9001). Willke (2006, p. 68) cites examples from Jacques Gansler (1989, p. 191) on military procurement: Military specification MIL-F-1499 F describes, in 18 pages exactly what must go into a fruitcake [...] there is a 20-page specification for towels, a 20-page specification for underwear, and a 16-page specification for plastic whistles, a 17-page specification for olives, a 20-page specification for hat chocolate [...].

At some point, all of this takes on dimensions that are no longer manageable and can no longer be handled efficiently and effectively, no matter how systematic the system and even with the help of the best software. In general, it can be said that from a certain extent, hierarchy no longer only helps to manage complexity, but becomes so complex itself that it ties up capacity that is only needed to manage its own complexity and is therefore no

[18] Systems theory explanations describe the limits of the possibility of hierarchical structures much more precisely, but also much more abstractly (e.g. Willke 2006). The latter is the reason why it is not used here. Nevertheless, further considerations are based on the insights of systems theory.

longer available to manage the complexity for which the hierarchy was actually created. At the end of such a development, according to Niklas Luhmann (1996, p. 56), systems perish from their own complexity; to put it more simply: at some point, they are only concerned with themselves.

A first, albeit too short-sighted, departure from hierarchy is teamwork, which was developed at the beginning of the 1960s and is now widespread, as teams do not structure their collaboration internally in a strictly hierarchical manner, but rather partly autonomously (Stock-Homburg 2013, p. 375 ff.). Whereas in a pure hierarchy, each task or problem solution is assigned to a specific, specialized position, in teams there are overlapping competencies, which can be observed particularly well in sports, when in soccer a defender occasionally scores a goal, while the striker sometimes helps to take the ball from the opponent. The players only follow hierarchical instructions to a very limited extent, especially during the game, even if some coaches sometimes wish they could while trying to intervene in the game from their coaching zone. Even the team captain is primus inter pares at best. Coordination is mainly achieved through situational coordination between the players in conjunction with practiced actions, especially in so-called standard situations. This can be applied to any type of teamwork, even in organizations.

However, the implementation of teamwork in an operational context proves to be difficult (e.g. Goeudevert 1996, p. 160 ff.). It can complement hierarchical organizational structures, especially in areas where problems need to be solved. In this respect, it can also and especially be found in Systems Engineering & Management. However, hierarchies are also often formed within teams, for example by appointing team leaders. However, this contradicts the actual team concept (Haug 2009, p. 103 f.). This once again demonstrates the great importance of hierarchy in organizations, which gives reason to take it up again in Section 6.1.1 and to go into it in more detail.

Now that the great importance of hierarchy has already been explained here in some detail, it may come as a surprise that a supplementary structure of socio-technical systems will nevertheless be introduced later on, which relativizes the widespread understanding of the apparent connection between hierarchy and domination, because after all, it was also and precisely the limits of hierarchical coordination that initially gave rise to working out the strengths of hierarchy. These limits and how to overcome them will be discussed later on (Sect. 6.1.2). In order to understand this discussion, however, it is first necessary to clarify in more detail than before what socio-technical systems consist of.

Action Systems

The answer to the question of what socio-technical systems consist of seems obvious: acting people, i.e. actors (Simon 2015, p. 16 ff.). After all, people become employees of a company through an employment contract or members of an association through an application for membership and its acceptance. In companies, they are assigned to specific departments according to their qualifications, where they solve certain problems or perform certain tasks. The departments are located in specific places where people have more or less permanently assigned jobs. All of this usually shows a very clear structure.

Against this background, it seems logical that most of the works cited here formulate a theory of action (e.g. Latour 2017, p. 76 ff.; Parsons 1986, p. 14 ff.; Ropohl 2009, p. 89 ff.; Weick 2015, p. 53 f.) However, Niklas Luhmann (1996, p. 1991 ff.) has demonstrated that much deeper insights can be gained if the theory of action is further developed into a theory of communication. In this, action remains significant, but in a relativized form. Fig. 2.56 serves to classify the terms that become relevant in this context and their interrelationships.

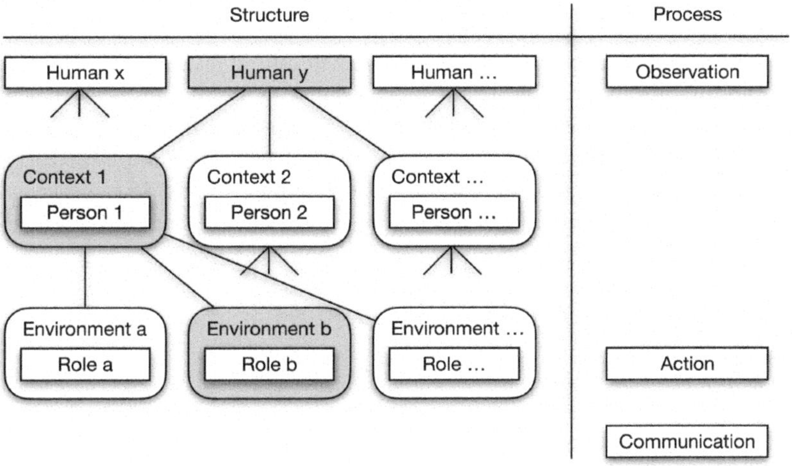

Fig. 2.56 Difference between human, person and role

People in organizations come and go – the company remains. It often even outlasts its founders. This applies not only to hiring and firing. Even in a shift operation, one person on the second shift replaces the other person who has just worked the first shift at the same workplace, and if a person is on vacation or sick, they are replaced by another person. If an organization were shaped by the specific people who are currently working in it, its nature would constantly change. It would be very unstable.

However, if people are not part of organizations themselves, they belong to their environment (Simon 2015, p. 13 f.). This makes them interchangeable, and this is exactly what everyday experience shows. Organizations decouple themselves from people and virtually lead their own, independent lives (Weick 1979, p. 53 f.). Successful founders in particular experience this when their companies grow over time. Like children, companies of a certain size and after a certain period of time also enter puberty, so to speak, develop their own character and finally emancipate themselves. If and to the extent that a founder cannot let go in such a phase, like a parent, but continues to try to impose his will on his pupil, this regularly results in a crisis for a company comparable to that of a child growing up (adolescence crisis).

The fact that people are not part of organizations makes them very different from other institutions such as marriage: if one of the partners in a marriage leaves, the marriage is divorced. A marriage is therefore not an organization, although married couples undoubtedly have to organize their life together. Also, people usually work for several organizations:

during the day they work in a company and spend their evenings in a sports club. Marriage, on the other hand, is based on exclusivity, at least officially.

Everyday experience also shows that people do not allow themselves to be incorporated into an organization as a biological or psychological system. They only owe an organization the contractually agreed work performance, while other abilities and skills remain unused. The fact that he can play the piano well, for example, will be of little use to him when assembling gearboxes at work. The fact that he has taken on the responsible position of treasurer in the sports club will not help him either. All that is required of him here is that he assembles gearboxes day in, day out.

This is a prerequisite for people being prepared to follow instructions within the legal, contractual and morally justifiable framework, even if they themselves do not appear to make sense. Chester Irving Barnard (1938, p. 167)[19] refers to this externally determined part of behavior as the *indifference zone*. As long as employees of a company feel adequately paid and valued, they can largely be indifferent to what they should do in return within the framework of the applicable regulations and agreements. If this is not the case and leads to complaints or even a refusal to work, they risk being dismissed.

Conceptually, this is expressed in systems theory through the distinction between *people* and *persons*. While a modern marriage (Luhmann 1994) is based on people contributing their entire physical and mental selves, in an organization they merely take on a role as documented in a job description, for example. The fact that this can be expressed by the term person can be explained etymologically, as in ancient times a mask was referred to as a person (Latin: per through and sonare to sound), which was worn in the theater to play a corresponding role. The human body is concealed behind the mask (person), as is the psyche (Luhmann 1996a, p. 429 f.). The person is what is relevant of a person in an organization (Luhmann 2011, p. 89); it is a fictitious quality constructed through communication that is attributed to a person. In Fig. 2.55, person y is therefore perceived as person a in a certain context a, in this case in the context of a company: as a fitter. In context b, in this case in a sports club, he represents person b: the treasurer.

A person can also take on different *roles*. Roles are to be understood as a bundle of interrelated *expectations*, as a system of rules that configure a certain game and, within this game, certain bundles of tasks based on the division of labor (Willke 2005, p. 151). For example, in addition to his main role b as a fitter, person 1 can also have the role a of a safety officer[20]. This means that he would be the contact person for his colleagues for questions on occupational health and safety and would have the voluntary task of supporting his superiors in the implementation of measures to prevent accidents at work and occupational illnesses. While in his role b he is only required to remain in the immediate vicinity b of the assembly workstation assigned to him, the environment a relevant to his role a as safety officer extends to the entire assembly department. In this role a, he is authorized to leave his workplace, which he is only permitted to do in his role b for certain purposes, for example as part of his distribution time.

All of the aspects explained so far are of a structural nature. In organizations, expectations of a person are documented in *employment contracts*, for example, and expectations of roles are documented in *job descriptions* or laid down by law. The structural similarity

[19] Quoted from Simon 2015, p. 14.

[20] In Germany in accordance with § 22 of the Seventh Social Security Code (SGB VII).

of organizations forms the prerequisite for what was referred to as *structural coupling* in connection with the basic constructivist understanding (Sect. 2.1). Even if a system cannot see through another system from the outside, it can still count on a certain probability of successful interaction. For example, every company will be able to create offers in response to inquiries and ultimately accept orders, however this is organized internally in detail. Interactions refer to what is categorized as processes in Fig. 2.54. This already indicates that organizations only fulfil their purpose when processes are created on the basis of their structures.

Now that the terms human, person and role have been introduced with the help of Fig. 2.56, further considerations should lead to the insight that and why it is helpful to differentiate between the terms observation, action and communication also used in the figure.

It has already been established that people are not regarded as elements of socio-technical systems because they are not absorbed in their wholeness, which is expressed by their humanity, but rather sometimes turn to this system, sometimes to that system and only contribute to it in specific respects. This is why people are seen as observers, i.e. as outsiders who, among other things, observe themselves. If they were part of systems, these would be weak in character and unstable. In fact, systems develop their own, system-specific culture; in organizations, for example, we speak of a corporate culture.

And even people are not elements of socio-technical systems, because – as already mentioned – employees come and go, the company remains. It often even outlasts its founders. What initially remains are the actions of individuals. These are also carried out by successors. For example, a new employee who is hired to drive the forklift truck in the incoming goods warehouse will do this in more or less the same way as his predecessor. Action theorists see this as confirmation that it is ultimately actions that ensure the continued existence of social systems.

Karl Edward Weick (2015, p. 163 ff.) describes actions in detail as *double interactions*. One person acts and a second reacts by acting: after a successful incoming goods inspection, the quality inspector points with his index finger to the goods that are now ready for storage and the forklift driver loads the pallet onto the forks. The quality inspector acknowledges this by nodding his head and turning to the goods to be inspected next. The double interaction takes on a cyclical form in that the behavior of the first person is interpreted as a communication of information and the behavior of the second person as a sign of understanding. It is also interpreted as a communication of information and is in turn answered by the behavior of the first person as a sign of understanding (the completed reaction) (Simon 2015, p. 20).

From this observation, Niklas Luhmann (1996a, p. 191 ff.) draws the conclusion that it is not the action as such, but its resolution into *notification*, *information* and *understanding* that ensures the existence of social systems. They interlock actions with each other. He describes the unity of these three selections, which always follow one another and therefore belong together, as *communication*. This in turn forms the basic unit of social systems, i.e. their elements. Unlike in data processing and communication technology, this is not about the transmission of signals, where it is assumed that a sender causally determines the states of the receiver through such a transmission. Under the assumption of operational closure and self-reference (autonomy), this is not possible, as such systems can at best be irritated from the outside, while the continuation of the processing of such irritations takes place within the system according to its own rules, which are not sufficiently visible from the outside.

Communication as an Element of Social Systems

Elements of socio-technical systems, i.e. units that cannot be meaningfully broken down further from the perspective of the system, are therefore individual acts of communication (Luhmann 1996a, p. 191 ff.). Furthermore, each individual act of communication is formed from the unity of information, notification and understanding. Since these insights are as counterintuitive as they are far-reaching, they will be discussed in more detail below.

In a first step, the sender obtains information by interpreting selected data, from which he selects those that he considers to be worth communicating. In the second step, the sender selects the receiver[21] with whom he wants to share his information and communicates it to him. In the third step, the receiver receives the information and tries to understand it.

First of all, it should be noted that communication understood in this way is contingent in many respects, i.e. also possible in other ways, and in this respect is by no means as unambiguous as is assumed in a technological interpretation. In sociology, this is even referred to as double contingency (Luhmann 1996a, p. 148 ff.), because it affects both the sender and the receiver. In detail, the selections on the part of the sender, which are always possible in other ways, relate to (1) the selection of data and (2) the way they are interpreted, to (3) the information deemed worthy of communication and (4) to the receiver. The receiver, in turn, will interpret the information according to his own ideas (5) and decide freely what his (6) reaction to it will be.

In view of the numerous selections as well as the possible errors that can occur, it is understandable why Friedrich Schleiermacher (1977, p. 75) considers understanding to be improbable. This can be calculated mathematically. As is well known, probabilities are multiplied along a path. If, for example, the probability of doing the right or wrong thing in each of the six steps of a communication act (see above) is 50 %, the probability that the correct reaction will occur at the end is 0.56 = 0.0156 = 1.56 %. So out of 1/0.0156 = 64 communication results, only one is successful. With three possibilities in each step, the probability of success of the communication act is reduced to 0.14 %. Now only one out of 1/0.0014 = 729 communication results is successful[22]. In addition, conversations usually consist of several, often even many communication acts. Even two consecutive acts of communication, for example an answer to a question, reduce the chance of success in the favorable case of a 50:50 probability to 0.24 ‰.

There are three main reasons why conversations nevertheless succeed more often than these calculations suggest:

- In his empirical studies on evolutionary anthropology, Michael Tomasello (2022, p. 13) identifies pointing and gesture play as a psychological platform for linguistic commu-

---

[21] Following Luhmann (1996, p. 148 ff.), the explanations here refer generally to a sender-receiver pair, but also apply to cases in which several participants communicate with each other.

[22] The probabilities decrease with the power -6 as the number of possibilities increases.

nication, specifically as a socio-cognitive and socio-motivational infrastructure. Mutual understanding of such a simple form of communication is possible due to shared attention, shared experience and shared cultural knowledge (ibid., p. 15 ff.). In addition, people have a pro-social motivation to help and share, what Michael Tomasello calls shared intentionality. Pointing is based on the natural tendency of humans to follow the line of sight of others to external objects, and gesturing is based on the natural tendency to interpret the actions of others as intentional (ibid., p. 20). Finally, humans have the skills of shared cultural learning and imitation in order to create and pass on commonly understood communicative conventions and constructions (ibid., p. 23). On this basis, language eventually replaces the naturalness of pointing and signing with a shared history of social learning (which everyone mutually knows is shared) (ibid., p. 20) and thus increases the efficiency of communication.
- Furthermore, social systems are able to form structures that make mutual understanding more likely. The word drinking can therefore be used neutrally to describe the process in which a body actively absorbs liquid through the mouth. In another context, however, it can also mean that someone is an alcoholic. This is the reason why it is so important to pay great attention to the design of structures: The more clearly they are shaped, the more likely they make communication successful and the more stable the corresponding social system is. Although processes also contribute to this, strictly speaking it is primarily their structures that are important in the selection of communication.
- After all, conversations are generally very tolerant of errors and misunderstandings. Natural language is an example of this, as it includes many terms, metaphors, allegories and symbols that are so vaguely defined that they leave great scope for interpretation (Lakoff and Johnson 2021). If the communication partners also encounter each other with the goodwill, empathy and interest in each other that is often present within social systems (Goleman 1996), everyone usually finds themselves in a conversation to such an extent that conflicts can be avoided. This is one of the main reasons why it is so important that companies do not take on their employees lock, stock and barrel, but instead demand and promote a limitation to the assigned role, because, according to Reinhold Niebuhr, this results in the serenity of individuals to accept circumstances that they cannot and should not change, the courage to tackle the projects that they can and should manage, and the wisdom to distinguish one from the other.

From a constructivist understanding, the insight was derived in the previous Section that participants in communication are not actors (doers), but observers (Simon 2015, p. 21). In the course of an act of communication, observers observe each other while at the same time observing that they themselves are being observed. For communication to be successful, they must recognize a meaning in it, or more precisely, invent a meaning that they attribute to their observations. It follows that communication cannot be observed directly, but can only be inferred. In order to make communication observable nonetheless, systems theorists use a trick and pretend that it is an action. After all, communication can be classified as successful if one can recognize from the subsequent action whether one has been understood or not (Luhmann 1996a, p. 226).

So far it has been worked out that social systems and thus also organizations (1) consist of individual acts of communication, that communication (2) is contingent and (3) originates from (4) changing persons. These are already four factors that make an organization appear unstable. This impression is reinforced when you consider that an act of communication (5) is fleeting, as it only ever takes place in the present. Karl Edward Weick (2015, 67) explains: We prefer a view of organization that assumes that organizations are constantly falling apart and therefore must be constantly rebuilt. Processes need to be constantly re-implemented. Most administrators know this; most organizational researchers need to be reminded of it. Against this background, the question of how organizations gain stability in the face of such a large number of destabilizing influences will be answered below. As the structural model of socio-technical systems explained below will show, this is achieved by what is known in the US political system as *checks and balances*: Different sub-systems with different functions control each other and compensate for their respective weaknesses with their own strengths, so that a fluid equilibrium can emerge overall.

2.6.2.3 Structural Model of Socio-Technical Systems

The hierarchical arrangement of the three sub-systems of the functional model of socio-technical systems can be misunderstood to mean that actions within an organization primarily follow a more or less strict chain of command from hierarchically superior to hierarchically subordinate instances. In fact, hierarchy makes a very significant contribution to mastering complexity. However, in order to understand the possibilities and, in particular, the limits of SEM, it is important to understand that the solutions ultimately realized are based on a more complex logic than the hierarchy would suggest. Such an understanding can help to avoid disappointment and the resulting loss of motivation, which can be observed in practice more regularly than exceptionally when what is ultimately implemented deviates significantly from what experts have systematically developed with a great deal of expertise and commitment.

To this end, the structural model for modeling socio-technical systems shown in Fig. 2.52 should be applied. The starting point is Talcott Parsons' AGIL scheme (1986, p. 50; 1996, p. 20), which has already been outlined to some extent in Fig. 2.4. Fig. 2.57 extends this diagram to include the essential relationships between the four part-systems (Münch 1992, p. 63).

It has already been explained above that the part-systems each fulfill specific functions for the overall system. In sociology, this is referred to as functional differentiation (Parsons 1986, p. 19 ff.). With the allocation of resources, the economic part-system ensures adaptability to the environment, with the absorption of contingency through decisions, the political system places the use of resources at the service of achieving set goals, the community negotiates collectively valid values among themselves and the culture ensures legitimacy through reflection. The part-systems initially do this autonomously by using specific semantic codes within their system boundaries: money in the economy, power in politics, morality in the community and arguments in culture.

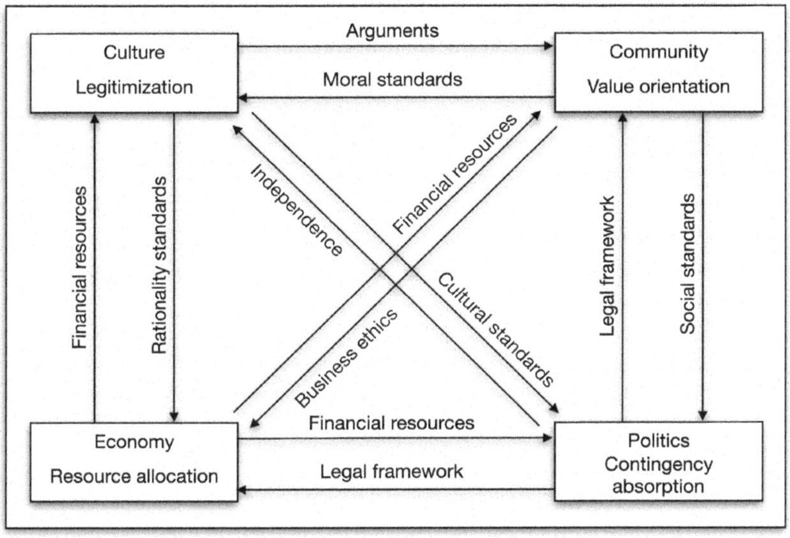

Fig. 2.57 AGIL scheme (Münch 1992, p. 63)

Relationships Between the Parts of Socio-Technical Systems

However, this would mean that the part-systems would not be viable on their own. In fact, they are dependent on the services of the other part-systems (Münch 1992, p. 119).

- The function of the economic part-system is to *allocate* resources. However, pure economism would result in a battle of everyone against everyone else, from which all participants would ultimately emerge as losers. Therefore, the goal of profit maximization only serves to motivate market participants, but not the private accumulation of wealth on any scale. The fact that economic globalization in conjunction with digitalization is currently leading to such distortions is proof that the structure of social systems explained here enables equilibrium, but can also lead to imbalances if not properly designed (Ahrens 1999). The economic part-system contributes to equilibrium when surplus profits are siphoned off and made available to the other part-systems so that they can fulfill their functions, even though they do not follow any economic logic. However, while too little tends to be siphoned off at the level of societies, those parts that are responsible for value creation within companies tend to be left with too few resources, which affects their adaptability.
- A dominance of the political part-system would lead to a dictatorship. In a state of equilibrium, however, this part-system limits itself to reducing contingency through decisions where this cannot be directly and sufficiently reduced by other logics, such as those of the economy. For example, it is possible to calculate from a business perspective whether an investment in certain operating resources promises to be profitable. It would be redundant and therefore superfluous to decide this again politically. However, whether a new market segment should be developed is a strategic question and can at best only be

answered to some extent in terms of costing (Mintzberg 2013). Therefore, the remaining knowledge gaps must first be closed by a decision. This is then the basis on which calculations can be made in the economic part-system[23].

- In a community that sees itself as a unit, it must be possible to relate actions to one another (*conformity*). This requires norms to which all actions are oriented. They provide a structure for communal action by selecting from the multitude of possible causes for action only those that should be relevant for the community in order to link these actions to certain expectations. The criterion for success here is the degree of integration of members, i.e. their understanding of each other and their commitment to the community.
- For the continued existence of a system, it is necessary to ensure the *consistency* (coherence, consistency, consistency) of actions. To this end, basic patterns of action are generalized. These generalizations then open up new scope for action (contingency). The criterion for success here is the meaningfulness of the action within a cultural frame of reference.

As already partially indicated, each of the four part-systems can contribute to the balance, but also to the imbalance of the overall system. This can be illustrated using Fig. 2.58. The functions shown here are already familiar and are only provided for orientation purposes. What is new is the connection to what Talcott Parsons (1986, p. 50) calls milieus of action.

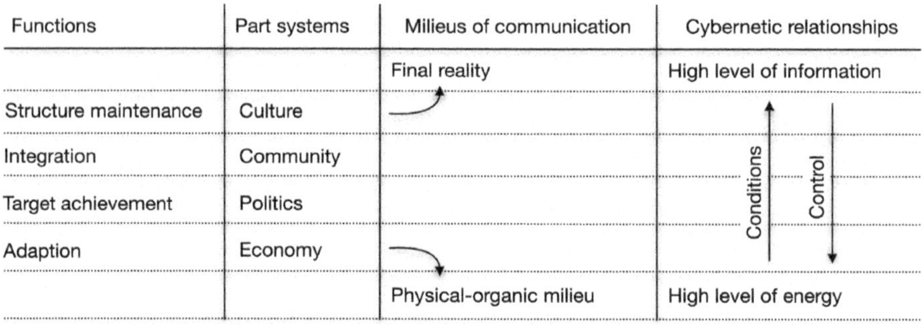

Fig. 2.58 Cybernetic relationships between the spheres of social communication (Parsons 1986, pp. 50 ff.)

- At the lower end of the scale, this is the physical-organic milieu, i.e. the natural and social environment of a system. Resources such as energy and raw materials are obtained from this environment and systems must ensure their survival under the prevailing conditions. As a system can only influence these conditions to a very limited extent, it must adapt to them. This function is primarily performed by the economic part-system, to which technology is also assigned as a means.
- At the upper end of the scale, social systems can be linked to what Talcott Parsons calls ultimate reality. Religions, for example, refer to this area as the afterlife or the

[23] This means that the underlying systems theory findings on the function of decisions are diametrically opposed to views in connection with agile methods. Joachim Goll and Daniel Hommel, for example, state that decisions must not be made out of ignorance (2015, p. 16). In fact, however, decisions are precisely there to close knowledge gaps. What is certain, e.g. that 1 + 1 = 2, does not require a decision.

kingdom of heaven. Establishing the connection between immanence and transcendence (eschatology) is one of the functions of culture.

The extent to which the four part-systems are subject to conditions and have to adapt to them decreases towards the top. At the same time, controlling factors increase. For a better understanding, this will be illustrated using operational phenomena as examples.

- As a representative of the economic part-system of a company, a salesperson is focused on selling as many products as possible. In doing so, they are directly exposed to the demands of their customers (social environment of the overall system). As a rule, he must react flexibly to this, for example by granting price discounts or promising short delivery times.
- However, if the company management, as the representative of the political part-system, does not pursue cost leadership as a *competitive strategy* (Porter 2013) but, for example, quality leadership (differentiation), it will restrict (control) the scope of the sales department when granting discounts. From the multitude of all possible sales measures (contingency), some are selected and others are prevented.
- It can also be assumed, for example, that integrity has a high priority in the workforce (common part-system). Even if it were to occur to the company management, due to its power (decision-making authority), to allow or even demand unfair measures in sales, this could hardly be enforced, as the corporate culture shaped by the community part-system exerts a restrictive (controlling) influence on company policy. This is particularly true because *corporate culture* cannot be shaped in a targeted manner, but emerges emergently (Simon 2015, p. 96 ff.).
- While values are largely lived without reflection in the community part-system, the cultural part-system develops a narrative that lends *legitimacy* to the values reflected in it. In traditional companies in particular, such narratives are also linked to the personality of their founders, for example, which can lend them enormous stability. One example of this is the history of Siemens, which is primarily associated with Werner von Siemens (e.g. Bähr 2016).

However, a very strong connection to a corporate tradition can hinder the adaptability of the overall system to changing environmental conditions, such as those that can arise in the course of globalization from the associated change in the competitive situation or from technical progress (e.g. digitalization), to such an extent that the continued existence of the company is at stake. In such a case, one speaks of the *dominance* of a controlling sub-part-system over dynamic part-systems at the lower end of the scale of cybernetic relationships shown in Fig. 2.58 (Münch 1992, p. 23). Such dominance can also occur, for example, in the case of a strong hierarchy, if the political leadership of a company misunderstands its decision-making authority in such a way that it can interfere with the competencies of the specialist staff without causing significant damage.

Conversely, controlling part-system can be *dynamized* by dynamic part-system at the lower end of the scale of cybernetic relationships, forcing controlling part-systems to *adapt* (ibid.). For example, a personnel policy that makes extensive use of temporary workers or commissions subcontractors to perform internal tasks can lead to a loss of solidarity among the workforce. Consequences such as a lack of motivation, limited commitment or high fluctuation can considerably destabilize the corporate culture and thus the overall system (Pfeffer 2002, p. 61 ff.).

Another destabilizing effect can come from *isolation* between part-systems. Daniel Goeudevert, for example, reports from his time as a board member at Ford: Particularly in large companies, board members often have the problem that their activities take them further and further away from real life (1996, p. 159). If you rise in the hierarchy of a company to the position of chairman, then you are usually on the last floor of the company building. And the further up you go, the more the windows turn into mirrors. Finally, at the last level of the hierarchy, you are not only alone, but also have no more windows. The view of the outside world is blocked. You only see yourself. [...] Medicine describes autism as follows: Mental disorder that manifests itself in pathological self-centeredness, affective apathy, loss of contact with the environment, withdrawal into one's own world and is sometimes accompanied by paranoia. The mental state of a top manager could be described in this way, or at least in many respects similarly (ibid., p. 180). Whether and, if so, to what extent the generalization contained in this individual assessment is tenable remains to be seen, but the risk of isolation between part-systems of a social system, to which reference is to be made here, is demonstrated in any case.

Before finally introducing the only form of interaction that enables a balance, the possibility of conflict between part-systems should be pointed out, as this can be observed in many organizations. Particularly between the political and economic part-systems, disputes repeatedly arise over the question of what results from factual considerations and what should be decided politically. However, this distinction is too crucial for Systems Engineering & Management to remain a mere reference to the insights of the former top manager Daniel Goeudevert (1996, p. 160 ff.), which are also revealing in this respect. This important topic will be discussed in more detail in Section 2.6.2.3.

Before that, however, it remains to explain the previously mentioned form of interaction, which is the only one that can bring about a stable balance between the part-systems of a social system: *interpenetration*. This refers to the mutual interpenetration of different part-systems, which nevertheless preserves the respective intrinsic laws. An example of this, if applied in the sense intended here, can be leadership through targets[24]. Here, political leadership merely formulates its objectives and the associated boundary conditions. What is often overlooked in practice, but is particularly important for the executive bodies, is the need to provide the necessary resources. In terms of interpenetration, time and money are particularly suitable for this, while material resources, for example, can already involve an encroachment on the autonomy of the executing body. How the resources are then actually used and what specific measures are taken to achieve the goal under the given framework conditions remains a matter for the implementing bodies. Furthermore, interim reports to political bodies, such as project progress reports, do not contain any detailed content-related aspects, but only those that are relevant in the political part-system. In practice, such information is usually condensed into key figures or very simple visualizations such as a traffic light, in which the color red symbolizes a critical situation and green a promising one. A more in-depth explanation of this rather superficial introduction to interpenetration

[24] In practice, the term target agreement is more common, but often misleading, because subordinates do not always have realistic opportunities (power) to effectively contradict the targets set by their superiors and enforce their own ideas to a significant extent.

as a form of interaction between different, autonomous part-systems can be found in Section 6.1.2.3.

Decisions

On the basis of the structural model of socio-technical systems (AGIL scheme), the function of the decision will now be explained in more detail, as it is also of great importance in Systems Engineering & Management, but at the same time gives rise to numerous misunderstandings. It is already known from the previous references that decisions are primarily made in the political part-system and that their function is essentially to absorb contingency that could not be eliminated in other part-systems. For the sake of simplicity and because this is of particular importance for the SEM, only the economic and political part-systems will be considered below.

The SEM is part of the economic part-system because it serves to solve problems properly. To this end, experts are at work who, on the basis of implicit knowledge, i.e. experience and the resulting skills, and explicit knowledge[25] , as documented in textbooks, standards and guidelines, for example, have determined and *calculated* solutions and demonstrated their probable effectiveness, if necessary also taking into account the state of research and with the help of scientifically sound and practically proven methods, procedures and tools[26]. However, despite the generally high level of expertise that flows into problem solutions in this way, uncertainties (contingency) always remain that are not accessible to any calculation, however skillful, simply because every problem solution is made *for the future*, which in principle cannot be accurately predicted, no matter how sophisticated the underlying forecasts (Sect. 4.2.4.11) may be, and because there are always relevant aspects that cannot be included due to their *spatial distance* (problem of time and space).

A decision is required to transform such residual uncertainties into certainty (March and Simon 1993, p. 165; Simon 2015, p. 66 ff.; Wirdemann 2017, p. 39 f.) – and only for this! Niklas Luhmann (1988, p. 170) explains that every action can take place in one way or another, but not arbitrarily. The function of the decision is to transform this this way, but also differently into a this way and not differently.

Now that the experts have contributed all their expertise, it has been used up and is no longer available to absorb the remaining contingency. Instead, managers are now required to apply other, additional criteria in order to enable the application of the results. They do this by deciding, not by calculating, because the fact that, for example, 1 + 1 equals 2 according to Adam Riese is just as important as all other problems of this kind.

The basis of decisions is power. Power, in turn, is not tied to professional qualifications (Wirdemann 2017, p. 40). On the contrary: professional qualifications can even stand in the way of decisions, as they can obscure the fact that the criteria on which decisions are to be based go beyond purely factual criteria. It becomes particularly problematic when managers believe themselves to be the best experts, because they easily overestimate themselves in professional terms and often underestimate experts simply because of their hierarchically

---

[25] On the distinction between explicit and implicit knowledge Nonaka and Takeuchi 2012, p. 74 ff.

[26] Decisions that are referred to in decision theory as decisions with certainty (e.g. Göbel 2018, p. 56; Laux et al. 2018, p. 63 ff.) are not decisions in this sense, but calculations.

subordinate positions[27]. Good managers who have understood what their function is and what it is not, know that they are only called upon where logical considerations reach their limits. They should therefore complement, not replace, professional insights.

As it would go too far here to expand on the entire decision theory, including the underlying forms of rule (Weber 1972, p. 122 ff.; Wolf 2011, p. 63 ff.), this statement is simply illustrated by the example that managers can also come to power by being the sons or daughters of entrepreneurs whose successors they take over. The basis for this is the right of disposal resulting from ownership, but not necessarily professional competence or even management competence. The different functions of specialists and technical managers on the one hand and political leaders on the other result, among other things, from different forms of responsibility.

- Specialists and technical managers must take responsibility for what is attributable to their technical qualifications. For example, engineers who specialize in the structural design of components must master the relevant calculation methods of technical mechanics and apply them correctly. If they miscalculate, they have to take responsibility for the consequences.
- Instead, political managers assume *political responsibility*. This does not necessarily require personal fault. The only criterion is success. To achieve this, they must prevail against the competition, i.e. come to power and stay in power by making decisions that lead to success. One of the instruments at their disposal is the selection of suitable specialists. Responsibility for personnel is an essential characteristic for distinguishing between salaried employees and executive employees (§ 5 Para. 3 Sentence 1 BetrVG). They are also responsible for organizational duties (e.g. § 43 Para. 1 GmbHG).

This of course contradicts the widespread conviction that decisions made by political leaders in particular must be made rationally according to objective criteria and must therefore also be explained causally (e.g. Goll and Hommel 2015, p. 16; Pfeffer and Sutton 1999). Contingency means that every decision can also be made differently and is therefore not objectively clear, i.e. causally conclusive.

In addition, this is usually observed by those who are affected by political decisions; the possibility that a different decision could have been made remains in consciousness even after a decision has been made (Martens and Ortmann 2006, p. 437 f.) and leads, among other things, to decisions still being questioned in retrospect, i.e. they can be used for later attributions of responsibility. In this respect, the absorption of contingency is usually only successful to a limited extent. This is the reason why attempts are often made, at least in appearance, to justify decisions causally. It can be exploited that only a few people are usually aware of the illusory nature of causal explanations. In this respect, objectification also helps to reduce contingency. It serves what Theresa Zimmermann (2004, p. 315 ff.) calls the *illusion of causality*. Richard Michael Cyert and James Gardner March (1963, p. 237) speak of the *competence illusion* in this context: it is not uncommon for chance

[27] However, there is a widespread misconception that you become a manager by being successful as a specialist. This is also suggested by universities, which mainly train students to become specialists, but also and especially offer the prospect of leadership positions. However, political leadership requires a completely different qualification, namely a political qualification.

to determine success or failure (Kahneman 2012, p. 268) and not the competence of decision-makers.

The distribution of technical and political responsibility among different functionaries has the advantage that experts remain in the respective organization if errors occur that were not logically foreseeable, while political leaders assume this risk and leave if damage occurs as a result. The latter are generally easier to replace than specialists, who often have qualifications that are relevant to the company's unique selling points. Leadership, on the other hand, is largely the same across organizations[28]. This is reflected, among other things, in the fact that political leaders in the business world enjoy no de facto protection against dismissal, i.e. they can be dismissed at any time without giving reasons. And in democratic states, management personnel are elected for a fixed term from the outset, whereas in public administration, which is predominantly staffed by experts, civil servants are not dismissed (Weber 1972).

This does not mean that political leaders cannot also have important specific qualifications. However, these also and above all consist of recognizing when it is time to leave. The extent to which political leaders who know how to do this are valued can be seen, for example, in the fact that they often take up a new, often even better paid position shortly afterwards. Smart owners or supervisory boards appreciate it when someone knows how to lead, i.e. is willing and able to take on political responsibility.

However, it is helpful if political leaders have at least enough technical expertise to be able to assess whether solutions to problems have been developed to the best of the knowledge and belief of the experts involved and what significance the solutions have in the context of the overall situation. So it is not always possible to do without specialist knowledge, but this is not comparable to what is known in specialist circles as specialist expertise and what is required to solve problems properly. Rather, political leadership is primarily a matter of *intelligence*, especially a *good grasp* (Grandke et al. 1998) and *assertiveness*. The concept of Lean Production also requires political leaders to always get a picture of the situation on site through Genchi Genbutsu or a Gemba Walk (Dombrowski and Mielke 2015, p. 244 ff.).

Evidence-based management (Pfeffer and Sutton 1999) therefore requires that decisions are prepared by experts in a professional and generally understandable way. In systems engineering, this includes in detail

- to solve the problem professionally, i.e. in particular
  - by incorporating implicit knowledge (experience) and skills,
  - by incorporating explicit knowledge (knowledge documented in textbooks, standards, guidelines, patents, etc.) patents etc., also: state of the art),
  - where appropriate and possible, including the current state of research (documented in scientific publications, discussed at specialist conferences, etc.),
  - using scientifically sound procedures, methods and tools,

[28] Laypeople sometimes wonder how a female doctor, as an executive in the German government, can first head the Ministry of Family Affairs, then the Ministry of Labor and then the Ministry of Defense in order to qualify for the presidency of the European Commission. From a systems theory perspective, however, this is evident: leadership only requires very limited professional qualifications and follows similar rules everywhere.

  - with the involvement of all relevant stakeholders and
  - on the basis of an explicit calculation of opportunities and risks in the light of strengths and weaknesses (SWOT analysis).

- Identify alternatives whose selection requires a decision, which also includes the presentation of possible consequences, insofar as these are recognizable on the basis of technical expertise,
- document the problem-solving cycle and its results, including any remaining gaps, thereby ensuring reproducibility and intersubjective verifiability. To this end, the documentation must on the one hand do justice to the subject matter, but at the same time present it in such a generally understandable way that it is also accessible to laypersons[29].

Systems engineering representatives, i.e. experts, are also involved in decision-making processes by providing technical advice to the decision-making bodies. However, they are not involved in the decisions themselves. These are exclusively the responsibility of the political authorities. However, they are required to incorporate the technical aspects as far as possible in the sense of evidence-based management (ibid.). As mentioned above, this does not relieve the experts of all responsibility. They are and remain responsible for the technical accuracy of the results, for example in the event that they have miscalculated. However, involving them in political decisions would go beyond their area of responsibility.

This clear separation between the areas of responsibility of the economic sub-system on the one hand and the political sub-system on the other could be called into question by the fact that uncertainties constantly arise in the course of factual work. And indeed, this requires interpenetration between the political and economic part-systems (Fig. 2.56). It is therefore incumbent on political decision-makers to provide a framework for risk calculations within which the experts in the economic part-system can calculate. Such a framework can specify the handling of risks, for example, by categorically specifying risk preferences such as risk appetite, risk neutrality and risk aversion. The methods used in the economic part-system include suitable parameters for calculating this.

The expected utility theory, also known as *Bernoulli's principle* (Göbel 2018, p. 110 ff.), provides an example of this. According to this, the expected value of the total utility $\Phi(a_i)$ for the choice a_i of a variant i results from the sum of the utility values n_{ij} weighted (multiplied) by the probability p_j of the respective environmental state u_j. The expected utility function, also known as the Von Neumann-Morgenstern function (VNM), is as follows

$$\Phi(a_i) = \sum_{j=1}^{U} n_{ij} \cdot pj \tag{2.41}$$

A concrete risk-benefit function is set up on the basis of a survey. The decision-maker is presented with the alternative of either a_1 a certain payout of a low amount e with $e_{max} > e > e_{min}$ (e.g. salary) or a_2 an uncertain payout of the maximum amount e_{max} (e.g. lottery winnings). In the second case, the probability of winning (e_{max}) is p and the probability of receiving nothing (e_{min}) is $1 - p$. The risky choice a_2 is called the basic

[29] From a professional perspective, decision-making bodies are to be classified as laypersons.

reference lottery (BRL), the safe choice a_2, which provides the decision-maker with the same benefit as the BRL, is called the safety equivalent (SE). Indifference between a_1 and a_2 exists if the expected utility (EUT) of the BRL corresponds to that of the SA.

$$EUT(BRL) = p \cdot n(e_{max}) + (1 - p) \cdot n(e_{min} = n(CE) \tag{2.42}$$

With $n(e_{min}) = 0$ and $n(e_{max}) = 1$, this results in

$$EUT(BRL) = p = n(CE) \tag{2.43}$$

Four relevant parameters must therefore be determined by the survey: the probability p, the two events e_{max} and emin and the certainty equivalent CE. A possible question to the decision-maker could therefore be what certain amount SE he would equate to a lottery win that would yield an amount of $e_{max} = \$1,000$ with a probability of $p = 40\%$, but an amount of $e_{min} = \$0$ with a probability of $(1 - p) = 60\%$. In this case, the expected value EUT(BRL) is \$400.

- If the decision-maker were to compare this expected value with a certainty equivalent of 400 e, he would behave in a risk-neutral manner; ultimately, it would not matter whether he chose one or the other (indifference).
- If the decision-maker were to specify \$300 as the certainty equivalent, he would be risk-averse. In this case, he would have to calculate a so-called risk premium of \$100 for foregoing the chance of a correspondingly higher amount of \$400 for security reasons.
- If, on the other hand, the safety equivalent were set at \$600, for example, this would correspond to a risk-averse decision. Only at this amount would he forgo the chance of receiving \$1,000 in the best case scenario by participating in the lottery. In this case, the risk premium would be negative.

Fig. 2.59 visualizes the risk-benefit functions for the case that the decision-maker is only guided by one target value. The function of a risk-neutral decision-maker corresponds to a straight line. In contrast, the function of a risk-taking (rf) decision-maker is convex. The marginal utility increases with increasing events. In contrast, the function of a risk-averse (ra) decision-maker is concave; the marginal utility decreases as the event increases (Laux et al. 2018, p. 134).

From this example, it can be deduced that decision-makers absorb contingency that is to be expected during a design project by defining the framework for risk-taking. The experts can then use this to calculate. Since the methods that are available on the basis of so-called decision theory to support the rational choice between variants (Göbel 2018, p. 48 ff.) are consistently comparable to the example of Bernoulli's principle outlined here, they are therefore not actually decisions in the sense intended here, but calculations.

This also applies to situations that are classified as uncertain in decision theory and are distinguished from risk situations in that it is only foreseeable which states will occur, but not the probabilities with which this will happen (Knight 1921). This can be countered by the fact that there does not necessarily have to be a situation in which no probabilities of occurrence can be specified. Rather, the only question that arises is how likely the

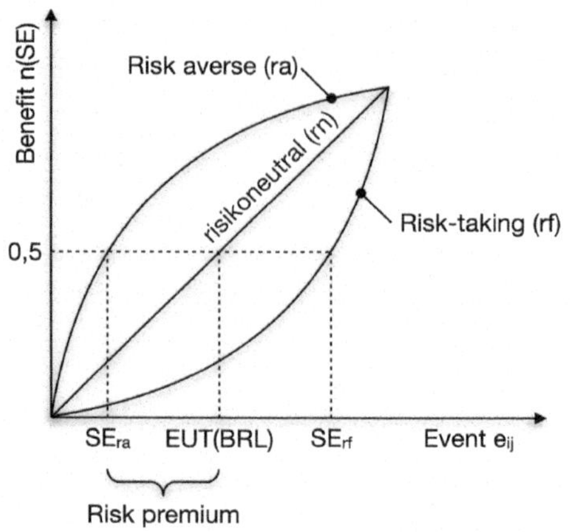

Fig. 2.59 Risk-utility function (Göbel 2018, p. 117; Laux et al. 2018, p. 134)

probabilities come close to reality (Esenführ et al. 2010, p. 23), i.e. how great the risk is that is inherent in every risk calculation.

Limited Rationality of Purpose

The fact-oriented work in the economic part-system can give the impression that organizations are very purpose-driven[30]. For example, companies are assumed to primarily maximize economic benefit. However, when founding a company, some founders will also and above all pursue an idea whose realization in the sense of self-realization may be more motivating than the hope of earning a lot of money with it at some point. It is not uncommon for subsequent generations of owners to have the maximization of their income in mind. But whatever purpose was associated with the founding of an organization, over time they emancipate themselves from it. Fritz B. Simon (2015, p. 28) explains: What remains is the goal to survive: The meaning of life is life – this also applies to organizations (also Probst 1987, p. 48). Of course, this does not exclude the possibility that they look for goals and solve problems, for example. But this is then a means to the end of survival. If you look at organizations as problem-solving tools, they are vagabond solutions in search of suitable problems.

The assumption of purposeful rationality of organizations is closely linked to the mechanistic interpretation pattern. One starting point is the IPO model (Fig. 2.6), according to which a system processes (P) input variables (I) into output variables (O) (e.g. Wiendahl 2014, p. 24). This is linked to the assumption that an observer knows which function is used to transform inputs into outputs and what state the system is in before, during and after this process. In this respect, this model assumes that a system is open in every respect.

[30] For example, the ISO/IEC/IEEE 15.288:2023 standard binds the definition of systems to the pursuit of predefined goals.

If one does not want to take objections such as those raised by Fritz B. Simon (2015, p. 28) as a reason to throw the idea of a purpose orientation completely overboard, one will nevertheless not be able to avoid the insight that organizations should generally not only fulfil one purpose, but many, sometimes even competing purposes. This results in the *stakeholder value approach*, as required by the globally widespread Quality Management standard DIN EN ISO 9001:2015 and, more generally, by DIN ISO 26000:2011 on social responsibility. Approaches such as this take into account the fact that organizations are exposed to a whole range of demands from different stakeholder groups: Not only owners want to earn money, but also employees, suppliers and service providers. In addition, the state demands taxes and levies, the social community expects contributions to general prosperity (*corporate citizenship*), the natural environment is to be protected and so on.

Niklas Luhmann (2000, p. 445) therefore describes an emphasis on a single purpose as a decision-related equivalent for what one would call 'objectivity' in epistemology. This is merely a rhetorical achievement with which one attempts to give certain positions the appearance of indisputability. This is a political maneuver whose paradoxical goal would be to prevent questions about the meaningfulness of a decision (Simon 2015, p. 30 f.). Fritz B. Simon (ibid.) summarizes this insight with the sentence that it is not the common goal of different stakeholder groups that ensures the survival of an organization, but the fact that the organization is able to serve as a common means for different *goals*.

In business administration, so-called behavior-oriented approaches such as that of Herbert Alexander Simon (1997, p. 33 ff.) seem to correspond to such a rationality concept in that they assume bounded rationality. On closer inspection, however, such approaches aim in a different direction in that they do not question the goal orientation of organizations in general, but merely regret their limitations due to a lack of qualification, capacity or motivation. As a result, organizations do not strive for the objectively best solution, but for one that appears satisfactory in light of limited resources and capabilities. If necessary, meaningful explanations for such solutions are often only constructed afterwards (Weick 1995).

The systems theory approach goes beyond this in that, firstly, it understands goal orientation as a political means to the end of securing existence and thus gives priority to securing existence over goal orientation and, secondly, it assumes a multitude of partially competing goals. Apart from this, it agrees with behavioural approaches to the effect that systems also do not choose the best path in their interaction with their environments such as owners, employees, other organizations and so on, but are satisfied with viable solutions due to limitations such as their limited capacity to process complexity (Simon 2015, p. 32).

Insights such as these have far-reaching consequences for systems engineering. This includes the selection of the problems whose solution is to be tackled in the first place. In fact, a large number of problems can usually be identified almost everywhere, but most of them remain unsolved for the moment, and many remain unsolved in the long term. Assuming rationality, it could be assumed that the problems to be solved are selected according to defined criteria and in the light of a calculated prioritization. This is, for example, the claim of *Project Portfolio Management* (DIN 69909-4:2015-11). In fact, the selection of relevant topics follows a more complex logic.

Fritz B. Simon (2015, p. 116) explains this using the *garbage can model*, citing its originators Michael D. Cohen, James Gardner March and Johan Peder Olsen (1972): To understand processes in organizations, one can view a decision-making situation as a garbage can into which various types of problems and solutions are thrown as members make decisions. The mix of waste in a single garbage can depends on the mix of available garbage cans, the labels stuck on the available garbage cans, the waste being produced and the speed at which the bins are emptied and the waste removed. He explains: The organization is confronted with problems and their descriptions, proven solutions and new ideas for solutions, with different people who provide different perspectives, and with opportunities and necessities to make decisions. Confusion is the normal state to be expected here.

In this mixed situation, it depends on which problems receive *attention*. This can come from a wide variety of sources:

- Hierarchically superior authorities can, for example, succeed in drawing attention to certain topics by being able to use formal structures. For example, they can convene meetings, determine the group of participants and set the agenda. However, the obligatory agenda item Miscellaneous on a standard agenda shows that subordinate bodies can also draw attention to topics that are not planned by superiors. Whether and, if so, how the topics then develop, i.e. whether they immediately lose attention again or are prioritized, can no longer be reliably controlled, even by those who have power. In addition, the motives for putting issues on the agenda can be very diverse: political profiling, factual necessity, personal preference and much more.
- Attention can also arise as a result of events in a system's environment. However, this presupposes that these events find their way into the system's internal communication processes. In a company, for example, this can take place via the sales department, which has specific access to the sales market and can derive a particular relevance from the topics it introduces.
- Generating attention within the company is also not tied to management responsibility. For example, the janitor can also be successful by pointing out that it is raining on the roof.
- Ultimately, all stakeholder groups can generate attention for certain issues on the basis of their respective threat potential, for example the works council on the basis of the Works Constitution Act for wages and salaries, health and safety officers with the backing of the employers' liability insurance associations for working conditions and so on. Conflicts between different claims arise time and again, and how these conflicts are resolved can neither be reliably controlled nor fully explained in retrospect. In fact, a large part of communication is a continuous struggle for attention for the most diverse topics, from which sometimes one topic emerges victorious, sometimes the other, and attempts at rationalization such as Project Portfolio Management (DIN 69909-4:2015-11) are only one of many means of gaining an advantage in this struggle.

Anyone expecting rationality, objectivity, factual orientation and the like will inevitably be disappointed. Even the selection of topics and thus the selection of problems to be solved only meets such requirements to a very limited extent. And the implementation of the solutions found is also subject to this restriction. This gives everyone cause for criticism, dissatisfaction and sometimes even frustration at any time. However, those who are dissatisfied with the choice of relevant topics and with the way solutions are handled are just as unlikely as others to have universally valid criteria that would allow them to make a better choice. They too can only participate in the struggle for attention and hope for at least occasional success.

For many, it is easier to remain emotionally calm and rationally level-headed in this mixed situation if they recognize and understand the observable behaviours. Prospect theory, for which Daniel Kahneman (2012) was awarded the Nobel Prize in Economic Sciences, provides a suitable basis for this, as it even allows mathematical modeling. The starting point for criticism of models of rational decision-making (Homo Oeconomicus, Commander Spock) is the expected utility theory (Sect. 2.6.2.3), which goes back to Johann I Bernoulli and is reflected in the Von Neumann-Morgenstern utility function (Eq. 2.41), among other things (ibid., p. 331 ff.). Prospect theory counters this with a value function that describes how people evaluate opportunities and risks, which are calculated as the product of the probability of occurrence and the extent of a gain or loss.

This involves going through two phases: (1) In the processing phase, all possible outcomes are ranked. (2) Subsequently, each outcome is assigned a corresponding benefit in order to finally select the outcome with the highest benefit (Eq. 2.44).

$$U = v(x_1) \cdot w(p_1) + v(x_2) \cdot w(p_2) + \ldots \tag{2.44}$$

In it, x_i denotes the potential outcome dimensions and p_i the corresponding probabilities of occurrence. v is a value function that assigns a respective benefit to an outcome (Fig. 2.60). It intersects the zero point, has an asymmetrical S-shape and weights losses more heavily than gains (*risk aversion*) with the same variance in absolute values. w is referred to as a probability weighting function and expresses the fact that people value unlikely events more highly than events with a medium or high probability of occurrence. In the area of gains, the S-curve is concave, in the area of losses it is convex and in the area of losses it is steeper than in the area of gains (Kahneman 2012, p. 348). In the case of positive events, safe payments are preferred to higher but uncertain gains. In the case of negative events, on the other hand, people are risk-averse and prefer an uncertain, high loss to a certain, lower loss.

The so-called *endowment effect* can also be observed: People value what they already own more than something that does not belong to them (ibid., p. 356 ff.). This is caused by *cognitive distortions*, some of which are explained in Section 2.1.2.

Finally, it can often be observed that groups have a higher willingness to take risks than each group member individually. This effect is referred to as a *risky shift* (Göbel 2018, p. 167). This is explained by the fact that (1) responsibility in groups can be spread across several shoulders so that each individual is less worried about having to take responsibility for negative consequences alone, and that (2) in a group, those who have fewer inhibitions are more likely to speak out anyway and therefore tend to be more willing to take risks, so that they can assert themselves over more reserved group members, while the latter can at least demonstrate through their voting behavior that they are not afraid, and that (3) more information comes together in a group than is available to an individual group member, so that people feel safer as a result.

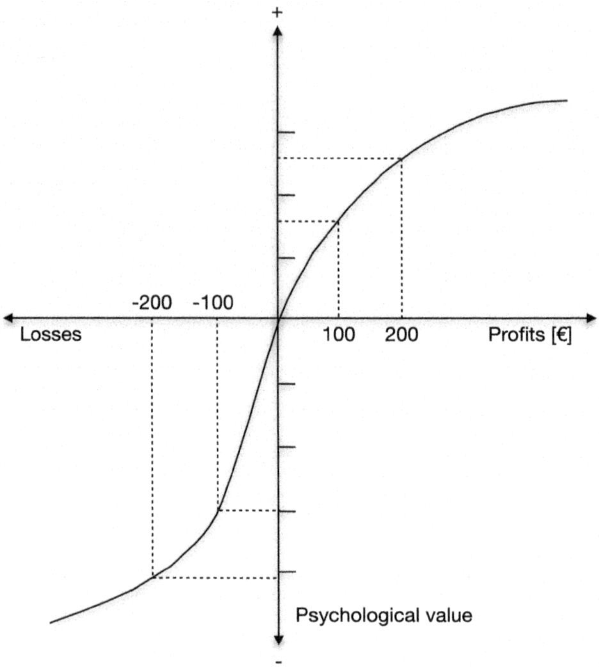

Fig. 2.60 Loss aversion (Kahneman 2012, p. 348)

Chapter 3
Life Cycle Model

The starting point and quasi unifying bracket of all process models is the life cycle model shown in Fig. 3.1 (e.g. Naefe 2018, p. 13 ff.; Thommen and Achleitner 2006, p. 60 f.; Wiendahl 2014, p. 99 f.). This describes the observation that all systems are virtually born, grow up, live for a while and finally die. This does not happen in a linear fashion, but in definable *phases*. There are recognizable *transitions* from one phase to the next. The first transition is *birth*. This is followed by a phase of development in which systems still have to develop their full potential. The duration and subject of this development can vary greatly from system to system: while humans, for example, need around a year to learn to walk, flight animals in particular succeed in doing so just a few minutes after birth. And the development of an airplane differs considerably from that of a rice stove, not only in terms of duration but also in terms of content. However, the basic course of the life cycles is always largely comparable.

The development phase is completed when *maturity* is reached. But just as the birth of a system is a decisive event in its life, this also applies to the transition from *development* (genesis) to maturity (homeostasis). In a person's life, such transitions are referred to as *initiations* and are often ritualized: in the Christian context, for example, initiation rituals are common in the form of baptism (birth) and Protestant confirmation or Catholic communion (transition into Christian adulthood). In Judaism, these are bar mitzvahs for boys and bat mitzvahs for girls, and those who do not ascribe to any religion sometimes celebrate the youth consecration. In the case of man-made things, such initiation rituals include, for example, the transition from development to ongoing operation, known as the *Start of Production* (SOP), and such transitions are also quasi-ritualized, for example, in Quality Management in the German automotive industry through the *Production Part Approval Process* (PPAP) in accordance with IATF 16949 or in the USA through the *Advanced Product Quality Planning* Process (APQP) according to the Automotive Industry Action Group (AIAG). In addition, many other terms have become established in science and practice for such transitions: *Milestones* (DIN 69901- 2:2009-01, p. 20), *Decision Gates* (Walden et al. 2023, p. 29 ff.), *Gates* (Cooper 2002, p. 128 ff.), *Control Points* (Ebel 2011)

V. Ahrens, *Systems Engineering & Management*, https://doi.org/10.1007/978-3-658-46875-0_3

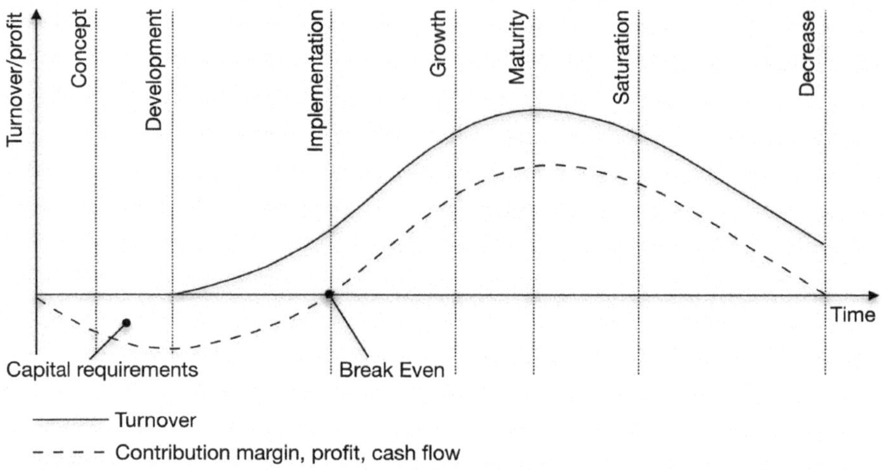

Fig. 3.1 Model of the system life cycle (Wiendahl 2014, p. 100)

et cetera. Unambiguous definitions and clear distinctions from one another or references to one another can only be found in part, but instead there is a lot of imagination in the invention of ever new terms, which in this context, however, contributes more to confusion than to progress. Essentially, they all mean the same thing, namely transitions between phases of the life cycle – however these can be meaningfully defined in a specific case.

In the maturity phase, systems in full possession of their abilities and characteristics do what they were created to do: People live self-determined (mature) lives, work, father children and raise them. Cars are driven, houses are lived in, balls are kicked at goals. As will be shown in more detail later on, the main aim in this phase is to maintain a *steady state* (homeostasis) for as long as possible. For example, while the developmental period in a person's life lasts around 20 years, the phase of mature life lasts around 50 years, based on the current retirement age in Germany after the age of 67. In contrast, the lifespan of man-made systems varies greatly: modern bridges often have to be replaced after just 30 years, while cars and even trucks are still driving over some Roman bridges after two millennia[1] .

From the perspective of the SEM, there are two key things to do in this phase of mature life:

- On the one hand, systems generally have to be managed in this phase. This is the reason why the title of this volume includes not only engineering, but also management. However, managing ongoing operations is often not part of the core of Systems Engineering (e.g. Walden et al. 2023, p. 35 f.). In the case of products, for example, this is done by users when they drive cars, live in houses or shoot balls at goals. In production, Production Planning and Control (PPC) experts, among others, are responsible for this, using specific methods, processes and tools (e.g. Lödding 2016). Nevertheless, it is relevant

[1] One example is the bridge over the Ouvèze in Vaison-la-Romaine, one of the author's favorite vacation destinations.

for systems engineers how the systems they design are subsequently operated, as this results in significant requirements for the design. With this in mind, at least a general (generic) understanding of the management of ongoing operations is created below.
- On the other hand, there is also work to be done during operation for which Systems Engineering can provide a good basis. This includes maintenance, which counteracts wear and tear and thus extends the service life (DIN EN 13.306:2018-02; DIN 31051:2012-09). Cars have to undergo vehicle inspections and sometimes require minor repairs, houses need a new coat of paint from time to time and footballs need to be pumped up from time to time. This also means that after the actual development has been completed, problems can still arise that make subsequent changes to the systems in use appear sensible. In production, this is described with the process of continuous improvement (CIP) or, in the frequently used Japanese expression, kaizen (e.g. Belz et al. 2015, p. 50 ff.) and stands in contrast to Kaikaku, which in Japanese refers to the design of something fundamentally new, as it is at the beginning of the product life cycle. In the case of cars, the tuning industry in particular benefits from kaizen, while the cosmetics industry helps with people. In this respect, systems engineers generally also have a lot to do in the operating phase.

In this way, Systems Engineering & Management (SEM) contributes to the overarching *Life Cycle Management* (LCM), which goes beyond technological aspects to include other aspects such as Business Management. In the course of *product policy*, for example, it is important to counteract *exogenous causes* of a shortened product life cycle. These can be changes in the market that suggest *product differentiation*, for example. In its horizontal form, it extends the properties of existing products, for example by adding additional colors or characteristics (e.g. the off-road capability of a road vehicle), thereby increasing *variety*. In its vertical form, it includes material and technical changes and is therefore aimed at *quality*. Further possibilities for differentiation arise, for example, from the addition of product-related services (value-added services) or from marketing, if it is possible to emotionalize the offer, for example by evoking associations with positive aspects of consumers' lives such as freedom or fitness. An important prerequisite for the profitability of such measures is a modular structure of the products (modular system).

Furthermore, pioneers, i.e. those who have launched a new or significantly modified product on the market (first-to-market strategy), may need to react to competitors (market followers) who are targeting the same market with copycat products (generics or analog preparations in the case of pharmaceuticals). Preventing or at least hindering this in the course of product development through industrial property rights (patents, utility models, design patents) is the subject of the SE (Sect. 4.4.4.1). If and to the extent that this does not succeed, which is particularly the case after the expiry of property rights, the LCM is responsible for reacting to competition by adjusting prices and conditions, among other things.

Moreover, pioneers are often forced to protect their innovative products against *counterfeiting* and *product piracy* (anti-counterfeiting). To this end, technical, organizational, economic and legal measures are taken, for which numerous associations such as the European umbrella organization for the engineering industries of mechanical and plant engineering, the electrical industry and metal processing (ORGALIME) or the Federation of German Industries (BDI) publish guidelines and checklists. Technically, for example, it is about the preventive prevention or avoidance of an analysis of the product in the sense of *reverse engineering* (e.g. Lunkeit and Zimmer 2021). This includes techniques for concealing key features and properties. Products are also given security features such as security labels (VOID films), infrared or ultraviolet coatings, holograms or kinegrams. The ORGALIME guidelines call for the use of methods that are as exclusive as possible (which may also be protected by intellectual property rights), the strongest possible connection between the security features and the product, a combination of visible and invisible elements, simple recognizability and controllability, and the avoidance of any possibility of copying, removal or alteration. In all of this, however, it is important to weigh up the relationship between cost and benefit. Organizational measures aim, among other things, to minimize the risk of betrayal of secrets or espionage, legal measures primarily include the enforcement of property rights and economic measures are intended to make counterfeiting and product piracy unattractive for competitors, for example through an appropriate pricing policy.

From the perspective of those pursuing a strategy of imitation (second-to-market strategy, late-to-market strategy), the LCM initially comprises above all corresponding market observation. If and to the extent that this makes the imitation of products appear possible and attractive, it is necessary to take precisely those measures that the pioneers' anti-counterfeiting attempts to prevent – naturally only to the extent that these are legally and morally permissible. In addition to the above-mentioned reverse engineering, which aims for a detailed reconstruction, functional re-creation can also be considered, which merely imitates the functions of the product to be imitated in the sense of a black box view. In particular, the now advanced 3D scanners or industrial computer tomography (iCT) offer far-reaching possibilities for the reproduction of geometries, especially since the data obtained can be processed by CAD systems and realized using additive manufacturing processes, among other things, especially if production using the original process is uneconomical, for example due to small quantities. The latter can often be the case in the manufacture of spare parts, for example.

In the course of what is initially still a rather rough consideration here, the second major phase of mature life or ongoing operation is followed by a third major phase, that of *decline* until death. In this phase, wear and tear is so great that even the most costly maintenance measures or continuous improvement measures can no longer compensate for it – at least not economically – and changes in the market mean that at some point products are no longer in sufficient demand. Sooner or later, all systems become a case for *palliation*.

However, this is where living systems and man-made systems (artefacts) differ, as the latter often offer the proverbial opportunity for retreading , which aims to develop a *successor product* in the sense of product variation or *product modification*. The basis for this is the observation that not all sub-systems and part-systems or elements of a system are equally affected by wear and tear or are no longer up to date. In this case, the replacement or supplementary renewal of the system can extend the life cycle and ultimately initiate a new life cycle. One example of this is the VW Golf, which has been in production since 1974 and is now successfully in its 8th generation.

A more detailed look at product variation leads to *revival*, which distinguishes between a *relaunch* and a *facelift*. A relaunch refers to the reintroduction of a product following a revision, for example in newly designed packaging. A facelift goes beyond this by also making technical enhancements to the product. In addition, a pure revival enables the revival of an old product in its original form, as can be observed with vinyl records, for example.

Finally, *Obsolescence Management* serves to design products in such a way that they can be used without restriction throughout their entire life cycle until the end. This is to be distinguished from so-called planned obsolescence, which serves to artificially shorten the useful life in order to increase sales. In view of the need for economic (consumer protection) and *ecological sustainability* (environmental protection), this form of planned obsolescence is counterproductive.

However, the launch of each new generation requires a new development phase, which is then followed by a corresponding maturity phase and so on. This can be depicted in different ways in the life cycle model: On the one hand, you can jump back from the end of the curve to the beginning, albeit at the cost of breaking the link to the time axis. On the other hand, you can continue the curve on the time axis and reassign the respective phases to it. However, this makes the representation either very long or very compressed over time. Dispensing with a linear time axis also makes a circular representation possible (e.g. Vajna 2022, p. 95 ff.), the advantage of which is that it emphasizes the *cycle of life*, i.e. the cyclical nature of the life cycle, after which new life emerges from death, which goes through the same phases as any life that does not come to a premature end due to an unexpected event.

A particularly popular form of circular representation of the system life cycle is the *PDCA cycle* (Fig. 3.2) introduced by Walter Andrew Shewhart (1939, p. 45) and further developed by William Edwards Deming (1950) (Taguchi et al. 2004, p. 1426), which was initially adapted by Quality Management in the Japanese automotive industry and has since become widespread worldwide as the concept of Lean Production, originally developed by Toyota, has spread (e.g. Belz et al. 2015, p. 54). However, it is so often misunderstood that a separate section is devoted to it later (Sect. 3.8) in order to dispel this misunderstanding. However, this first requires an understanding of the other process models.

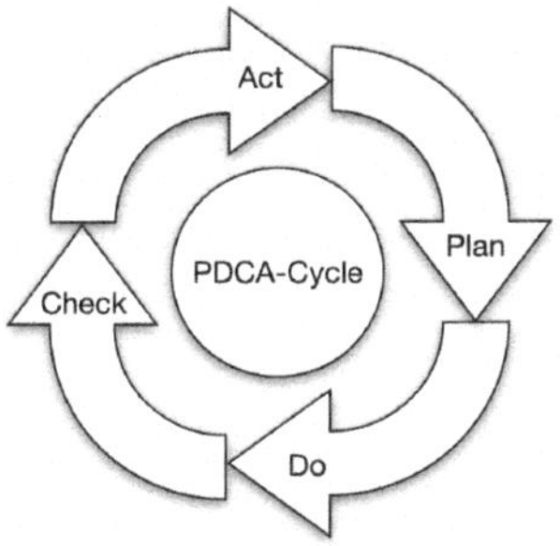

Fig. 3.2 PDCA cycle (Belz et al. 2015, p. 54)

Another variation of the system life cycle is the two-dimensional four-field portfolio shown in Fig. 4.21, which was developed by the Boston Consulting Group and is therefore also known as the *BCG matrix* (e.g. Kirchner 2020, p. 29 ff.). This model establishes the relationship between the relative market share of a product and the growth of the relevant market, which has not been reliably proven either theoretically or empirically, but is often sufficient for practical application, and derives standard strategies from this (Sect. 4.2.4.8). While the expected future market growth on the ordinate represents the environmental influences, the relative market share based on the *experience value concept* on the abscissa represents the company's internal perspective and expresses the assessment that a company with higher sales compared to its competitors will gain more experience and can therefore reduce its costs and risk (Henderson 1968). The applicability of the BCG matrix is generally limited to oligopoly markets. For polypoles, McKinsey has therefore developed this matrix into a nine-field matrix, which depicts market attractiveness (market growth, market size, market quality, energy and raw material supply, etc.) as dimensions on the ordinate and relative competitive advantages (relative market position or market share, relative production and/or research and development potential, relative qualifications of employees, etc.) on the abscissa.

3.1 Quantitative Models of the System Life Cycle

In Systems Engineering (e.g. Walden et al. 2023, p. 26 ff.) and in related fields such as design methodology (e.g. Ehrlenspiel 2020, p. 123 ff.; Naefe 2018, p. 13 f.) or marketing (e.g. Bruhn 2022, p. 45 ff.; Kreutzer 2022, p. 262 ff.), the system life cycle model is widely used, but is predominantly outlined in qualitative terms. Mathematical models that would allow a deeper understanding of the underlying laws are rarely found in these fields of knowledge. Furthermore, the presentation of the system life cycle is generally limited to one product and neglects, for example, the influence of competing products.

More far-reaching insights can be found in theoretical ecology in particular (e.g. Bungartz et al. 2009, p. 251 ff.; Hofbauer and Sigmund 1984). There, mathematical models are used to describe the dynamics of biological populations. Assuming that the laws underlying the system life cycle and *population dynamics* are sufficiently similar, some of the ecological models are outlined below. First of all, however, it should be explicitly pointed out that *evolutionary economic approaches* (e.g. Lehmann-Waffenschmidt and Peneder 2022) such as this one differ significantly from those of the more widespread *neoclassical microeconomics*. While the latter assumes that economic equilibria are formed on markets through supply and demand, i.e. that the whole is inferred from the multitude of decisions made by individuals (bottom-up), evolutionary economics assumes that the decisions of individuals also depend to a large extent on the whole (top-down).

Like all models (Sect. 2.2), these are not intended to capture reality in all possible respects. In view of the very large number of influencing factors, however, real product life cycles will be so specific that they can only be empirically determined and interpreted in specific cases. However, models can promote basic understanding and thus support application-specific interpretation.

The factors usually recorded by ecological models that influence the density of a population, which can be compared with the market penetration of a product (market volume), are divided into density-independent and density-dependent factors. The former include *interspecific competition* in which different animal or plant species share a biotope and have similar demands on the resources it contains (food, retreat areas, etc.). The limitation of a population then results not only from its own density, but also from that of other populations. By analogy, such a situation can be assumed for products if different products are to be sold in the same market segment, such as kitchen utensils, for which the average private household has a limited budget. In response to this, the inferior population can adapt by finding an *ecological niche*, for example by choosing alternative resources (niching), in which it can avoid the immediate competition. For example, the ANNA G corkscrew from Italian design company Alessi is designed in such a way that it can also be used as decoration and can therefore be allocated to the budget earmarked for this purpose.

In ecology, *non-infectious diseases* are also counted as density-independent factors, as they occur randomly, meaning that their relative frequency does not increase with increasing density, unlike infectious diseases. In the case of products, this corresponds to the random failures that occur in the middle section of the so-called bathtub curve shown in Fig. 4.56.

Another density-independent factor that influences population dynamics in ecological systems results from the *food chain* (trophic cascade). For example, it was observed that fishing for lobsters on the shallow rocky coast of St. Margaret's Bay on the Canadian Atlantic coast led to the increasing number of sea cucumbers, which were no longer eaten by the lobsters, completely grazing the originally extensive kelp forests (Pinnegar et al. 2000, p. 179 ff.). Similarly, a decline in sales of compact discs (audio CDs) as a result of the increasing spread of streaming services leads to a decline in sales of corresponding players, without this being attributable to the market penetration (population density) of the devices.

Density-dependent factors include *intraspecific competition*, i.e. the access of individuals in a population to the same resources. In contrast to interspecific competition, it is more difficult for similar animals or plants than for different species to escape competition by moving into niches. Another factor in this category is social stress (also known as the crowding factor) as a result of encounters, which often trigger behavioral changes such as aggression in animals and humans. Furthermore, *contagious diseases* (infections, parasite infestations) can spread more easily with increasing population density. Similarly, a platform strategy, according to which ostensibly different products are actually built largely in the same way, results in significantly greater damage in the event of a recall than a product strategy that relies on substantial diversification so that only small numbers of units are affected in the event of damage.

3.1.1 Model of the Utilization Phase without Competition

Theoretical ecology uses a whole range of mathematical models to understand the effect of such influencing factors on the dynamics of populations. This in turn allows the identification of measures to influence the dynamics. However, the life cycle cannot be fully modeled using a single equation. Instead, different models are used for the various phases, from which the complete cycle can be assembled. Some of these models are outlined below, while others can be found in the relevant literature.

The growth and maturity phase is often modeled using the logistic equation derived in Section 2.4.5.2 based on Fig. 2.30 (e.g. Tattersall et al. 2021, p. 9113 ff.). However, this is an analogous model, while products occur in integer form. Therefore, a discrete form of this equation will be used below. Furthermore, only one product is to be modeled initially, so that density-dependent influences such as intraspecific competition are not taken into account for the time being. Bernulli's differential equation (Eq. 2.39) now takes the form of a non-trivial recursion rule (e.g. Argyris et al. 2017, p. 69[2]; Bungartz et al. 2009, p. 301 ff.; Hofbauer and Sigmund 1984, p. 61 ff.; Schuster 1994, p. 37):

$$y_{n+1} = ky_n(1 - y_n) \tag{3.1}$$

y stands for the relative population size, i.e. the ratio of the current size to the maximum possible size, whereby the latter is limited by the available resources. In economics, this corresponds to the ratio in which the total available market potential is exhausted by the actual market volume in the face of intraspecific competition. The index n stands for the period under consideration and k indicates the rate of change per period, which in ecology in turn results from influencing factors such as food resources, climatic conditions and so on: $k = 1.03$, for example, means growth of 3 % per period, $k = 0.95$ a shrinkage of 5 % per period. In the economy, for example, this can be used to model the effect of market penetration through marketing. The model is first introduced below before it is interpreted in terms of its practical usefulness.

$f(y) = ky(1 - y)$ describes a parabola in the interval $0 \leq y \leq 1$, which has its vertex at $y = 1/2$ with the function value $f(1/2) = k/4$, is open at the bottom and intersects the x-axis at $y = 0$ and $y = 1$ (Fig. 3.3 above for $k = 2.00$). To ensure that y does not leave the specified interval, k must remain in the interval $0 \leq k \leq 4$. Possible effects are now investigated by varying the rate of change k.

With a starting value of $y_0 = 0.01$ and a rate of change of $k = 2.00$, an S-shaped curve of the population development is obtained according to Fig. 3.3 below, which is comparable to the curve in Fig. 2.30. The upper part of Fig. 3.3 shows the graph of the logistic mapping and the angle bisector. Furthermore, the staircase curve between these two curves shows the way in which the maximum value is approached. At the beginning of the development (lag phase), the population is still small, so that density-dependent influencing factors do not play a role. As density increases, influences such as word-of-mouth advertising, which is explained using the diffusion model by Frank M. Bass (1967) and shown in Fig. 2.32, increasingly occur and cause exponential growth (log phase). However, the closer the population size gets to the saturation point, the more its growth is slowed down by competition for increasingly scarce resources, until it finally comes to a complete standstill. From now on, births and deaths balance each other out; the population size remains constant at half the maximum size in this parameter constellation: $y_{n>17} = 0.5$.

If the rate of change is increased to $k = 2.10$, the maximum population size also increases to $y_{n>19} = 0.5238$. This, together with the still relatively low utilization of the maximum relative population size of 100 %, suggests that the rate of change should be increased further. However, this leads to remarkable effects. Fig. 3.4 below shows the course of the

[2] The following explanations largely follow this source, without this being stated repeatedly.

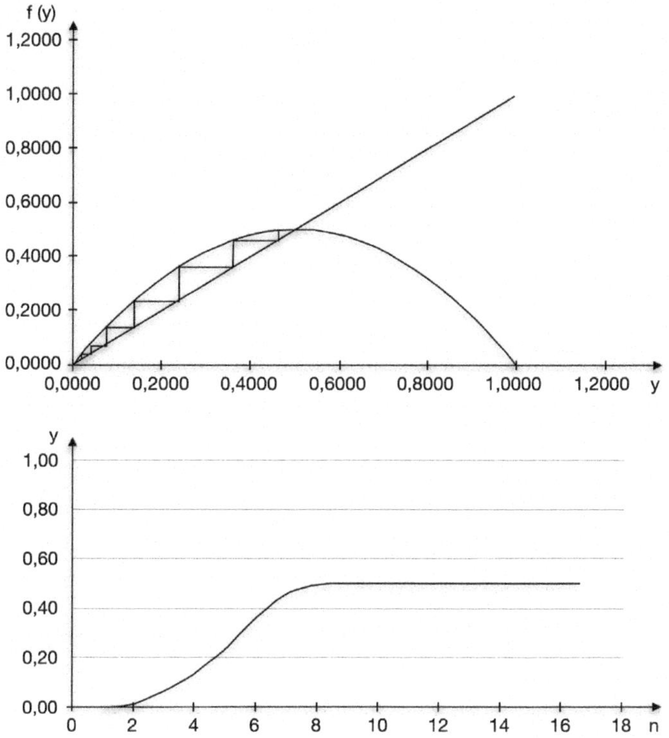

Fig. 3.3 Iterative approximation of the logistic function (top) and graphical construction of the iteration (bottom) for k = 2.0

population size for $k = 2.99$. As can be seen, the population size now fluctuates between two values, whereby the oscillation decreases only slightly in the further course. Nevertheless, the two values lie between $y = 0.7203$ and $y = 0.6044$, on average at $\overline{y} \approx 0.6667$ and thus significantly above the previously achieved value. However, this higher yield comes at the price of unstable behavior.

With a further increase in the rate of change to $k = 3.50$, the instability increases: the population size fluctuates between four values, the mean value $\overline{y} \approx 0.6470$ becomes smaller (Fig. 3.5 below). Finally, at a rate of change of $k = 3.99$, Fig. 3.6 below shows the transition to a deterministic-chaotic (non-linear dynamic) behavior in the sense of the word. Deterministic means that the system behaves in exactly the same way under exactly the same initial conditions. It is chaotic in the sense that even the smallest changes in the initial conditions can lead to completely different reactions.

Mathematically, this non-trivial behavior, which is caused by bifurcations or period doublings, can be explained by searching for possible fixed points (point attractors). The following condition must be met for the development to aim for a fixed point y_s:

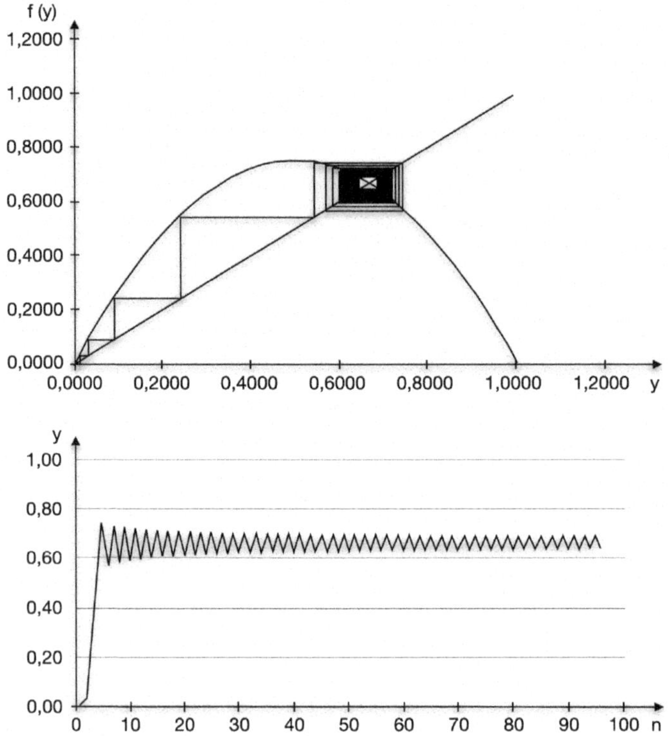

Fig. 3.4 Iterative approximation of the logistic function (top) and graphical construction of the iteration (bottom) for k = 2.99

$$y_s = f(y_s) \tag{3.2}$$

The parabola therefore has the two fixed points

$$y_s = 0 \qquad y_s = 1 - \frac{1}{k} \tag{3.3}$$

Next, the closer surroundings of the two fixed points are examined. For this purpose, the parameter ϵ is introduced, which differs slightly from the respective y_s. If the starting point $y_n = y_s + \epsilon$ is inserted into the 1st order difference equation $y_n + 1 = f(y_n)$, from which Eq. 3.1 was developed, this results in the following point

$$y_{n+1} = f(y_n) = f(y_s + \epsilon) \approx f(y_s) + \epsilon f'(y_s) \tag{3.4}$$

The disturbance therefore subsides when $|f'(y_s)| < 1$. This means that the fixed point is stable (point attractor). For $|f'(y_s)| > 1$, on the other hand, the fixed point is unstable. For the parabola, the derivative of the function is specifically

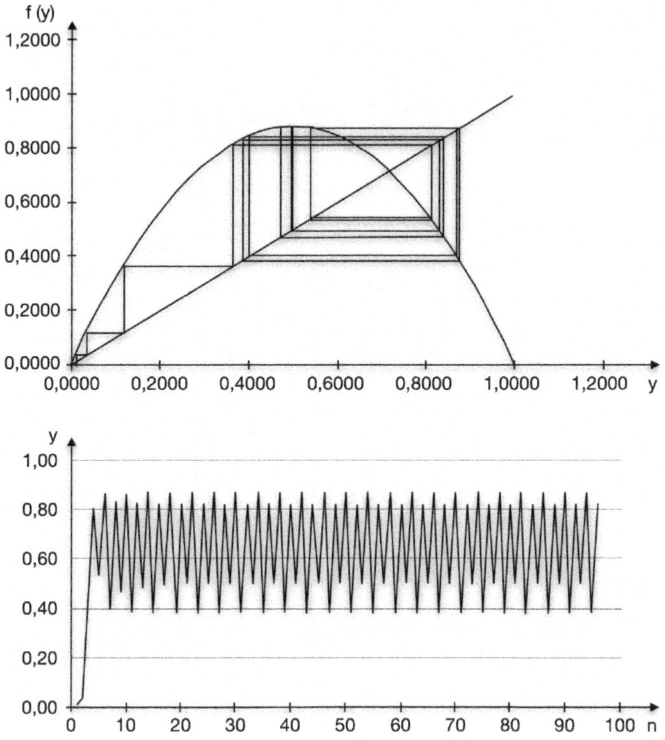

Fig. 3.5 Iterative approximation of the logistic function (top) and graphical construction of the iteration (bottom) for k = 3.5

$$f'(y_n) = k(1 - 2y_n) \tag{3.5}$$

The fixed points are therefore

$$f'(0) = kf'\left(1 - \frac{1}{k}\right) = 2 - k \tag{3.6}$$

It follows that for $0 < k < 1$ every development ends in $y_s = 0$. For $1 < k < 3$, the point attractor $y_s = 1 - 1/k$ is effective. This is illustrated in Fig. 3.3 with the fixed point $y_s = 1 - 1/2 = 0.5$ and Fig. 3.4 with the fixed point $y_s = 1 - 1/3 = 0.6667$. For $k > 3$, neither of the two fixed points is attractive according to Eq. 3.6 Instead, the population sizes oscillate between $y_{s1} = f(y_{s2})$ and $y_{s2} = f(y_{s1})$. Neither of the two points forms an attractor on its own. Rather, they form a cycle in their entirety. The further iteration process for y_{s1} and y_{s2} results in

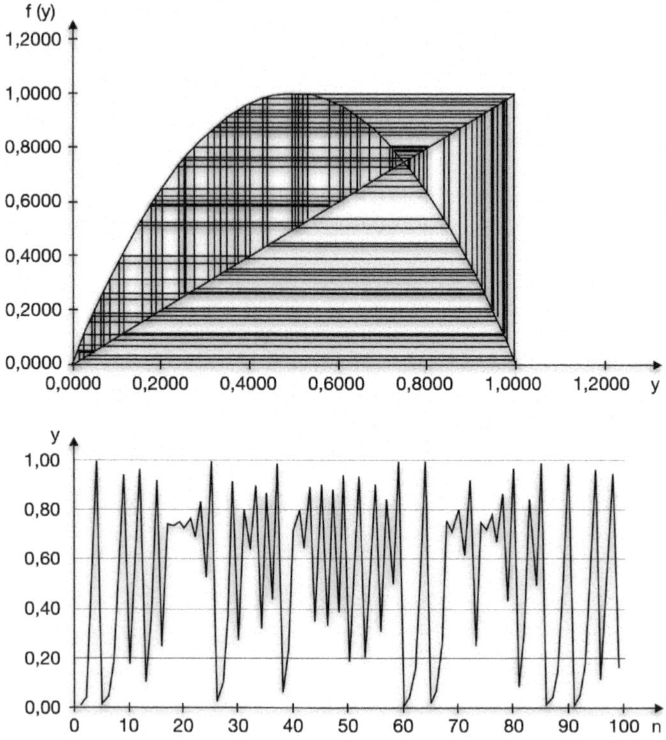

Fig. 3.6 Iterative approximation of the logistic function (top) and graphical construction of the iteration (bottom) for k = 3.99

$$y_{s1} = f(f(y_{s1})) \qquad y_{s2} = f(f(y_{s2})) \tag{3.7}$$

Both points are not fixed points of the function $f(y_n)$, but they are fixed points of the mapping function $f^2 = f(f(y_n))$, which describes a movement in the event that two iteration steps are carried out at once (doubling transformation). The following applies to the parabola relevant here

$$f^2(y_n) = kf(y_n)[1 - f(y_n)] = k^2 y_n(1 - y_n)[1 - ky_n(1 - y_n)] \tag{3.8}$$

The attractor condition for this Eq. 3.8 is

$$y_s = f^2(y_s) \qquad |f^{2'}(y_s)| < 1 \tag{3.9}$$

Fig. 3.7 qualitatively shows two graphs of the doubling transformations $f^2(y_n)$ for different rates of change k, on the left for $k < 3$ and on the right for $k > 3$. The curve shown on the left intersects the first bisector – apart from the coordinate origin $y_s = 0$ – in

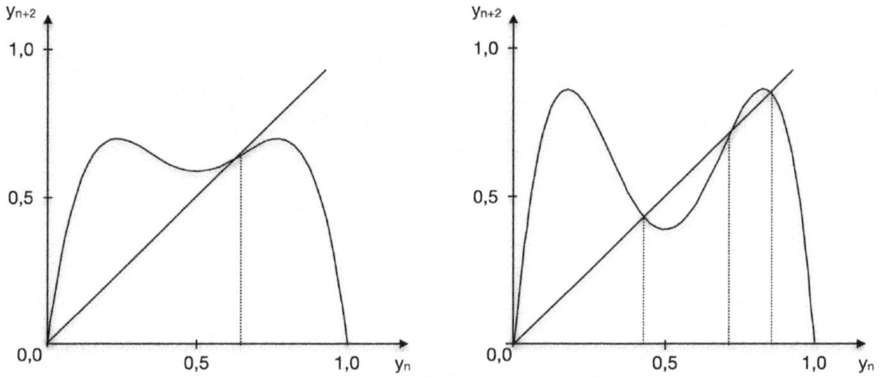

Fig. 3.7 Graph of the doubling transformations $f^2(y_n)$ for various rates of change (Hofbauer and Sigmund 1984, p. 67)

exactly one point attractor $y_s = 1 - 1/k$ (Eq. 3.8). This explains the stable behavior of the population development according to Fig. 3.3. As the rate of change k increases, the curves of the doubling transforms become steeper and steeper; at $k = 3$, the curve only touches the angle bisector, after which two further intersection points are formed in addition to the previous intersection point y_s (Fig. 3.7, right), the two point attractors y_{s1} and y_{s2}. The fixed point y_s is now *instabil*. If the rate of change k exceeds the value $k = 1 + \sqrt{6} \approx 3.449$, the same effect occurs for the two new point attractors y_{s1} and y_{s2} as for the original fixed point y_s: this effect is referred to as *period doubling*. The result is a cycle of period 4. The intersection points of the graph $f^2(y_n)$ are now fixed points of the function

$$f^4(y_n) = f^2\left(f^2(y_n)\right) \tag{3.10}$$

This development continues as the rate of change k increases: further instabilities lead to $2n$ cycles, whereby the intervals between the period doublings decrease rapidly. The sequence k_i tends towards an accumulation point k_∞. For $k > k_\infty$, irregular (chaotic) behavior is observed, which, however, is interrupted from time to time by areas of regular behavior (periodic windows). The exact value from which chaotic behavior sets in is $k_\infty = 3.5699456...$ The distances $k_i + 1 - k_i$ decrease like a geometric sequence.This gives rise to the Feigenbaum constant δ, named after its discoverer, Mitchell Feigenbaum, for which it can be shown that the corresponding laws apply to all discrete distance functions with a quadratic maximum. This results in the Feigenbaum constant δ, named after its discoverer, Mitchell Feigenbaum, for which it can be shown that the corresponding laws apply to all discrete mapping functions with a quadratic maximum. Fig. 3.8 illustrates the scenario just explained in the form of the so-called *bifurcation diagram*.

In order to interpret the logistical illustration as a model of the system life cycle, it is first necessary to recall the underlying assumptions. These include the fact that it is a single product for which there is no competition on the market. Furthermore, key influencing factors are assumed to be constant, such as the price. In practice, this is approximately

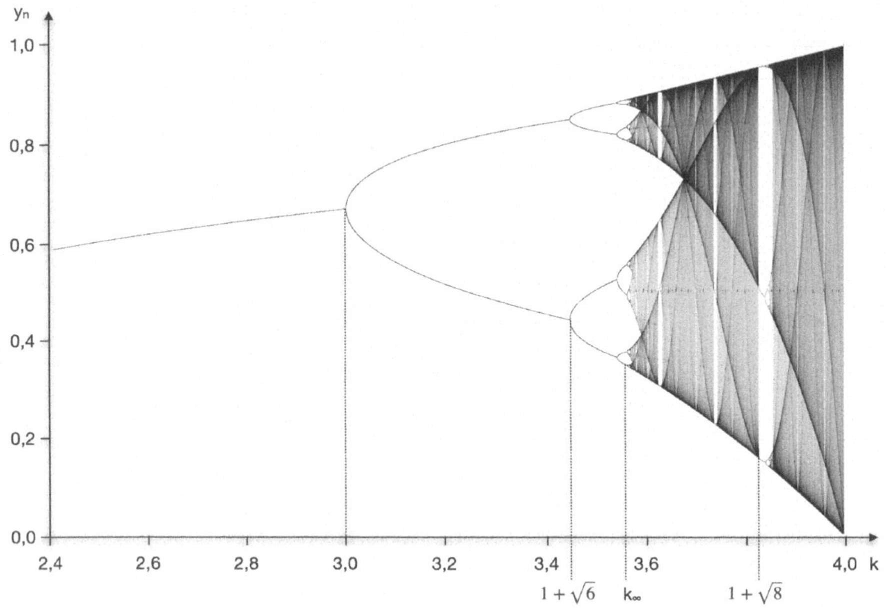

Fig. 3.8 Bifurcation diagram of the logistic figure (Bungartz et al. 2009, p. 310)

possible in monopoly-like markets, for example. Again, the automotive industry is an example of this in cases where customers are loyal to a certain brand and possibly even a certain vehicle model. Furthermore, the capacity limit, i.e. the market potential in economic terms, is assumed to be constant in the long term. The market volume that can be exploited, which in the parameter constellation shown in Fig. 3.2 is 50 % of the market potential, depends in practice on the solvency of the interested customers, among other things: If the real price is above the natural price, it limits the possibility of exploiting the entire sales potential.

What can be learned from the model is that the speed at which a new product is introduced to the market can have a significant influence on the development of sales. If it is very low, the market launch can fail ($k < 1$). Up to a certain limit (in the model up to around $k < 2.5$), a constant trend emerges – with higher values after an increasingly longer settling phase – but at the price of a relatively low exploitation of the market potential. This potential can be further exploited if the speed of change is further increased to approach the limit of chaos, but at the price of instability in the product life cycle. Increased sales are now offset by costly adjustment measures in production, for example by having to adapt capacity to sales fluctuations. Finally, an excessive rate of change can lead to an irregular (chaotic) course of the system life cycle. For example, marketing measures for market penetration such as special offers, installment purchase offers or consumer loans can also reach groups of buyers who cannot actually afford the product. If payment defaults or premature returns occur as a result, this can lead to a collapse in sales. Highly motivated or even over-motivated marketing will also come up with something to boost sales again, and it may even succeed, but if and when it overshoots the mark, the fluctuations can continue. Sustainable and controllable growth should therefore be approached ambitiously in order to exploit the maximum sensible market potential on the edge of chaos. At the same time, however, it must not be overdone, because too much ambition (greed) leads to irregular life cycles that are no longer economically manageable (Dittes 2021, p. 98).

3.1.2 Model of the Utilization Phase with Competition

The model developed so far only takes into account inter-specific competition. However, since intraspecific competition plays a major role in practice, it is taken into account in the extension outlined below. For this purpose, the capacity limits K_a and K_b for two competing products a and b (ecologically: species) are no longer specified in relative terms, but in absolute terms. Furthermore, each product has its own rate of change k_a or k_b. In this respect, the information is essentially the same as before. What is new, however, are the competition coefficients α and β, which express the extent to which one product threatens the existence of the other. α therefore expresses how strong the influence of product a is on product b and vice versa.

$$y_{a,n+1} = y_{a,n}k_a\left(1 - \frac{y_{a,n} + \beta y_{b,n}}{K_a}\right) \tag{3.11}$$

$$y_{b,n+1} = y_{b,n}k_b\left(1 - \frac{y_{b,n} + \alpha y_{b,n}}{K_b}\right) \tag{3.12}$$

For a first experiment, let us assume that both products have almost the same prerequisites: They enter the market at the same time, their sales increase at the same rate $k_a = k_b = 0.1$, their market potential is the same at $K_a = K_b = 500$ units and the competition coefficients are also identical: $\alpha = \beta = 1.5$. Only in terms of initial quantity is product a with 21 units 5 % higher than that of product b with 20 units. As Fig. 3.9 shows, this small lead for product a is enough to completely eliminate product b from the market. This is also known as the *competitive exclusion principle*.

If, in a second experiment, product b with $\beta = 1.54$ is slightly stronger than product a with $\alpha = 1.5$ under otherwise identical conditions, Fig. 3.10 shows that although product a can initially hold its own on the market due to its better initial conditions, it is then overtaken by product b after around 80 periods and finally driven out of the market completely.

In a final experiment, based on the conditions of the first experiment, which also remain the same here with one exception, competition coefficients are selected that are less than one, whereby $\alpha = 0.8$ is slightly greater than $\beta = 0.7$. As a result of this now relatively weak mutual influence, coexistence occurs according to Fig. 3.11, whereby the stronger product a receives a higher market share than the weaker product b, as expected. If both competition coefficients were set to zero, indicating that the two products do not compete with each other, both products would fully exploit their respective market potential. For the sake of

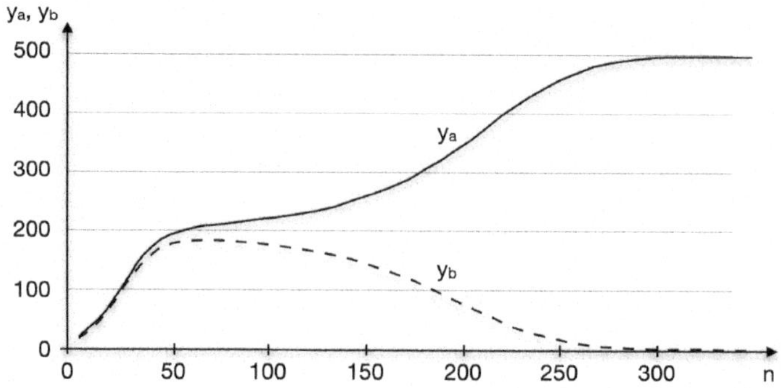

Fig. 3.9 Life cycles of two competing products a and b, with a slight initial lead of product a over product b, with the same growth rate

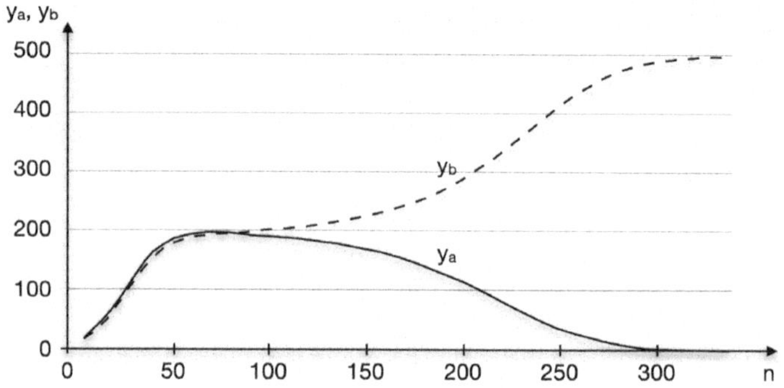

Fig. 3.10 Life cycles of two competing products, a and b, with a slight initial advantage for product a and a slightly greater growth rate for product b

completeness, it should also be mentioned that the rates of change in these experiments were deliberately kept low so as not to obscure the illustration of the effects demonstrated by other effects. If the rate of change were to increase significantly, the curves would be just as unstable as previously explained. Finally, it should be noted that these are very simple models. Josef Hofbauer and Karl Sigmund (1984, p. 75 ff.) are among those who show that models such as Vito Volterra's can be used to formulate more differentiated competition models. In addition, so-called predator-prey models (ibid., p. 68 ff.) such as that of Vito Volterra and Alfred James Lotka are available, which can also be used to model phenomena in domains other than ecology. Since the presentation of all modeling approaches would go beyond the scope of this paper, there is no way around reading further literature if necessary. With regard to a comparison with empirical data, however, a final modeling approach will be outlined below, which has already been used in an economic context. While the previous models focused primarily on the *introduction and growth phase* (Fig. 3.1), the following model also and above all depicts the further course of the system life cycle in the *maturity phase*. In these phases, pricing policy plays an increasingly important role. Particularly in view of competing products, whose market entry becomes more likely as the product life cycle increases, prices that are initially above the natural price are usually reduced continuously. In terms of competition, many markets are characterized by the presence of many small and a few large suppliers; the sales figures of different suppliers of a product are assumed to be lognormally distributed (Kalbach 2013, p. 3).

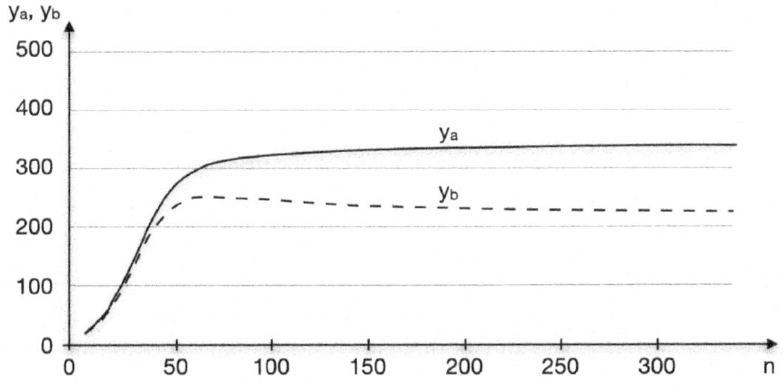

Fig. 3.11 Life cycles of two competing products a and b with a competition coefficient of less than 1: coexistence

Furthermore, the model should be able to depict an initially comparatively rapid spread of a product through *multiplication* as well as a subsequent, slower adaptation to market conditions. The latter can occur, for example, through a *symbiosis*, in which the product virtually comes to terms with competing products, or by *displacing* competing products. This can increase the original capacity limit (market potential). The rapid increase occurs primarily in the course of the market launch and can be modeled, for example, with the help of the *bass diffusion* already explained in Section 2.4.5.2. This initially takes into account the influencing factors of making a new product known through marketing measures (innovation effect) and, following on from this, word-of-mouth advertising (imitation effect) in the event that the product is accepted by the market. However, total sales result from the sum of initial and repeat sales, whereas the Bass diffusion model only takes initial sales into account. With regard to the real introductory price μ_0, it is assumed, as expected, that this is above the natural price μ_n. As not all interested buyers can or want to afford the product in this case, the market potential (capacity limit) is higher than the tappable market volume.

In the introductory phase of a product, which, as mentioned, is represented by the Bass diffusion model, the introductory price determines the initial market volume. Once this volume has been reached comparatively quickly, the system life cycle continues through evolutionary diffusion by means of slower adaptation, which can be brought about by varying the product to be adapted (ibid., p. 5). This, in turn, is essentially achieved by reducing the real price. This decreases exponentially and asymptotically approaches the natural price: $\mu(t) \sim e^{-\varepsilon t}$. The market volume then corresponds to the market potential (saturation). The parameter ε is proportional to the width of the price distribution. The greater the competition, the faster the average real price falls. In this model, $\varepsilon = 0$ describes the special case of monopoly, in which evolutionary diffusion is absent, so that only bass diffusion comes into play.

Sales as a result of evolutionary diffusion now describe an S-curve, just like the Bass diffusion (Fig. 2.32). However, this curve is so deformed by the exponential price decline that it must be described by a different model. The *Gompertz function* is suitable for this purpose

$$y(t) = ae^{-be^{-ct}} \tag{3.13}$$

In this, a describes the upper asymptote, as the exponent assumes the value zero due to its negative sign for large values (after a long time: $t \rightarrow \infty$). $b > 0$ describes the temporal shift and $c > 0$ is the gradient. Fig. 3.12 shows the corresponding curve at the bottom right.

This model approach can be extended to include repeat purchases, which in turn result from both multiple purchases and replacement purchases. The latter depends on the lifespan of the product, which is discussed in Section 4.4.5.3 and is based on the so-called bathtub curve (Fig. 4.56). In this respect, the times of the first purchase and the replacement purchase are more or less deterministically linked to each other. This results in a wave-shaped curve in the saturation phase, which is synchronized with both the increase in sales figures due to the Bass diffusion and the further increase due to the Gompertz diffusion (Fig. 3.12, bottom right).

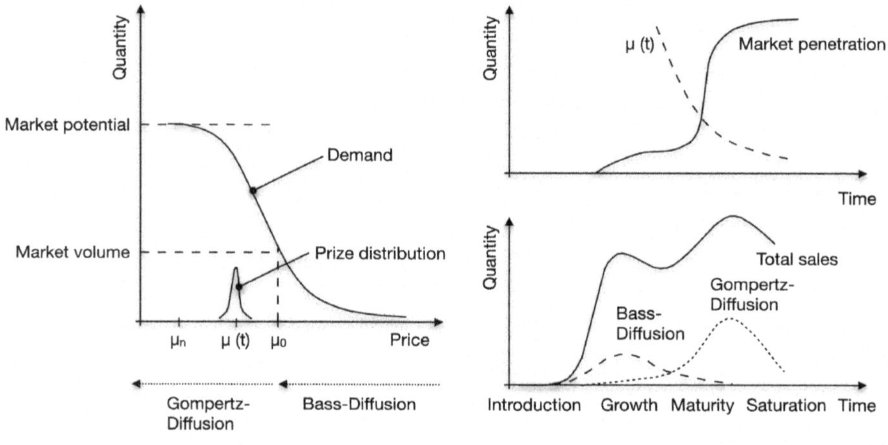

Fig. 3.12 Fitness of a durable good as a function of the real price (Kaldasch 2013, p. 11)

The fact that models such as these correlate with empirical data and are therefore suitable for practical application is discussed by Joachim Kaldasch (2013, p. 7 f.) using the example of the Mercedes-Benz S-Class, among others. Fig. 3.13 shows both the actual sales figures and the model-based curve. The initial conclusion that can be drawn from this is that the market volume up to 1964 was too low to achieve a breakthrough. From 1964 onwards, however, sales figures increased to the extent suggested by the bass diffusion. It can now be assumed that the subsequent wave-like progression correlates with the respective model changes. The model periods shown in Fig. 3.13 thus allow a comparison to be made. And indeed, both the introduction of the W126 and the W126 model are obviously associated with an increase in sales figures. However, the situation is different with the two successor models W140 and W220 in the 1990s. The introduction of the W140 was accompanied by a sharp rise in prices. This may at least partly explain the decline in sales figures. However, the introduction of the next model, the W220, shortly afterwards was not accompanied by another price increase. Despite this, sales figures continued to fall. It was not until the next model, the W463, that sales rose again significantly. Although the previous decline could be explained by the low attractiveness of the models, for example, the models used as a basis here also allow a different interpretation. For example, the average lifespan of an S-Class vehicle is around 15 years, while the model changes were made in shorter succession. If one assumes that vehicles in this price segment must first pay for themselves with at least some of the customers before an investment in a new vehicle is considered, it can be argued that the model changes have not taken sufficient account of this. From this, Joachim Kaldasch (ibid.) draws the conclusion that suppliers of goods can only influence the associated life cycle to a limited extent according to their own ideas. Although so-called *planned obsolescence* offers more possibilities in this respect by deliberately limiting the service life through sometimes even rather clumsy technical measures such as the installation of meters that prevent further operation after a defined number of usage cycles, this appears legally questionable, is morally unjustifiable and contributes significantly to the burden on the natural environment.

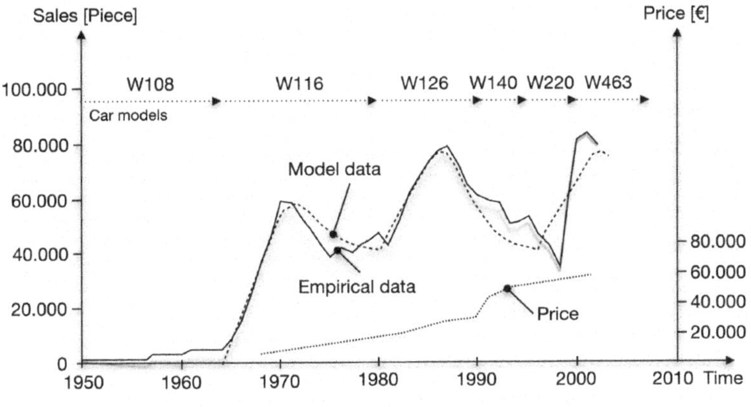

Fig. 3.13 Empirical sales figures for the Mercedes-Benz S-Class compared to model data and entry prices (Kaldasch 2013, p. 15)

The models and their comparison with empirical data have made it clear that trivializing representations of the system life cycle such as those in Fig. 2.60 do not do justice to the complexity of the actual phenomena. Models such as those mentioned here offer much more potential for a not only theoretical, but also practice-oriented understanding of the influencing variables and their effects. Once again, it should be noted that the aim is not to reproduce reality by deriving quantifying sales forecasts from it, for example. The familiar statistical forecasting methods discussed in Section 4.2.4.11 are better suited to this. Here, on the other hand, the focus is on the use of models for the purpose explained in Section 2.2, in particular using the model-based knowledge process shown in Fig. 2.5: Models make it possible to arrive at insights that allow for more competent action in practice than would be possible without the apparent detour via modeling.

3.1.3 Model of the Development Phase

The SI model is a similar approach for modeling the development phase. This model is the simplest of the models used to try to understand epidemics and pandemics such as the COVID-19 pandemic. However, it is not chosen here because it is simple, but because it can best be applied to modeling the system life cycle.

The S stands for suspectible individuals, i.e. healthy individuals, and the I for infectious individuals, i.e. individuals who can infect other individuals as a result of their own illness. For the intended use case here, however, these interpretations must be reinterpreted. For this purpose, $S(t)$ is used to describe the total number of work packages to be completed in a development project, i.e. the time required to complete them. $I(t)$ stands for the work packages that have already been identified and are infectious in that they trigger the processing of subsequent work packages. The sum of the work packages that have not yet been completed and those that have already been completed is constant and corresponds to the total number of work packages N to be completed: $S(t) + I(t) = N = const$. So if the number of completed work packages $I(t)$ increases, the number of uncompleted work packages $S(t)$ decreases to the same extent.

Similar to the rate of change k used above, the succession rate is used here, which is also abbreviated to k for reasons of consistency, while in epidemiology it is usually referred to as the infection rate and in biology as the replication or reproduction rate and is then abbreviated to r. The numerical example may help with understanding, according to which a completed work package triggers two subsequent work packages, so that three work packages are then completed if the two most recently triggered work packages each trigger two work packages after their completion, so that seven work packages are completed after their completion, four of which each trigger two new work packages, and so on. Mathematically, this is expressed by $2t - 1$ or $e^{kt} - 1$ with $k = ln(2) = 0.693$.

A linear functional response with a growth term c, which is similar to the law of mass action known from chemistry, can be formulated as a solution approach:

$$\frac{dS}{dt} = -c \cdot I(t) \cdot S(t) \tag{3.14}$$

$$\frac{dI}{dt} = c \cdot I(t) \cdot S(t) \tag{3.15}$$

The product $I(t) \cdot S(t)$ gives the number of constants if all work packages that have not yet been started can in principle be triggered by any work package that has already been completed. This is associated with the model-like simplifying assumption that content dependencies do not entail any significant restrictions, which undoubtedly does not quite do justice to reality, but simplifies the calculations considerably and provides sufficiently meaningful results for a basic understanding. Up to this point, the factor c still determines the successor rate and is only replaced by the already introduced rate k later on. From the conservation law mentioned above, the following now follows

$$\frac{dI}{dt} = -c \cdot I(t) \cdot (N - I(t)) = c \cdot N \cdot I(t) \left(1 - \frac{I(t)}{N}\right) \tag{3.16}$$

The proportionality factor $c - N$ results from the initial exponential increase in the processed work packages. In this situation, $I(t) \ll N$, so that the differential equation 3.15 becomes

$$\frac{dI}{dt} = c \cdot N \cdot I(t) = k \cdot I(t) \tag{3.17}$$

This now contains the actual successor rate r, so that Eq. 3.17 is independent of the total number of work packages N. From this follows

$$\frac{dI}{dt} = c \cdot N \cdot I \cdot \left(1 - \frac{I}{N}\right) \tag{3.18}$$

If you write the two differential equations as follows:

$$\frac{dS}{dt} = -[k \cdot S(t)] \cdot I(t) \tag{3.19}$$

$$\frac{dI}{dt} = [k \cdot I(t)] \cdot S(t) \tag{3.20}$$

reveals a similarity to predator-prey systems, but without showing the oscillation behavior often associated with them.

The solution of the differential equation 3.17 is obtained as follows with the aid of relevant integral tables:

$$I(t) = \frac{N}{1 + \left(\frac{N}{I(0)} - 1\right) \cdot e^{-k(t-t_w)}} \tag{3.21}$$

In this use case, t_w maps the average processing time of a work package, which is calculated as follows:

$$t_w = \frac{ln\left(\frac{N}{I(0)} - 1\right)}{k} \tag{3.22}$$

The first time derivative of the cumulative number of completed work packages (Eq. 3.23) is shown in Fig. 3.14 and shows the expected course of an S-curve, which initially increases exponentially until a few remaining work packages drag on for a comparatively long time towards the end of development. This corresponds to common experience. Linked to this is the number of work packages processed per time, which is described by the first time derivative of Eq. 3.21 and is also shown in Fig. 3.14, albeit on a different scale. After starting with just a few work packages, their number increases rapidly and reaches its maximum around the middle of a development project, only to decrease exponentially again afterwards.

$$\frac{dI}{dt} = \frac{k \cdot N \cdot \left(\frac{N}{I(0)} - 1\right) \cdot e^{-kt}}{\left[1 + \left(\frac{N}{I(0)} - 1\right) \cdot e^{-kt}\right]^2} \tag{3.23}$$

The costs, which are often used in sketches of the system life cycle such as the example shown in Fig. 3.1, can be derived from this model as a good approximation proportional to the number of work packages. In the development phase as well as in the use phase, it becomes clear that the actual life cycles of products are likely to look very different from those that are usually published.

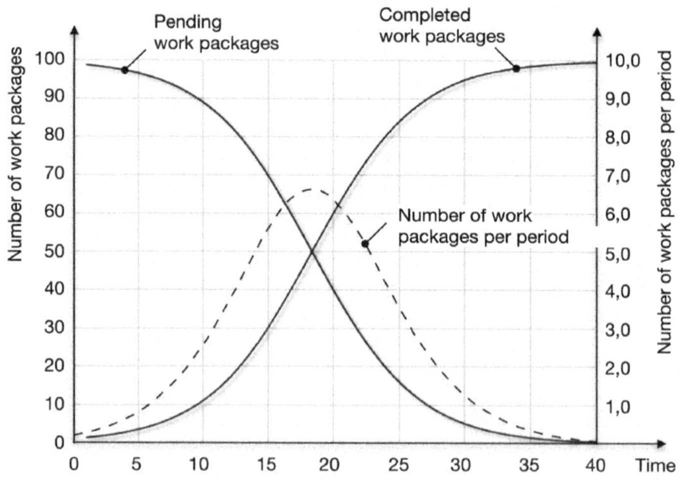

Fig. 3.14 Timing of the workload in development phases

3.2 Problems and Tasks

The following considerations lead from the cognitive science-based distinction between problems and tasks (e.g. Dewey 1997; Dörner 1979; Duncker 1974; Lüer and Spada 1990; Sell and Schimweg 2002) to the contrast (dichotomy) between projects and processes. In addition, this distinction also established the separation of *mental and manual work*, which goes back to Frederick Winslow Taylor (1919), and inspired the development of *artificial intelligence* with the differentiation between well-structured and poorly structured problems proposed by John McCarthy (Dörner 1979, p. 13). Finally, the distinction has also found its way into *design methodology* (Ehrlenspiel and Meerkamm 2013, p. 56 f.) and Project Management (Jakoby 2021, p. 15 ff.).

However, the insight into the necessity of differentiating between projects and processes is hindered by the fact that *process orientation* is a paradigm (Sect. 2.4) that Michael Hammer and James Champy (2003) established at the beginning of the 1990s and which still prevails today. Thilo Knuppertz and Uwe Feddern (2015, p. V), for example, associate it with an almost epochal development. For Karl Werner Wagner and Gerold Patzak (2015, p. VII), running a company without Process Management is no longer conceivable and Gerd Nanz (2012, p. 210) claims that Business Process Management is an integral part of the day-to-day work of most companies and organizations. The standard DIN EN ISO 9001:2015 for Quality Management, which is probably the most widely used worldwide, requires all companies certified according to it to have a process organization, and ISO/IEC/IEEE 15.288:2023 for Systems and Software Engineering also maps all activities that occur during the system life cycle as processes (Walden et al. 2023, p. 39 ff.).

This could be countered by the fact that increasing importance is also sometimes attached to project organization. For example, a HAYS study (Rump et al. 2010, p. 4) comes to the conclusion that traditional organizational structures and procedures with their formalized processes are increasingly proving to be an obstacle in view of current and expected developments in the economy. Project-based organizational structures, on the other hand, would bring much greater agility, with work firmly anchored in projects. However, such comparisons of process organization on the one hand and project organization on the other suggest that one can easily be replaced by the other, i.e. that both forms of organization are in principle suitable for dealing with all types of tasks or problems and only differ in their respective advantages and disadvantages. In addition, protagonists of Process Management simply equalize differences between the two forms of organization by also interpreting projects as processes.

- For example, Robert G. Cooper describes his widely known process model for product development as a stage-gate process (2010, p. 146).
- Walter Jakoby explains that every project is a special problem-solving process (2021, p. 1).
- According to Manfred Burghard, the process characterizes the actual procedure in the project (2013, p. 19).
- The international standard DIN ISO 21500:2016-02 on Project Management states that Project Management is realized through processes, and the national series of standards DIN 69901 even dedicates a separate sheet 2 to the underlying Process Management.

In fact, a look at reliable sources unclouded by the prevailing paradigm reveals that Project and Process Management differ significantly:

- Michael Gaitanides (2012, p. 35), for example, explicitly limits Process Management to processing procedures that are accessible to routine. This is to be distinguished from the implementation of one-off projects.[3] Only activities with a certain repetitive character should therefore be subject to process organizational design considerations.
- Gerold Patzak and Günther Rattay (2014, p. 19) describe processes as a frequently repeated, rather sequential chain of activities, whereby the initial situation and the desired result are defined and the necessary measures are categorized or specified. There are only insignificant uncertainties with regard to the achievement of objectives. In contrast, projects are a one-off, parallel and sequential network of activities, whereby the initial situation is defined, the desired result is specified and the necessary measures are still completely open in some cases, so that there is significant uncertainty regarding the achievement of objectives.
- Matthias Geirhos (2017, p. 15) emphasizes that projects stand out from routine activities and that a key characteristic of projects is a project-specific organization.
- The latter also emerges from DIN standard 69.901-5:2009-01 (p. 11) for Project Management, in which project-specific organization is named as a characteristic feature of projects.

In view of this mixed situation, a return to the basics seems instructive. As already announced, *cognitive science insights* will be used here. These take behavior as their starting point (Fig. 3.15), which becomes action through an orientation towards a goal (e.g. Dörner 1976, p. 10 ff.; Funke 2003, p. 17 ff.). A goal, in turn, describes the difference between a given situation and a desired situation. Cognitive scientists refer to this difference as a barrier. If the means to overcome this *barrier* are known and available, they speak of tasks; if the means are unknown or unavailable, they are problems (Dörner 1979, p. 19; Duncker 1974, p. 1; Ehrlenspiel and Meerkamm 2013, p. 56 f.; Funcke 2003, p. 25; Lüer and Spada 1990, p. 256; Naefe 2018, p. 19).

Fig. 3.16 illustrates the relationship between problems and tasks in the notation of an Event-driven Process Chain (EPC) (Sect. 2.4.5.2). First of all, every non-trivial barrier represents a problem when it is first overcome. The attempt to solve a problem can lead to three events:

- The attempt may fail. As long as the problem is perceived as such, this leads to another attempt to solve it.
- The problem can be solved once and for all. In this case, there is nothing more to do.
- Although the problem can be solved as such by acquiring the necessary knowledge and all the necessary resources, it often happens that further situations arise in which the barrier has to be overcome. This is then no longer a problem, but a task. After all, overcoming the barrier can be done routinely. In this respect, the relationship between problems and tasks can be described as *complementary*. They relate to the same concern, namely the transformation of an actual state into a deviating target state, but are based on fundamentally different situations and cannot be reduced to one another in this respect.

[3] The latter is a tautology, as projects are by definition unique.

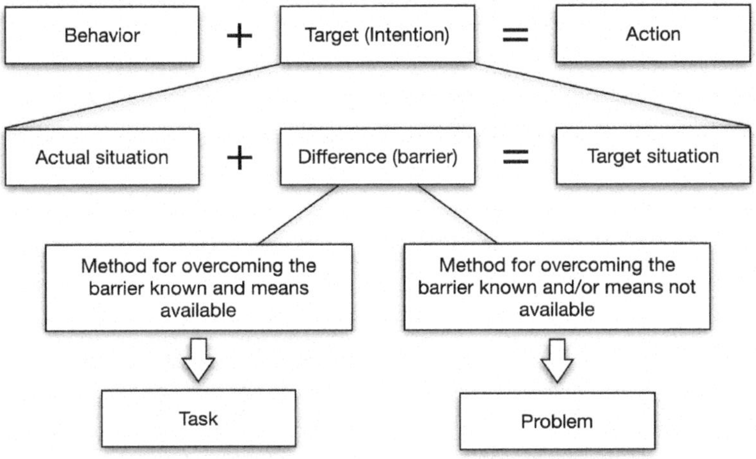

Fig. 3.15 Differentiating between problems and tasks

At the same time, both categories are always relevant for coping with life, so that they complement each other in this respect, i.e. they cannot do without each other.

The fact that the two categories are based on fundamentally different situations and cannot be reduced to one another in this respect can be described as a *dichotomy*. This can be seen, among other things, in the dynamics of the respective approach: While this is cyclical and in this respect discontinuous (discrete) when solving problems, the routine and thus always recurring completion of tasks is continuous (analogous). The latter leads to a steady state that always delivers the same or at least sufficiently similar results, while the result of one problem solution differs significantly from that of any other problem solution. The steady state is therefore contrasted with a genesis, a change. These aspects, which have only been introduced here, will be discussed in more detail below.

3.2.1 Continuity versus Discontinuity

The distinction between the continuity sought in routine activities and the discontinuity required to solve one-off activities has already been outlined above, but will be dealt with in more detail below in order to justify the necessity of this distinction not only by listing the relevant sources, but also in terms of content.

One reason for interruptions in the workflow are checks of interim results. The number and duration of such checks increase with increasing uncertainty (Patzak and Rattay 2014, p. 19). In the course of routine activities, they can be kept within reasonable limits and integrated into the workflow to such an extent that they do not significantly impair quasi-continuous processing. Up to this level, one can speak of a continuity of value-adding processing, as aimed for by Process Management. For example, by applying the poka-yoke principle, error avoidance or detection through technical measures (Liker 2016, p. 196),

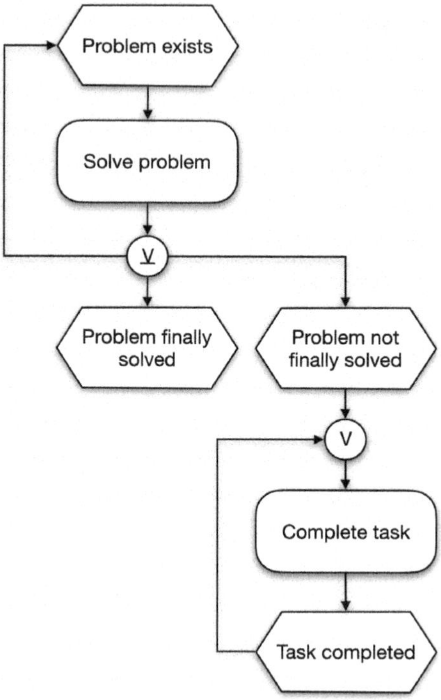

Fig. 3.16 Relationship between problems and tasks

possibly supplemented by worker self-inspection, the inspection effort in many areas of industrial series and mass production can be minimized to a very large extent. This fulfills a prerequisite for flow production, which in turn can be considered a prime example of a process in the sense intended here. To put it simply, one can say that Interruptions are avoidable.

When working on one-off projects, uncertainty is naturally much greater than when carrying out routine activities. Audits therefore usually extend not only to interim results, but also, for example, to such far-reaching questions as whether the continuation of the project is possible at all. Furthermore, the audits themselves often have to be developed with the necessary methods and tools. Finally, many more stakeholders are usually involved in such audits than in routine audits. As a result, audits usually lead to significant interruptions in the course of projects (so-called milestones), which can even result in the entire project being terminated. To put it simply, one can say that Interruptions are obligatory.

3.2.2 Degree of Automation

In view of the frequent repetition of the same or at least similar activities over a long period of time, investments in measures to identify, design, implement and maintain rules for handling business processes through to automation, for example with the help of *Workflow Management Systems* or *Business Process Management Systems*, pay for themselves (Brocke and Rosemann 2015). Protagonists of Process Management, such as Michael Gaitanides (2012, p. 7), are critical of the associated information technology Taylorism, but cannot avoid admitting that the benefits cannot be ruled out in individual cases. Alfred Kieser (2006, p. 99) confirms that such management doctrines, which he characterizes as simple, are indeed widespread today.

One-off projects resist automation of their processing because they are usually too different from one another, so that corresponding measures can only be amortized in some areas at best. Although computer programs have long been offered to support Project Management, their consistent use usually falls short of promises or expectations.

Ultimately, there is only a small overlap between project and Process Management, in which processes are also possible and useful in projects. For example, invoices also have to be paid in projects, which firstly takes place in a similar way in most companies and secondly involves several work processes such as checking for factual accuracy by a specialist department, checking for arithmetical correctness by the accounting department, approval by a superior department, posting, payment instructions and archiving and is usually carried out based on a division of labor. The processing of vacation requests or the handling of business trips also fall into this category of processes, which are not managed any differently in projects than in routine organization. In such cases, projects can often even fall back on existing processes from the routine organization. In this respect, it is questionable whether such processes still need to be referred to as project processes.

The fact that the paradigmatic pressure to ultimately identify everything as a process that seems even rudimentarily suitable for this purpose can clearly overshoot the mark is shown, for example, by DIN standard 69.901-2:2009-01 on Project Management, when it even describes the holding of a final meeting as a process (p. 48). It is true that time also passes during such a meeting, and there is no doubt that various things have to be done from the presentation, discussion and documentation of findings to the publication of lessons learned. However, it is doubtful whether this really means what the protagonists of process organization originally imagined it to mean.

3.2.3 Sequence versus Circularity

Due to the objective of designing processes consistently, these are usually modeled as consecutive activities. Examples such as the spaghetti diagram (Fig. 2.26) have already been introduced in Section 2.4.5.2.

The situation is completely different in projects. For example, the Systems Engineering process model for problem solving (Haberfellner et al. 2018, p. 72) explicitly sees itself as a cycle (Ahrens 2020, p. 85). What should be an exception in Process Management becomes the rule here. This can be explained by the significantly greater uncertainty involved in solving one-off projects compared to completing routine tasks. The need to take the unforeseen into account and to return to previous phases of problem solving or even abort the project is unavoidable when dealing with previously unknown challenges.

3.2.4 Consistent Approach versus Consistent Results

The diametricality of project and process also becomes clear when you realize what their respective regularity is. Routine tasks should always lead to structurally identical results, i.e. inquiries to offers and ultimately to orders, vacation and travel requests to approvals or rejections, the assembly of gearboxes to functioning gearboxes and so on. Here, the regularity refers to the respective results. The procedures required in each case, on the other hand, are always specific: the process for assembling a gearbox is fundamentally different from the preparation of a quotation. To put it simply, the results of processes are always largely the same, but the procedure is always different.

With projects, the situation is exactly the opposite: while the results of problem solutions differ significantly, the approach is always roughly the same for every project (Haberfellner et al. 2018): The micro-cycle is always the problem-solving cycle, regardless of whether it is labeled as such or otherwise, or whether it has been specified with regard to certain areas of application. Furthermore, macrology follows the life cycle of the design object from planning through implementation and operation to disposal, and there is also a whole series of design principles, such as an approach from rough to detailed (Sect. 3.1.2). All in all, the system for handling projects is of course too complex to be explained here in even the slightest detail, but the aspects mentioned should be sufficient to make it clear that it can be simplified: The approach to projects is always largely the same, but the results vary from project to project.

3.2.5 Complementarity of Project and Process

If we summarize the insights gained so far, we come to the conclusion that project and process are complementary in the sense of the word: on the one hand, they are opposed to each other and cannot be reduced to each other (dichotomous), and on the other hand, every organization is dependent on both forms of organization. In this respect, they complement each other. In addition, both forms of organization can be substantiated by systems theory:

- In the 1940s, Norbert Wiener (1961) founded *cybernetics*, a sub-discipline of systems theory that deals primarily with homeostatic processes. This initially purely technical approach was subsequently adapted by very different scientific disciplines, including business administration. For example, Anthony Stafford Beer (1981) developed his well-known *Viable Systems Model* (VSM) on this basis in the early 1950s (Fig. 3.17), which inspired, among other things, the *St. Gallen Management Model* (Ulrich 2001) and Fredmund Malik's *Management Cybernetics* (2015, p. 24), which are still widely known and recognized today.
- At around the same time that cybernetics was being developed with regard to homeostatic processes, Systems Engineering, which provides a process model as well as methods and techniques for the design (genesis) of complex systems, emerged in large organizations such as Bell Laboratories (Ropohl 2009, p. 74; Walden et al. 2023, p. 3 ff.) or the National Aeronautics and Space Administration (NASA). Project Management is closely linked to this (Haberfellner et al. 2018, p. 31).

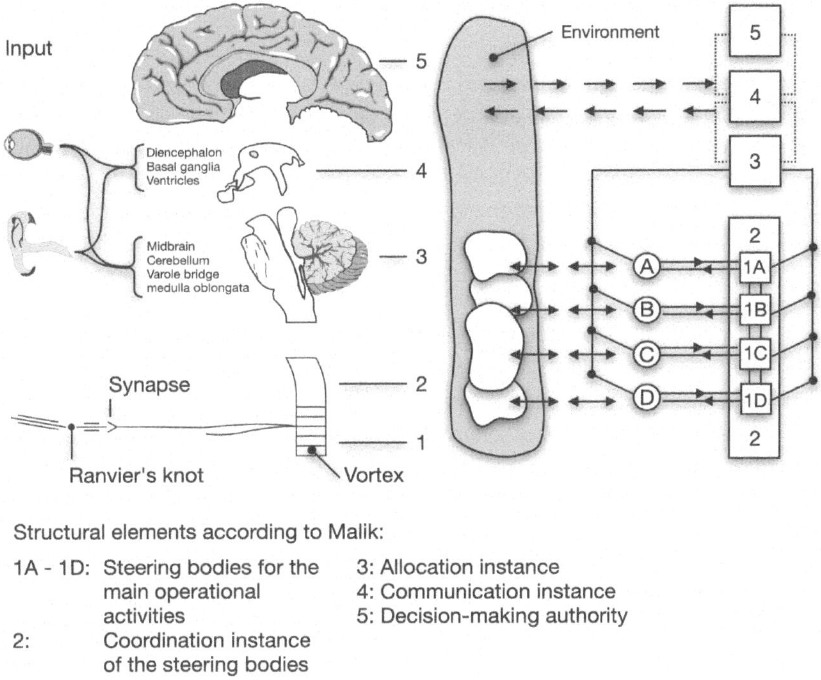

Fig. 3.17 Viable Systems Model (Malik 2015, S. 24)

Against this background, the two *generic process models* for solving problems (Sect. 3.3) and for completing (routine) tasks (Sect. 3.4) are introduced in the following two sections, but initially only outlined in an overview in order to bring them into a manageable context with the previous explanations.

3.3 L 1: Development Phase

An initial, still rough breakdown of the life cycle has shown that three main phases can be distinguished almost by nature: the (L 1) development phase, the (L 2) utilization phase and the (L 3) return phase. These phases are separated from each other by four transitions: (L 0 - L 1) birth, (L 1 - L 2) commissioning (SOP), (L 2 - L 3) decommissioning and (L 3 - L 0) death or (L 3 - L 1) retreading. Further phases can be added to these natural phases. In the development phase in particular, it is common to include further phases depending on the scope and complexity of the project, for example an initial preliminary study, a subsequent concept, design, prototyping and so on.

3.3.1 From the Rough to the Detail

The division of the development phase into further phases usually follows the simple logic of proceeding from the rough to the detailed (Haberfellner et al. 2018, p. 54 f.), which is also referred to as the *top-down principle*. From a generic perspective, it is not very important how many phases are planned and how they are designated in detail. This depends on numerous specific factors, including the type of project, the temporal (duration) and spatial extent, traditions of certain industries and the personal tastes of individual participants.

Walden et al. (2023, p. 27) provide an overview that includes the phase classification of the international standard ISO/IEC/IEEE 15.288:2023, that of the US Department of Defense and NASA. This is supplemented by our own proposal, which is based on the I CAN/CSA-ISO/IEC/IEEE 24.748-1:2019-06-01 standard. This diversity raises doubts as to whether universality can be claimed in this respect, which is an essential prerequisite for the presentation of models that claim to be generic. Therefore, no further proposal for a phase classification will be made here. Instead, the following information should be understood as examples. What remains generally valid is the procedure from the rough to the detailed.

In the humanities and social sciences, this approach finds its equivalent in *hermeneutics* (Gadamer 2010). For a long time, and still today, the scientific and epistemological discussion about suitable methods was and is strongly influenced by scientific and now also technical disciplines. This is attributed to the visible success expressed above all in technical progress. In the 19th century, however, a counter-movement developed in the humanities, initiated primarily by Friedrich Schleiermacher and Wilhelm Dilthey, which methodologically distanced itself from the natural sciences and technology and emphasized their fundamental differences. Among other things, this gave rise to the method of hermeneutics as a doctrine of interpersonal understanding.

The difference between the natural sciences and the humanities is usually emphasized in the statement that the natural sciences are about explaining, whereas the humanities are about understanding. While scientific and therefore also technical explanations are based on laws, hermeneuticists believe that explanations in the humanities cannot be traced back to laws. This initially resulted in the model of the *hermeneutic circle* (Ast 1808, p. 179 f.). Wolfgang Stegmüller (1974, p. 64), however, rightly pointed out that a circular movement does not model progress, and alternatively proposed the form of the *spiral*, in which the changing radius represents the progress of knowledge. What has remained is the question of whether the circle is traversed from the inside out or from the outside in; both can be found in the literature. In this case, the outward-to-inward progression is preferred in order to show that although understanding begins at different starting points, it subsequently converges to a more or less common understanding. Furthermore, the *Fibonacci spiral* (Lausch 2009) is chosen as the model here, as it has proven to be particularly good for illustrating natural phenomena, which also include cycles of human knowledge acquisition. In addition, this spiral is scale-invariant and therefore suitable for capturing complexity in the sense intended here (Sect. 2.4.2).

Against this background, Fig. 3.18[4] shows that the starting point (1.1) is characterized by a prior understanding. For example, when it comes to understanding cultural expressions such as literary texts or works of art, such a pre-understanding can consist, among other things, of already having information about the author and the circumstances of their work (epoch, social conditions, living conditions, etc.). The initial examination of the object under consideration (1.2) creates an initial understanding, which in a third step (1.3) is compared with the already existing understanding of the world and classified within it. The further developed understanding of the world is internalized in a fourth step (1.4). With regard to the object under consideration, this results (2.1) in a new, expanded pre-understanding. A renewed examination of this (2.2) leads to an expanded understanding, which in turn (2.3) is compared with the already developed understanding of the world. The now further developed understanding of the world is internalized a second time (2.4) and leads (3.1) to an even deeper pre-understanding of the object under consideration, and so on.

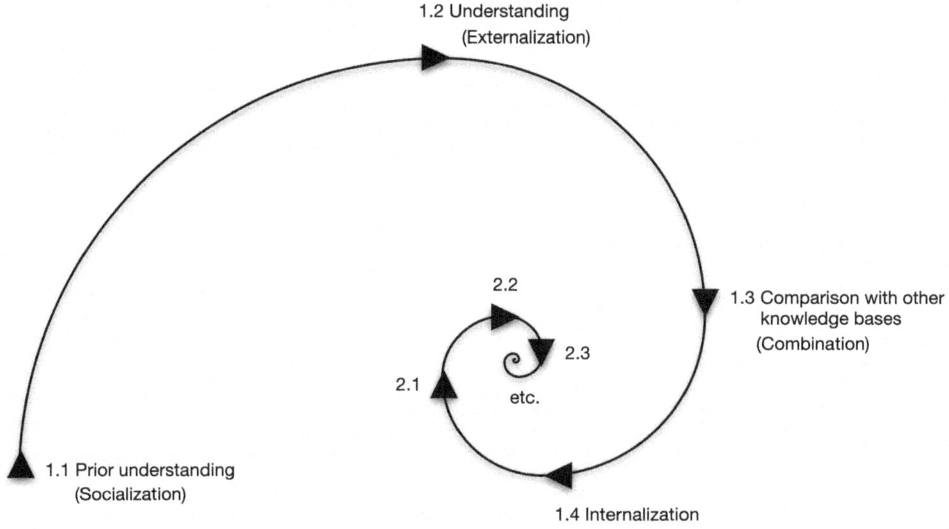

Fig. 3.18 Hermeneutic spiral

However, the convergence of the common understanding at one point in the core of the spiral is in conflict with the constructivism on which this is based (Sect. 2.1). In order to resolve this conflict, Leonard Bachman (2007, p. 10) interprets the center of the spiral, albeit with a certain amount of tongue-in-cheek, as a *strange attractor*, i.e. as a location in a phase space that describes the final state of a dynamic process. Such attractors are *fractals* (Mandelbrot 1987) that cannot be described geometrically in a closed form. An attractive

[4] In order to ensure compatibility with the knowledge spiral shown in Fig. 3.33, the phases of socialization, externalization, combination and internalization differentiated there are already added in brackets in Fig. 3.19.

feature of such an interpretation is that systems that converge in this way react sensitively to small differences in the initial situation and are therefore not subject to strict *causality*. However, no concrete attractor can be calculated for the present epistemological case, as is quite possible for other systems (Peitgen et al. 1994). In this respect, only a metaphorical interpretation is possible here.

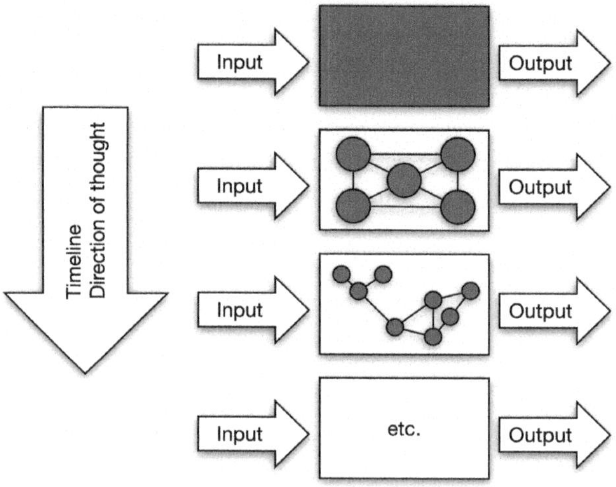

Fig. 3.19 Approach from rough to detailed (Haberfellner et al. 2018, p. 54)

Another variant of the procedure from rough to detailed is based on a so-called *black box* (Fig. 3.19), developed from gray boxes, whose shade of gray becomes lighter from step to step until finally the so-called *white box* is available (Haberfellner et al. 2018, p. 55). This means that the first step is primarily about narrowing down the relevant system and examining its relationship to its environment. In the subsequent steps, sub-relationships or sub-systems and elements and their relationships are then included in increasing detail. However, the widespread form of representation shown in Fig. 3.19 can be misunderstood to mean that the approach from the rough to the detailed is based on linear thinking. Fig. 3.20 should therefore make it clear that, as with the hermeneutic circle (Fig. 3.18), the object being observed is the same each time a new loop of knowledge is passed through, but provides increasingly detailed insights. However, the basic constructivist understanding on which this is based must also be taken into account here, which means that any observation of reality, no matter how detailed and exact, remains model-like and hides the vast majority of actual reality.

Factory planning (e.g. Wiendahl 2014, p. 230 ff.) provides an illustrative example of the procedure from the rough to the detailed. It begins with planning the associated *concept*. This is a rough draft of the project, which essentially derives the initial situation and the resulting potentials and deficits (purpose of the project). At this early stage, the underlying information is often only based on estimates and qualitative statements. This is followed by a rough outline of the idea and the means to be used to achieve the goal of tapping the potential or eliminating the deficits. In the context of factory planning, for example,

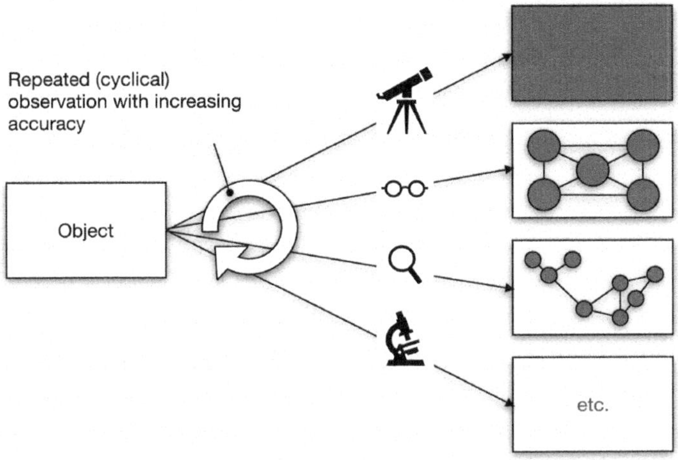

Fig. 3.20 A cyclical approach from the general to the specific

the initial situation may be characterized by a significant increase in the sales volume of manufactured products, from which the goal is derived to counter the resulting lack of space in the existing factory by building a new factory hall. For this purpose, sketches are often made, often by hand, to develop a rough idea of how the project can be implemented spatially.

At the next detailed stage, which is referred to here as *rough planning*, the data for analyzing the initial situation is already collected quantitatively. For example, a representative production program and the associated parts descriptions are extracted from an existing Merchandise Management System in order to derive the capacity requirements (Wiendahl 2014, p. 236). The available capacity is determined from the machine descriptions and the production potential, so that the future demand for personnel, operating resources and space can be calculated by comparing these two analysis results. On this basis, a rough layout is developed for the new factory building, which shows the production areas in blocks and indicates their logistical links (ibid., p. 237). Simple computer programs are often used to produce drawings.

In the course of *detailed planning* in the subsequent third planning phase, everything is planned in as much detail as is necessary for the subsequent realization. Drawings are now usually produced using CAD software and contain, among other things, the layout of the individual machines, including the necessary connections to electricity, compressed air, water and so on. Furthermore, the workstations must be designed according to ergonomic aspects (Bokranz and Landau 2012). The trade association regulations on occupational health and safety must be observed (Kohte et al. 2017) and environmental protection aspects must be taken into account (Förtsch and Meinholz 2018).

Detailed planning is usually followed by *implementation planning*, which involves making all the necessary preparations for the implementation of the plan. This includes, for example, submitting applications to the authorities and concluding contracts with contractors.

Finally, the plan is put into practice in a separate *realization phase*. In the example of factory planning, the factory hall is built, the machines are installed, the connections are made and so on. The subsequent commissioning can be part of the realization phase, but can also be planned as an independent *commissioning phase*. In any case, it includes trial operation. In addition, the acceptance of trades by the responsible authorities, such as the employers' liability insurance association, can take a long time, meaning that it cannot be carried out at the exact time of the start of production (SOP).

Formal transitions (milestones, decision thresholds, gates, control points, etc.) are usually provided between all phases to check whether and, if so, to what extent the results of the previous phase appear suitable to provide a promising basis for the following phase. Since developments in the cognitive science sense are problem solutions (Sell and Schimweg 2002), where it is not known from the outset whether there is a solution at all and, if there is a solution, what it looks like, an examination may well lead to the conclusion that a continuation of the project does not appear to be sufficiently promising. In such cases, it may be necessary to revert to previous phases and even abandon the project.

In practice, the possibility of backtracking or abandoning a project is associated with serious problems, as such a necessity is often not justified by the project, but attributed to the supposed incompetence of those responsible. For understandable reasons, they are therefore keen to bring projects to a straightforward conclusion at all costs. Instead of an end with horror, however, one is then confronted with a horror without end. One solution to this problem may lie in dealing openly with this conflict of objectives and avoiding apportioning blame (Liker 2016, p. 358 ff.).

A strictly sequential, quasi-waterfall-like progression through development phases often leads to a long development process. In addition, in later phases, such as the phase in which the production process required to manufacture the previously developed product is developed, knowledge is gained that could have been incorporated in previous phases in order to improve the overall result. For example, certain product features may be irrelevant to the product itself, so that they are not initially considered by product developers. However, such features can result in considerable simplifications for production, for example through *Poka Yoke* (Niemann et al. 2021, p. 63 ff.). However, this often only becomes apparent during the development of the production process. However, incorporating such features into the products can be time-consuming or even impossible.

Against this background, *simultaneous engineering* (also known as concurrent engineering) aims to partially parallelize development phases (Bullinger and Warschat 2013, p. 15 ff.). Fig. 3.21 shows that this can be useful not only for the product and production process development phases, but also for other phases. The disadvantage, however, is that this can lead to additional work, as not all the information required for the subsequent phases is available due to the fact that the preceding phase has not yet been completed. This can result in the so-called *simultaneous engineering paradox*, according to which the individual phases can be extended while the overall development time is shortened at the same time.

An essential component of cross-phase collaboration is *design for production and assembly* (Bender and Gericke 2021, p. 704 ff.). The MTM Association, for example, offers PROKON, a method and a supporting computer program to support assembly-oriented design (Bokranz and Landau 2012, vol. 2, p. 74 ff.; Schlick et al. 2018, p. 594).

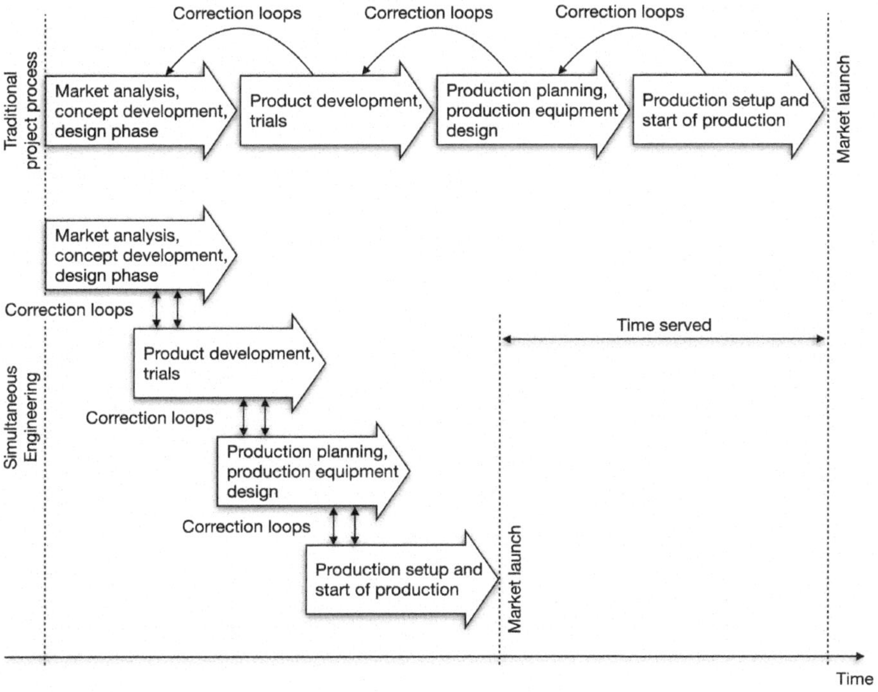

Fig. 3.21 Simultaneous Engineering (Holger Weihe)

This involves identifying influencing variables for individual parts and assemblies during the design phase in order to derive assembly difficulties. Quantifying these difficulties allows a statement to be made about the suitability for assembly and provides starting points for their improvement during product development.

Furthermore, *Product Lifecycle Management* (PLM) supports the parallelization of development phases (Scheer et al. 2006) by allowing systematic management of all product-relevant data during the entire system life cycle with the help of appropriate computer programs (Fig. 3.22). In practice, however, it is not always easy to link such computer programs with other programs such as *Manufacturing Execution Systems* (MES) or *Enterprise Resource Planning* (ERP) as well as with programs from the context of the *digital factory* (VDI 4499 Sheet 1:2008-02).

Finally, cross-phase collaboration is also supported by using the *Systems Modeling Language* (SysML) (Alt 2012; Weilkiens 2014).

With the help of graphical representations, technical systems can be modeled in an object-oriented way. So-called diagrams offer application-specific views of the overall model (Sect. 2.4.6). For example, structure diagrams such as the block definition diagram (Fig. 2.18) or the package diagram allow the structure of a system to be defined, analyzed and tested. A requirements diagram not only makes it possible to manage all the requirements of a system, but also to relate the requirements to each other and to the model elements. Behavioral diagrams such as the sequence diagram (Fig. 2.46) are used to integrate the

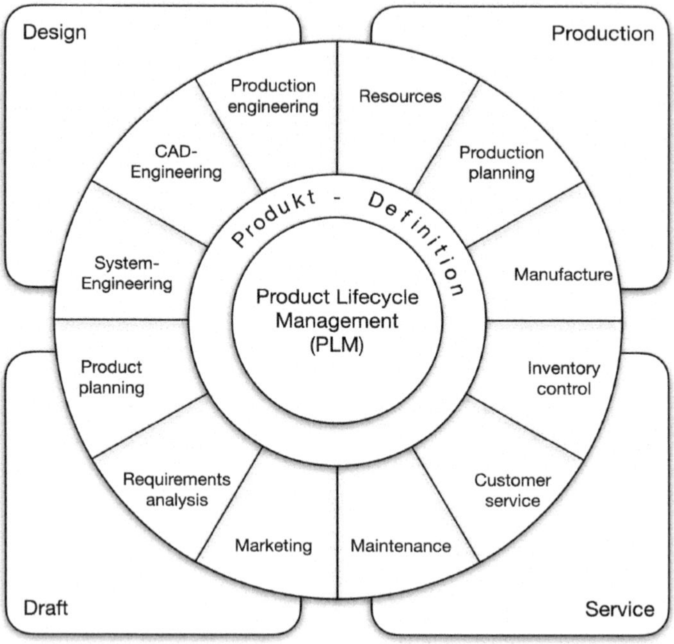

Fig. 3.22 Product Lifecycle Management

focal system into a higher-level context. This allows different stakeholder groups to make specific contributions in different phases of system design and control on the basis of a common, consistent database.

3.3.2 Problem-solving Cycle

Systems Engineering follows the generic process model of the problem-solving cycle shown in Fig. 3.23 (Haberfellner et al. 2018, p. 71 ff.; Thommen and Achleitner 2006, p. 45 f.). This cycle goes through 6 phases (Ahrens 2020, p. 85 ff.), which are numbered consecutively here for the sake of simplicity. In addition, the numbers are preceded by the letter P for problem-solving cycle in order to distinguish them from other process models (L for product life cycle, R for control loop, etc.).

(P 1) The initiation phase begins the first time through the problem-solving cycle with the perception of a problem and ends with the decision to tackle the solution to this problem. In subsequent runs of the problem-solving cycle, the perception of the problem is replaced by the decision to continue the project that has been started in the course of a procedure from rough to detailed (Sect. 3.1.2) in a further level of detail. In detail, the perception of a problem results from the difference between an actual state and a target state (Fig. 3.1). However, problems are defined by the fact that the difference (barrier) cannot be

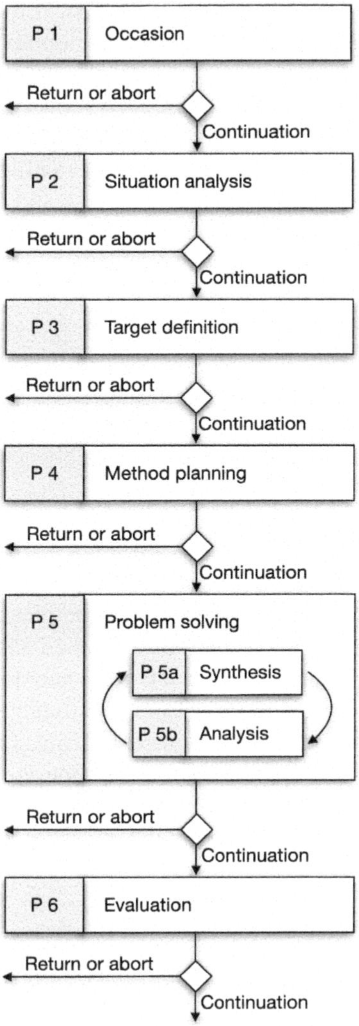

Fig. 3.23 Problem-solving cycle (Ahrens 2020, p. 85; Haberfellner et al. 2018, p. 70)

identified with sufficient accuracy straight away and subsequently eliminated immediately. In a professional context, this requires a systematic, methodically supported approach.

(P 2) Against this background, the situation analysis[5] serves to systematically examine the current situation. This leads to a sufficiently comprehensive and detailed knowledge of the initial situation, in particular to identify the opportunities and/or risks or deficits that have led to the situation being perceived as a problem, as well as the company's own strengths and weaknesses. In this phase, it is also important to identify the requirements

[5] The names of the phases may vary depending on the respective area of application. For example, situation analysis is referred to as diagnosis in medicine and as situation assessment in the military.

of all stakeholder groups (stakeholders, interested parties) for the problem solution to be developed and to forecast which solution-relevant developments are to be expected.

(P 3) In the light of the now transparent initial situation, the target state is defined in the course of formulating the objective. As Fig. 3.14 shows, this results logically from the knowledge of the actual situation and the barrier. Furthermore, the target formulation phase includes an examination of its achievability. If there are conflicting objectives, this can also include finding a compromise or setting priorities. The result of this phase is, in particular, the criteria that can be used in the further course of the problem-solving process to assess whether the interim solutions and, in particular, the final solution meet the requirements. Tests are defined for this purpose.

(P 4) Now that the barrier between the actual and target state is known, the next phase is to identify and make available the methods and tools with which the barrier is to be overcome. The results of this phase include, in particular, determining how the tests are to be carried out, whether and to what extent the requirements for the interim solutions and the final solution are met. The sum of all reasonably selected methods results in the methodology.

The two phases of problem solving (P 5) and solution evaluation (P 6) will now be explained in context. For this purpose, *systematic search strategies* outlined by Reinhard Haberfellner et al. (ibid., p. 253 ff.), partly with reference to Horst Wilhelm Jakob Rittel (1992), are chosen as a starting point. For the sake of completeness, the trial and error strategy should be mentioned here, which is certainly justified and will therefore be discussed in more detail later (Sect. 4.4.5.2), but will be left aside here for the time being. Instead, the following considerations will be based on the first-fit strategy visualized in Fig. 3.24. The authors mentioned describe this strategy as non-optimizing. This is contradicted by the fact that this strategy does optimize by examining several options over several levels of detail. However, the *first best solution* (first fit) is selected at each level. For this to be classified as the optimum, one condition must be met: Time pressure. It is therefore mainly used in *exceptional situations*, the management of which is explained in detail in Section 3.7.

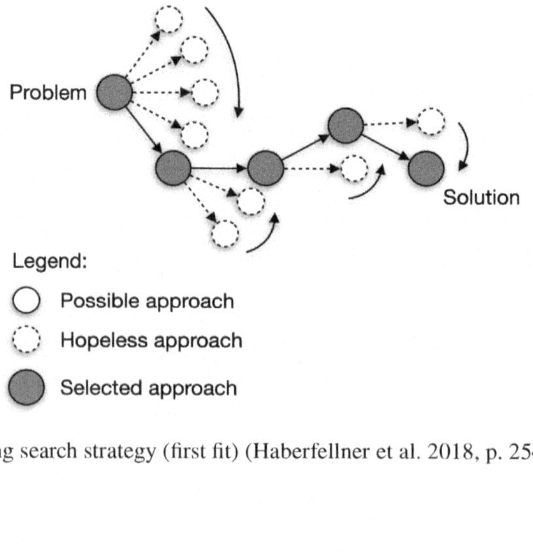

Fig. 3.24 Non-optimizing search strategy (first fit) (Haberfellner et al. 2018, p. 254)

A more detailed consideration of the search strategy now reveals the interaction of problem solving (P 5) and solution evaluation (P 6). To do this, however, it is first necessary to divide the problem-solving phase into two phases, synthesis (P 5a) and analysis (P 5b). While synthesis serves to generate creative and therefore constructive solution ideas, these are checked for suitability in the course of critical and therefore destructive analysis. Reinhard Haberfellner et al. (ibid., p. 245 ff.) justify this with the principle of deferred judgment (ibid., p. 75; Sell and Schimweg 2002, p. 77; Traut-Mattausch and Kerschreiter 2018, p. 274). This is based on the insight that with almost every idea of how something could work, an idea is immediately born as to why it probably won't work after all. As a result, many ideas are lost that may prove to be worthwhile upon closer examination. However, as this type of analysis is usually intuitive, it can neither be avoided nor guided intuitively. This requires a conscious and methodically supported approach and the methods for supporting creative work, which are explained in detail in Section 4.4.4.2, also explicitly contribute to this.

Furthermore, the distinction between verification and validation already introduced in Section 2.2 is necessary for understanding the further explanations. To reiterate, it should be briefly explained once again that verification serves to check whether a solution approach works in itself, i.e. whether, to put it simply, 1 + 1 really make 2. Validation, on the other hand, is intended to clarify whether a solution that is fundamentally functional is desirable in the light of the interests of all those involved and affected (stakeholders) and in view of alternative possible solutions. With the help of this distinction, verification can now be assigned to analysis (P 5b), while validation takes place in the course of evaluation (P 6). Fig. 3.25 thus visualizes the overall context: Solution ideas are developed in the course of synthesis (P 5a). In this phase, it is assumed that every idea can be realized in some way. For this reason, all arrows and circles in the diagram are shown with solid lines. The analysis (P 5b) is then used to test the ideas for their suitability[6]. Since Fig. 3.25 is based on Fig. 3.24 and thus a search strategy that aims for the first suitable solution, the y-analysis ends with this variant in this case; all others are now represented by dashed lines and thus as hopeless approaches. Only with this search strategy is there nothing left for the evaluation but to choose the first best suitable variant. The search strategies explained below will show that the evaluation normally selects the relatively best variant from several suitable variants, using the objectives formulated in phase P 2 as criteria. In analogy to mathematics, one could say in this respect that the bracket that is opened with the formulation of objectives (P 2) is closed again with the evaluation (P 6).

The explanation of the evaluation already anticipates what is explained in detail in Section 4.4.5.2, namely that it only really makes sense if there are several ways of proceeding. The simplest possibility is to evaluate whether a solution found is better than the previous solution (actual state). As this is a comparison of exactly two possibilities, it is also referred to as an alternative. As explained in detail in Section 4.4.5.2, in terms of optimization it is generally more expedient to develop several options, from which the relatively best option can then be selected. If there are more than two options, this is no longer referred to as an alternative, but as a solution variant.

[6] The distinction between synthesis and analysis was adopted from Haberfellner et al. (2018, p. 71 ff.). Of course, strictly speaking, the complete problem-solving cycle is always followed.

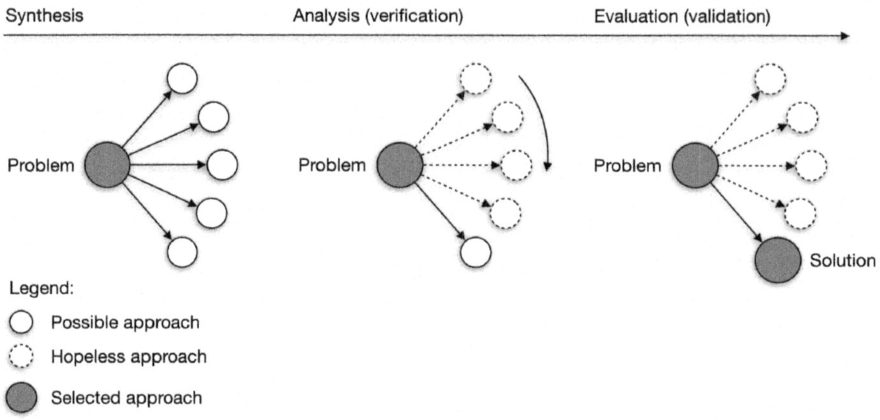

Fig. 3.25 Sequence of synthesis, analysis (verification) and evaluation (validation)

Such variants are present in the *one-step optimizing search strategy* visualized in Fig. 3.26. It is therefore obvious that this figure is basically the same as Fig. 4.52, which is used to explain variant planning in more detail. At each level of detail, several fundamentally possible solution variants are developed in order to select the relatively most promising variant through a comparative evaluation, which is further developed at the next level of detail. This also results in new variants and so on. However, the search is virtually doomed to success at each stage, as there is no way back. This can help to limit the development effort, but carries the risk of suboptimal solutions.

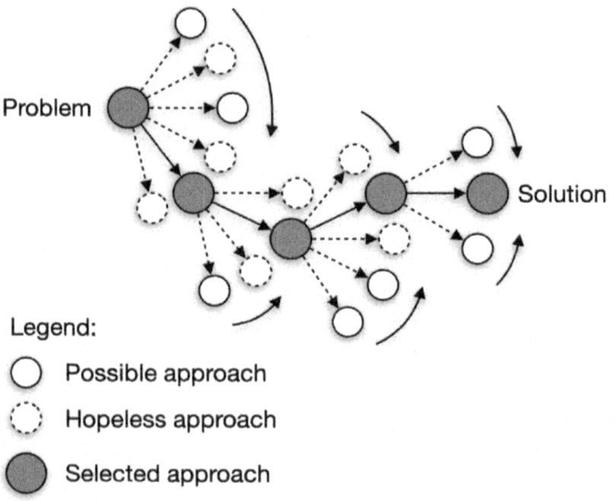

Fig. 3.26 One-step optimizing search strategy (Haberfellner et al. 2018, p. 255)

To minimize this risk, the *cyclical search strategy* shown in Fig. 3.27 is the best way to proceed, as it corresponds best with the cyclical approach to problem solving recommended here. If an initially chosen solution path turns out to be a dead end later on, a return to an earlier stage is made in order to take a different path from there. Although this potentially increases the development effort and the associated uncertainty, it is more likely to lead to better solutions than the search strategies mentioned above.

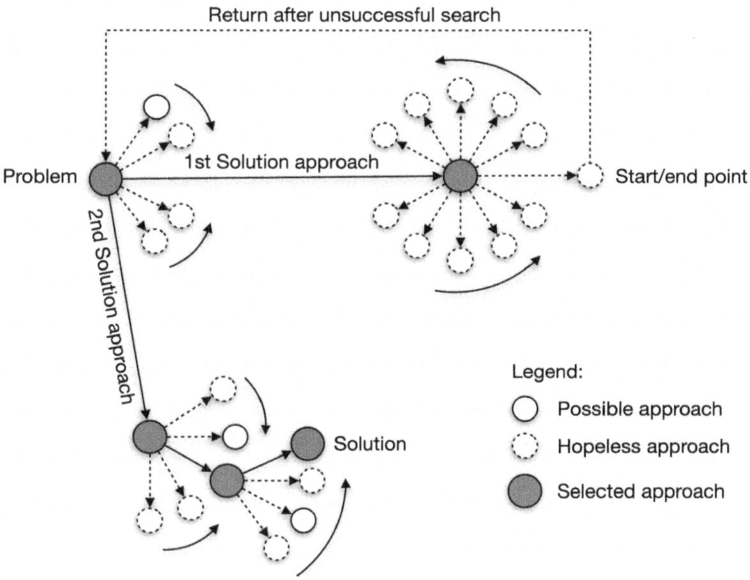

Fig. 3.27 Cyclical search strategy (Haberfellner et al. 2018, p. 256)

With the help of the findings now available, it can be clarified why synthesis (P 5a) and analysis (P 5b) are combined into the solution development phase (P 5), while solution evaluation (P 6) forms a separate phase at the top level of the problem-solving cycle. Two reasons can be given for this: (1) The principle of deferred judgment was given as the reason for the strict separation of synthesis and analysis. In this respect, this is also correct and important. On the other hand, the two phases naturally belong very closely together and not every test that results in the conclusion that a solution approach cannot be implemented in this way must immediately result in a return to early phases of the superordinate problem-solving cycle, especially as this cycle, as will be shown in Section 4.1, is self-similar anyway and is run through in a similar way on a small scale as on a large scale (Fig. 4.1). In fact, solution attempts and tests, whose findings lead to new, often only slightly modified solution attempts, usually alternate in quick succession. The strict separation of synthesis and analysis is so important precisely because they are in danger of merging and mixing with each other. If this were not observed, the two would probably be perceived as one. However, this could be to the detriment of creative solutions, i.e. limit the courage to give solutions that are initially perceived as outlandish, unusual or unpromising a chance to prove themselves.

(2) A second reason for classifying the analysis (P 5b) differently from the evaluation (P 6) is that an analysis does not require a decision, but an evaluation certainly does. As already explained above, the analysis serves to verify, i.e. to check whether – to put it very simply once again – 1 + 1 actually add up to 2. This cannot and does not have to be decided; it can be calculated, and calculations are the domain of specialists (experts). Assessments also include value judgments, and in this respect the term is quite aptly chosen. Value judgments should not be left to experts alone, even if they are of course just as entitled as others to have interests, a legal opinion and a moral concept beyond their expertise, on the basis of which they can and may also evaluate facts. Beyond this, however, all stakeholder groups must be involved in assessments. The choice they have is the result of the problem-solving process and experts play a key role in this from a technical perspective, but the final choice, the validation, is the responsibility of all those involved and affected. However, since even such a choice cannot usually eliminate all uncertainties, a decision must ultimately be made, and a decision only makes sense if there is something to decide, i.e. if there are several options, each of which is feasible in principle, but only one of which can and should be implemented. Decisions are also made not only on the basis of technical considerations, but also and above all on the basis of political calculations (Sect. 2.6.2.3). Therefore, the evaluation phase (P 6) leads to a decision and an analysis (P 5a) does not.

There are phase transitions between the phases and at the end of the last phase, which are also referred to here as *milestones* in line with Project Management. These are intended interruptions that serve to check whether the results expected up to the previous phase have been achieved (Gessler and Kaestner 2011, p. 353). If this is not sufficiently the case, a decision must be made as to whether to return to one of the previous phases in the expectation that a satisfactory result can be achieved by going through the corresponding phases again or whether the project should be discontinued for the time being or permanently.

Such setbacks are common in the course of solving problems (Haberfellner et al. 2018, p. 153 ff.; Wirdemann 2017, p. 26), as by definition not all the necessary knowledge and/or not all the necessary resources are available to overcome the associated barrier. In the context of agile methods such as Scrum, the need for so-called *empirical management* is derived from this (Wirdemann 2017, p. 46 f.). Beate Bender and Kilian Gericke (2021, p. 39) refer to this with reference to other authors and in agreement with them as the *co-evolution of problem and solution*. When trying to solve a problem and thus overcome the corresponding barrier, new problems usually arise that initially remained unrecognized or were not given their actual importance. This results in a new, usually more detailed description of the problem, which leads to new attempts to solve it, which in turn lead to new problems and so on. This inevitably means that the phases of the problem-solving cycle have to be run through again and again until a satisfactory solution is finally found. It is the rule rather than the exception that such a solution does not always meet all requirements to the complete satisfaction of all stakeholders, because not all problems can be fully solved, especially not those problems that were only identified during the development of the solution. This is in the nature of things.

In practice, too little attention is paid to this insight. Setbacks or discontinuations are often justified by a lack of competence on the part of those involved in solving the problem. This in turn leads to those responsible pursuing solutions to problems at all costs and often trying to cover up failures. Instead of an end with a scare, the result is a scare without an end. In fact, in many cases it would be more economical to terminate projects as soon as the lack of prospects for success is recognized or to deliberately prolong them by jumping backwards in the face of unsatisfactory interim results. In view of the usual expectations, however, this requires a high degree of rationality on the part of all those involved and courage on the part of those who take such initiatives.

In this respect, conventional projects can learn from *agile approaches* such as Scrum. They emphasize the *cyclical nature* of the approach. A so-called *sprint*, which corresponds to the problem-solving phase P 5 of the problem-solving cycle (Fig. 3.22), lasts a maximum of 30 days. This is followed by an evaluation of the development progress achieved up to that point (Wirdemann 2017, p. 20 ff.), which corresponds to evaluation phase P 6 of the problem-solving cycle. In addition, so-called *daily scrums* are carried out: daily runs of feedback loops that reveal problems and coordination requirements at short notice. *Timeboxing* is used to adjust the work content if necessary rather than the completion dates (ibid., p. 33). If changes are required after the completion of a sprint or a daily scrum, these are incorporated into subsequent sprints and daily scrums. Due to the short cycle of the sprints and the daily feedback loops, the *transparency* of the *development progress* is maintained just as well as the *flexibility*.

In line with its name, the problem-solving cycle also provides for a cyclical approach that includes regressions and even complete abortions. And there is no reason anchored in Systems Engineering not to run through the cycles just as quickly, flexibly and transparently as is propagated in agile approaches. In this respect, there is therefore no reason why the cyclical approach should not be advocated more aggressively in SE.

This is all the more true as SE is superior to agile approaches in many other respects. For example, Joachim Goll and Daniel Hommel (2015, p. 82), two protagonists of Scrum, name clear limits to agile approaches. In general, these are only suitable for small projects[7] and, above all, only for those that are not safety-relevant. Ursula Kusay-Merkle (2021) only deals with agile Project Management for small and medium-sized projects from the outset. However, since most projects, including many small and medium-sized ones, include safety-relevant aspects, the scope of application of agile methods generally seems manageable. In addition, SE offers a complete set of instruments for projects of all types and sizes, from which the set of instruments required for a specific project can be compiled and configured (tailoring). According to Joachim Goll and Daniel Hommel (2015, p. 79), this is not always the case with agile approaches such as Scrum. They explain that the associated bottom-up approach does not always lead to success, without naming the examples they cite to back up their statement. What they criticize in particular is the lack of an initial idea of the overall system to be developed. In fact, agile approaches assume that this arises more or less automatically from the incrementally developed partial solutions (fragments). This deficiency should be alleviated by the fact that the participants should at least develop a shared vision of the desired result at the start of a design project.

In this respect, too, SE has a more comprehensive set of instruments to offer, in particular through a methodically supported situation analysis (P 2), through an objective derived from this (P 3) and through an explicit plan of how the actual state can and should be

[7] The statement by the two authors mentioned above that agile approaches are suitable for complex tasks (actually problems) seems to contradict this, at least if one can assume that small projects do not usually deal with problems that could be described as exceptionally complex.

transformed into the target state (P 4: Methodology, in particular work program). Against this background, what Joachim Goll and Daniel Hommel (2015, p. 73) say about the benefits of agile approaches is certainly true: although agility is not precisely defined, it sharpens the senses for flexibility in the project and good cooperation with the customer.

However, from an SE perspective, it should also be critically noted that agile approaches give the customer a primary role, while SE attaches no less importance to customers, but at the same time also has all other stakeholder groups in mind, because their approval is just as crucial for the success of projects. Even the rejection of a single stakeholder group is enough to cause a project to fail (Sect. 4.2.4.6).

Finally, the statement found in this context that agile approaches help to reduce bureaucracy also appears questionable (ibid.), as many projects are carried out precisely in order to create bureaucracy, for example in the form of business processes or technologies such as software that are to be routinely used in the course of such processes. Reducing bureaucracy would therefore mean removing the results of one's own work from circulation immediately afterwards. This in turn would be an admission of poor work. In fact, reducing bureaucracy is not an end in itself and does not make sense across the board, as regular operations are not possible without bureaucracy. It must therefore be a declared goal of design projects to develop good bureaucracy and to dose it correctly to avoid bureaucracy.

3.4 L 2: Utilization Phase

In the utilization phase, the aim is to maintain a steady state (homeostasis) for as long as possible. *Control loops*, which act as bureaucracy in organizations, serve this purpose.

Fig. 3.28 first shows the model of a technical control loop. In it, the reference variable $w(t)$ specifies how the output variable $x(t)$ should behave: $w(t) = x(t)$. A time-invariant reference variable $w(t) = const$ is also referred to as the setpoint or setpoint state. The controlled system F_S has the task of producing the output variable $x(t)$. This is not always possible due to disturbances $z(t)$. Although disturbances can be triggered at many points and affect all parts of the control loop, all disturbances are summarized in the disturbance element F_Z of the control loop for simplicity's sake. The output variable $x(t)$, which may be disturbed, is continuously recorded by a measuring element F_M and compared with the reference variable $w(t)$ at the reference junction as a feedback variable $r(t)$, possibly in a processed form. The resulting control difference $e(t) = w(t) - r(t)$ must then be minimized by the controller F_R using suitable measures. The manipulated variable $y(t)$ serves as the output variable of the controller for this purpose.

As control technology is mathematically quite demanding (e.g. Zacher and Reuter 2017, p. 15 ff.), the control loop will only be illustrated here using a simple example. This involves controlling the temperature of a living room (ibid., p. 3 f.). The room forms the controlled system F_S. The setpoint, a specific temperature $w(t) = const$, is set using the setpoint setting of a radiator thermostatic valve (thermostat for short) in accordance with DIN EN 215-1. Disturbances $z(t)$ that cause a deviation from the actual room temperature $x(t)$ can be, for example, windows and doors that are opened and closed. A sensor (measuring element F_M) built into the thermostat continuously measures the room temperature $x(t)$.

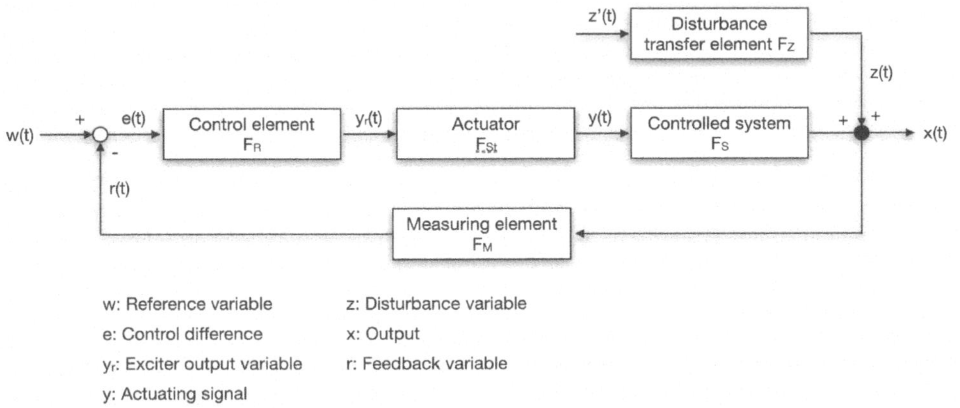

Fig. 3.28 Technological model of the control loop

Wax, liquid or gas is usually used for this purpose, which expands as the temperature increases (ibid., p. 9). If the room temperature increases due to disturbing influences, for example, the expanding medium in a closed cylinder designed as a metal bellows causes longitudinal expansion, which is transferred to a pin. The pin presses on the valve, causing it to close. As the room temperature decreases, the volume of the medium in the cylinder is reduced so that a spring can push the pin back again, closing the valve. In this way, the volume flow (flow rate) of the hot water flowing through the radiator (radiator, convector) is varied. This in turn influences the room temperature x(t) with the aim of automatically maintaining a flow equilibrium $x(t) \approx const$ (homeostasis).

Since this is not only about the design of technical systems, but also about the organizational process model for carrying out routine activities (tasks) as the basis of Systems Management, the model of the technical control loop in Fig. 3.29 is now shown as a management control loop. The organizational structure is usually structured hierarchically (Sect. 6.1.1). For this reason, the technical model of the control loop (Fig. 3.28) has been rotated by 90 degrees to represent the management control loop (Fig. 3.29), so that the control variable $w(t)$ now comes from the hierarchically superior position at the top. Middle management assumes the role of the controller F_R by taking measures on the basis of the most clearly defined scope of action possible in order to steer the control loop F_S, i.e. the value-creating area, in such a way that it fulfills the expectations placed on it in terms of productivity, quality and so on. The results (products, services, rights) are continuously monitored. To this end, organizations have various institutions that are generally oriented towards the stakeholder groups whose interests must be taken into account. For example, business controlling serves to protect the interests of investors, Quality Management (e.g. DIN EN ISO 9001:2015) serves the interests of customers, occupational Health and Safety Management (DIN EN ISO 45001:2018) serves the interests of employees and Environmental Protection Management (DIN EN ISO 14001:2015) to the natural environment.

From an engineering perspective, this model is compatible due to its mechanistic nature and is therefore initially justified. From a psychological and social science perspective, however, it quickly reaches its limits, because as effective as hierarchy is in coping with complexity (Sect. 6.1), it quickly fails when systems become larger. In fact, in the *AGIL scheme* (Fig. 2.57), corporate management is assigned to the political part-system (G-system), while middle management and the value-adding area are assigned to the economic part-system (A-system). They therefore follow different logics by using different semantic codes. Controlling, which is mentioned here as an abbreviation for all the other control institutions mentioned above, has the function of mediating between the political and economic spheres of action, i.e. ensuring structural coupling. However, explaining this in detail here would take us too far away from the line of thought pursued here. It will therefore be discussed in more detail in Section 6.1, so that the mechanistic argumentation can be developed further below.

The design of control loops follows the process model of the problem-solving cycle (Sect. 3.3.2). Therefore, systems engineers must not only master this system of their own approach, but also thinking in control loops, as this must always be an essential part of their problem solving when the result of their work according to Fig. 3.15 serves the routine execution of tasks (Fig. 3.29) or automation (Fig. 3.28) from an organizational point of view.

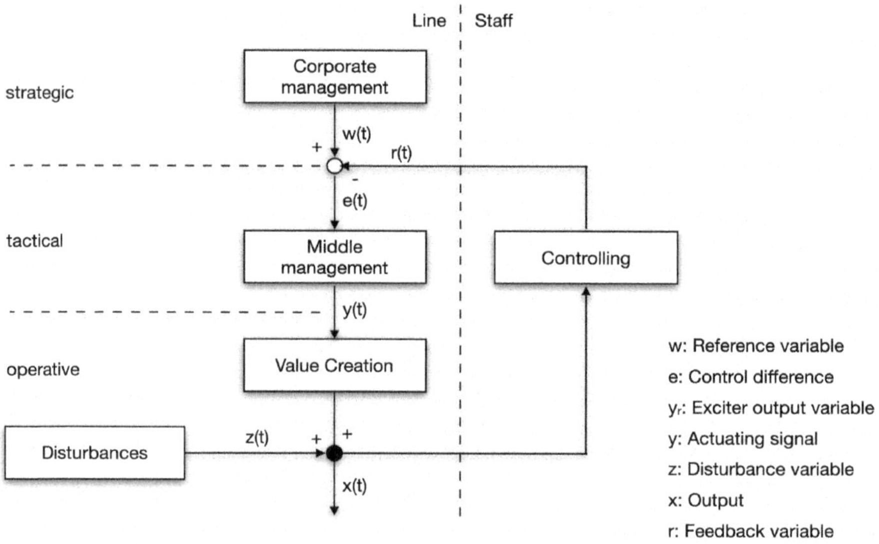

Fig. 3.29 Management control loop

In the case of technical control loops, the physical or chemical behaviour of the systems to be controlled (controlled system FS) must first be formulated mathematically (ibid., p. 51 ff.) in order to then bring them into a calculable relationship with the mathematical relationships of the control devices that are also to be technically implemented (ibid., p. 99 ff.). This is the subject of *engineering systems theory*, which focuses on electrical engineering (e.g. Rennert and Bundschuh 2013). Particular challenges include ensuring the stability of control loops (Zacher and Reuter 2017, p. 167 ff.). In particular, the Hurwitz and Nyquist criteria must be fulfilled. Furthermore, control loops should exhibit good control behavior if and insofar as the control variable is variable over time: $w(t) = const.$ This means that the limits within which a control loop fulfills its function should not be set too narrowly. Today, comprehensive and largely mature models, methods and tools are available for all aspects of this type. Nevertheless, the design of technical control loops is and remains a very challenging problem.

The requirements placed on technical control loops, such as management behaviour and stability, also apply in a figurative sense to organizational control loops (Fig. 3.29). However, due to human influences, organizational aspects can only be modelled mathematically to a very limited extent. Instead, sociological and psychological findings must be used as a basis.

Paul Anton Watzlawick et al. (2013, p. 32) show that such approaches can also be based on mathematical logic, at least as a model, by developing their psychological analysis of the relationship between continuity (regular operation) and change (design, Change Management) on the basis of *group theory* (Glosauer 2016) in order to understand continuity, and by using *logical type theory* (Whitehead and Russell 1910-1913) to identify the possibilities and limits of change (Sect. 2.4.4). For this reason, the findings derived from group theory on the existence of systems are first reproduced below and harmonized with the control loop, while the explanation of the possibility of change is initially only included to the extent that it appears necessary to understand the existence, in order to then examine it in more detail in the following Section on Change Management.

As already explained in Section 2.4.4, we can speak of groups if the elements of a system fulfill certain conditions (axioms). The first condition mentioned was the *closed nature*, which results from the fact that the combination of two elements always produces an element of the same group (Glosauer 2016, p. 21). The Bible quotes Jesus as saying: A prophet is nowhere less valid than in his fatherland and at home with his own (Mark 6:4). In practice, the inability of a system to bring about changes from its own specific mode of operation that affect the system as a whole, which is derived from the seclusion of groups, can be observed, following a comparable example by Paul Anton Watzlawick et al. (2013, p. 39), in the so far unsuccessful effort to reduce the very large number of MPs in the German Bundestag compared to other countries. Corresponding laws would have to be approved by a majority of those MPs, many of whom would lose their seats as a result. Practical experiences such as these are confirmed in theory by the *incompleteness theorems* (Hoffmann 2012), which Kurt Friedrich Gödel formulated on the basis of the Principia Mathematica by Alfred North Whitehead and Bertrand Russell (1910-1913) and which prove that mathematics cannot formally prove the validity of its statements by itself. This would require meta-mathematics.

The second condition stated in Section 2.4.4 is the validity of the *associative law*, according to which the order of execution of the corresponding operations is not important when linking elements (Glosauer 2016, p. 21). Paul Anton Watzlawick et al. (2013, p. 41) give the example of a problem that is commonly referred to as the *chicken and egg problem*: the arms race. The circular nature of the interaction between hostile states or communities of states often makes it practically impossible to decide whether a particular action a is the cause or the consequence of an action by the other side. [...] In its *own view*, each party naturally perceives its own actions as caused and provoked by the actions of the other side.

Seen from the *outside* and in the overall context, however, every action by one party proves to be the trigger for a reaction by the other, and this reaction then becomes the cause of what the first side merely regards as its reaction. Although the participants change through their continuous adaptation to what they observe in the other party, changes of this kind are described by Paul Anton Watzlawick et al. (ibid., p. 33) as first-order changes that do not change the system as a whole. As in a so-called *vicious circle*, the opponents cannot break out of the spiral of conflict. This always requires an external intervention, what Paul Anton Watzlawick et al. (ibid.) call a second-order change. In this assessment, they see themselves confirmed by William Ross Ashby (1956), whom they (Watzlawick et al. 2013, p. 31) quote as saying: 'So you can see that the word change can mean two very different things. There is the change from one (internal) state to another..., which corresponds to the behavior of the machine due to its internal dynamics, and there is the change from transformation to transformation..., which is a *change in its overall behavior* and which is brought about arbitrarily by the experimenter or some other external factor. This distinction is fundamental and must not be neglected under any circumstances (Ashby 1956, p. 43).

With regard to Change Management, which in the sense of Paul Anton Watzlawick et al. (2013, p. 33) aims at second-order changes and in the sense of Helmut Willke (2005) requires interventions, another phenomenon appears relevant that concerns so-called group dynamics (social systems) and therefore also and especially has a stabilizing effect in organizations. Based on an example given by Paul Anton Watzlawick et al. (2013, p. 41), the following case is often observed in a similar way: a member of a working group regularly contributes innovative ideas, but is just as regularly met with rejection by one of his colleagues. However, if the always skeptical colleague suddenly slips into the role of innovator, it is not uncommon for the normally innovative colleague to switch to the role of skeptic, so that a change of the first order occurs, while the system as a whole retains its stability. Both before and after the role reversal, the combination of the two colleagues' modes of operation results in the *unit element*.

Paul Anton Watzlawick et al. (2013, p. 43) derive the insight from the fact that in groups, each element produces the unit element by linking with its *inverse element* (Glosauer 2016, p. 21) that all knowledge results from the perception of opposites. They quote Lao Tzu (1921, chapter 2): If on earth all recognize the beautiful as beautiful, the ugly is thereby already set. If everyone on earth recognizes the good as good, then the non-good is already established. This gives rise to the question, which is relevant for marketing among other things, of whether names are derived from the named or whether the named first develops its concrete effectiveness through its naming. Paul Anton Watzlawick et al. (2013, p. 43 f.) explain this using the example of an incident during the Chinese Cultural Revolution: the Red Guards replaced all public signs and inscriptions on streets, buildings, stores and so on that referred to the bourgeois past with names that conformed to the revolution. What they probably did not realize, however, was that this measure was in line with a very old basic rule of Chinese culture, which Confucius called the rectification of names. The related assumption, which can also be found in George Orwell's novel 1984, in which harmful terms of everyday language are replaced by a new language, is that language shapes thought and thus also perception. Against this background, however, Paul Anton Watzlawick et al. (ibid.) only recognize first-order changes in the actions of the Red Guards, as they gave new validity to the old Chinese culture instead of revolutionizing it in the sense of a second-order change.

The insights developed so far from a psychological perspective are confirmed by systemic sociology. At this level, it is the contexts of meaning that are constituted by social systems (Luhmann 1996, p. 92 ff.; Willke 2005, p. 50) and subsequently have a stabilizing effect. If, for example, a department in a manufacturing company has found its purpose in carrying out the routine activities assigned to it without errors, i.e. without deviations from the specified standard (reference variable $w(t)$), it will not give this up easily. Fritz B. Simon (2015, p. 28) explains this by saying that the purpose of life, including that of an organization or an organizational unit (department), is simply to live. This life is threatened by attacks on the constitutive context of meaning, and social systems know how to defend themselves against this – often with great success, even under the conditions of hierarchically structured power.

3.5 L 3: Change Management

The considerations in the following Section will be preceded by three core statements, which will form the basis for the rest of the discussion. (1) As the reasons for the first core statement are too diverse to be listed here in any detail, and as the reasons should also be immediately obvious to everyone, general experience is used to point out the need for systems to be changed from time to time, for example in the course of *continuous improvement* (CIP, Kaizen), as envisaged in the concept of Lean Production (Toyota Production System (TPS)) (e.g. Belz et al. 2015, p. 50 ff.). This refers to the second-order changes referred to in the previous Section (Watzlawick et al. 2013, p. 48 ff.), which affect the overall behavior of the corresponding system (Ashby 1956, p. 43).

(2) The second core statement picks up on the insight derived in the previous Section from group theory (Glosauer 2016) that systems cannot bring about second-order changes on their own, but are dependent on external interventions (Watzlawick et al. 2013, p. 33). This is explained in more detail in Section 3.5.1.

(3) The view derived in Section 2.1 and modeled in Section 2.6.2 that psycho- and socio-technical systems are operationally closed and create openness to external irritations according to their own conditions and rules that are not directly visible from the outside stands in the way of this (e.g. Willke 2013, p. 24 f.). Systems organize themselves. The stabilizing effect of first-order changes leads to systems defending themselves against second-order changes in the sense of control loops. Attempts of this kind are perceived as disturbances and eliminated by means of the control loop. The better the Systems Engineering has designed a control loop, the more difficult it is for designers to change this control loop during operation. This is particularly the case when it comes to changing its own substance. But even if changes take place in its environment, for example by the elimination or addition of other control loops or by their fundamental change, the focal control loop reacts to this and its reaction will be negative if its own mode of operation is substantially affected. Against this background, Section 3.5.2 will deal with the question of how second-order changes are nevertheless possible.

3.5.1 Second-order Changes

The following considerations tie in with the second core statement (see above) and thus also with the explanations in Section 3.4, which, analogous to group theory (Glosauer 2016), have led to the insight that systems stabilize themselves in regular operation and cannot bring about changes to their fundamental mode of operation on their own. Although first-order changes are constantly occurring within the systems, these contribute to stabilization rather than changing the system as a whole.

From the perspective of a focal system, second-order changes are therefore always seen as something involuntary, even incomprehensible, a quantum leap, a sudden enlightenment that occurs unexpectedly after a long and often discouraging mental effort, sometimes like a dream, sometimes almost as an act of grace in the theological sense. In his book Der göttliche Funke (The Divine Spark), Koestler compiled an encyclopaedic collection of examples of this phenomenon and introduced the term *bisociation* for it. He understands it as the sudden flash of insight that shows a familiar situation or event in a different light and evokes a new attitude towards it[8] (Watzlawick et al. 2013, p. 49).

However, this only applies within the focal system. Second-order changes can only be categorized if you step outside the system and look at it from the outside. Paul Anton Watzlawick et al. (2013, p. 50 f.) illustrate this using the nine-point problem, which is also used in the cognitive sciences to investigate human problem-solving behavior (Sell and Schimweg 2002, p. 31). The nine dots shown on the left in Fig. 3.30 must be connected by four straight, contiguous lines without lifting the pencil. Observations repeatedly show that many test subjects fail to solve this problem because they make the unfounded assumption that the solution can only be found within the area spanned by the dots. In fact, however, such a condition is not part of the problem at all. All attempts to solve the problem under this self-supplemented condition only result in first-order changes. A second-order change and, in this respect, the solution to the problem requires stepping out of the assumed boundaries of the system, because as Fig. 3.30 in the right-hand section shows, the lines must be drawn beyond them. The ability to step out of the original frame of reference and adopt new perspectives is an essential prerequisite for *innovation*, especially in the context of Systems Engineering.

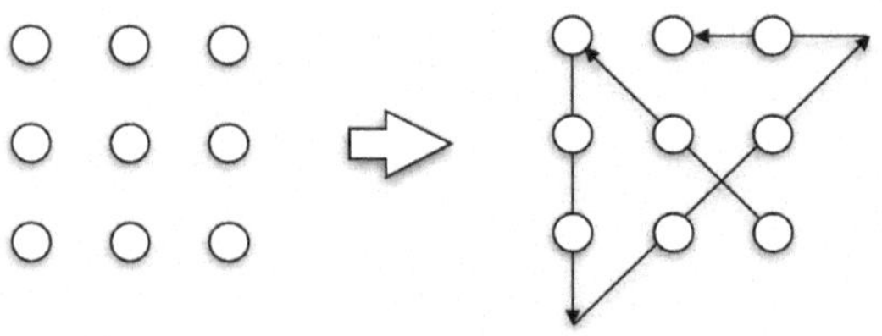

Fig. 3.30 Nine-point problem (Sell and Schimweg 2002, p. 31)

[8] Kostler 1966, p. 36.

In terms of group theory, a second-order change can be explained as a change in the linking of the elements (Watzlawick et al. 2013, p. 50). It follows from the logical theory of types that you cannot talk about a set (class, system) in the language of its elements (ibid., p. 54). Psychologists working in therapy know this and devote a considerable amount of their attention to preventing themselves from being taken over by the system to be treated, because they would no longer be able to help from the moment this happened. Conversely, it is part of the subconscious or unconscious repertoire of systems in need of therapy to continually and often with remarkable skill attempt to draw the therapist into the logic of their system in order to maintain their own stability, however pathological this may be.

The logical type theory assumes a hierarchy of logical types with regard to the effect of second-order changes (Watzlawick et al. 2013, p. 28). On this basis, a change of this magnitude always starts from a higher level of logical abstraction, i.e. from a meta-level. However, it would be too simplistic to equate this hierarchy with the chain of command in organizations, which is usually also organized hierarchically.

Bruno Lauer (2019, p. 158) provides an example of such an error with the widespread view that initiatives for change come from the leadership of an organization and, following the hierarchical structure, should be implemented in a cascade from top to bottom. Nevertheless, he recognizes what he sees as an apparent contradiction with the equally widespread demand that all those involved in and affected by change should be fully involved as early as possible. As a solution, he suggests publishing at least one plan from the outset, from which everyone can see that they will be involved and from when and in what form. This can succeed if the decisions made in the early phases leave enough leeway for those who will only be involved in later phases and if they are kept informed at all times.

From a systems theory perspective, however, the orientation towards hierarchical chains of command appears to have limited effectiveness at best. It is true that power-conscious people will attach importance to their respective positions of power being respected. In this respect, such an orientation cannot be completely avoided. However, the boundaries of social sub- and part-systems are different and, for the reasons outlined in Section 2.6, overcoming these boundaries is generally far more difficult than safeguarding the power interests of managers derived from their respective status. For example, a department in a company is made up of a head of department (management personnel) and employees. Together they form a social system. It is therefore not only important that the management personnel find ways to come to terms with change and make constructive contributions to it. What is also and above all important is that the social systems such as departments, project groups and so on cooperate constructively within their system boundaries. This can lead to difficulties, for example, if the system boundaries themselves or affiliations to systems are to be changed, because social systems are, as mentioned, generally very stable and highly interested in maintaining boundaries, affiliations and identity-forming structures and processes (e.g. tasks, responsibilities).

Even a manager will not easily be able to avoid this within the system they manage. This can result in conflicts of interest that cannot simply be resolved. If a manager introduces the idea of change into the system from the outside, there is a risk of encountering resistance within the system and losing acceptance. Conversely, if a manager advocates the desire for stability outside of the social system, for example within the management circle, there is a risk of the same thing happening there. Ultimately, there is a risk of sitting between two stools, so to speak. In fact, influencing individual people is not enough (Watzlawick et al. 2013, p. 48 ff.; Willke 2005, p. 51), because the order of a social system is emergent. As stated, this means that interventions must affect a system as a whole (ibid.). The solutions offered by systems theory are not simple, but may be somewhat more effective than conventional solutions, whose effects are so limited that authors who propagate such solutions estimate their failure rate at around 75 % (e.g. Lauer 2019, p. 49 ff.). However, *systemic intervention theory* in particular recognizes the inherent limits of its effectiveness. Helmut Willke (2005, p. 4) comments: What can be achieved in a first step is [...] a central prerequisite for a revised practice of intervention: It is about creating an understanding of the fundamental difficulties of intervention in complex systems and making it clear that in the context of non-trivial systems it is no longer enough to press a button, make a law, give an order, administer a drug or issue a new regulation. Rather, the *improbability of successful intervention*, the failure of trivializing strategies of change, the exacerbation of problems through the usual reaction of more of the same becomes the normal case.

A first, brute option for enforcing change against the resistance of social systems is to destroy the relevant systems, for example by simply dissolving a department in a company and possibly even dismissing the employees concerned. Apart from the limits placed on such a possibility by laws, agreements and morals, however, this is often not sensible simply because the services provided by the department are still needed and because a complete reconstruction after a complete dissolution is unlikely to pay off in such a case. In addition, measures such as these are also observed by other social systems and can then also have effects there that do more harm than good. Although this does not fundamentally speak against the option of occasionally dissolving systems, it does speak in favor of doing so on the basis of suitable Change Management.

If a social system is to remain fundamentally intact and continue to provide essential parts of its services for the whole, there is no other option from a systems theory perspective than for the system to ultimately change itself (Willke 2005, p. 119). This, in turn, cannot be left solely to the managers who bear responsibility in and for the respective system. Rather, the system as a whole must find its way into a new *modus operandi* (Bourdieu 2003).

The *indifference zone* introduced in Section 2.6.2.2 (Barnard 1938, p. 167)[9] provides a further starting point for bringing about second-order changes: as long as the members of an organization feel adequately paid and valued, they can largely be indifferent as to what they should do in return within the framework of the applicable regulations and agreements. For Change Management, this means that too much commitment on the part of employees to company matters that go beyond their immediate area of work can hinder change. However, since too little commitment can also hinder the development of an organization, it is important to promote the basic attitude of *composure* in the sense of Reinhold Niebuhr, i.e. to accept without complaint what cannot be changed, but at the same time to have the courage to change what can be changed and to have the wisdom to distinguish one from the other. The conscious *distinction between human being and person* (Sect. 2.6.2.2) can contribute to this. This in turn requires people to be in a state of inner balance so that they can live out those parts of their humanity that they cannot bring to bear as individuals in their respective roles in an organization in other ways. Promoting a good *work-life balance*

[9] Quoted from Simon 2015, p. 14.

is therefore not only in the interests of employees, but also of employers (Willke 2005, p. 190 ff.).

Achieving and maintaining such a balance can be supported by *supervision* or *coaching*, for example (Belardi 2018). As outsiders, consultants can contribute to reconciliation between the individual and the person by encouraging those who develop a narrow view in everyday life to reflect on their roles and offering perspectives, methods, time and space to do so, without interfering in their clients' scope of action themselves and while maintaining confidentiality.

3.5.2 Intervention

Advanced methods for stimulating (irritation, perturbation) systems for *self-organized change* have already proven their worth in family therapy, among other areas (e.g. Willke 2005, p. 1). However, they are not yet established in other areas, particularly in the organizational theory relevant here. Therefore, the following explanations can only be of a provisional nature. In addition, systemic intervention theory is so comprehensive that it can at best only be introduced in a rudimentary way here. It is therefore strongly recommended that further literature (especially Willke 2005) be studied in depth.

Fig. 3.31 provides an ideal-typical overview of possible ways of controlling socio-technical systems depending on the complexity of the environment and the complexity of the system to be controlled. Repressive[10] control was found at the societal level in archaic (pre-modern) societies (ibid., p. 46). Today, this type of governance can be found in small organizations such as start-up companies, in which the relationship between members is essentially based on their personal relationships with one another, while a formal organization is of only secondary importance, if at all. The founder of such a company is usually also the owner and, as a result, usually also the decision-maker. As the originator of the business idea, he also has extensive specialist expertise. Both of these qualities give him a natural authority with employees, so that he hardly has to expect any resistance from the workforce in everything he does. Finally, the need for change arises from the small size of the company, which usually results in a desire for growth, which in turn is associated positively and is therefore also welcomed rather than rejected. Against this background, the management of change in such organizations is usually quite effortless. The main focus in this respect is on establishing the business idea.

- If the complexity of the environment remains low, such organizational forms develop into self-directed (liberalistic) organizations with increasing size and, in this respect, also with increasing internal complexity. One example of this is the company oose in Hamburg, which is well-known in the Systems Engineering community and has adopted the organizational form of a cooperative. All that needs to be done is to formulate general framework conditions under which functionally differentiated sub- and part-systems perform their respective functions in a self-organized manner. Therefore, the manage-

[10] This term may evoke negative associations, but is meant to be value-neutral and is based on a corresponding term by David Émile Durkheim (1992).

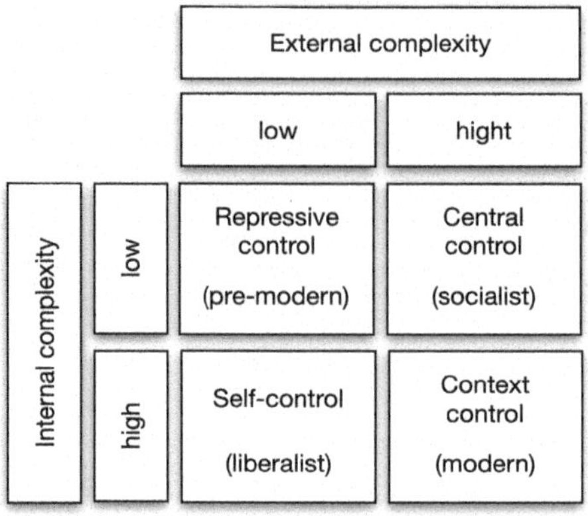

Fig. 3.31 Relationship between system complexity and form of control (Willke 2005, p. 48)

ment of change also takes place in the self-determined departments and thus fulfills the requirement formulated above that social systems can only change themselves, while the probability of success of external interventions in this regard is usually relatively low.

- If, on the other hand, the complexity of the environment increases while the internal complexity remains low, social systems develop a central (hierarchical) control. As will be explained in detail in Section 6, this is an exceptionally effective way of controlling complexity, although it comes at the cost of a considerable limitation of adaptive capacity. As a result, it comes into conflict with the basic *law of requisite variety* formulated by William Ross Ashby (1956, p. 206 ff.) (Willke 2005, p. 47), according to which a controlling system (e.g. corporate management) must have at least the same degree of complexity as the system to be controlled (e.g. complex company in a complex environment: market, society). Examples such as the demise of the once very successful company Kodak show that hierarchical management of change fails as soon as the external pressure to change (business complexity) takes on serious proportions (Lucas and Goh 2009, p. 46 ff.). Precisely because the company was so successful in the past, there was no need to increase internal complexity through more complex forms of management.
- As a rule, many large organizations have developed a high level of internal complexity in the course of *functional differentiation* (Münch 1992, p. 34 ff.) and at the same time face a high level of environmental complexity. To control such systems, Helmut Willke (2005, p. 49 f.) recommends *decentralized context control*. In organizations, interventions in this sense are expressed as *patterns of expectation*. This requires *interaction patterns* between individuals and mixed forms between expectation and interaction patterns for groups.

In light of the insight explained above (Sect. 2.6.2.2) that social systems consist of communication (elements), the context in which a social system is integrated consists of the communications in its social environment that appear relevant to the system (ibid.,

p. 40). Social systems[11] must and want to prove themselves in this context. To do this, they compare themselves with other social systems with which they share their context, in the same way that a fashion-conscious man (psychological system) orients himself to the clothing style of other fashion-conscious men in his environment and, for example, wears a tie from Monday to Thursday, but not on Casual Friday. In addition, social systems usually find an oversized range of possible survival strategies in their respective context (ibid., p. 124 f.), in an economic context, for example, management concepts (Simon 2009) such as Six Sigma or agile Project Management, buzzwords such as Internet of Things (Kaufmann and Servatius 2020) or Industry 4.0 (Mertens et al. 2017), fads such as Design Thinking (Schallmo and Lang 2020) or robust theories such as intervention theory (Willke 2005). Of these options, social systems select those that promise to maintain their *connectivity* to their respective context. Strictly speaking, they try out *self-descriptions* (Weick 1995) that make sense (Sect.s 2.1 and 2.6.2.2), like the fashion-conscious man who looks in the mirror before leaving the house to make sure he is competitive with other men.

Helmut Willke (2005, p. 126) describes this kind of mutual observation of social systems in a shared context as *circular intervention*: since it is fundamentally unclear today, and in highly complex systems it probably remains unclear, which activities actually have which effects, there is massive internal pressure to respond to the (real or imagined) external pressure, i.e. to join the new wave. After all, the others are doing it too. I join a change because the others are changing. And the others are changing because they assume and expect me to change.

For those who want to actively manage change, this means that they have to become part of the context of the social systems that they want to encourage to change. In this context, they can then make offers that are perceived as attractive (useful, fashionable, opportune) by the other social systems. In addition, it can be helpful to be perceived as exemplary (worthy of imitation) as an intervening system. In turn, measurable success, but also prestige, for example, can be helpful. This may be one reason why managers are sometimes expected to act in an exemplary manner (e.g. Liker 2016, p. 247 ff.). However, as the explanations show, this is at best necessary, but by no means sufficient for successful Change Management.

Helmut Willke (2005, p. 127 ff.) mentions *paradoxical intervention* as a second possibility. In his view, it is first necessary to understand that social systems that exhibit behaviours that are considered worthy of change (pathologies) have developed strategies around them, so to speak, that minimize the system-internal problems that result from them. As a result, everything often seems to be in perfect order for the system in question and this perfect order is defended against attacks (change initiatives) from outside (every piece of advice is also a blow). The paradox of such situations is that the system-internal problems are recognized as such and that this can also result in the insight that these or their causes (the behaviours worthy of change) must be eliminated. At the same time, however, the system as a whole considers the internally established order to be worth preserving because, as a result of the cushioning of the pathologies through the relevant strategies, everything seems to be in perfect order overall.

[11] The description of social systems is given here by way of example. The same also applies to psychological systems, which are not discussed in detail here, e.g. Willke 2005, p. 157 ff.

Mara Selvini Palazzoli et al. (2011, p. 42) formulate the paradox as follows: It is not that you should *do* something else – you must *be* what you are not, only in this way can you help me to be who I am not, but who I could be if you were what you are not. In such situations, they recommend countering with a counter-paradox and advising the social system not to change and to continue living with the recognized pathology: By positively evaluating the homeostatic tendency, we paradoxically initiate the ability to change, because the positive evaluation opens our eyes to the paradox of how it can be possible that the cohesion of the group, which the therapists describe as good and desirable, can only be maintained at the cost of one person becoming a patient (ibid, S. 66). A positive evaluation of the pathology as a contribution to the stabilization of the system by those who attempt to intervene from the outside can be understood by the system to mean that there is no longer a threat from the outside. Freed from the burden of having to defend itself, the system can then try to look for solutions itself. Offers of help from outside in this regard may now be seen as offers and no longer as attacks.

Helmut Willke (2005, p. 136 ff.) mentions crisis intervention as a third possibility for intervention. However, as will be shown in Section 3.5, the term crisis, especially in the form of the acute crisis explicitly mentioned by Willke, does not describe what is meant, so that we will speak here of *emergency intervention* instead. In doing so, we take advantage of the fact that parts of the way social systems operate are particularly evident in emergency situations. It can therefore be expedient to deliberately initiate an emergency. The system, which has been thrown out of balance as a result, is supposed to reinvent itself to a certain extent when it regains its equilibrium.

What all three intervention strategies have in common is that they harbor considerable risks, as all of these strategies are dependent on the way the system to be changed operates and cannot be influenced from the outside. The probability that an intervention or change project will succeed is therefore low. It depends largely on whether and, if so, to what extent it is possible to deduce the mode of operation from the outside and, as it were, the sore point that can be addressed. Helmut Willke (2005, p. 7) cites Odysseus' cunning in the Trojan War as a historical example of a successful intervention. After all attempts at direct intervention with battering rams and storm ladders had bounced off Troy's autonomy and operational unity for years, Odysseus finally succeeded in taking the city overnight by no longer basing his own ideas of the desired change on its own logic. And so he realized that the Trojans would only see the horse from their own perspective, as a sign of the Greek attackers' understanding of their defeat and as praise for the victor. As a result, Troy ultimately defeated itself.

3.5.3 Participation

Although the methods for implementing change presented below are based on the insights into the operation of social systems explained in the previous Section, they will be much easier to understand and put into practice than the intervention strategies explained so far. However, the following explanations cannot completely avoid further theoretical foundations. One reason why Change Management has probably only had limited success so far (e.g. Lauer 2019, p. 49 ff.) is that the desire for the simplest possible approaches clashes with the complexity of the problems to be solved. So if you like it simple, you should refrain from Change Management. Simple attempts often make projects even more complicated than they already are, because once fronts have been established and hardened, they can hardly be resolved peacefully and with little loss.

A comparatively simple basis for the participatory intervention explained below is the so-called *IKEA effect* (Norton 2009). In *behavioral economics*, this refers to the insight, named after the Swedish furniture store, that something you make yourself is more highly valued than something that is merely consumed. This effect can motivate those who are affected by change and are actively involved in it to accept and implement what has finally been changed. However, those who routinely apply solutions to problems in day-to-day operations are usually laypeople when it comes to design activities. As a result, their involvement in the implementation can lead to amateurish and therefore not always optimal solutions. Anyone who has ever taken a closer look at the Billy shelves in the homes of friends and acquaintances (with the exception of their own masterpieces, of course) will know what we are talking about here, given the many assembly errors or carelessness.

In order to address problems of this kind, the process model shown in Fig. 3.32 is introduced here, which is based on the design phase of the system life cycle shown in Fig. 3.1 and in particular implements the procedure explained in Section 3.3.1 from rough to detailed using the example of technical product development (this applies analogously to developments in other domains) (here and in the following Ahrens 2019). Comprehensive projects usually start with a *concept*[12]. In this phase, solutions are initially only roughly *sketched out*. At the same time, the final solution is already very largely defined in this most creative of all phases, while the subsequent phases of increasing detailing are increasingly limited to the pure implementation of the solution through to its realization and commissioning (Ehrlenspiel and Meerkamm 2017, p. 1). Participation must therefore not only begin with the conception of a problem solution, it must also be at its strongest in this phase so that those who are to use the solution later have the greatest possible influence on it.

This is facilitated by the fact that professional problem-solving skills are not yet required in this phase. For example, methods and tools that are difficult to learn are not yet used. This is only increasingly the case in later design phases. Instead, sketches are made with paper and pencil, for example. Even non-experts can handle this.

In the *rough planning* that follows the conception, the concrete form of participation can be described as *involvement*, because in this phase users no longer have to and can no longer contribute to the development of the solution to the same extent as was the case in the first phase. In this phase, for example, decisions are made as to which part of the desired problem solution should be developed in-house and which part should be purchased from business partners (make-or-buy decisions). However, this is a matter for specialists. In the course of *variant planning* (Sect. 4.4.5.2), different solution variants are usually designed. In the case of a buy decision, for example, offers from different providers are obtained. Against this background, participation can now consist of involving the users of the relevant solution or their representatives, such as a works council, in the selection of variants.

[12] The number and names of the phases mentioned here may vary from project to project.

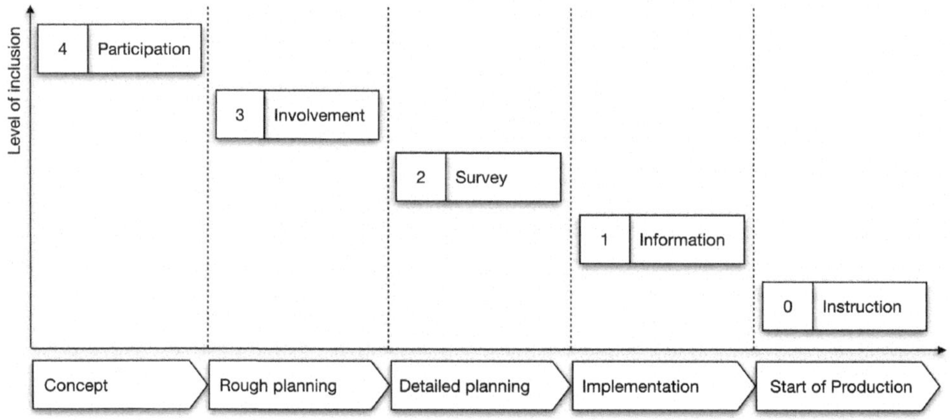

Fig. 3.32 Participation in the course of Change Management

In the detailed planning phase, complex methods and tools such as CAD programs are generally used to create technical drawings. The operation of such computer programs alone requires extensive specialist skills. In addition, extensive engineering skills from the respective areas of product and production process development are required. Prototypes are also often built in this phase, which requires professional manufacturing know-how. As a result, users can usually only contribute a little to the development of solutions from a technical perspective. Nevertheless, participation is also required in this phase as part of Change Management, which can take the form of a *survey*, for example. This can ensure that the users can still make their original contributions to the concept to a sufficient extent, as there are regularly numerous changes to the original approaches on the way to the realization of a solution, as every development is always accompanied by a learning process, which means that not everything that was imagined in a rough planning phase always turns out to be feasible. However, moving too far away from the concept into which the users have incorporated their ideas and expectations almost inevitably leads to a loss of acceptance.

In the course of *execution planning* and *implementation* of the problem solution, it is often a matter of very detailed questions of realization, in which the requirement for technical expertise reaches its maximum, while users as laypersons can hardly contribute anything significant to the solution. Nevertheless, participation in the course of Change Management remains important in this phase too. However, it can usually be limited to *information* about the progress of work. Furthermore, the information usually also includes *training* in dealing with the upcoming changes. One of the reasons why many change projects fail is that they only start with participation now and limit it to information and training. However, the latter can only contribute to the success of a change project if the preceding phases of participation have laid the foundations for this.

Before the final step of a change project, the *start of production* (SOP), an *instruction* from a superior authority is required. With the associated *decision*, the experts working up to this point are relieved of their responsibility in those respects in which uncertainties (contingency) remain that could not be eliminated on the basis of specialist knowledge with the aid of specialist methods, procedures and tools. These uncertainties are a political responsibility and specific (management) personnel are generally responsible for them. Details on this have already been explained in Paragraph 2.6.2.3.

Practical applications of this approach can be found, for example, in *co-creation*, which provides for the participation of consumers in the development of products or the involvement of stakeholders in research projects (e.g. Deserti et al. 2019, p. 17 f.). While traditional development projects only attempt to ensure that the requirements of those affected are taken into account through requirements engineering (e.g. Schlick et al. 2018, p. 51 f.), they are actively involved in the sense intended here. The procedure explained here is almost in contradiction to approaches according to which the conception is to be carried out by managers, while those affected are only involved at the end of a development project in the course of detailing (Schlick et al. 2018, p. 718). On the one hand, this collides with the fact that there is hardly any room for maneuver left in this phase of development and that methods and tools are used for detailed planning that can only be mastered by experts (Ehrlenspiel and Meerkamm 2017, p. 1). Furthermore, such approaches disregard the role of managers as decision-makers, as explained in Paragraph 2.6.2.3: Managers are not so because they are the best experts, but because they are the best at leading.

3.5.4 Mediation between Implicit and Explicit Knowledge

A major advantage of participation is that it not only promotes the acceptance of changes by those who are to adopt them into their everyday routines. In addition, solutions in which their users participate are generally better than those that are only developed by experts.

This is due to the fact that users inevitably have experience from routine operations that is not available to those who specialize in the development of solutions, at least not to the same extent.

However, empirical knowledge is usually available in a form that cannot be utilized immediately in the course of solving a problem. Ikujiro Nonaka and Hirotaka Takeuchi (1997, p. 74 ff.) refer to this as implicit knowledge and distinguish it from the explicit knowledge of experts. For example, an assembler can assemble gearboxes very skilfully, effortlessly using a working environment optimized according to ergonomic criteria, but he will usually not be able to explain how exactly and why his work is so easy. Conversely, the ergonomist will be able to explain very precisely and with reference to relevant sources why the work system is designed in exactly the same way and not differently, without being able to carry out the work in it as skillfully as the fitter.

Fig. 3.33 shows the widespread SECI model of corporate *Knowledge Management* (ibid.), which, however, is not specific to this, but essentially corresponds to the epistemological *inductive-deductive scheme* known since antiquity, which is explained in Section 4.1.4 using Fig. 4.2. According to the SECI model, implicit knowledge (ability) arises through socialization from existing implicit knowledge. Specifically, this takes place through the training of a new employee, for example through demonstration by an already trained employee and imitation by the new employee, or through an exchange of experience between employees with different levels of experience. Knowledge is primarily transferred through observation, imitation and practice (ibid., p. 75; Tomasello 2022, p. 20 f.). This results in experience that is located in the area of application.

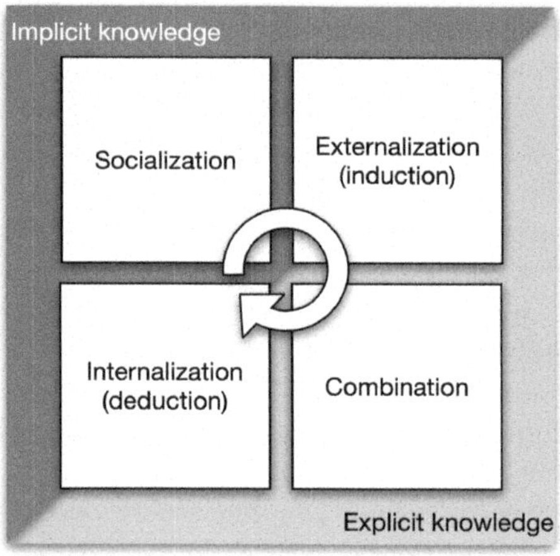

Fig. 3.33 SECI Model (Nonaka and Takeuchi 1997, p. 74 ff.)

Surveys, participation and, above all, participation are ways of converting implicit application knowledge into explicit expert knowledge in the course of *externalization*. The implicit knowledge takes the form of metaphors, analogies, models or hypotheses (ibid., p. 77). In terms of scientific theory, this can be categorized as *induction*, i.e. as an attempt to infer generally valid laws from observed phenomena. The following Section builds on this and explains how this *conversion* can be carried out in concrete terms.

The explicit knowledge gained from externalization is subsequently combined with existing explicit knowledge within the development process by the experts working there and, if necessary, with the involvement of the scientific community.[13]

By applying the resulting innovations in routine operational work, this expert knowledge is converted back into implicit knowledge in the course of *internalization*. In science, deduction is primarily used in this phase, i.e. the application of generally valid laws to specific individual cases. The knowledge spiral thus begins a new cycle at a higher level of knowledge (ibid., p. 84).

3.5.5 Intermediary Configuration of the Conversion of Skills into Knowledge

Since those who are primarily involved in development (problem solving) belong to different social systems than those who routinely apply problem solutions, communication

[13] To differentiate between research and science, the former can be assigned to externalization and the latter to combination.

between the systems is regularly associated with considerable difficulties (Sect. 2.5). However, the approaches cited below have shown ways to increase the probability of mutual understanding and an understanding of the concerns of the other systems. This will be explained using the interface between implicit and explicit knowledge (Fig. 3.33).

To this end, we will draw on insights gained with the help of *actor-network theory* (ANT) (Latour 2005). Unlike other social science theories, it relates social interaction not only to human communication[14], but also to communication emanating from matter (natural and technical (artificial) objects). To avoid complicated theoretical explanations, let us illustrate this with a thought experiment: a ball (quasi-object, Latour 2005, p. 238) is placed in the middle of a sports hall, i.e. clearly visible. First-year pupils are then asked in turn to enter the hall after changing and immediately take a seat on the bench (the ball is placed in the middle of the hall again before each first-year pupil enters). It is easy to imagine that virtually no first-year pupil will follow this instruction without reservation. In fact, almost every first-year pupil will kick the ball.

Comparable experiments prove that matter also communicates, which is linked to interpersonal communication. In the example outlined above, the ball, especially in the environment of a sports hall (the experiment would hopefully be different in a china store), emits a request (*affordance*, Gibson 1973) that has a stronger effect on most pupils than the PE teacher's request to take a seat on the bench, although there may even be a threat of sanctions if the teacher's instruction is ignored.

If you modify the thought experiment and move the time and place of the event to the break of a scientific conference, a carelessly discarded drinks can can spontaneously form two soccer teams out of conference participants who have actually only gathered outside the door of the conference hotel to catch their breath, while two posts of the door railing serve as goal boundaries. Even fuzzy rules such as where the boundary of the playing field runs are quickly agreed upon and understood by all players (with the exception of the drinks can).

Considerations such as these make it clear that matter plays an active role in social systems, although it cannot express itself in the same way as human actors. According to ANT, matter brings about transformations (Latour 2005, p. 37 ff.): The conference attendees were briefly transformed into soccer teams, the two posts of the door railing into a soccer goal. Rules were agreed and the knowledge of these rules was easily communicated to all participants.

In the context of occupational science, objects such as the ball are referred to as *boundary objects* (e.g. Boujut and Blanco 2003, p. 205 ff.; Conceição et al. 2012, p. 127 ff.; Sachse 2002; Vinck et al. 2009, p. 297 ff.). However, Latour (2005, p. 37 ff.) points out that this ascribes a rather passive role to the objects. In reality, however, the objects would have an active effect, so that they take on a role more akin to that of a mediator. However, mediators also limit themselves to ensuring compliance with the procedural rules (§ 2 MediationsG). They are only active in this respect. Otherwise, they act like *catalysts* who neither intervene in terms of content nor change themselves as a result of their mediation activities. However, this is precisely what the intended role of objects is: They also influence the content of communication in social systems and can also change themselves in the process.

[14] ANT views the actors themselves as elements of social systems. In contrast, Niklas Luhmann's (1996, p. 194) insight that social systems consist of communication is taken as a basis here.

The latter takes into account the systems-theoretical insight that communication acts as elements of social systems are very fleeting and must be constantly regenerated in order to maintain the system (Luhmann 1996, p. 28). However, communicative contributions from a quasi-static object would quickly become boring in a constantly changing context, so that their effect would soon diminish. If, on the other hand, they change themselves, they also actively contribute to maintaining communication. In soccer in particular, the constantly changing location of the ball in relation to the pitch and the players is crucial for the progress of the game. For example, shortly before the final whistle, you can observe that a team that has been narrowly losing up to that point has an increased interest in the longest possible stoppage time in order to preserve a last chance to equalize. If the goal difference is greater, this interest is significantly lower.

A term that not only describes the active participation of matter in social systems, but also its impact in terms of content and the resulting change in itself, has not yet been established. The objects should therefore be referred to here as *converting objects*. This takes into account the insight that different spheres of activity, such as development and application, use different semantic codes or, to put it simply, follow different inherent logics, so that a translation service between the codes is required in the course of interpenetration (Sect. 2.6.2.3).

As a rule, a translator cannot avoid adding something of his own to his translations in view of the wide scope for interpretation and thus making not only an active but also a content-related contribution. In addition, he will have to change over time as a result of possible reactions to his own contributions. Bible translations are an example of this. In the new German Luther translation of the Holy Scriptures, which has been in circulation since January 2019, the Greek word Adelphoi is no longer translated with its original meaning, i.e. brothers, but with brothers and sisters (e.g. 1 Corinthians 7:29). Against this background, it can be argued that the concept of converting objects includes all aspects that are to be addressed: (1) active communication, (2) influencing the content and (3) changing themselves.

The practical application will be demonstrated below using the example of *cardboard engineering* (Gorecki and Pautsch 2012, p. 29 ff.; Hagen 2009, p. 10 ff.), an ergonomic method for designing work systems with the participation of those who are to routinely carry out work in them during operation. This method provides for the construction of mock-ups (dummies) made of corrugated cardboard. In addition, edge protection angles or tubes made of cardboard, wooden strips, simple connecting elements such as screws and nuts, etc. are used. The material is processed with simple tools such as cutter knives and glue guns. As an extension to two-dimensional sketches, such as those drawn with pencil on paper, such mock-ups can be described as *three-dimensional sketches*. They are usually built on a scale of 1:1 and are so stable that they allow work processes to be simulated. They therefore have the properties that converting objects must have in order to be able to mediate between the implicit knowledge of users and the explicit knowledge of ergonomists.

- Just like the ball in the sports hall (although certainly not with the same intensity), materials and tools in a workshop also represent an invitation (Gibson 1973) to use them to produce something. In this respect, active communication emanates from the converting objects.

- The content-related contribution to communication initially consists in the fact that the converting objects support the translation of knowledge into skills and vice versa. Experts can realize their specialist knowledge in the model just as users are able to express their practical experience in the course of joint model construction. Discussions in which both sides are involved do not get bogged down in abstract patterns of thought and argumentation on the part of developers or in unspecific information on the part of users; the contributions of both sides are manifested in the jointly produced mockup. And different approaches also come together: a systematic approach, an orientation towards norms and standards, a consideration of proven ergonomic findings, as is to be expected from the development side, does not collide with trial and error, which is likely to prevail on the user side. On the contrary: the latter can promote creativity and contribute to innovative solutions. This is one of the main advantages of participation.
- The mockup is constantly changing during the course of development. The progress of the work is not linear, but cyclical (Sect. 3.3) and is therefore characterized by numerous regressions, which is particularly characteristic of the search for solutions based on the principle of trial and error. This approach is supported by the fact that the mockup is easy to modify and, in terms of the material used, the costs involved are negligible. As a result, the progress of knowledge is visible to all those involved at all times in the truest sense of the word and constantly encourages them to continue working. Experts, users and the mockup are virtually in a triangular relationship with each other and are in a continuous dialog in which the mockup, as a converting object, actively participates through its prompting effect, which it renews again and again through its constant change. Without the mockup, experts and users would have considerably greater difficulties in communicating with each other, because explicit knowledge, which suddenly encounters implicit knowledge (ability), largely encounters a lack of understanding and comprehension and vice versa.

It is obvious that this method is used at the beginning of a design project in the conception phase, as it allows for nothing other than the creation of a rough three-dimensional sketch. This sketching requires no knowledge or skills on the part of the user that cannot be assumed to be known and mastered by them: crafting with cardboard and glue is usually learned at kindergarten age and is therefore playfully easy in the sense of the word. Nevertheless, this type of work should not be underestimated. The ease with which the materials are handled is deliberately chosen in order to bring experts and users on an equal footing, so to speak, and to support work on an equal footing. And by the way, the costs for the materials used are very affordable.

3.6 L 4: Palliation Phase

Particular attention must be paid to what should happen to a system after its death. Among other things, this involves *protecting the environment*. The fact that this question is also of economic interest shows the great importance that the waste disposal industry now has. And since this is also a very innovative sector in which many solutions still need to be found, and will continue to be found in the long term, in order to return systems to the cycle of life at the end of their service life, this sector is also a large and interesting area of application for Systems Engineering & Management. Against this backdrop, the term *disposal* seems unfortunate, as it suggests that it is simply a matter of getting rid of systems at the end of their life. In the following, this term will therefore be replaced by the term *recycling* in the sense of a *circular economy*.

Today, there is widespread agreement that the return of systems must be designed in the same way as their use (e.g. Walden et al. 2023, p. 151 ff.). As an object of design, it should therefore be assigned to the first phase of the life cycle. One of the difficulties associated with this, however, is having to think about death right at the beginning of life and make all the necessary arrangements for it. This is tantamount to giving a newborn a coffin at its christening. Although this may be less emotionally charged in the case of artifacts, those who create new things with joy and zeal will not fare so differently. What's more, the requirements that result from a later return to the design of an artifact are numerous and far-reaching, significantly limiting the creativity of solutions. Among other things, difficult trade-offs have to be made between advantages and disadvantages as well as opportunities and risks, which often necessitate compromises. This entails the risk of giving more weight to obvious advantages than to disadvantages that are far away in terms of time and space.

These difficulties often mean that the design of the return is not carried out at all or not sufficiently in the first phase of the life cycle, but in the worst case only after death has occurred. One of the most obvious examples of this is the handling of radioactive waste from nuclear power plants. If the associated recycling problems had already been taken into account in the design phase, not a single nuclear power plant of today's design would ever have been built - not only for environmental reasons, but also for economic reasons. This does not speak against nuclear power plants in principle, but it does speak in favor of thinking technologies through to the end, including their feedback, before using them, if you don't want to end up like Goethe's sorcerer's apprentice: Only the master masters the brooms!

A further difficulty in including traceability in the life cycle is which systems this relates to. It is relatively clear, for example, that a manufacturing company is responsible for the whereabouts of its own product-specific production system after the last series product has been manufactured with it. However, it is not quite as clear whether the same company is also responsible for the return of each individual product that was manufactured using the production system. It is also conceivable that those who own the products at the end of their useful life, the so-called consumers, are responsible for this. This term already suggests that the products are consumed through their use, so that nothing is left over at the end. And indeed, there are products to which this applies, for example a waffle ice cream purchased at an ice cream stand on Duke Kahanamoku Beach. However, these tend to be exceptions. Most products are not used up completely, even if it is only because their packaging is left over, for example.

One of the many possibilities that make it difficult to classify the design of recycling in the life cycle in general terms is that what is left over at the end of the use phase of systems finds its way into a completely new life cycle, for example when old packaging is used as a raw material for completely different products. A well-known example of such recycling is used PET bottles that are used to make fleece fabric. In such a case, the responsibility for designing the recycling process can be transferred to another organization, thereby relieving the previously responsible organization in this respect. In contrast, reuse such as that of returnable bottles involves returning unused artifacts to the product manufacturer to be used repeatedly for the same purpose. This means that the final return of systems that are no longer viable often remains the responsibility of the manufacturer. Another option is to reuse used items elsewhere, possibly after prior reconditioning or refurbishment. One example of this is cars used for business purposes, which are usually used as new vehicles and are bought by private individuals as used vehicles after just a few years. Finally, in the course of recycling, valuable materials such as metal can be reused in their original condition as metallic materials. In contrast to reuse, however, reconditioning is not sufficient. Instead, a new manufacturing process is required, for example by using scrap metal to produce metal in a blast furnace (Arnold 2013, p. 156 f.).

In view of these difficulties in classifying the return to the life cycle, the following proposal is intended to simplify matters. To this end, the generality that has been deliberately emphasized so far by the parallels between the life cycle of living beings and that of artefacts must be abandoned. In the case of a living being, repatriation usually only begins after death. Organs that are donated are only removed after they have been clearly established. This is followed by burial. For artifacts, this sequence is now to be shifted in that the repatriation phase ends with the quasi-final death. As long as the repatriation has not yet been completed, someone remains responsible for completing it. It must be managed for this length of time, and if design activities are still required, as is currently the case, for example, with regard to the fate of radioactive waste from nuclear power plants, then these must also be attributed to the repatriation phase if and insofar as they were not already completed in the initial design phase. However, this does not correspond to the principles of systems engineering: according to these principles, essential design activities must be completed before an artifact is put into operation.

3.7 L 5: Management of Exceptional Situations

The coronavirus pandemic has made it clear that the organization and infrastructure of states and companies are not sufficiently suitable for coping with exceptional situations of this magnitude (here and in the following Ahrens 2022, p. 119 ff.; Timchenko 2021, p. 50). This is problematic insofar as an increasing number of such exceptional situations are predicted for the future. So far, one cause has dominated in this respect: war. This is said to have occurred around 14,400 times in the documented history of mankind (Baylis et al. 2008, p. 212). Against this background, it is understandable that states and communities of states have established and continuously maintain a specific infrastructure: the military, and a specific form of organization: in particular emergency laws, in order to be prepared for the case of defence. However, there is now an increasing likelihood that other causes will also lead to exceptional situations: Natural disasters in the wake of climate change, major damage situations in technical facilities such as nuclear power plants, hacker attacks on data processing and communication systems, especially in the area of system-relevant infrastructure such as energy or water supply, the collapse of the international financial system or terrorist attacks, for example with so-called radiological weapons (Müller 2015,

p. 1 ff.). Against this backdrop, it is also necessary to create a suitable organization and a specific infrastructure for this purpose. This no longer only affects states and communities of states, but also companies (Ahrens 2022, p. 119 ff.).

Approaches such as Risk Management (DIN EN 31.010:2010-11) or the associated Business Continuity Management (BCM) in accordance with DIN EN ISO 22301:2020-06 as well as a number of other approaches for managing specific risks, for example in data processing and communication technology (ISO/IEC 27.001:2013-10), are intended for this purpose (Müller 2015, p. 93 ff.), but have obviously only had a limited impact in coping with the most recent disasters. There may be two reasons for this: The procedures, methods and tools available

- are not consistently applied (ibid., p. 17 ff.) and/or
- are not up to the scale of such disasters.

Risk Management serves to make potential losses manageable through prevention (Kersten and Klett 2017, p. 41 ff.). Systematic identification and assessment (Rühl 2021, p. 45 ff.) is followed by measures to avoid, reduce and pass on risks, the latter to insurance companies, for example. Residual risks are borne by the company itself (Horst and Saitz 2000, p. 102). However, this is based on the assumption that residual risks are small and therefore bearable. However, there are risks for which this does not apply, i.e. which are not consistently covered by conventional management systems. For example, technical plants are designed for the *Maximum Credible Accident* (MCA). By contrast, only rudimentary precautions are taken, if at all, for risks that go beyond this (ultimate MCA). However, situations such as those mentioned above, for example the corona pandemic, are to be classified as a ultimate MCA. The following explanations therefore focus on the question of how such beyond-design-basis risks can be managed better than before.

Business Continuity Management provides for Crisis Management (Rühl 2021, p. 79 ff.). However, it does not always seem clear what is meant by this and what it involves. For example, Heinrich Kersten and Gerhard Klett (2017, p. 152) explain that a crisis can also develop into a catastrophe (disaster). While they recommend reducing the number of people involved in managing crises, this should be expanded again in the event of a disaster. Furthermore, the reduced group of people should form a crisis team. However, while a Crisis Management team is usually a body that merely assists managers in an advisory capacity (Heimann and Hofinger 2022, p. 3), Heinrich Kersten and Gerhard Klett (2017, p. 152) explain that a Crisis Management team is not only a planning, coordinating and informing advisory body, but also the central management body in Emergency Management.

Such ambiguities are to be clarified below by differentiating between crisis and disaster (synonymously: exceptional situation, disaster, emergency, ultimate MCA). Crises are understood to be all naturally transient phases in the life cycle of systems, for example the transition from the change of a business process to its routine application. In such cases, *Crisis Management* corresponds to the commonly known Change Management (Lauer 2019). The following considerations, on the other hand, focus on the management of exceptional situations, which is to be distinguished from this and, in contrast, is not yet sufficiently mature.

On the basis of a value-neutral concept of crisis, it can be concluded that projects are actually designed to manage crises, namely to shape and steer the transition from an actual state to a deviating target state. The term crisis therefore refers to everyday situations that can be managed using familiar procedures, methods, processes and tools. In addition, although Crisis Management can fail, it does not have to. On the contrary: good Project Management, particularly in conjunction with Change Management (Sect. 3.6), makes a significant contribution to increasing the probability of successful Crisis Management. In this respect, this does not lead any further here.

The situation could be different with terms such as incident, emergency and disaster (hereinafter referred to collectively as exceptional situations (synonymous: emergency)), as relevant sources such as the German Disaster Prevention Act (KatSG), which was passed in 1999 and last amended in 2004, defines *disasters* as major events that lead to a current danger to the life or health of a large number of people, to the environment or to other significant legal assets and which cannot be adequately managed by the authorities responsible for averting danger with their own forces and resources.

- Crises are predictable. They announce themselves and develop over a longer period of time, whether they are natural, such as an adolescent crisis, deliberately caused, such as the development of a new product, or unintentional, such as a project that has fallen into crisis (e.g. Neubauer 1999, p. 20). Although exceptional situations can generally be calculated in terms of their probability and potential extent of damage as part of Risk Management, it is generally impossible to predict exactly when they will occur. At the moment they occur, they appear to be sudden.
- Furthermore, the consequences of a crisis do not necessarily have to be negative. On the contrary: as already shown above with the examples of the adolescent crisis and product development, people generally even expect a desirable state. What links crises with negative associations is merely the experience that overcoming them is usually associated with difficulties and that a negative outcome cannot be completely ruled out, especially as gains are often associated with losses, for example the final loss of childhood and the light-heartedness of life often associated with it during the transition to adulthood. In contrast, the consequences of exceptional situations are always negative.
- It follows from the always negative consequences of exceptional situations that the reaction to them must be as quick as possible (e.g. Bortal 2016, p. 116 ff.). In contrast, the most important thing in crises is to lead them to a positive outcome and secure this in the long term. To this end, it can be important to take time to overcome them.
- In both planned (e.g. product development) and natural (e.g. adolescence) crises, it is particularly obvious that the declared goal is to reach a new state. This is less obvious in the case of unintended crises, for example when a project itself has fallen into a crisis. However, the fact that the aim in such cases must also be to strive for a different state than the previous one can be explained by the fact that the previous state was the cause of the crisis, so that a return to the same state would not help to overcome the crisis. The situation is different in exceptional situations, where the aim is usually to restore the original state or return to the previous plan. Against this background, the question can now be asked as to whether such a rapid response and the combating of an emergency situation can be ensured by Project Management. Some of the key characteristics of projects, such as their uniqueness, time, financial, personnel and other limitations as well as their differentiation from other projects (DIN 69901-5:2009-01, p. 11), speak

in favor of this. Karim Bortal (ibid.), for example, explains that exceptional situations[15] can indeed be managed according to the rules of Project Management, only faster.

However, the need for re-action could stand in the way of such a classification, as this puts management in a defensive situation. In contrast, projects are also defined by the fact that they follow a specific objective, i.e. they require and enable action. They are based on a plan. Although it could be assumed that the goal in exceptional situations is to overcome them and return to a controlled situation, this seems too unspecific as a target for a project in the conventional sense.

The only remedy in this respect is Risk Management (DIN ISO 31000:2018), which identifies and calculates risks in advance in terms of their probability and potential extent of damage and plans measures to be taken if a foreseeable loss event occurs. This turns re-action into action; management is put in a position to proactively tackle damage control. As Risk Management is mandatory in well-managed companies (e.g. Section 91 (2) AktG), it should be assumed here that exceptional situations always occur suddenly, but not unexpectedly as a rule, so that there is nothing to prevent them from being dealt with using conventional Project Management – just faster.

If we go further into the characteristics of projects, there is still their specific organization (DIN 69901-5:2009-01, p. 11), which differs from the organization of regular operations (Ahrens 2018). In this respect, however, there is a difference to what the German Civil Protection Act says, for example. According to this, exceptional situations are also characterized in particular by the fact that they [...] cannot be adequately managed by the authorities responsible for emergency response with their own forces and resources.[16] Assuming that well-managed companies have defined responsibilities for both regular operations and Project Management, established procedures, methods, processes and tools and provided the necessary resources, it is precisely these responsibilities (authorities, forces) and resources that allegedly cannot be used to adequately manage exceptional situations. This leads to the hypothesis that, in addition to routine organization and project organization, a different form of organization is required to manage exceptional situations.

And there is a second indication that we should look for this specific form of organization, which complements the organization of regular operations and projects, because basic research in systems theory has identified three organizational principles with (1) programs, (2) communication channels and (3) people (Simon 2015, p. 70 ff.), which, as will be shown below, cannot be fully reconciled with the two organizational forms of routine organization and project organization that are widespread in business administration.

(1) Luhmann (1997a, p. 842) defines programs as expectations that apply to more than one decision. A distinction is made between conditional programs and purpose programs (Simon 2019, p. 71).

(1.1) Conditional programs are based on the distinction between conditions and their respective consequences (Luhmann 2011, p. 261) and can be interpreted as what is referred to in business administration as process organization (Nordistik 1931, p. 158 ff.; Kosiol 1934, p. 81 ff.), since the 1990s also synonymously as business Process Management

[15] However, Karim Bortal (2016, p. 3 ff.) defines crises and disasters differently to this article.

[16] Authorities are to be transformed to the operational level and interpreted there, for example, as departments.

(Gaitanides 2012) and serves the organization of routine activities (cognitive science: tasks). Programs of this type follow a simple if-then logic and thus serve technology as the basis for automation.

(1.2) Purpose programs are based on the distinction between goals and the means required to achieve them (Luhmann 2011, p. 261). Like conditional programs, they correspond to the process organization known in business administration, but differ from these in that they are not part of Process Management, but of Project Management (Ahrens 2018). From a cognitive science perspective, this is about solving problems (e.g. Sell and Schimweg 2002). Since the development of routines that can be used repeatedly for this purpose, for example on the basis of artificial intelligence, is still in its infancy, the automation of functional programs is at best only rudimentary.

(2) Sociological systems theory has identified communication channels as a second organizational principle (Simon 2015, p. 71 f.). These correspond to the so-called organizational structure in business administration (Nordsieck 1931, p. 158 ff.; Kosiol 1934, p. 81 ff.). As a result, competencies are distributed across different hierarchical levels. However, since both the delegation of decisions to higher levels and the consultation of subordinate levels before decisions are made also follow defined programs, namely official channels (Weber 1972, p. 125 f.), the distinction between programs and communication channels is not entirely clear (Simon 2015, p. 73). However, this is irrelevant for the considerations made here. In this respect, it is only important that conditional and functional programmes are the two main forms of organization that have already found expression in business administration.

(3) The third organizational principle named by systems theory is persons or groups of persons (Simon 2015, p. 73 f.). In this context, a person is understood to be what is relevant to an organization from a person as a psychological system (Luhmann 2011, p. 89). People are therefore interchangeable for an organization, but persons are not[17] (Simon 2015, p. 73). For the considerations made here, the distinction between people and persons is just as important as their coupling to each other, because whenever an organization is confronted with exceptional situations, programs prove to be not flexible enough, while people with sufficient trustworthiness, competence, intelligence, creativity and judgement may be able to make significant contributions to coping with such situations (ibid., p. 74).

While programmes and communication channels thus find their equivalent in the usual company organization and at state level in *bureaucracy* (Weber 1972, p. 126 ff.), a form of organization based on people or groups of people does not appear to be formally (sometimes informally) clearly identifiable. In addition to the two organizational forms for handling routine activities (tasks) and for solving problems (Project Management), the specific form of organization sought here can be based on these.

The procedure for developing a form of organization for coping with exceptional situations can be compared to some extent with the experimental proof of the Higgs boson in elementary particle physics (Higgs 1964, p. 132), which was initially only postulated on the basis of models. Here, too, a missing piece of the puzzle is being sought, so to speak, to close a gap that has so far only become visible through the structure of other pieces of the puzzle, and at best the contours of the missing piece can only be recognized, namely a

[17] In a marriage based on love, things are different: in such a social system, the spouses have to contribute as human beings and are therefore not interchangeable (Luhmann 1994).

form of organization that is significantly different from the previously established forms of organization.

3.7.1 Approaches to Managing Exceptional Situations

The starting point for developing an approach to managing exceptional situations is the organizational principle introduced above, which is derived from sociological systems theory and is based on a person or groups of people (Simon 2015, p. 73 f.). It appears to be advantageous that this approach correlates with widespread views and regulations in practice:

- In Project Management, for example, the assumption is widespread that in difficult situations it is primarily the managerial staff that is important (e.g. Graën 2014; Neubauer 1999, p. 15; Weßels and Winters 2010, p. 1 ff.).
- The emergency laws of the Federal Republic of Germany provide for the executive to be strengthened in a state of defense (Art. 115 f. and i GG) and for the power of command and control over the armed forces to be transferred to the Federal Chancellor (Art. 115b GG). In such exceptional situations, parliaments are only involved as a scaled-down emergency parliament through the Joint Committee (Art. 53a GG).
- In airplanes, the autopilot takes over regular operations. In irregular situations, on the other hand, pilots tend to rely on their expertise.

Fritz B. Simon (2015, p. 74) explains the advantage of linking the organization to people by the fact that these in turn are linked to psychological systems, i.e. to people. As mentioned above, in systems theory, a person is understood to be that which is relevant from a psychological system in a social system. This can be derived etymologically, as in ancient times a mask was referred to as a person (Latin: per through and sonare to sound), which was worn in the theater to play a corresponding role. The human body is concealed behind the mask (person), as is the psyche (Luhmann 1996, p. 429 f.).

In organizations, the mask corresponds, among other things, to a job description, i.e. a fictitious characteristic constructed through communication that is attributed to a person. In situations in which an organization is bound to programmes and communication channels, people cannot get out of their personal skin, so to speak. Their options for action (in systems theory terms, actually communication options) are limited to the roles they play as persons. For example, the company electrician may be able to play the piano as well as he likes, but nobody will want to take any notice of this in everyday company life, except at the company Christmas party.

Nevertheless, people naturally retain their entire repertoire of possible actions. If the organization is tied to one person or to comparatively small groups of people, such as a *task force* (Bortal 2016, p. 128 ff.), these additional, normally hidden competencies should be included as an exception if the system to be organized is predominantly confronted with ignorance (Simon 2015, p. 74), i.e. if it is a matter of managing exceptional situations in the sense intended here. According to Fritz B. Simon (ibid., p. 75), what takes place in the psyche of a member [of the organization] is unpredictable (non-trivial). Therefore, psychological environments can provide the organization with the necessary potential for irritation. If they feed their perceptions, opinions etc. into communication or make them the basis for decisions, this can break up routines. Wherever people act as decision-making premises, unpredictability is introduced. This is because the person does not get the organization without the autonomous mental systems that are firmly linked to them. Persons always bring a certain degree of anarchy into the organization (ibid., p. 73).

From these considerations, we can distil the insight that the unpredictability of the actions of influential people, or what Fritz B. Simon (ibid.) calls anarchy, can be useful for the management of exceptional situations. However, *anarchy* is not to be equated with *anomie*, i.e. with an extensive or even complete absence of order, as is commonly the case. In fact, it merely refers to a different form of rule.

The situation is similar with *improvisation*, which should also be included in the considerations at this point as possibly further-reaching. It too does not mean arbitrary action, but one that opens up degrees of freedom within certain limits that are closed off by programs and communication channels. In commedia dell' arte, for example, the actors' contributions are virtually framed by their respective roles, such as that of Harlequin, Pierrot or Colombine, as well as by the theme of the play being performed, but develop impromptu, i.e. in their concrete form spontaneously out of the respective situation as a reaction to the contributions of the other participants, sometimes even including reactions from the audience (Riha 1990, p. 53). In this respect, creativity, coupled with spontaneity, but also a firm grasp of the rules of improvisation and the roles of the other participants, is of particular importance.

From these divergent considerations, it should now be deduced in a converging manner that, on the one hand, people play a central role in exceptional situations and that, on the other hand, a set of rules must be developed that is geared towards this, which enables creativity and spontaneity, but at the same time steers them in a goal-oriented direction. And since we are talking about exceptional situations, it will be necessary to clarify when such a situation should be considered to have occurred, who decides when such a situation has ended and who in turn has to decide this.

In this respect, too, the considerations do not have to begin ab ovo. If, for example, we take the idea formulated by Felix Timchenko (2021, p. 54) that democratic decisions and the long rounds of debate that often accompany them are not expedient in dealing with exceptional situations, *military organization* is the obvious choice. In view of the millennia-old tradition of settling conflicts by force, it is the most developed form of managing exceptional situations. However, there is currently a widespread view that the prevailing forms of organization, i.e. bureaucracy in particular at state level, need to be changed as a result of the experience gained in the fight against the pandemic. There is no doubt that bureaucracy has not reached its optimum, neither at state level nor in companies. In fact, bureaucratic deficits are omnipresent. Nevertheless, the corona pandemic was neither a suitable trigger for the further development of bureaucracy nor can such further development serve to better manage future major emergencies, because bureaucracy is created for normal situations. Exceptional situations require a completely different form of organization.

This is demonstrated by the organization in the event of defence:

- Decision-making powers are restricted to a smaller group of people (Federal Chancellor, Joint Committee) in accordance with clearly defined rules (Art. 115a to 115l GG).
- Rules (laws) that apply in times of peace are partially replaced by other rules in times of defense, for example with regard to the free choice of profession (Art. 12a GG) or freedom of movement (Art. 11 GG).
- To cope with the exceptional situation, specially trained personnel are available in the form of soldiers, who have no other task in peacetime than to prepare for the case of defense in order to be fully operational immediately if necessary.
- The Bundeswehr has a specific infrastructure (bases, vehicles, means of communication, energy supply, weapons, etc.) for dealing with exceptional situations, which has no other function in peacetime than to be kept available and kept up to date for the case of defense in order to be fully operational immediately if required.

In principle, a similar organization and infrastructure are available to combat a pandemic, albeit with significant deficits:

- Although decision-making powers were limited to a smaller group of people, the Prime Ministers' Conference, in which this group of people met, was merely an unofficial body. As a result, the rules according to which decisions were made there were so weak that lengthy decision-making processes and numerous deviations from jointly reached agreements were an inevitable consequence.
- With the Infection Protection Act (IfSG), the Federal Republic of Germany does have a specific set of rules that allows measures to be adapted to the specific exceptional situation, but practice has shown, for example in the implementation of rapid tests and in the procurement and distribution of protective equipment, that this is far from sufficient to deal with such an exceptional situation. Where bureaucracy was relied upon, it was found to be unable to cope with the situation in many cases, which was to be expected insofar as it is created for the standard case and identifies and eliminates exceptions to this as malfunctions, like a technical control loop (e.g. Heinrich and Schneider 2019, p. 4 f.). As a solution, the so-called unbureaucratic path was sometimes chosen, which, however, did not replace the bureaucratic path with other rules, but by largely dispensing with rules. The foreseeable consequences were mismanagement and a license for corruption and crime.
- Although the Robert Koch Institute (RKI) is an institutionalized epidemiological center in Germany that provides staff and infrastructure in cooperation with virologists and epidemiologists from hospitals and research institutions, practical experience has shown, for example, that the equipment and capacity of health authorities is far from sufficient to cope with such an exceptional situation.
- In addition, the German Federal Armed Forces are available with their personnel and infrastructure in the course of administrative assistance in disaster emergencies (Art. 35, para. 1 GG) and in the event of an internal emergency (Art. 87, para. 4 GG). However, their already notoriously poor equipment is hardly suitable for exceptional situations other than defense, and the education and training of soldiers is also primarily aimed at military national defense. However, the state can also fall back on other organizations such as the German Federal Agency for Technical Relief (THW) or the German Red

Cross (DRK), so that the conditions for coping with exceptional situations at state level are generally good for further expansion.

In companies, the organization and infrastructure for coping with exceptional situations are even weaker. Risk Management is based on the relevant demands of the various stakeholder groups and their power to assert their respective claims:

- The claims of capital investors (shareholders) are primarily regulated by law. Public limited companies[18] are obliged to take appropriate measures, in particular to set up a monitoring system, to ensure that developments that could jeopardize the continued existence of the company are identified at an early stage (Sect. 91 (2) AktG). This refers to Risk Management and the establishment of an internal audit function with a view to risky transactions, accounting inaccuracies and breaches of statutory provisions that have a significant impact on the net assets, financial position and results of operations of the company or group (BT-Drs. 13/9712, p. 15). As part of the audit, it must be examined whether the monitoring system can fulfill its tasks (Sect. 317 (4) HGB). However, it is neither required nor systematically checked whether and, if so, which specific measures are taken to manage risks (IDW PS 340). In fact, it is essentially only a question of greater transparency of the economic situation of companies vis-à-vis investors in the sense of the German Corporate Control and Transparency Act (KonTraG). The effectiveness of measures to manage risks is only examined indirectly to the extent that they have an impact on the extent of the risks to be disclosed, i.e. in particular on the potential amount of damage and/or the probability of damage occurring (Keim 2004, p. 274). Against this background, specific rules that deviate significantly from normal operations only become effective when insolvency occurs in the course of the then applicable Insolvency Code (InsO) in order to minimize the actual damage that has occurred.
- Customers enforce their Risk Management requirements in particular on the basis of the international Quality Management standard DIN EN ISO 9001 (in particular Section 6.1) by obliging their suppliers to be certified in accordance with this standard by independent third parties (DIN EN ISO 19011). This also includes measures to manage risks, which in turn include organization and infrastructure. In addition, customers are protected by the statutory warranty obligation (Sect. 365 BGB).
- In Germany, employees are primarily represented by the statutory accident insurance institutions with regard to the risks of injury or illness (Sect. 15 SGB VII). In addition, agreements on occupational health and safety may also be made in works agreements and collective agreements (Kern and Schmauder 2005, p. 222 ff.). Finally, the international standard DIN ISO 45001 for occupational Health and Safety Management systems has been in place since 2018, similar to Quality Management.
- The operational management of ecological risks is regulated by law. Corresponding regulations can be found for the various areas of environmental protection such as air, water, soil, noise, etc. in a large number of relevant laws such as the Immission Control Act (BimSchG), the Water Resources Act (WHG), the Chemicals Act (ChemG) or the Closed Substance Cycle and Waste Management Act (KrW-/AbfG). On the other hand,

---

[18] For limited liability companies (GmbH), a spill-over effect is assumed (BT-Drs. 13/9712, p. 15), so that there is no explicit equivalent in the GmbHG.

companies can organize their operational Environmental Protection Management independently on the basis of standards such as DIN EN ISO 14001 or the Eco-Management and Audit Scheme (EMAS).

Although there are now signs of uniformity and cross-divisional standardization (Ahrens and Hofmann-Kamensky 2000), operational Risk Management lacks consistency and a systematic approach. In contrast to the state level, there is no uniform company-wide organization and infrastructure that would enable a specific approach in exceptional situations. This limits the effectiveness of the relevant management and is inefficient in that resources are distributed unsystematically across many areas of a company and are therefore often unnecessarily redundant. In addition, standards such as DIN EN ISO 9001 on Quality Management recommend the integration of Risk Management into the process organization (Sect. 4.1) and thus into the organization of routine activities (Ahrens 2018). This contradicts the approach intended here of creating a separate organization and infrastructure for the management of exceptional situations, which differs significantly from routine organization and Project Management.

3.7.2 Elements of a Management System for Exceptional Situations

From the previous considerations, concrete elements and relationships between these elements are now to be derived, which can be part of a system for the management of takeover situations. For this purpose, parallels to existing systems will continue to be used as analogies, albeit in greater detail, so as not to reinvent the wheel. However, this can only be a first draft, which will subsequently need to be corrected and supplemented.

3.7.2.1 System Boundaries

Exceptional situations should remain exceptions, in line with their designation (Timchenko 2021, p. 54). Therefore, similar to projects (DIN 69901-5:2009-01, p. 11), they must be limited in time and space. To do this, it is again necessary to determine which bodies have the authority to set these limits. In turn, it would seem sensible to define criteria that can be used as a basis for such decisions.

In principle, the boundaries should be set as narrowly as possible. It follows from this that the management of exceptional situations can be found at all levels and that escalation should start from the smallest units and only continue as far as appears necessary to manage the situation.

- At state level, for example, chief administrative officers such as district administrators or mayors in independent cities can declare a disaster. In such cases, the state disaster control laws (LKatSG) apply in the Federal Republic of Germany. At the same time, as outlined above, there are corresponding regulations at federal level, which also includes the Federal Office of Civil Protection and Disaster Assistance (BBK), which reports to the Federal Ministry of the Interior (BMI).

- This cascading can be adopted for companies. Here too, formal regulations should be established at all hierarchical levels that allow for the declaration of a state of emergency and any necessary expansion.

The *Jidōka* (Dombrowski and Mielke 2015, p. 16), which is well-known in Lean Management, can serve as a model, which empowers employees to stop assembly at any time by activating a pull cord (andon cord) as soon as a problem arises. This takes place in several escalation steps:

1. In the first step, the employee who has identified the problem tries to find a solution independently. Depending on the assembly cycle, a certain amount of time is available for this.
2. If the problem cannot be solved within this period, help is requested in a second step. This is associated with
   a. an increase in capacity,
   b. an expansion of the qualifications available to solve the problem and
   c. an expansion of powers.
3. If the problem cannot be solved in this way within a given time, the entire assembly is stopped in the third step. All available resources are then mobilized to solve the problem. This in turn is accompanied by
   a. a further increase in capacity,
   b. a further expansion of the skills available to solve the problem and
   c. an even greater expansion of powers.

The aim is to solve problems immediately and sustainably according to the principle of *do it right the first time* (DIRFT), so that they do not lead to further problems of greater magnitude as they drag on. This may seem very expensive in the current situation, but Toyota and many other companies that are now imitating this approach are convinced that it pays off in the long term. At Amazon, for example, every employee in complaints processing is authorized to immediately remove a product completely from the sales program due to a customer complaint. Only when the problem has been reliably and sustainably resolved is the affected product offered for sale again.

3.7.2.2 Role of Persons

The strengthening of the role of individuals or small groups of people in exceptional situations, which has proven to be a guiding principle in the course of the previous considerations, can now be worked out in more detail using the example of military leadership. According to Army Service Regulation (HDv) 100/200 No. 1001 from 2009, leadership in the Bundeswehr is understood as a process of controlling the behavior of people in order to achieve a goal. It comprises the targeted deployment of forces and resources as well as information in space and time. The characteristics of military leadership are as follows:

- the unity of leadership, i.e. the definition of clear subordination relationships (hierarchy, chain of command),

- the interaction of command and obedience as well as leading by order, which can be understood as a prerequisite for the speed of reactions and serves the unity of will (synchronization),
- the indivisible, personal responsibility of military leaders and
- the enforcement of their will.

With regard to a transfer to companies, however, it must be taken into account that the military is an independent entity at state level that takes over control in exceptional situations and thus overrides those who bear responsibility in regular situations. The analogy therefore suggests that an organizational unit should also be created in companies, from which personnel are recruited in exceptional situations to take charge in the sense of a *task force* (Bortal 2016; Timchenko 2021, p. 53 ff.). This can be justified by the fact that this personnel must also be equipped with special powers and qualifications in exceptional operational situations that neither the personnel deployed in routine operations nor in Project Management have. Furthermore, such an organizational unit provides additional capacity.

It is not yet possible to provide a definitive list of specific powers at the current stage of consideration, but initial contours are emerging if the analogy to military leadership is retained.

- For example, it may make sense to grant the right to spontaneously *recruit* staff. It will need to be clarified whether and, if so, to what extent this must be agreed in advance in contracts (employment contracts, works agreements, collective agreements, etc.) and/or whether such contracts may be deviated from in exceptional situations. The same applies to official requirements. Only disregarding the law appears to be excluded.
- The specific powers may also include the right to spontaneous *requisition* (tools, materials, space, etc.) and *contribution* (e.g. use of parts of third-party budgets). In order to ensure responsiveness (speed, flexibility), bureaucratic processes such as posting in Merchandise Management Systems will have to be dispensed with in such cases. However, these processes will have to be replaced by processes that are fast and flexible, but limit the resulting damage in the event of a later return to regular operations. As mentioned above, improvisation must also follow rules.
- It will also be necessary to grant people who become active in exceptional situations the right to deviate from regular processes. However, it will need to be clarified whether and, if so, to what extent this must be agreed in advance with business partners, with whom certain processes are often agreed on the basis of certified management systems (DIN EN ISO 9001, DIN EN ISO 14001, etc.), for example, and with authorities, which often impose requirements for compliance with certain procedures, or whether temporary deviations in exceptional situations are permissible even without explicit agreements.

Although the strengthening of individuals or small groups of people promotes the speed and effectiveness of measures, there is a risk that specialist expertise will fall by the wayside. To counteract this, it is common practice in the military in particular, but also in the police and fire department, to support command personnel with staff. In the German Armed Forces, for example, these include the following so-called basic command areas (FGG), the names of which have been slightly modified with a view to transferring them to the operational organization:

- FGG 1: Human resources, external communication
- FGG 2: Internal communication
- FGG 3: Management, organization, training
- FGG 4: Logistics
- FGG 6: Command service (control center).

In principle, it seems sensible to also support operational personnel for the command of special operations forces through staffs. However, which aspects are relevant and how they are distributed among different staffs is of secondary importance at this point, especially as this is also handled differently in the armies of different countries. In addition, this may also depend on the specific requirements of the respective companies.

3.7.2.3 Infrastructure

Just as the military has its own infrastructure, which is only intended for defence purposes and is only used in peacetime to practice and be ready for use at any time, it can also make sense for companies to maintain a specialized infrastructure for dealing with exceptional situations (e.g. also Timchenko 2021, p. 63). However, it is to be expected that different infrastructures will be required, at least in part, for different exceptional situations, just as the military was only able to contribute to a small extent to overcoming the coronavirus pandemic, for example by providing personnel to carry out tests and vaccinations.

One example of this is the communication system.

- An exceptional situation can arise, for example, if the Merchandise Management System fails. In such a case, a suitable infrastructure can consist of a redundant (mirrored) IT system or an emergency system limited to essential functions.
- For another exceptional situation in which, for example, production comes to a standstill while the ERP system remains fully functional, it may make sense to provide the task force with its own application environment within the IT system in order, for example, to be able to carry out postings of material movements in such a way that they later enable the smoothest possible return to regular operations, but to skip or at least significantly shorten certain process steps that normally take time, such as approvals, releases and the like, in order to be able to act quickly and flexibly.

The personnel who are kept on hand to deal with exceptional situations must, among other things, be familiar with this infrastructure, which differs from regular operations. Since speed is also required above all in the event of an emergency, appropriate training is also required.

3.7.2.4 Procedure Model for Dealing with Exceptional Situations

The procedure for dealing with exceptional situations cannot be fundamentally different from the procedure for dealing with problems, since, in cognitive science terms (Sell and Schimweg 2013), the aim in both cases is to solve problems. This always follows the same logic of the problem-solving cycle (Sect. 3.3) as shown in Fig. 3.23.

Phase	Designation	Explanation
F	Facts	Situation assessment
O	Options	Identification of options for action
R	Risks	Determination and assessment of the associated prospects of success (opportunities) and risks
–		Pause briefly to reflect on what you have learned so far
D	Decision	Decision in favor of the option with the greatest prospects of success and the lowest risks
E	Execution	Implementation of the measures
C	Check	Evaluation of the success of the measures

Fig. 3.34 FOR-DEC procedure model for managing exceptional situations (Hofinger 2022, p. 186 ff.)

- The declaration of a state of emergency provides the impetus
- This is followed by an analysis of the situation.
- This is followed by the formulation of concrete goals. However, while the target state concretized in this way is still unknown in many respects in the context of projects, it is already known in the management of exceptional situations to the extent that it is essentially a matter of returning to the original state. Similar to the goal formulation phase, the subsequent phase of identifying suitable methods, procedures and tools to solve the problem is also a phase in Project Management, but there are significant differences in terms of content. In both situations, the aim is to achieve an optimal result, so that the methods to be selected in this phase are basically optimization methods. When dealing with crises, i.e. in the course of Project Management, you can, and often must, take your time. Optimality refers to the result of the design (Haberfellner et al. 2018, p. 253 ff.) and its sustainability. When dealing with exceptional situations, on the other hand, flexibility and speed are generally important. A solution is therefore optimal if it is a quick solution, a first-fit solution (ibid., p. 254). The methods to be selected must be designed for this.
- The same applies to solution development. In contrast to projects, where this usually takes place from the rough to the detailed and initially provides for a concept, for example, which is then implemented as a prototype before the final realization takes place, a final realization is sought immediately when dealing with exceptional situations. In contrast to projects, variants are not developed in the same detail from which the relatively best one is finally selected. This would also take too much time. Instead, the first best solution is often the best solution.
- Finally, the final assessment will only clarify whether the solution found is suitable for ending the exceptional situation. At this stage, it is important to bear in mind that in such situations, it is not possible to find an optimal solution.

Key aspects of this special approach to dealing with exceptional situations can be found, for example, in the FOR-DEC model (Hofinger 2022, p. 186 ff.) outlined in Fig. 3.34, which is familiar from the aviation industry. In this model, the fact, option and risk phases in the problem-solving cycle correspond to the situation analysis as shown in Fig. 3.23. The brief pause takes account of the fact that this is a phase model which, in the event that the intended reflection comes to the conclusion that not everything worth considering has been taken into account, can lead to setbacks or even the abandonment of the project. In common parlance, this corresponds to the insight that it can be useful to sleep on it for a night before making far-reaching decisions. In the problem-solving cycle, the decision phase corresponds to goal setting. What is missing here, as in the original problem-solving cycle according to Reinhard Haberfellner et al. (2018, p. 71 ff.), is a methodical phase, which is also necessary in exceptional situations and argues in favor of ultimately proceeding according to the problem-solving cycle instead of the FOR-DEC model. The execution phase is comparable to the problem-solving phase and both process models end with an evaluation phase.

3.7.2.5 Conclusion on the Management of Exceptional Situations

In summary, it can be concluded that it would fall short of the mark to blame the established forms of organization for the suboptimal management of the coronavirus pandemic.

- for the completion of (routine) tasks, i.e.
  - at the state level in bureaucracy and
  - at company level in Process Management,
- or to solve problems, i.e. in Project Management,

to look for. Rather, a specific organizational form is required.

The elements of a management system for exceptional situations are (1) the transfer of decision-making powers to individuals or small groups of people, (2) a staff to support the decision-making bodies, (3) specific (improvisation) rules, (4) a dedicated infrastructure and (5) an adapted procedural model. model.

In view of the early stage of development, however, the outlines of such a system can at best be recognized. If and insofar as the development intended here continues to prove meaningful and possible, it will be necessary to further specify the still very general elements. In doing so, it may prove necessary to differentiate between certain classes of exceptional situations in order to align the concretization in each case. In view of the variety of possible exceptional situations (at state level, e.g. illnesses, major damage to technical facilities, terrorist attacks, ecological disasters, at company level, e.g. loss of important suppliers, long-term loss of essential resources), standardization, which is desirable for reasons of efficiency, is likely to reach its limits for reasons of effectiveness. However, insights based on systems theory in particular can help to promote efficiency and effectiveness in equal measure by applying a systematic approach.

3.8 PDCA Cycle

Now that all the generic process models required to design and manage the system life cycle have been introduced, the PDCA cycle will be revisited in order to categorize it as a specific form of the system life cycle (Sect. 3.1). This is due to its great popularity and the observation that it is often misunderstood.

For example, the Quality Management standard DIN EN ISO 9001:2015- 11 (Sect. 0.3) claims that the control of processes and Quality Management Systems can be achieved through the PDCA cycle. The importance it attaches to this process model for solving problems is so great that it devotes an entire subsection to it (ibid., section 0.3.2) and uses one of only two illustrations to show that and how the entire structure of the standard is based on it.

In other places, the PDCA cycle is often located at the operational company level and associated with the continuous improvement process, for example in Bokranz and Landau (2015, vol. 2, p. 48 ff.). In such cases, the planning of improvements is carried out regularly at short intervals of, for example, 14 days in small groups made up of people who are involved in the immediate value creation process or are more or less directly entrusted with its management. The main focus is on identifying deficiencies and their causes on site during ongoing operations (Plan), taking measures to eliminate these deficiencies (Do), measuring the success of these measures (Check) and ensuring the sustainability of the new performance level through standardization (Act).

However, this interpretation is not compatible with that of DIN EN ISO 9001:2015-11, as the standard locates the PDCA cycle at the strategic company level. This can be seen in the example of the planning phase, in which the organization should plan the entire Quality Management System (ibid., chap. 6). This also includes the definition of quality objectives, which is usually a task for top management. All other fields of action assigned to the PDCA cycle (ibid., section 0.3.2) contain corresponding references.

Although objections based on the assumption that the PDCA cycle can be applied at all company levels are to be expected at this point, this appears problematic insofar as a continuous improvement process, at least from a pragmatic point of view, can be integrated into the organization of routine activities for the ongoing operation of systems and is thus itself part of the routine organization, whereas, for example, the planning of a company-wide Quality Management System should normally be the subject of a corresponding project and must therefore be organized quite differently from a routine activity.

In light of the distinction between problems and tasks (Sect. 3.2), the PDCA cycle will now be examined to determine whether and, if so, to what extent it represents either a process model for the organization of routine activities or one for the management of projects. A planning-based approach is taken as the basis for the investigation with regard to Project Management. Agile approaches such as Scrum, Feature Driven Development (FDD), eXtreme Programming (XP) or Crystal (Haberfellner et al. 2018, p. 99 ff.), which largely attempt to dispense with comprehensive planning, are not taken into account here because their compatibility with the PDCA cycle appears questionable from the outset due to the fact that it provides for an explicit planning phase.

Without referring here to a concrete process model for Project Management, the PDCA cycle should be roughly interpreted as a model for problem solving. In this case, it is immediately clear that a corresponding procedure begins with a plan, which, following Rainer Bokranz and Kurt Landau (2012, p. 549), includes clarifying the task, surveying the current situation, identifying deficiencies, setting improvement goals and developing suitable measures. It is equally clear that the measures are to be implemented (Do) and then checked for effectiveness (Check).

However, the interpretation of the Act phase is now problematic. Rainer Bokranz and Kurt Landau (ibid.) explain that deviations identified in the Check phase are to be corrected in this phase. However, this raises the question of why a plan-based approach refrains from clarifying the associated task, surveying the situation, identifying the deficiencies, setting improvement targets, planning suitable measures, implementing them and measuring success again – in short, going through the Plan, Do and Check phases again, immediately after completing the previous Check phase. Similarly problematic is the requirement of DIN EN ISO 9001:2015-11 (Sect. 10.2) to not only make continuous improvements in the course of the Act phase, but also to make groundbreaking changes, innovations and reorganizations (ibid., section 10.1), as well as to extend corrective measures to the change of the Quality Management System itself if necessary. For this, too, it seems necessary to go through the plan, do and check phases.

To counter problems like these, it would be conceivable to apply the PDCA cycle in the Act phase of a higher-level PDCA cycle in such cases, i.e. to nest cycles inside each other like matryoshkas. However, the argument against this is that the problems are repeated in the subordinate cycles and that it is no longer clear whether, for example, a change to the Quality Management System is to be attributed to the superordinate or subordinate cycle. A further interpretation problem arises from the fact that the standard formulates requirements within the framework of the improvement, which it assigns to the Act phase, that are actually to be assigned to the other phases, but does not do so consistently. For example, one requirement is to take measures for monitoring and correction, whereby monitoring should be assigned to the Check phase and correction to the Do phase. A Plan phase is missing in this context, as is an Act phase. Elsewhere, there is a requirement to initiate measures, which can be attributed to the Plan phase. This requirement is then followed by the requirement to check the effectiveness of the measures, which would be part of the check phase, although the do and act phases are missing in this context. Overall, the requirements listed therefore seem too arbitrary to be able to be classified in a process model based on the PDCA cycle, as the standard claims (Sect. 03).

Against this background, it can be stated that the Act phase is actually obsolete in the execution of projects. In this respect, the PDCA cycle is not suitable as a process model for projects in every respect. It would have to be shortened to a PDC cycle.

For the second examination with regard to the suitability of the PDCA cycle as a process model for handling routine activities, the model of a control loop (Sect. 3.4) should be used as a benchmark, because this originally cybernetic model has proven itself in technical applications when it comes to the continuous achievement of predefined goals (management variables) (e.g. Heinrich and Schneider 2019). Such an approach is sometimes criticized because it is based on a mechanistic view of the world. However, it should be remembered that it was Stafford Beer, a business economist, who transferred the cybernetic insights gained by Norbert Wiener in the course of the technical development of automatic target tracking systems for anti-aircraft guns (Galison 1997, p. 282) to Business Management applications, and that such applications are still relevant today (e.g. Adam 2000; Malik 2015). Moreover, the use of control loop models in connection with business organization has also been firmly anchored in Industrial Engineering for decades (e.g. REFA 1991, p. 44 ff.).

Against this background, the phases of the PDCA cycle should be mapped to a typical control loop. Fig. 3.35 shows that this is successful for the Plan, Do and Check phases, while the Act phase has no equivalent here either. In addition, only the Do and Check phases are run through cyclically, while the Plan phase provides the reference variable, which does not have to be changed every time the Do and Check phases are run through, but can also be kept constant in the longer term. In fact, this is the rule rather than the exception. In this respect, the PDCA cycle also appears to be unsuitable as a process model for the organization of routine tasks. It should essentially be reduced to a DC cycle, which should be supplemented by a plan phase on a case-by-case basis. However, the latter does not have to be run through chronologically, but could also run in parallel.

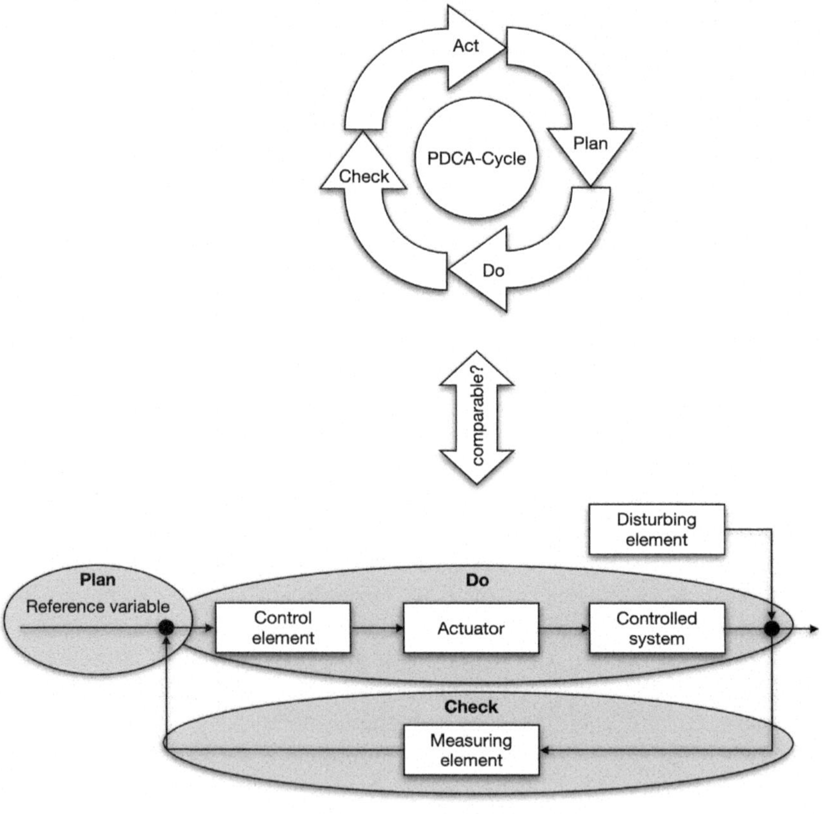

Fig. 3.35 Unsuccessful attempt to compare the PDCA cycle with the control loop model

If the PDCA cycle is not to be completely discarded, especially in view of its popularity, the only remaining option is to interpret it as a specific form of the system life cycle (Sect. 3.1). The PDC phases can then be understood as at least a rudimentary approach to problem solving (Sect. 3.3), while the Act phase depicts ongoing operations. The problem-solving cycle is then applied during the design phase, while control is achieved with the help of control loops. However, as this contributes to confusion rather than clarity, the use of the PDCA cycle is not recommended here.

Chapter 4
L 1: Systems Design

In Section 3.2, the problem-solving cycle introduced in Section 3.3, was derived from cognitive science findings (e.g. Dewey 1997; Dörner 1979; Duncker 1974; Lüer and Spada 1990; Sell and Schimweg 2002). As this approach is of central importance for the design (genesis) of systems, i.e. for Systems Engineering, it will now be explained here in the necessary detail. As will be shown, it is also closely linked to scientific work (Ahrens 2020), because both in practice and in theory, the aim is to come as close as possible to reality or the truth. Although this is countered by the constructivist insight (Sect. 2.1) that, in principle, there can be no complete correspondence between the human imagination and actual reality, evolution has equipped humans with abilities and characteristics that enable them to act more or less successfully. They are *viable* (von Glaserfeld 1996, p. 43). Experience and the theoretical basis of methodological constructivism show that the probability of success can be increased by using suitable methods, procedures and tools (Gehlen 2016). One of these promising methods is the problem-solving cycle (Haberfellner et al. 2018, p. 71 ff.).

Parallel to the actual solution to the problem, this must be organized: Work packages must be put together and coordinated, roles such as managers or controllers must be filled with people, schedules must be drawn up and checked for compliance, contracts must be concluded and applications submitted, and much more. However, this belongs to the area of *Project Management* (Haberfellner et al. 2018, p. 10), which is closely linked to the actual problem solving, but forms a separate domain that will not be discussed in detail here. There are numerous good sources on this (e.g. Burghardt 2013; DIN 69901-1:2009-01 ff.; Gloger 2016; Hirzel et al. 2019; Jakoby 2021), which we recommend reading at this point.

V. Ahrens, *Systems Engineering & Management*, https://doi.org/10.1007/978-3-658-46875-0_4

4.1 P 1: Initiation of a Design Project

According to Karl Raimund Popper (1987, p. 80), all knowledge begins with problems: No knowledge without problems – but also no problem without knowledge. Every *problem*, we are told, arises from the discovery that something in our supposed knowledge is not in order, or logically speaking, from the discovery of an internal contradiction in our supposed knowledge, or a contradiction between our supposed knowledge and the facts. Speaking of problems here and in the following, however, carries the risk of a negative interpretation. To avoid this, some prefer to speak synonymously of *challenges* (e.g. Jakoby 2021, p. 15). In fact, however, problems are understood here in a completely value-neutral way as the difference between what is and what should be, if it is not known how this difference can be minimized and/or if not all the necessary means are available (Fig. 3.15). For this reason, the term problem should be retained here to delimit tasks. Anyone who relies on talking up problems will not deal with the aspects of problem solving discussed here in the breadth and depth presented here anyway.

There are numerous hurdles to overcome before a problem can be solved. The first hurdle is that a problem does not always manifest itself clearly and unambiguously at first. It often appears diffuse over a longer period of time and may only express itself unconsciously in an indeterminable discomfort (Haberfellner et al. 2018, p. 62; Jakoby 2021, p. 37 ff.). Therefore, it is not always possible to pinpoint the exact point in time at which a problem perception enters the consciousness, and the authorship is often no longer traceable. The *theory of science* refers to this phase as the *context of discovery* (Reichenbach 1938, p. 382).

A second hurdle is that not everyone has the right or the opportunity to name a problem as such in every context – one of the reasons why authorship is often no longer traceable. For example, if employees at lower hierarchical levels of a large corporation were to problematize the corporate strategy formulated by the Management Board, this would not have much effect in most cases. Communicating about problems, like all communication, forms the elements of social systems (Sect. 2.6.2.2). Within the respective social system, communicating about an issue that is perceived as problematic only has a chance of being perceived collectively if it can be connected to the prevailing communication. For example, if the general mood is good, communication about a risk is less likely to be accepted than in a situation in which the mood is characterized by worries. It is even more difficult to communicate across system boundaries – especially about problems. As explained in Section 2.1, psychological and social systems only perceive stimuli from their environment as disturbances (irritations, perturbations), to which they only react occasionally, but even then according to their own laws, which are difficult to recognize or even understand from the outside. A very large number of problems therefore do not even make it from individual perception to collective awareness.

A third hurdle is that although many problems are perceived as such, they are not pursued further because initial, at this stage still mostly unconscious or only partially conscious examinations lead to the assessment that it cannot be solved at all or not here or not now or not under the given circumstances – even without student procrastination (Leach 2004, p. 71). To this end, the problem-solving cycle (Fig. 3.23) is already run through unconsciously at one time or another. Siegmund Freud (1911, p. 233) coined the term pro-treatment for this

(Sect. 2.1). Methodologically, this is based on the *hermeneutic circle*, which has already been introduced in Section 3.3.1 and is taken up again in Section 4.1.4.3, in order to then explicitly link it to the initiation of a design project.

- If the situation analysis then leads to the realization that there is no chance of finding out sufficient or precise information about the current situation, thinking about the problem is already interrupted in this phase.
- If, on the other hand, one comes to the conclusion that the given situation can be clarified with a reasonable amount of effort, but then realizes that it is not at all clear what should actually change, the reflection ends in this phase of goal formulation.
- If, on the other hand, the target state can be outlined with sufficient prospects of success, the next step is to assess the chances of obtaining the knowledge and resources required to solve the problem. If the chances in this regard are classified as too low, thinking about the problem ends in this phase.
- If, on the other hand, this also seems achievable and worthwhile, the next step is to draw up and evaluate rough sketches of a solution. Only when these phases have also been successfully completed does the problem reach higher levels of awareness and concretization until it is finally officially placed on an agenda. In DIN 69901-2:2009-01, this corresponds to the so-called processes[1] I.3.1, D.3.3 and P.3.2 *Granting approval*, depending on the respective degree of concretization of the design project.

It is difficult to give an order of magnitude of how many identifiable problems actually reach awareness or how many problems that make it there are actually tackled in the further course, or how many problems that were tackled could ultimately be solved. Klaus Ehrlenspiel and Harald Meerkamm (2013, p. 368) estimate that the number of products successfully launched on the market is vanishingly small, accounting for just a few percent of the original product ideas. However, it also seems possible that the order of magnitude should be in the per mille range rather than the percentage range. Nevertheless, the number of problems that ultimately end up on *to-do lists* is often so large that they cannot be dealt with simultaneously with the available resources. As a result, problems usually have to be organized and prioritized in terms of *Project Portfolio Management* (DIN 69909-4:2015-11).

Once a design project has made it onto the agenda, professional Project Management usually involves appointing a person to arrange everything else. In DIN 69901-2:2009-01, this corresponds to Process I.5.1 *Clarify responsibility*. This is usually also associated with the transfer of responsibility. However, it is just as regularly not clarified what responsibility entails. It should therefore be added here that this is a multi-digit relational concept (Ropohl 1987, p. 155 ff.): who (actor) is responsible when (before/during/after) before which instance (court/client/third party/own conscience) what (action/omission/product/service/right) for what (immediate/direct consequences, delayed/distant consequences), how (compensation/punishment) and why (laws/norms/values).

[1] Since the term process in the context of Project Management was classified as misleading in Section 3.2.5, it is only used in the context of Project Management standards to show the cross-references to them in a comprehensible way. In fact, the process orientation of the standards reveals that and how much they are trapped in the corresponding paradigm (Section 2.5).

Furthermore, it is important to note that projects are naturally (Sect. 3.2) characterized by a barrier that the necessary knowledge and/or resources are not available to overcome. If and to the extent that a project fails, the person to whom responsibility has been transferred cannot be held responsible for this without further ado, as it is often not possible to clearly determine whether and, if so, to what extent a failure must be attributed to the insurmountability of the barrier. The concern of those responsible that they will be held responsible even if they had no chance of preventing the failure can lead to them wanting to make the project a success at all costs, even if it is only an apparent success. It is therefore important to make it clear that a project that is terminated prematurely or one that includes relapses into previous phases can also be a success in justified cases and is then attributed as such to the person responsible.

A second important aspect of the delegation of responsibility is the *principle of congruence*. According to this principle, delegation must always be accompanied by the transfer of the necessary powers and the provision of the required resources. For example, a matrix organization in which the disciplinary responsibility lies with the line organization (parent organization) and the technical responsibility with the project can mean at least a partial violation of the congruence principle for the project manager.

Depending on the expected scope of further activities, the project manager will *organize* the project. DIN 69901-2-2009:01 provides for processes such as I.5.2 Select PM processes, P.1.3 Create project plan, P.5.1 Plan project organization and many more. The more advanced the degree of concretization of a design project, the more detailed these processes will need to be. This will not always be discussed here or in the following, as the focus is not on Project Management but on solving the problem. Nevertheless, references to the standard will be made from time to time, as many of the processes mentioned there are the subject of Systems Engineering anyway.

When initiating a project, it is often necessary to carry out an initial explicit, rough but systematic and methodically supported *Feasibility Analysis* (DIN 69901-2:2009-01, Process D.8.3). The questions to be clarified and the procedure (problem-solving cycle) have already been outlined. Methodologically, it is essentially a matter of calculating the *opportunities and risks* (DIN 69901-2:2009-01, Process D.8 .1) in the light of the company's own *strengths and weaknesses* (SWOT analysis, see Section 4.2.4.8), whereby, in accordance with the general understanding, a risk is understood here as the possibility of failing to achieve the intended in an unfavorable way, while an opportunity refers to the possibility of failing to achieve the intended in an advantageous way, i.e. achieving more than intended, for example (Romeike and Stallinger 2021, p. 26; Wälder and Wälder 2017, p. 1). Both can be calculated according to the so-called *insurance formula* as the product of probability of occurrence and extent (e.g. Andler 2015, p. 410 ff.), risks with a negative sign, so to speak, and opportunities with a positive sign.

DIN ISO 31000:2018-10 and, guided by it, numerous authors such as Stefan Bracke (2022, p. 41 ff.) define a risk as the impact of *uncertainty* on objectives. However, this results in two possible misunderstandings: (1) Opportunities are also effects of uncertainties. The resulting difficulty leads, for example, to opportunities being subsumed under risks, with the difference that they lead to positive deviations from the desired outcome. However, this contradicts the general understanding according to which opportunities and risks are virtually on an equal footing (Romeike and Stallinger 2021, p. 26; Wälder and Wälder

2017, p. 1). (2) Furthermore, the reference to uncertainty often leads to the assumption that, on the contrary, there is such a thing as certainty (e.g. Göbel 2018, p. 73 ff.; Hoffmann 2017, p. 2). However, the only certainty seems to be death and even in this respect, many believe in an afterlife, be it in heaven or hell. In this respect, there are uncertainties even in this regard, which can have a significant impact on the way people lead their lives. For this reason, it seems to be more far-reaching to understand *security* at best as a practically unattainable borderline case of uncertainty and, moreover, to assume that every pursuit of an intention harbors more or less great opportunities and risks.

Furthermore, there are definitions such as that of Frank Romeike and Manfred Stallinger (2021, p. 26 ff.), which agree with the definition recommended here with regard to the equal importance of opportunities and risks, but concretize uncertainty to the effect that it results from the fundamental unpredictability of the future. Although this is not entirely incorrect, it is incomplete, as the spatial dimension must always be taken into account in addition to the time dimension (Fig. 4.26). If one wants to make such a concretization at all, one could generally refer to the necessity, derived from the basic constructivist understanding (Sect. 2.1), of always having to create models of reality in order to deal with it as if it were the model. However, since it is actually much more than that, some of the many influences that were not included in the model in the course of abstraction have an effect in reality during concrete action. Once again, this insight leads to the conclusion that safety as a complete absence of uncertainty cannot exist in practice for reasons of principle. What then remains is the question of whether the model-like (theoretical) assumption of certainty can be taken any further. The magnitude of numerous anthropogenic problems that have arisen in recent decades in particular (instability of the international financial system (Deutsche Bundesbank 2022), climate change, etc.) may cast doubt on this (Beck 1986; Jonas 1984).

In addition to the concept of risk, there are a number of other related terms that are sometimes differentiated from each other or simply used synonymously, while in a professional context it can be helpful to make full use of the linguistic differentiation options. For example, a *threat* can be distinguished from a risk by merely assessing the extent of the damage regardless of its probability of occurrence (Romeike and Stallinger 2021, p. 26). *Hazards* can in turn be distinguished from risks by the fact that their causes cannot be directly influenced by humans, as is the case with an approaching storm, for example, while the causes of risks can be attributed to humans (Luhmann 1997b, p. 327 ff.), which – to stay with the example – would be the case if you forget your umbrella in an approaching storm. Against this background, a *venture* corresponds to a risk, even if one exposes oneself to a danger, since the exposure could be avoided and is therefore the responsibility of the person. The underlying differentiation between risk and danger makes it clear why Ulrich Beck (1986) describes today's society as a risk society and sees it on the way to a different modern age, because more and more of what was previously perceived as a danger is becoming a risk, as people are claiming the right to shape things, to which Systems Engineering in turn attempts to contribute. In this respect, awareness of this appears relevant in this context.

In contrast to danger, *force majeure* is also referred to, particularly in the legal context and in the insurance industry, when events that cause damage occur by chance and cannot be averted even with the utmost, reasonable precautions. This includes natural disasters such as storms, earthquakes or pandemics, but also man-made losses that occur by chance (casus fortuitus), such as wars, revolutions, terrorism, sabotage, hostage-taking or arson. In this respect, too, risks that are consciously taken and calculable, such as going to sea, are not classified as force majeure in the event of damage.

Finally, for the sake of completeness, reference should also be made in this context to the Maximum Credible Accident (MCA), which is to be counted as a risk because it can and must be assumed to be possible, but in view of the potential for catastrophe it sets a higher limit for the reasonableness of prevention and response measures. The term ultimate MCA is then used to describe what goes beyond this level of damage. Examples such as the peaceful use of nuclear energy and, in this respect, particularly in view of the ultimate MCA that has already occurred in Chernobyl and Fukushima, are the subject of controversial debate in societies as to whether the use of technology with disaster potential is at all responsible. Irrespective of the answer to this question, however, the use of nuclear energy is currently prohibited simply because there is no responsible solution for the final disposal of radioactive waste.

Methods used to manage risks include *Failure Mode and Effects Analysis* (FMEA) or *Fault Tree Analysis* in Quality Management (e.g. Brüggemann and Bremer 2015, p. 38 ff.; Edler et al. 2015; Wälder and Wälder 2017) and *Risk Assessment* in occupational Health and Safety Management (e.g. ifaa 2017, p. 17 ff.). The management of opportunities, i.e. *Innovation Management*, takes place primarily in the course of *Research & Development* (R&D) of products and production processes or of services and service provision processes, and numerous methods are also used for this purpose, although even a rudimentary list of these would go beyond the scope here (e.g. Ehrlenspiel and Meerkamm 2013, p. 367 ff.; Wiendahl 2014, p. 63 ff.).

The methods of Innovation and Risk Management often take other factors into account in addition to the probability of occurrence and the extent. For example, the FMEA supplements the two core factors with a probability of detection and calculates the risk priority number (RPN) by multiplying the three influencing variables, while in the selection of product ideas, for example, factors can be added to take future viability or compliance with the corporate strategy into account (Ehrlenspiel and Meerkamm 2013, p. 387). There is nothing to be said against this as long as a sufficiently comprehensible quantifiability is maintained, because the purpose of such calculations is ultimately to *prioritize* problems as rationally as possible if not all problems can be tackled at the same time due to an almost always existing lack of resources. In addition, the quantification gives an indication of which lever should be prioritized: if the probability of occurrence is high, this should be reduced; if it is high, this should be reduced.

More extensive methods for coordinating and prioritizing a (too) large number of problems can be found in *Project Portfolio Management* (synonym: Multi-Project Management) (DIN 69909-4:2015-11; Hirzel et al. 2019). In detail, this also concerns the form in which projects are applied for and commissioned, the financial, personnel, material, time and spatial resources that are made available to the individual projects and the framework conditions according to which projects are to be handled uniformly (standardized) within an organization. This concerns, for example, cooperation with the parent organization and with other projects, reporting and much more. However, as this book is not about Project Management, but about Systems Engineering & Management, please refer to the relevant literature (e.g. Jakoby 2021, p. 72 ff.), while the following will primarily focus on generic aspects of *initiating projects*.

We will now take up what has already been explained above, namely that initial checks regarding the solvability of problems initially take place more or less unconsciously, but that such checks also follow the problem-solving cycle. From a point in time that cannot be determined in more detail, such an implicit check (pre-treatment) turns into an explicit check. This is unavoidable at the latest with the explicit intention to actually tackle a problem solution. Therefore, the procedure for initiating a problem solution follows the same problem-solving cycle on which the entire project is based. In this respect, the problem-solving cycle has the property of *self-similarity* (Fig. 2.15) already explained in Section 2.4.2 as a characteristic of complexity, i.e. it follows the principle of *recursion* (Chapter 1). According to an admittedly somewhat free interpretation of the law of necessary variety (Ashby 1956, p. 206 ff.) in this context, this is also necessary if the procedure is to contribute to mastering complexity. Fig. 4.1 shows this characteristic at the level of the overall project as an example for the formulation of objectives (P 3), which is divided into the phases of initiation (P 3.1) to evaluation (P 3.6). Depending on the project, the structure can go even deeper. Fig. 4.1 shows an example of this for the problem-solving phase (P 3.5), which in turn can be divided into the phases P 3.5.1 to P 3.5.6. This can be, for example, the procedure model for the application of a method. This makes it clear that the same process model can always be used to solve any problem, regardless of the level of detail. Against this background, the widespread development of new process models for specific areas of application is obsolete and confuses more than it contributes to the efficiency and effectiveness of problem solutions.

For the structure of this volume, however, this brings with it the difficulty that the problem-solving cycle would actually have to be dealt with in detail on the basis of the phase of initiation just explained, which would make a detailed explanation of the further phases of the superordinate cycle superfluous or lead to redundancies. In order to avoid this, the main contents are only briefly mentioned here in order to then go into them in more detail when explaining the superordinate phases.

In a very simplified way, the following is about answering the well-known *W questions*, i.e. the questions about the who (stakeholders), the how (existing approaches), the when (before, i.e. the current situation, and after, i.e. the target situation or objective), the what (requirements) and the why (opportunities, risks). In view of the definition of a problem as the difference between a given state and a desired state, it seems obvious at first glance that of these questions only the question of how must initially remain open, because the presence of a barrier (lack of knowledge and/or lack of resources) is the essential difference between a problem and a task (e.g. Dewey 1997; Dörner 1997). e.g. Dewey 1997; Dörner 1979; Duncker 1974; Ehrlenspiel and Meerkamm 2013, p. 56 f.; Jakoby 2021, p. 15 ff.; Lüer and Spada 1990; Sell and Schimweg 2002). However, there are also problems that are characterized by an unclear initial situation and/or an unclear objective (Dörner 1976, p. 14). This applies in particular to so-called dialectical problems (Sell and Schimweg 2002, p. 25). It follows that both the implementation of a situation analysis and the formulation of an objective can be the subject of a problem solution and can therefore be carried out as an independent project.

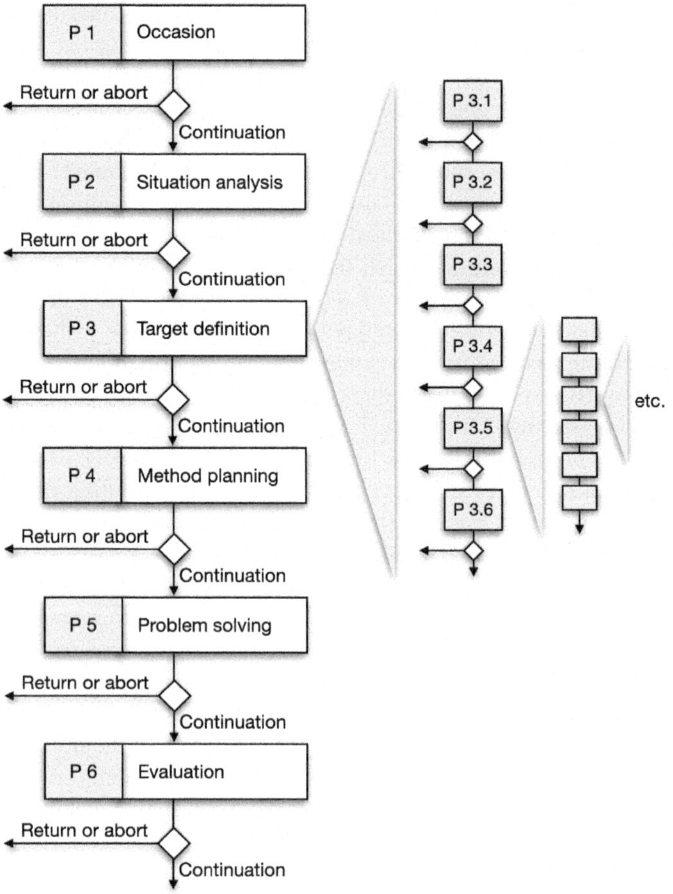

Fig. 4.1 Self-similarity of the problem-solving cycle

Furthermore, before explaining the individual phases of initiating a problem solution, it should be noted that it can be useful to structure not only the procedure itself, but also the documentation of a problem solution according to the problem-solving cycle (Ahrens 2020). This makes the problem solution comprehensible in the sense of the word.

4.1.1 P 1.1: Occasion

When initiating a problem solution, the first step is to identify the cause as clearly as possible. The first step is to provide a convincing answer to the question of what exactly the problem is. If, for example, a risk calculation such as the FMEA forms the starting point, it is important to state which of the risks mentioned therein is to be reduced; if the cause comes from Innovation Management, it must be stated which of the product ideas contained in the list of all product ideas is to be implemented; if Quality Management is confronted with customer complaints that give rise to a systematic elimination of the causes, these must be stated.

Furthermore, it is fair to name the discoverer (originator), as they deserve respect for having identified the problem and overcome all the hurdles mentioned in the previous Section. In fact, many people who point out problems repeatedly experience the fate of Cassandra, the tragic heroine of ancient mythology who had the gift of foreseeing disaster but was never listened to. However, when it comes to eliminating the disaster (solving the problem), it is not uncommon for others to claim authorship, i.e. fame and honor[2].

4.1.2 P 1.2: Situation Analysis

In order to justify that the problem is indeed a problem, at least a rough idea of the initial situation characterized by this problem is required in the course of initiating the corresponding problem solution. The difference between the actual situation and the target situation should emerge as clearly as possible at the given time from the comparison with the objective to be outlined subsequently (P 1.3). Since the problem has not yet been solved in this phase, this project will regularly reach its limits (Haberfellner et al. 2018, p. 62), but the previous implicit checks (see above) have shown that the actual situation can be understood clearly enough to be able to tackle the problem solution on this basis. At least everything that belongs to the findings of these implicit checks should now be expressed explicitly.

This also requires a *Stakeholder Analysis* (Göbel 2010, p. 125 ff.; Krips 2017) in order to clarify which stakeholder groups are involved in and/or affected by the project. On this basis, it is then necessary to determine how the problem presents itself from the respective perspectives of the different stakeholder groups. This is particularly important because a solution to a problem is often initially initiated by just one stakeholder group. For example, a project to reduce costs will primarily be initiated by investors or their internal representatives, while the reduction of injury risks is likely to be primarily in the interests of employees. However, employees will also be involved in an intended cost reduction, at least to the extent that this affects their interests, and conversely, investors will also be interested in occupational health and safety measures if, for example, costs are incurred as a result. Stakeholder Value Management thus merges with *Requirements Engineering* (Ebert 2019).

On this basis, a *SWOT analysis* provides a relatively comprehensive impression of the current situation by comparing the *opportunities and risks* associated with the project with the *strengths and weaknesses* of those carrying out the project and/ This also requires a Stakeholder Analysis (Göbel 2010, p. 125 ff.; Krips 2017) in order to clarify which

[2] Remarkably depicted in the science fiction drama Contact (1997) in the scene in the White House in which Dr. David Drumlin degrades the author Dr. Eleanor Arroway to the role of the presenter of the lecture slides, while he takes the presentation of her research results, which were obtained against his resistance, out of her hands and passes them off as his own.

stakeholder groups are involved in and/or affected by the project. On this basis, it is then necessary to determine how the problem presents itself from the respective perspectives of the different stakeholder groups. This is particularly important because a solution to a problem is often initially initiated by just one stakeholder group. For example, a project to reduce costs will primarily be initiated by investors or their internal representatives, while the reduction of injury risks is likely to be primarily in the interests of employees. However, employees will also be involved in an intended cost reduction, at least to the extent that this affects their interests, and conversely, investors will also be interested in occupational health and safety measures if, for example, costs are incurred as a result. Stakeholder Value Management thus merges with Requirements Engineering (Ebert 2019).

On this basis, a SWOT analysis provides a relatively comprehensive impression of the current situation by comparing the opportunities and risks associated with the project with the strengths and weaknesses of those carrying out the project and/ or the company.

Finally, the situation analysis includes a search for already known approaches to solving the problem. This can be based on a *literature review* or *benchmarking* (e.g. Kairies 2017), but *observations* and *surveys* can also provide information in this regard (Atteslander 2010). Only if and to the extent that no suitable solutions can be found to solve the defined problem is it actually still a problem.

4.1.3 P 1.3: Target Definition

In the course of formulating the target, the aim is now to describe the target state. However, Reinhard Haberfellner et al. (2018, p. 62 f.) rightly point out that no *promises* should be made that are too ambitious. Since a problem is defined by the existence of a barrier, any solution to the problem is naturally associated with the risk of not achieving the desired goal or only achieving it to a limited extent. In addition, problems are often so large that individual projects can only make a contribution to their solution. However, the resulting need for a certain *humility* (Hartmann 1949, p. 476) is often more likely to be recognized by those who commission problem solutions than by those who are commissioned to do so.

In detail, the formulation of objectives takes up certain aspects of the situation analysis and continues them. This applies, for example, to Stakeholder Value Management and the closely related Requirements Engineering, in which the requirements of the stakeholders are now combined into a target system (requirements catalog). In the higher-level phase of goal formulation (P 3), this can result in a *requirements specification* (DIN 69901-5:2009-01, p. 9), which documents what is required for what purpose. Large projects even require the use of specific software (databases) for the management of requirements. Such a comprehensive and detailed elaboration would only be appropriate for very large projects, but thinking in the categories of a specification can help to formulate the target system.

In view of the often quite different interests (claims, requirements) of different stakeholders, it is to be expected that conflicts between individual objectives will become apparent in the target system. In order to achieve a sustainable solution, it is important to make these conflicts transparent, rather than hiding them, playing them down or even ignoring or denying them, possibly out of fear of resistance to the project. Since stakeholders are defined in a narrow sense as those who have the potential to threaten the prevention or significant change of a project, they will usually. The later this happens, the more expensive it becomes. Therefore, it is a matter of common sense, especially of economic efficiency, to disclose *conflicting goals* (Haberfellner et al. 2018, p. 237 f.) at the earliest possible stage and to resolve them through *compromises* or *priorities*. This also applies, at least in a rough form, to the phase of initiating a problem solution.

The SWOT analysis, which was already started during the situation analysis (P 1.2), results in strategies that are assigned to the four fields of the matrix. For example, the generic recommendation for an SO constellation, where one's own strengths (S) encounter opportunities (O), is to use the strengths to seize new opportunities (Andler 2015, p. 287). General recommendations like this can be transferred to the specific goals of the problem solution.

Finally, the approaches to solving the problem that may have been identified in the course of the situation analysis can also be used as suggestions for the project to be initiated, even if they cannot contribute to the desired solution in the present form or not sufficiently. For example, a computer program that is already in use may not (yet) be used in this way, but it may be possible to expand it to this end or at least use it in part, or perhaps it can only serve as a model.

In the higher-level phase of goal formulation (P 3), a specification (DIN 69901-5:2009-01, p. 10) can be created that documents how and with what the requirements (goals) of the stakeholders are to be met. However, it should also be noted in this regard that this describes the barrier that distinguishes a problem from a task, in that the knowledge and/or means required to overcome it are not yet sufficiently known or available at the time of initiation. Furthermore, as with the above-mentioned specifications, not every project requires the development of a complete and detailed set of specifications, especially not in the early stages of the initiation. Nevertheless, thinking in the categories of a set of specifications can help to formulate the objectives of a project as clearly as possible.

In terms of *Change Management*, it is important to actively involve all stakeholders in this phase. These stakeholders usually make contributions that are beneficial to the project, and because better results are achieved as a result, this leads to broad acceptance, which is necessary for the problem solutions that are ultimately developed to be able to develop their full effect in everyday life (in routine application).

4.1.4 P 1.4: Method Planning

The phase of initiating a solution to a problem often tends to be rather unstructured (Haberfellner et al. 2018, p. 62). Against this background, the approaches presented here are intended to help give this phase as much structure as possible. This, in turn, can help to get projects that prove to be meaningful and promising off the ground as quickly as possible and to sort out all other projects just as quickly.

For this purpose, the consistent use of methods, procedures and tools is also recommended in this phase. However, a comprehensive introduction and explanation of methods will follow. This should not be anticipated too far. Therefore, the following considerations focus on methods that correspond to the often still unstructured approach in the initiation phase by being rather general. This is usually the case in this phase, especially for projects that are to be driven forward with little or medium effort. Only for larger and very large projects, for example in the context of a feasibility study, will the entire repertoire of methods be used even in the course of the initiation.

The methodological approaches that are introduced and briefly explained below are based on the *philosophy of science*. However, these approaches carry the risk that those who see themselves as practitioners may be only moderately enthusiastic about them. However, sceptics should be reminded of the words of Kurt Lewin, which were used at the beginning to justify the practical usefulness of systems thinking (Sect. 4.2), namely that nothing is more practical than a good theory (Wegener et al , 2011, p. 47), and Günter Ropohl (2009, p. 73) has rightly pointed out that the *scientization of practical problem solving* is one of the four roots of contemporary systems thinking. On the other hand, scientists who do not delve deeply enough into the theory are expected to take the needs of practitioners into account. Without such mutual accommodation, Systems Engineering & Management is hardly teachable. The same applies to natural scientists and engineers on the one hand and social scientists and humanists on the other. The intended promotion of interdisciplinarity also requires concessions from all sides in this respect[3]. Fig. 4.2 visualizes this four-sided field of tension, but also shows connections.

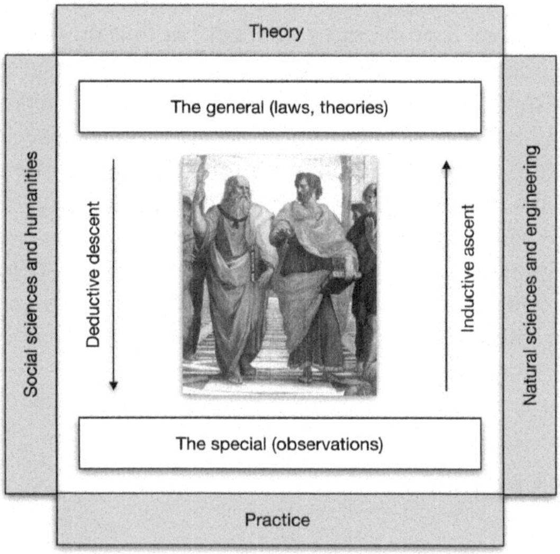

Fig. 4.2 Inductive-deductive scheme

[3] The following explanations are, to a not inconsiderable extent, partly also verbatim, taken from Volker Ahrens (2020, p. 87 ff.). To ensure readability, this is not always and everywhere indicated by appropriate citations.

In detail, the cause of a necessary problem solution can depend on the type of *argumentation* and *reasoning* used. The discussion in the philosophy of science about which type of argumentation is best suited is as old as science itself and has not yet been concluded. Therefore, it would be too far-reaching to reproduce this discussion here, let alone to develop one's own position. Instead, we would like to refer you to the large number of relevant publications on this topic, for example Gerhard Schurz (2014, p. 47 ff.) or Martin Carrier (2021, p. 9 ff.). The approaches are presented below on an equal footing, as they can all contribute to the methodological support of the initiation phase.

Fig. 4.2 shows the inductive-deductive scheme[4], which is used to explain *linear-logical argumentation*. The ancient origin of this scheme is indicated in the center of the figure by a detail from the painting "The School of Athens" by Raphael. In the center of this painting, Plato (left) and Aristotle (right) are depicted. From this, it can be deduced that of all the ancient philosophers, they are considered to be the most important. However, Socrates is also included. Although he is standing somewhat apart in the painting, he is brought into the center by the fact that his student Plato and his student Aristotle are doing what corresponds to his philosophical teachings: they are in dialogue with each other. This dialogue is taken up again in Section 4.1.4.4 In the context of the inductive-deductive scheme, however, it is important to note here that Plato points upwards with his hand, while Aristotle lowers his hand. This symbolism indicates that Plato, as the founder of *constructivism*, which is also the basis of this discussion, assumes that the idea, the model-like abstract concept, the image of the original or, as René Descartes called it, the rex cogitans (Dilthey 1924, p. 143 f.) is the starting point of all knowledge. Aristotle, on the other hand, held the opposite view, namely that all knowledge starts with what Immanuel Kant called the thing in itself, with phenomena, or, in the sense of René Descartes, with the res extensa. These two points of view and the consequences for Systems Engineering will be explained in detail below.

4.1.4.1 Deductive Approach

According to Plato, the first step is to formulate a *hypothesis* (assumption) (Haberfellner et al. 2018, p. 212), i.e. to develop a model-based, abstract idea of the problem, the systems involved and the systems affected (Fig. 4.2 above). Where the idea comes from and how it is justified is not important at first (Schurz 2014, p. 21). Therefore, it can also come about through induction, i.e. through the observation of individual cases and their generalization. What matters is how the hypothesis is justified in the further course. From the hypothetical starting point, deduction is used to arrive at the particular (the specific case of application): a new insight is derived from an existing one – independently of sensory experiences in reality. However, the theories obtained must have empirical consequences by which they can be tested (ibid., p. 14). In this respect, this approach corresponds to the *pragmatism* introduced above (Sect. 2.3).

Critical rationalism, which was essentially founded by Karl Raimund Popper (2005), now attempts to avoid the *induction problem*, which is still controversially discussed in

[4] The SECI model developed by Ikujiro Nonaka and Hirotaka Takeuchi (1997, p. 74 ff.), which is widely used in corporate Knowledge Management and is shown in Fig. 3.33, largely corresponds to this generic scheme.

the philosophy of science (e.g. Schurz 2014, p. 47 ff.; Keuth 2011, p. 3 ff.), and which is particularly widespread in practice-oriented journals. Again and again, authors attempt to prove the truth content (economic efficiency, attractiveness, etc.) of the facts they are discussing, however it is derived, by means of numerous arguments that speak in its favor. The more arguments that can be found, the more likely it is, so the assumption, to have come close to the truth (justification as a method). Especially in the case of facts that can be attributed to a currently prevailing *paradigm* (Sect. 2.4), this type of argumentation and reasoning usually meets with broad approval and is then often imitated as so-called *best practices*. One advantage of this type of argumentation, which should not be underestimated in either practice or science, is that it is easy and therefore convenient to generate a large number of publications. However, a significant disadvantage is that progress is only possible within the paradigms.

As early as the 18th century, the English philosopher David Hume pointed out that even the observation of a large number of individual cases does not allow general law. The observation of the color of swans is often used as an explanatory example. According to this, the fact that a large number of swans are white does not allow the generalization that all swans are white (Keuth 2011, p. 5 ff.). The discovery of a single black swan would already refute this law. From this, Karl Raimund Popper (2005) derives the insight that hypotheses are not to be verified, but rather to be *falsified*.

Although humans use science to search for the truth, i.e. for accurate statements about reality, and although it is indeed possible for them to make such statements, they can never be certain of them. The distinction between truth and certainty leads to *criticism as a method* (fallibilism). According to this, the main focus is not on trying to prove a statement, but on trying to disprove its validity. And the more often a statement has withstood criticism, the more likely it is to be valid. If, on the other hand, a statement is disproved even just once, it is considered to be disproved overall and must be discarded. An example of such an approach was the search for the Higgs particle, whose existence was postulated as early as the 1960s (Higgs 1964, p. 132 f.), but could only be proven by the European Organization for Nuclear Research (CERN) in July 2012. Examples of the application of criticism as a method are the *peer review* of scientific papers and the *disputation* for the attainment of academic degrees.

In the light of the theory of scientific revolutions published by Thomas Samuel Kuhn (2003) in 1962, Imre Lakatos (1974) objected to the hard condition that a single counter-example is sufficient to refute a hypothesis, arguing that a theory is not usually abandoned in practice on the basis of a single counter-example. Rather, it is initially immunized against contradictions by modifications. Only when the theory does not prove itself even after continued modification is it finally rejected. This objection is now widely recognized, but it does not fundamentally question critical rationalism.

4.1.4.2 Inductive Approach

Induction[5] results from the scientific point of view of *empiricism*, which is the exact opposite of rationalism in that it considers sensory perception, i.e. the particular, the observable, to be the most important source of human knowledge (Schurz 2014, p. 12 f.). Based on this, general statements, laws and theories are derived from the initial recognition of patterns, the subsequent *categorization* (concept formation) and by *projecting* the *propositions* found in the process onto previously unrecorded facts, general statements, laws and theories are arrived at (Fig. 4.2, below).

A very common application of inductive reasoning is *probability calculation*. In this case, a *sample* (partial quantity) is taken from a population that is usually too large to be examined in its entirety (full survey). Provided that the sample was selected at *random* (other aspects are also possible in principle), it is possible to draw conclusions about certain *characteristics* of the elements of the sample (observations, statistical units) with a certain probability (confidence interval) with regard to the characteristics of the population. For example, if a sample of 400 randomly selected parts is taken from a batch of 12,000 parts that has just been produced in a metal-processing company, and if 40 of these 400 parts, i.e. 10%, have a defective expression of the quality characteristic being examined quality characteristic, it is assumed that, taking into account the confidence interval of 95% in this case, the batch produced contains at least 600 and at most 1800 parts, on average 1200 parts, which have the same quality defect.

Furthermore, *artificial intelligence* (AI), for example in the context of *machine learning*, is concerned with *knowledge discovery in databases* and with *data mining* using inductive reasoning (e.g. Matzka 2021). Approaches such as these use *algorithms* to recognize patterns based on the previously mentioned statistical models using training data. The regularities recognized in this way are then applied to unknown data. In addition to such symbolic approaches, non-symbolic approaches are also being pursued, in which *neural networks*, comparable to the brain of living beings, implicitly represent their trained knowledge (e.g. Klüver et al. 2021, p. 173 ff.).

Finally, the generalization of human *experience* also corresponds to an inductive approach, so that this approach can be seen as the natural way of reasoning. In this context, deduction could be understood as an intellectual way of arguing. This, in turn, could lead to the assumption that deduction is superior to induction. However, this is not to be assumed here. In fact, both approaches are used here and in the following, especially against the background that hypotheses can also be inductively derived. What can be observed, however, is that deductive approaches are more frequently chosen in the humanities and social sciences than in the natural and engineering sciences. Fig. 4.2 may help to promote mutual understanding in this context.

---

[5] The well-known mathematical method of complete induction (Sect. 6.1) is actually not an inductive conclusion, but a deductive way of reasoning.

4.1.4.3 Hermeneutic Approach

While both deductive and inductive approaches, based on a linear-logical approach to argumentation and reasoning, claim to arrive at universally valid statements (universal propositions, laws), the hermeneutic approach abandons this claim (Gadamer 2010[6]). In the humanities and social sciences in particular, the view is widespread that whenever humans are involved, and this is also the case in connection with technology in the sense of socio-technical and psycho-technical systems, it is more or less always (Sect. 2.6.2), it is not a matter of *explaining* facts, as in the natural and technical sciences, but rather of *understanding* experiences and ideas that make *sense* (Dilthey 1924, p. 139 f.).

The recommended method for this is the hermeneutic spiral shown in Fig. 3.18, which, as already explained in Section 3.3.1, corresponds to the approach *from the general to the specific* in Systems Engineering. In a narrower sense, the method is used in the humanities to interpret human expressions in the form of literature, music or art. In this context, however, a broader understanding is used, according to which it is possible to understand psycho-technical and socio-technical systems with the help of the hermeneutic method. To do this, it is not based on a linear-logical approach, but on a circular approach, as the name suggests. Such an approach can also be observed in the perception of problems, right up to the concrete cause. A fundamental starting point of hermeneutics is the realization that there is no standpoint that can claim objectivity for itself. Every perception is subjective. Freedom from prejudice or impartiality are unattainable. In this respect, everyone perceives a problem differently, to the extent that some may see a problem where others cannot recognize one at all. Therefore, in a social system, it will always be necessary to first develop a collective perception of the problem, which at least partially includes compatible problem descriptions.

Based on an initial understanding of a problem, one will usually delve deeper, as the saying goes, take a closer look at the matter. As explained in Section 4.1, the problem-solving cycle is run through, although this is not always consciously done in the early stages. Often, other people (stakeholders) are also involved, who then contribute their subjective views or, as it is sometimes said, new perspectives. This results in an expanded and deeper understanding of the problem, which in turn can be the starting point for even more and/or deeper investigations and considerations, which take place in an increasingly conscious and formal manner until the problem is finally no longer pursued, postponed or concretely addressed.

In addition to this circular approach, which is basically just a different way of looking at the *problem-solving cycle*, the objective of understanding not only the scientific and technical facts of the problem, but also the meaning associated with the emergence and use of these facts, is of great importance. To achieve this, it is essential to understand the properties and functioning of mental and social systems, as explained above, at the rational level, as well as a scientific and technical understanding, and empathy at the emotional level. The latter also indicates that Change Management and, in particular, the associated participation already begins in this phase and, in this phase in particular, sets the course for whether and, if so, to what extent a design project will ultimately succeed.

---

[6] Hans-Georg Gadamer expressly did not want hermeneutics to be understood as a method. In this respect, the explanations at this point are simplistic.

4.1.4.4 Dialectical Approach

The discussion about the methodological distinction between the humanities and the natural sciences, which was led in the 19th century in particular by Friedrich Daniel Ernst Schleiermacher and Wilhelm Dilthey, was continued in the 20th century. One of the results of this was the *critical theory* developed by the Frankfurt School around Max Horkheimer, Theodor W. Adorno, Herbert Marcuse and Jürgen Habermas, which combines the hermeneutic approach with the insight that scientific knowledge is in principle not free of *interests*. This was associated with a departure from the previously fundamental claim of the value neutrality of scientific work.

The Frankfurt School recommends dialectics as a method, which, like the inductive and deductive approaches, has its roots in antiquity and is simplistically described as the "art of conversation". The starting point is the realization that reality is characterized by contradictions that cannot be resolved without further ado. In order to reveal the contradictions and thus make them fruitful for gaining knowledge, a *thesis* is formulated and contrasted with an *antithesis* (Fig. 4.3). This can happen simultaneously or in succession, in that the antithesis can develop from the criticism of the thesis. In a discourse, the thesis and antithesis are then compared with the arguments for and against them. The aim of the discourse is to develop a *synthesis* that preserves both the useful parts of the thesis and antithesis and adds new insights.

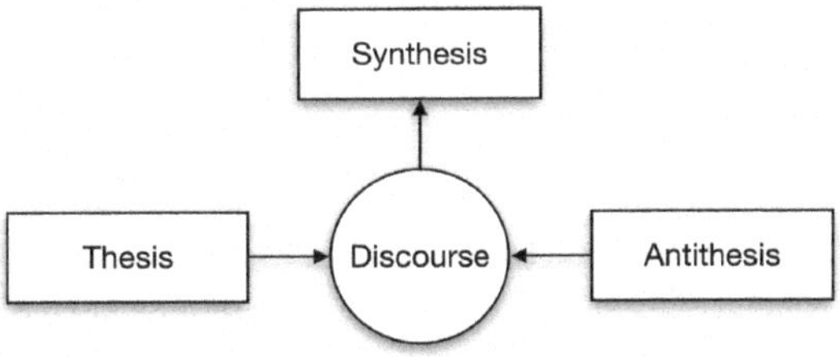

Fig. 4.3 Discourse

In a similar way to the hermeneutic spiral, which is comparable to the approach of moving from the general to the specific that is common in Systems Engineering, the dialectical way of arguing and concluding can be compared to *variant planning*, which will be explained in Section 4.4.5.2 In this process, various solution variants are developed in order to compare them with each other and select the best one. While one variant can be interpreted as a thesis, an alternative variant can be seen as an antithesis. A discussion of the respective advantages and disadvantages corresponds to the discourse, and the solution variant that is ultimately realized corresponds to the synthesis.

However, it is still much too early to do so in the course of initiating a design project. However, as explained in Section 4.1, the problem-solving cycle is already being implemented in this initial phase, which means that discussions can already take place at this stage, and these do not have to be limited to developing solution variants. Ultimately, all questions can be addressed in the sense of a discourse, whether it be the consensus-based interpretation of the given situation, aspects of the objective or suitable methods, procedures and tools for problem solving. In this context, it is important to reiterate that it is not only a matter of what is being argued about in a constructive sense, but also that there is an argument at all. In the light of the underlying constructivist understanding, an early consensus always seems suspicious.

4.1.4.5 Estimate of Expenditure

Now that methods for solving problems in terms of content have been explained, a few methods for organizational processing, i.e. Project Management, will be introduced below. A major difficulty in the course of the initiation is what Walter Jakoby (2021, p. 90) calls the *effort-order dilemma*. On the one hand, there is naturally uncertainty in every problem-solving process as to whether and, if so, how and to what extent the barrier between the actual and the target state can be overcome, and on the other hand, a client naturally expects a calculation of the effort (resources, time, resulting costs) that is as precise and binding as possible before placing an order. As the term "dilemma" suggests, there is no simple solution to this problem. Rather, concessions are required from all sides: a contractor will have to commit to a certain extent, depending on the degree of concretization of the project, and the client will have to admit that neither the success nor the contribution to be made by him can be guaranteed. The approach outlined here, in particular the approach from the rough to the detailed (Sect. 3.3.1 3.3.1) and the problem-solving cycle in conjunction with the iterative approach and the possibility of premature termination at any time, contributes significantly to dealing with this dilemma. And systematic, method-based work also increases the probability of success.

The term "estimation" used in Project Management to describe the calculation of costs (DIN 69901-2:2009-01, "Process" D.4.1 "Roughly estimate costs"; Jakoby 2021, p. 153) also points to the dilemma and at the same time indicates that all calculations based on it based calculations, such as the allocation of resources to solve the problem (ibid., "Process" P.7.1 "Create resource plan"), the creation of schedules, cost and financial plans (ibid., "Processes" P.1.2 "Create schedule" and P.4.1 "Create cost and financial plan") and so on, are equally uncertain. In this respect, the plans are not much more than a replacement for chance with error. Nevertheless, error is usually preferred to chance. Therefore, methods of estimating effort are briefly explained below. Further information can be found in the relevant literature on Project Management (e.g. Jakoby 2021, p. 153 ff.).

The starting point for estimating the effort is the structuring of the project. For this purpose, Project Management provides the *project structure plan*. This in turn is derived from the *product structure*, which is derived from the structure parts list (Wiendahl 2014, p. 143 ff.). The relevant principles are explained generically in Section 2.4.2. To develop each element and each sub- and part-system (hereinafter collectively referred to as elements), several tasks usually have to be completed. Therefore, a so-called process is created for each definable *work package* (Jakoby 2021, p. 148 ff.). The term "package" indicates that the work within the packages can be broken down further and further in the sense of the recursion principle, until individual basic movements are resolved in the sense of ergonomics (Bokranz and Landau 2012, Vol. 1, p. 102). Against this background, it is worth recalling the insight formulated in the context of the system-theoretical foundations that an element of a system always represents the smallest unit that can be meaningfully resolved from the perspective of the system, which does not mean that the element could not be resolved differently or further with regard to other cognitive purposes, for example into molecules, atoms, and so on. A process is usually represented as a functional model, as explained in Section 2.4.1.

Furthermore, the dependencies (relationships) between the processes are identified: certain tasks must be completed before subsequent tasks can begin. This can lead to parallelism (AND relationships) and alternatives (OR relationships) (Fig. 2.28). The corresponding principles can be found in Section 2.4.5.

In the further course, the required resources (personnel, material, machines/ equipment/ tools, energy, space (space requirements), etc.), the time required for processing and the resulting costs are to be estimated (DIN 69901-3:2009-01, p. 4 ff.). Depending on the degree of concretization and detail, this can be done for entire projects, but also for sub-projects down to individual work processes.

In the case of a very high level of uncertainty, which is common at the beginning of design projects in particular, a so-called *intuitive estimate* is often recommended (Jakoby 2021, p. 157 f.). However, intuition is generally understood to mean a subjective insight without discursive use of the mind, i.e. without conscious conclusions. This contradicts the systematic approach that is recommended throughout here – with the exception of some methods of creative thinking (Sect. 4.4.4.2). In connection with this type of estimation, reference is usually made to the use of experience, especially from previous and comparable projects. However, the evaluation of experience in a professional context is usually not intuitive, but is based on records that can be systematically evaluated in the sense of the data analysis explained in Section 4.2.4.4 or in a somewhat less quantifiable way in the form of the document study explained in Section 4.2.4.1. This is certainly useful, but not intuitive. Another experience-based method of estimation is the expert survey, which was explained in Section 4.2.4.2. This also involves systematically drawing on experience.

A *comparative estimate* can be made based on the experience of previous and comparable projects (DIN 69901-3:2009-01, p. 5; Jakoby 2021, p. 158 f.). To do this, the projects used for comparison with the new project are ranked in terms of the criterion to be estimated (resource requirements, time, etc.), which often makes it easier to classify the new project by classifying it as less expensive than more expensive projects and more expensive than less expensive projects.

In addition to qualitative estimates, *quantitative estimates* are also possible. However, the data required for this is specific to each application. One example is price indices for the construction industry, which are collected and published by the Federal Statistical Office. These show, for example, that a factory building in the former territory of the Federal Republic of Germany, excluding West Berlin, will cost an average of 1257 euros per square meter in 2021, based on the 2005 reporting year. Multiplying this figure by the planned area of a factory building allows the costs of the building to be estimated. Such estimates can be further detailed using corresponding key figures for individual trades.

More detailed quantified estimates can incorporate *probabilities*, as explained in Section 2.4.5.1 (ibid., p. 164 ff.). The three-point estimate explained in Section 2.4.5.2 (DIN 69901-3:2009-01, p. 5) is an example of this.

4.1.4.6 Scheduling and Capacity Planning

Starting with the rough outline and then moving on to the details (Sect. 3.3.1), the effort estimate and the subsequent schedule and capacity planning are usually carried out at the beginning of the system life cycle (Chapter 3) for the project as a whole. As the project becomes more concrete, the level of detail involved ultimately extends to the level of work processes. Since the effort required affects both the time needed and the resources required, while resources and time are interdependent, it is not possible to optimize both the time and resource requirements at the same time. Therefore, the respective optimizations are carried out sequentially and iteratively.

The network planning technique is suitable for scheduling, which combines the elements and relations of the project structure with the estimated effort (Jakoby 2021, p. 178 ff.). In addition to the PERT notation explained in Section 2.4.5.2 using Fig. 2.29, there are a whole range of other notations that differ, among other things, in that they focus on the processes in the case of activity-on-node networks (AON) or on the events in the case of event-on-node networks (EON). All notations, often with the support of corresponding software, allow the determination of the start and end dates for the elements (tasks), the resulting buffer times (time flexibilities) and the *critical path*, on which no buffer times are available, so that the postponement of a single task on this path affects the end date of the entire project. The network plans are visualized using Gantt charts.

For scheduling, it is usually assumed that the required resources (capacities) are available in a statically defined amount in the required period (Jakoby 2021, p. 195). These assumptions are then adjusted as far as possible to bring them into line with the associated conditions, such as the partial availability of a resource in the planned period, as soon as a draft schedule (network plan) is available. On this basis, a new schedule is then generated, to which the capacities must be adjusted again, and so on. This iterative process ultimately leads to a project plan that more or less matches the dates and capacities. The main measures used to adjust capacities are time shifts and changes in availability.

4.1.4.7 Planning the Approach to Instigation

When planning the specific procedure for initiating the solution to the problem, it may be necessary to observe formal requirements. These may include, for example, the formulation of applications, the submission of evidence or the preparation of offers. These may in turn be subject to requirements in terms of format and scope. Furthermore, deadlines and official channels may have to be observed, approvals obtained and much more.

For this purpose, systematic *Project Management* may be required, especially for large-scale projects, for the technical preparation (Fig. 2.57: economic sphere of action), because for all this, it is particularly important to clarify who does what, when, with the help of which resources, who coordinates what with whom, and who brings which competencies and receives which authorizations. Project Management offers a large number of methods, procedures and tools for this purpose, of which only a selection will be explained below in view of the somewhat different focus on the more generic Systems Engineering & Management. For the rest, reference is made to the relevant literature on Project Management (e.g. Jakoby 2021).

In addition to the technical preparation, it is always necessary to develop a *strategy for political implementation* (Fig. 2.57: Political sphere of action). The Stakeholder Analysis carried out in the previous phases can provide a suitable basis for this. However, from a political perspective, it is not the substantive requirements of the various stakeholders that are important, but rather their interests, i.e. the advantages that they expect to gain from the project to be initiated or from not carrying it out. In contrast to substantive requirements, these are advantages that do not have to be logically justified. They include *power, influence or prestige*, sometimes even *resentment*, and even the hope of an opportunity to gain *personal advantage*. The latter indicates that the transition to corruption and crime can be fluid in this context. It is therefore important to set limits on the basis of applicable laws and business and *corporate ethics* (e.g. Göbel 2010). *Compliance Management* (DIN ISO 19600) offers suitable methods, procedures and tools for this purpose.

Furthermore, it must be ensured that the solution sought can be socially anchored (Fig. 2.57: social sphere of action). To this end, it must be compatible with the values and norms of the mental and social systems involved and/or affected. One discipline that can provide support in this context is *technology assessment* (TA) (e.g. Simonis 2013).

Finally, the desired solution must be compatible with the culture of the society in which it is to be implemented (Fig. 2.57: cultural sphere of action). Appropriate starting points can be found in *intercultural marketing*, for example. Stefan Müller and Katja Gelbrich (2015, p. 3) cite the example of travelers in Saudi Arabia who prefer to stay with their extended families, so they are reluctant to share a large open-plan carriage with strangers on trains.

4.1.5 P 1.5: Implementation of the Order

Whenever a project is not carried out by the initiator itself, it requires a contract to be awarded. This can take place both within an organization and between different organizations. An assignment between organizations is usually preceded by an offer. This is usually accompanied by contract negotiations. Within an organization, an assignment usually requires fewer formalities, but at least in an analogous manner, the commissioned entity must also be willing (offer) to accept the assignment. This will also involve discussions (negotiations), which will also conclude with a more or less formal contract. To this extent, the differences are fluid, so that they will only be discussed below if and to the extent that they are of particular importance. Otherwise, the comments refer to the commissioning of one organization by another.

In the legal sense, the offer and the order must ultimately contain matching declarations of intent in order for a *contract* to come into existence (Felkai and Beiderwieden 2015, p. 119 f.). However, since the effort-contract dilemma explained in Section 4.1.4.5 cannot be clearly resolved, this must be expressly the subject of the contract by mutual agreement. Otherwise, the contract would be unworkable. This is all the more important to bear in mind because contracts for the execution of projects are usually *contracts for work and services*, in which the contractor is liable for the result in accordance with § 631 of the German Civil Code (BGB), i.e. the success of the project. Within an organization, by contrast, *service contracts* are usually valid in accordance with § 611 of the German Civil Code (BGB), according to which employees are obliged to perform the services specified in their respective employment contracts. Since they work according to instructions, they are not responsible for the success of their actions.

It is also important to bear in mind that the execution of a project is a *service*. While the responsibility for a product clearly lies with the manufacturer or dealer before the transfer of risk and with the buyer after the transfer of risk, the customer must always cooperate in the provision of a service. For example, if you go to the hairdresser but keep shaking your head while you are being styled, you will not be able to expect a decent haircut from even the best hairdresser. And because this applies to every service and thus also to the execution of projects, it is important to take this into account in the contracts. In detail, it must be agreed who is to provide which services. For the client, this may, for example, consist of providing the necessary information and granting access to relevant areas of his company.

The key parameters of project contracts are quality (performance), adherence to deadlines and costs (QTC circle) (Geiger and Kotte 2008, p. 29). In Project Management, the scope for interpretation of the QTC circle, known as the *magic triangle*, is often not fully utilized. It is generally agreed that the three influencing factors are fundamental requirements for problem solving and that they are in conflict with each other (e.g. Jakoby 2021, p. 80 f.): For example, quality in usually costs time and money. What is often overlooked is that the meanings of the three requirements change over time:

- The costs are the first thing to be considered, and are particularly relevant in the course of the initiation of an order and the associated conclusion of a contract for externally awarded orders. As a rule, the contract is awarded to the party that demands the lowest price for its services.
- Once the costs have been fixed, the time that passes until the project is completed becomes more important. Meeting deadlines is now the top priority.
- After the project has been completed, i.e. during operation, quality becomes more important, while the costs and the time to completion are soon forgotten. The importance of quality actually increases over time, because as the service life increases and the associated wear and tear, it becomes clear whether and, if so, to what extent the solution still functions with low maintenance and without errors. This, in turn, has a significant influence on the customer's decision to turn to the same problem solver again or to recommend them to others after the end of the useful life. While quality seems to be rather disposable in the course of the initiation, it turns out to be the most important factor in the long term. The conclusion to be drawn from this is that sustainable solutions often cost time and money in the beginning, but save both in the long run.

Ultimately, it is important to involve those stakeholders who are not directly involved in the contract negotiations but who will nevertheless be affected (Göbel 2010, p. 125 ff.). In this respect, the QTK circle is too short, since it only addresses the owners and other investors with the costs and only the customers with quality and adherence to schedules, i.e. only two of the usually much larger number of stakeholders. Pure project marketing, as envisaged by DIN 69901-2:2009-01 with the "process" D.3.2, is usually not enough for this, because according to the understanding conveyed there, it is simply a matter of informing stakeholders in such a way that they support the achievement of the project goals. The fact that Stakeholder Value Management, in the sense of successful Change Management, usually goes far beyond this was explained in detail in Section 3.5.

4.1.6 P 1.6: Evaluation

The result of the implementation should be a successful initiation. Nevertheless, it happens again and again that projects do not survive this first phase. In both cases, the result and, in particular, the way to get there, must be finalized. this phase constructively and critically. This is also referred to as *lessons learned*.

If a project is ultimately not initiated despite all the efforts made, this does not necessarily have to be seen as a failure. In particular, it does not have to be associated with an inability and/or a lack of commitment on the part of those responsible. This also applies to all terminations and setbacks in the subsequent phases of the problem-solving process. This widespread view carries a not inconsiderable risk, because those in charge who have to expect that the abandonment of a project or a return to earlier phases will be held against them as a shortcoming will be tempted to push a project through at almost any cost, to a possibly bitter end. Instead of a horrible end, a horrible end without end is to be feared, and that usually comes at a high price.

Good management recognizes this in good time. This does not mean that there is no bad management, but it is reasonable to assume that in most cases there are good reasons for not starting a project or for subsequently abandoning it or at least not continuing it in the originally intended manner. This is in the nature of things, because problem solutions, as has been explained many times before, but should be emphasized again here, are defined by a barrier, the overcoming of which cannot be known before it is overcome, whether it can be overcome at all and, if so, with a justifiable effort.

4.2 P 2: Situation Analysis

The recommendation here to also go through the problem-solving cycle within its individual phases can seem excessive for straightforward projects. However, since the recommendations apply to all types and sizes of design projects, it may help to imagine that a project is being carried out that is limited to a situation analysis alone. This could be part of a *feasibility study*, for example, to determine the conditions and calculate the expected costs before the detailed planning and concrete implementation of a very extensive project such as the construction of a new airport, an underground train station or a concert hall. Even a project like this can be divided into numerous sub-projects, which in turn can be divided into sub-projects and so on; all of them serve the situation analysis and all of them follow the problem-solving cycle. This makes it clear that the problem-solving cycle can be *scaled* to almost any depth. However, for small projects, it may be sufficient to follow the phases of the problem-solving cycle only in thought within the phases to be passed through, without explicitly identifying them, for example in documentation in the form of individual chapters.

4.2.1 P 2.1: Occasion

What has just been said in the introduction to this Section applies particularly to the initiation of a situation analysis. In a straightforward project that has already been initiated and encompasses all further phases up to the final evaluation, the reason for the situation analysis already arises from the preceding initiation of the overall project. In such a case, the transition to the situation analysis phase can be limited to the conscious determination of the beginning of this phase.

However, if a project is limited to a situation analysis, for example in the sense of the feasibility study mentioned in the introduction, which in turn involves a great deal of work and/or has far-reaching consequences, it may be necessary to take into account all the aspects explained in Section 4.1 on the initiation (P 1). Another reason for such a detailed preamble may be the commissioning of such a study to third parties. There may be gradations between these two poles depending on the respective circumstances (effort, scope, degree of commitment, etc.).

4.2.2 P 2.2: Situation Analysis

At the beginning of a situation analysis (P 2), the situation in which the analysis is to be carried out (P 2.2) must also be analyzed. Following the principle of recursion, this analysis (P 2.2) encompasses on a smaller scale what is the subject of the larger-scale superordinate situation analysis (P 2), the starting point of which the situation analysis (P 2.2) is intended to reveal. At this point, it should become clear what the benefit of coding the phases is. A purely verbal explanation could quickly lead to confusion, because as simple as the recursion principle is in principle, its consistent application is sometimes difficult. In addition to the code, the visualization in Fig. 4.1 can also help to promote understanding, even if the formulation of objectives has been chosen as an exemplary phase.

However, in order to avoid too much confusion, this Section 4.2.2 will not explain the situational analysis as a whole. This will be done in the superordinate Section 4.2 Instead, only those aspects that are specific to the situational analysis (P 2.2) of the situational analysis (P 2) will be addressed here.

It is essential to clarify the question of under which conditions the situation analysis (P 2) can and should be carried out.

- This concerns the available time, i.e. the point in time at which the analysis is to begin, as well as the period of time until it is to be completed (gross period). In addition, there are periods during which an analysis must be interrupted, for example weekends, holidays, phases of plant shutdown or periods during which the objects of the analysis are needed elsewhere and therefore cannot be analyzed (net period). Although the planning of the analysis period is generally the subject of a corresponding Project Management and good management will take aspects such as those mentioned into account so that the situation analysis (P 2) can be successfully carried out (this also applies to all further aspects addressed below), the fact remains that the time frame that is ultimately available should be consciously noted at the beginning of a situation analysis (P 2), at least to the extent that it is expected to affect the analysis results.

- If time is relevant in a context, then, for the sake of simplicity, this applies to the IPO model (Fig. 2.6)[7], mostly also for space. However, this systematic approach must always be interpreted in an application-specific way. Here, this can be done by asking whether and, if so, how those involved in the situation analysis will have access to all the rooms, places, facilities, machines, plants, etc. that need to be included in the necessary research. For example, their accessibility (distance, connection to transport infrastructure, environmental conditions such as climatic conditions) must be clarified. In addition, there may be access restrictions that require the granting of appropriate access authorizations. Furthermore, it may be necessary to provide accommodation at the locations where the situation is to be analyzed, including the provision of essentials, an internet connection and the like. As already indicated above, aspects such as these are generally the responsibility of the Project Management and will not be discussed in detail here. What is important, however, is whether and, if so, to what extent the conditions can influence the analysis results.
- Furthermore, the available resources for the conditions under which a situation analysis is to be carried out usually include financial resources in the first place, but also personnel (number, competencies in the sense of the competency model explained on the basis of Fig. 1.2, etc.) and material resources. The available or possibly unavailable resources can also have an influence on the analysis results, which should be clarified before the start of the actual situation analysis (P 2.5).
- If we continue to follow the IPO model in Fig. 2.6, then, in addition to space and time and the available resources, the three flow variables of *mass* (material), *energy* and *information* (data, commands) are relevant. Which of the variables are of particular interest and whether other flow variables can be neglected depends on the intended investigation. For example, when planning a factory, the material flow may be crucial (e.g. Burggräf and Schuh 2021). In the energy industry, the design of supply networks is more likely to be concerned with the energy flow (e.g. Konstantin 2017, p. 313 ff.). In the course of the situation analysis (P 2.2) of the situation analysis (P 2), it is necessary to clarify whether, and if so how and under what conditions, the relevant flow variables can be measured. However, information is almost always relevant. This can affect access to digital data sources as well as the availability of contacts (experts, participants, stakeholders, etc.).
- Furthermore, it should be noted at this point that it is not only a matter of clarifying what is accessible to the situation analysis, but also, and quite explicitly, what should remain

[7] Strictly speaking, the functional model of socio-technical systems according to Fig. 2.53 would have to be used here, which, however, is comparable from an external perspective with the IPO model according to Fig. 2.6.

hidden from it. This may, for example, concern confidentiality interests in general, but also, in particular, the protection of personal rights (data protection).

- Haberfellner et al. (2018, p. 214 ff.) recommend distinguishing between the problem area, the solution area, the intervention area and the impact area (Fig. 4.4), thus provide specific information on how the corresponding "process" D.11.2 "Delimit project content" of the Project Management standard DIN 69901-2:2009-01 can be implemented in detail.

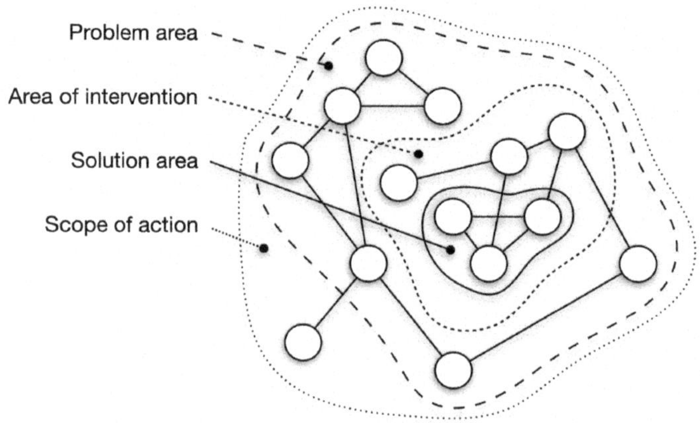

Fig. 4.4 Demarcation between the problem, intervention, solution and impact areas (Haberfellner et al. 2018, p. 213)

- They describe the area in which the problems are suspected as the *problem area*, which gives rise to the solution. To avoid misunderstandings that may arise from the fact that problems are often associated with something negative, it should be noted that the neutral understanding of the problem (Sect. 3.2) on which this is based also includes the perception of opportunities. Since at the beginning of a situation analysis it is often not yet clear how broadly the boundaries of the problem area need to be defined in order to capture all relevant aspects, the boundaries can still be changed in the course of the situation analysis. However, Haberfellner et al. (2018, p. 134) warn against following the often-observed tendency to expand the problem area too far. If we interpret what the authors state as a possible reason for this in the sense of the understanding of the problem on which this is based, then it can be observed that people often tend to look for problems outside their area of responsibility whenever they can be interpreted negatively. On the other hand, people like to ascribe opportunities to themselves.
- The *area of influence* within the problem area is the area in which solutions can be sought. This area is smaller than the problem area if limits are imposed on the design. These may result from restricted rights of disposal, from limited available resources, from lack of time or from other restrictions. An example of the distinction between the problem and intervention areas is the provision of services. These differ from

payments in kind, among other things, in the absence of a clearly definable transfer of risk in the sense of the law of obligations. Rather, services are characterized, among other things, by the fact that the customer always has to play a decisive role (Fritzsche 2001) due to the so-called una-actu principle. Thus, even if a hairdresser is one of the best in his profession, he will only be able to style a customer's hair properly if the customer keeps his head still during the process. Against this background, the problem area could now be formed by the fact that there are more customer complaints. However, since service providers can only exert a very limited influence on their customers (in the business sense, they are external factors), the area of intervention will have to be largely limited to the service provider's sphere of influence.

- Haberfellner et al. (2018, p. 213 f.) describe the *solution area* as an even narrower area than the intervention area, in which the problem is finally solved. For example, the hairdresser could possibly install monitors on which his customers can watch films or television programs while being styled, in the hope that they will then fix their eyes on the monitors in front of them and thus keep their heads still.
- Finally, the *scope of the solution* encompasses the area in which it takes effect. This can go beyond the areas mentioned so far. To stick with the rather simple example, word of the hairdresser's monitor solution spreading so much so that manufacturers of hairdressing salon furnishings come up with the idea of developing mirrors with monitors integrated into them from the outset, and then sell these to a wider customer base.

4.2.3 P 2.3: Target Definition

Within the scope and limits of the situation analysis (P 2), which were determined in the course of the situation analysis (P 2.2), the objectives of the situation analysis (P 2) must now be defined. These depend, among other things, on the level of detail in the course of the system life cycle (Fig. 3.1). In a conception phase, for example, the focus will tend to be on a rough analysis of the environment (Sect. 4.2.4.6), whereas in a more advanced planning phase the focus will tend to be on a detailed System Analysis (Sect. 4.2.4.5). Examples of specific goals of a situation analysis can be found in Reinhard Haberfellner et al. (2018, p. 202 f.).

The general aim is to make the actual situation transparent in order to identify the problem (syndromes) and, in particular, the associated barrier (Sect. 3.2). To this end, suitable evaluation criteria should already be defined in this phase, on the basis of which P 2.5 can be used to assess whether and, if so, to what extent the findings are further relevant after the situation analysis has been completed.

The required transparency primarily concerns the relevant structures and processes of the system to be designed, which should be presented at a level of detail that corresponds to the current phase in the system's life cycle (Fig. 3.1).

To do this, it is also necessary, among other things, to present the causes of the problem and their effects (Fig. 4.11), although the problems in the actual situation are often already known, so the causes must be sought. An evaluation of the success of this search can be made on the basis of the answer to the question of how many of the known effects have corresponding causes been found and whether these are well enough understood in each case to be confident of finding solutions.

Furthermore, it is helpful, especially at the beginning of a design project, to also take into account the broader environment of the focal system (Sect. 4.2.4.6), in order to incorporate factors that affect the system into the intended design. For very general questions such as these, it is naturally not easy to specify concrete evaluation criteria. However, what is important is the conclusiveness of the argumentation, which can be measured by whether it convinces as many people as possible or whether it tends to provoke widespread contradiction.

The requirements that the parties involved and affected place on a solution are of central importance. To this end, it is necessary to determine in detail which stakeholder groups must be taken into account (Sect. 4.2.4.7) and what expectations these groups associate with the project. The final agreement of all stakeholder groups to the list of requirements developed can serve as an evaluation criterion.

Furthermore, the opportunities and risks associated with the project must be calculated. If, at the same time, one's own strengths and weaknesses are also reflected, interesting strategies can be derived from the combination of these four influencing variables (Sect. 4.2.4.8). For example, the number of identified opportunities and risks, as well as strengths and weaknesses, can be used as a benchmark, which can be compared with corresponding lists of comparable projects.

With regard to possible solutions (therapy), the relevant state of research and technology should be determined as part of the situation analysis (Sect. 4.2.4.11). One benchmark can be the number and quality of the sources consulted, which can be supplemented by criteria such as the degree of novelty based on the publication dates or the internationality.

Finally, forecasts must be created that allow an assessment of how the historical analysis results will develop in the future (Sect. 4.2.4.11). Corresponding forecasting methods often include methods for assessing the quality of the forecast.

4.2.4 P 2.4: Method Planning

Now that the objectives of the situation analysis (P 2) are known, the next step is to clarify how these objectives are to be achieved. As explained in Section 3.2, based on Fig. 3.13, problems are characterized in contrast to tasks by the fact that all or at least some of the knowledge required for the solution is missing and/or that not all the necessary means are available. Nevertheless, problems should be solved as efficiently and effectively (economically) as possible, and as reliably and sustainably as possible, at least in a professional context. To this end, methods are used to replace the missing knowledge and/or the missing means (Haberfellner et al. 2018, p. 215 ff.), which, through the associated systematic, targeted approach, that has usually already proven itself in comparable situations, offer a prospect of success that cannot be expected from an amateur-intuitive approach.

In detail, it is about clarifying who (resources) should collect what information (what) when (time) and where (space) for what purpose (why) in what form (how) and to what extent (how much). In addition, each goal (requirement, expectation) should indicate how the degree of fulfillment is to be measured (assessed) after the situation analysis (P 2) has been completed in the course of the evaluation (P 2.6). Finally, the form in which the results are to be documented and presented should be specified.

The following example, which has been deliberately kept simple for the time being, is intended to illustrate how a goal of the situation analysis (P 2) can be formulated in concrete terms. For this purpose, it is assumed that additional quality assurance measures are to be taken in a manufacturing company after a series of customer complaints. To do this, employees of the department responsible for quality assurance (QA) are to take samples (how much) in accordance with DIN ISO 2859-1:2014-08 (how: method) after the next batch of the products complained about has been produced (when) in the punching shop (where) and check the quality characteristics complained about according to the inspection plan (how: method). The Clopper-Pearson confidence interval of the results should be 95% (requirement for the reliability of the analysis results). The results of the tests should be stored in the QS database, as should all test results from regular quality tests. In addition, the employee should summarize the results in a report of no more than one page, send it immediately to the Project Management by email, and report briefly and informally on the test results at the project meeting that immediately follows the last quality test.

However, not all methods are as easy to identify. In many cases, different methods that are fundamentally suitable have to be weighed against each other in order to select one or a few methods that are best suited for the respective application. In addition, many methods have to be specified before they can be applied. One example of this is surveying. Although textbooks document how surveys are to be conducted in principle (e.g. Diekmann 2020, p. 434 ff.), However, it is necessary to specify, for example, which questions are to be asked in which form (e.g. open versus closed questions) and in which form answers are expected. And there are usually numerous alternatives for this as well, which in turn have to be weighed against each other in order to select the most suitable variants for the application. The necessity to consider the methodology before the actual situation analysis (P 2.5) can go even further if methods are not yet available for a particular application, so that they first have to be developed. It may also be the case that methods are available in principle but still need to be modified for the intended application. Such cases are particularly common in science, but also in corporate research and development.

In this context, methods to support the situational analysis are to be introduced below. However, this is limited by the fact that there are a very large number of possible methods and an even larger number of variants of these methods, and that a lot can be explained about many of these methods. Specialized books on the subject often run to hundreds of pages (e.g. ibid.). Documenting all of this again here would not only go beyond the scope of this article; Above all, it seems unnecessary in view of the quality of the specialized works. Therefore, reading them is expressly recommended at this point. However, in order to at least provide an overview and to facilitate an introduction to the problem-solving cycle, selected methods are presented as examples.

These are, first and foremost, the methods of *empirical social research* (ibid.). These include the study of documents (often also referred to as literature research in a scientific context), surveys, observations (and experiments. In addition, some methods are recommended that relate to specific aspects of design. These include the *Stakeholder Analysis* with regard to the Stakeholder Value Management that follows in further phases, and the SWOT analysis, which correlates strengths and weaknesses with opportunities and risks. This is followed by the *Requirements Analysis* and the *determination of the state of research and technology*. Finally, the insight that all design projects are directed towards the future, while the situation analysis (P 2) can only be based on the past, is taken into account by the explanation of *forecasts*.

Before the methods mentioned are explained in detail, the comments in Section 2.1 should be recalled. The constructivist understanding that every mental and social system, and thus every socio-technical system, can only make an image (model) of the original (real reality), limits the possibilities of a comprehensive and objective analysis, as does the very limited willingness of the mental and social systems to be analyzed to be looked at.

4.2.4.1 Study of Documents

In many cases, documents are available that can provide information about the situation to be analyzed. What documents usually have in common is that they

- made for the purpose of providing data and/or information in such a way that it can be used by a specific target group,
- are thematically limited (Luhmann 1996, pp. 213 f.),
- remain unchanged over a longer period of time and
- can be found, for example, organized in archives.

Apart from that, documents can come in many different forms, for example

- as printed or handwritten documents, i.e. documents that mainly contain text, while other forms of presentation such as images or drawings merely supplement the text. Examples are books (textbooks, specialist books, conference proceedings, fiction, etc.), patents, newspaper and magazine articles, and websites. or posts on social media, records of observations and surveys or experiments, legal texts, standards and guidelines, letters (analog and digital as e-mail) and so on,
- as drawings, for example in the form of sketches, technical drawings, maps and so on,
- as images, for example in the form of photos, paintings or graphics, whereby the transition to drawings can be fluid,
- as films,
- as sound recordings, although films usually have sound added to them,
- as data collections, i.e. as collections of individual data, usually organized in tables, for example in electronic databases,
- as completed forms from the company suggestion scheme (Böhm 2005, p. 181), which usually not only identify problems but also suggest possible solutions,
- and so on.

Documents are a form of *communication*. In Section 2.6.2.2, the foundations were laid for a systems-theoretical understanding according to which acts of communication are the elements of social systems (Luhmann 1996, pp. 194 ff.). Therefore, a document study essentially corresponds to a communication analysis. Since communication serves to coordinate the actions[8] of social and thus also socio-technical systems, the analysis of this coordination focuses primarily on their *dynamics*. At first glance, this seems to require explanation in that documents, due to their aforementioned property of remaining unchanged over a longer period of time, appear static and thus structure-determining. It only becomes clear at second glance that they are uninteresting in this respect. What matters is what contribution they make to communication, in other words, by whom they are used, when, how and why (the simple questions in the German W-Fragen also help to clarify this once more).

The fact that they remain unchanged over a longer period of time gives documents a great advantage over other forms of communication, because communication is usually very fleeting: one word gives rise to another, as the saying goes. Individual acts of communication are "events that only ever take place in the present and have already passed by the moment they occur. In order to maintain social systems over time, communication must therefore be continued..." (Simon 2015, p. 22). In this context, documents are, so to speak, captured parts of a communication. They freeze parts of it for a certain period of time, preserve it, and represent a kind of memory that can be looked into. And the recipient has a certain amount of time to engage with it, can read, see or hear one thing or another multiple times and discuss it with others. However, if the recipient does not want to carry out purely historical analyses, the time they can take for this is not too long, because even documents lose their relevance over time for their use as part of communication acts: norms and guidelines are revised, laws reformed and so on.

Communication always takes place between at least two people. The person who initiates an act of communication is often referred to in the social sciences as *Alter* (Alter Ego), and *Ego* is the person at whom an act of communication is directed (Luhmann 1996, p. 95). In this respect, a person's role changes continually: sometimes they take on the role of Alter, sometimes that of Ego, depending on whether they are speaking or listening.

If documents are understood as parts of communication acts, however, it must be clarified how the statement that communication always takes place between at least two people can be reconciled with the possibility that one can also interact with documents one has created oneself. The answer to this question is based on the observation that documents remain unchanged over a longer period of time. As a rule, one will also have created one's own documents in the past and in a different place, so that a certain amount of time has passed and a spatial distance has been created between their creation and their analysis. Due to the knowledge and experiences gained in this time and in other places, you are no longer the same person after this time and at this spatial distance as you were at the time and place of the documentation. So you communicate with yourself and try to fathom what may have motivated you at that time and in that place to document what you find at the time and place of the analysis. This can be just as challenging as trying to understand what others were thinking when they created documents. If you want to experience this for yourself, you might look through the photos from your last vacation. The more there are, the more often you will wonder what prompted you to capture the scene in question.

[8] For the concept of trade, see Section 2.6.2.2

According to Fig. 4.5, an act of communication always forms a unit consisting of *information, message and understanding* (Luhmann 1996, p. 194 ff.). In this unit, documents only represent messages. That is why they were referred to above as parts of communication acts. These parts are, as also mentioned above, uninteresting on their own. In the context of a situation analysis, however, they become interesting if they are accessible to third parties (observers) and thus enable them to listen, watch or read while the people whose actions (communication) are to be analyzed are talking. However, it is important to realize that this is by no means enough to observe everything that is relevant in the course of a communication, because every message includes information that Alter wants to communicate and understanding from Ego.

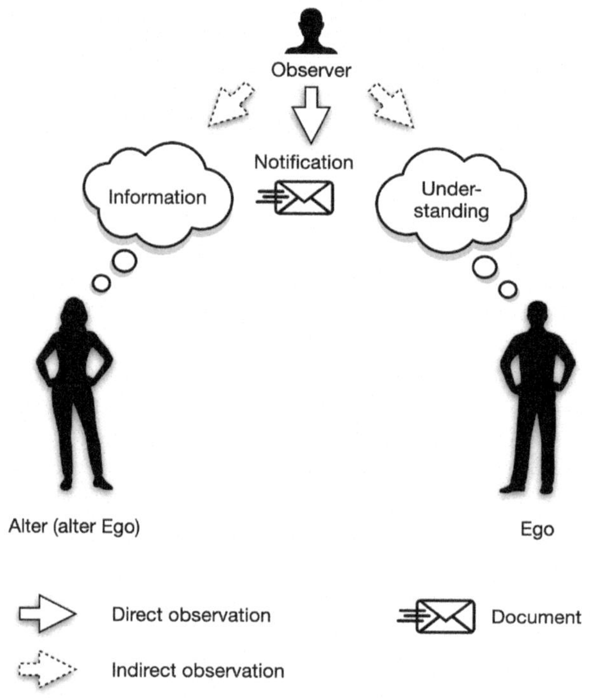

Fig. 4.5 Unit of information, message and understanding

From the constructivist basic understanding follows that both Alter and Ego are quasi-trapped in their respective *self-reference* – somewhat comparable to the ability of Zaphod Beeblebrox, which is the only one that enables him to survive the Total Perspective Vortex unscathed. In the social sciences, the resulting relationship is referred to as *double contingency* (ibid., p. 148 ff.). On the one hand, the *selection* of information that Alter considers worthy of communication is always also possible differently. Consequently, Alter must translate what presents itself as information to him on the basis of his self-image and world view into a message (ibid., p. 197), and this is also associated with considerable uncertainties. Even correct spelling and punctuation is difficult for many, not to mention printing. In addition, there is the necessity, especially in a professional context, to use technical languages, to make technical drawings, to develop mathematical equations and much more. The messages created with the help of such forms of expression, in this case documents, must subsequently be transmitted to Ego without errors. To do this, it is often only necessary to fulfill technical and organizational requirements: a reliable postal service, functioning computer hardware and software, a well-organized Document Management system. and so on. Finally, Ego must extract the information encoded in the message and understand the consequences that follow from it. To do this, the message must be connectable to what he already knows, can do and wants, and that is always possible in different ways (contingent). With the *selective perception* that inevitably results from his self-reference, he will only take note of information that he himself considers relevant. Whether the information has been understood by Ego can be deduced by Alter, on the one hand, from Ego's actions and, on the other hand, from possible answers.

In addition to these numerous opportunities for communication to fail, a situation analysis shows that the communication is observed by third parties (outsiders). They also bring their own self-reference with them, and in addition, in the case of the message discussed here, they can only take note of one of the three components of a communication act directly (Fig. 4.5). The other two parts of a communication act, i.e. the information and the understanding, can only be inferred indirectly at best. Nevertheless, the inference of these two parts of communication acts must always be taken into account. A situation analysis based only on documents falls short. However, even if all three parts of a communication are included in the analysis, the probability of understanding remains low both between those who communicate with each other and for the observers of this communication. Misunderstandings are more the rule than the exception.

Due to the fact that documents from the perspective of observers are often created by third parties and usually at an earlier point in time and/or in a different place, their analysis in a scientific context is referred to as *secondary research*, while methods that are applied by the analyzing subject directly to the analyzed object are referred to as *primary research*. This usually includes self-conducted observations, surveys and experiments. Strictly speaking, however, it is necessary to adhere to the three Aristotelian units, which were originally used in ancient times to construct dramas: the *unity of space, time and action* (the action here corresponds specifically to the survey and its analysis). If, on the other hand, you conduct a survey and document it in a protocol that is only evaluated at a later point in time, in a different place and under different circumstances, then, as already mentioned, it is strictly speaking already considered secondary research (document study).

One specific method that takes such an understanding of social systems into account is content analysis, which is particularly widespread in *communication and media studies* (Früh 2017; Mayring 2016, p. 114 ff.). It aims to deliver results that are objective and thus also intersubjectively comprehensible despite the high probability of misunderstanding. While qualitative variants of this method aim to understand the content of documents, quantitative variants limit themselves to describing and explaining documents. Therefore, qualitative content analysis is particularly relevant here, since it not only deals with the documents as messages, but also tries to explore what information Alter wanted to encode in them and what Ego finally understands (Fig. 4.5). Objectivity and, consequently, intersubjective verifiability should be achieved, among other things, by analyzing the data on the basis of previously defined categories (Kuckartz 2009, p. 57).

Another method of studying documents is *excerpting*, which is one of the basics of scientific work. The starting point is a central (research) question (this can also be a hypothesis) that guides interest. With a view to answering this question, the recipient notes the statements of a document that he considers essential in a condensed, summary way. These can be supplemented by the user's own considerations, taking care to keep them clearly separate from the statements of the received documents. The excerption can be supported by the use of Literature Management Programs.

An alternative form of documentation of excerpts is the *card index*[9]. In it, notes on analyzed documents, including source references, are stored in card index boxes and organized according to keywords. This makes it easy to link notes with each other. In electronic (virtual) card indexes (wikis), this can be done using hyperlinks, for example.

Finally, *knowledge maps* (iMaps) can be created in which records are spatially arranged and related to each other in order to support a visualization of the relationships. This corresponds to the structuring of systems explained in Section 2.4.2 using Fig. 2.10. The aim is to recognize patterns such as the topologies sketched in Fig. 2.11 from the arrangement of the sub- or partial systems and elements as well as from their relations. Specific software products are also available for this.

All the methods mentioned focus to a greater or lesser extent on the direct analysis of documents (Fig. 4.5). If the hidden parts of a communication act, the information at Alter and the understanding at Ego, are considered at all, then, for example, in the form of a certain amazement that company guidelines are not followed, or that some guidelines are not even known, or that supporting documents (e.g. forms) are not filled out, not filled out completely, or not filled out correctly. This quickly leads to the simple demand that this should be done consistently in the future. The fact that there are reasons why documents do not always have the desired effect is rarely questioned. If such analyses go further, they often lead to the assumption that the causes for limited use lie in a lack of user-friendliness or in insufficient availability. Other analyses conclude that there is a lack of skills among potential users, and in such contexts a lack of motivation is usually also identified.

Much of this will often not be entirely wrong, and so in many companies, usually well-trained and highly motivated professionals are repeatedly taking measures to eliminate the deficits associated with this[10]. All the more remarkable, then, is the finding that the problems outlined do not appear to become significantly smaller over time. This leads to two conclusions: (1) On the one hand, communication necessarily always remains uncertain, in particular due to the double contingency associated with it. From this, in turn, one can conclude that the lever should not be applied so much to ever more extensive attempts to improve communication, but rather to the fault tolerance of the systems that depend on communication, and that includes all systems. (2) On the other hand, the *hermeneutic method* (Sect. 4.1.4.3) can help to make mutual understanding between communication partners on the one hand, and understanding of communication between partners by an

---

[9] A recent example of this is the card index (in German: Zettelkasten) used by Niklas Luhmann, who is often quoted in this book.

[10] These include, among others, numerous students of the author, particularly in the course of writing their theses.

observer of that communication, a little more likely. Above all, *patience* and *empathy* are required for this. In addition, all participants need to be able to *reflect on themselves* in order to recognize their own self-reference and then to find and accept it in the other person as well. Finally, *mutual trust* and *goodwill* are important prerequisites for successful communication.

4.2.4.2 Survey

The survey is characterized by a whole series of specific features, the characteristics of which can largely influence (1) success or failure. In order to do justice to the importance of these characteristics, they will be discussed individually later. To do this, they will first be removed from a rough explanation. A relevant definition already provides initial clues. Surveys, for example, are understood as a planned procedure (method) with an objective, in which the test subject is to be prompted to provide verbal information through a series of targeted questions or stimuli[11] (Scheuch 1973, pp. 70 f., cited here from Diekmann 2020, p. 493).

Thus, surveys are also part of the communication between mental and social systems (Fig. 4.6), as is the study of documents, as explained above. However, they differ from other forms of communication, such as observation or the study of documents, in that the questioner and the respondent interact with each other more or less directly (2) verbally (3). In this regard, it will be discussed in more detail later that (3) a common language is important here (Diekmann 2020, p. 441) and that (2) the communication partners are more closely involved with each other than, for example, in an observation, which can also be done from a distance, or in a document study, which can be hidden from those whose communication is being observed at the moment of observation. Fig. 4.6 expresses this by showing that the observer and the questioner are the same person, each taking on different roles.

In the following, surveys will be distinguished from other forms of direct verbal communication (ibid., p. 439, Atteslander 2010, p. 110 f.). For example, interviews differ from everyday *conversations* in that the latter often take place between people who belong to a common social system (e.g. a private household, a department in a company, a singing club) or who (e.g. a family), while interviews take place in an *artificially created situation* in which the communication partners tacitly agree that no lasting relationship should arise from it. This suggests that the direct interaction (2) that has just been mentioned will need to be considered in even more detail.

Furthermore, conversations in everyday life do not have to be geared towards a specific goal. First and foremost, they serve the pure self-preservation of social systems, which, because their elements consist of acts of communication and because these are fleeting, must repeatedly produce new acts of communication in order to continue their existence. So pursuing goals is possible in the course of conversations, but not necessary. By contrast, surveys in the sense intended here are always geared towards a specific goal. Achieving this goal corresponds to the above-mentioned success (1) of an interview.

---

[11] Stimuli can be understood here as irritations or perturbations (Sect. 2.1)

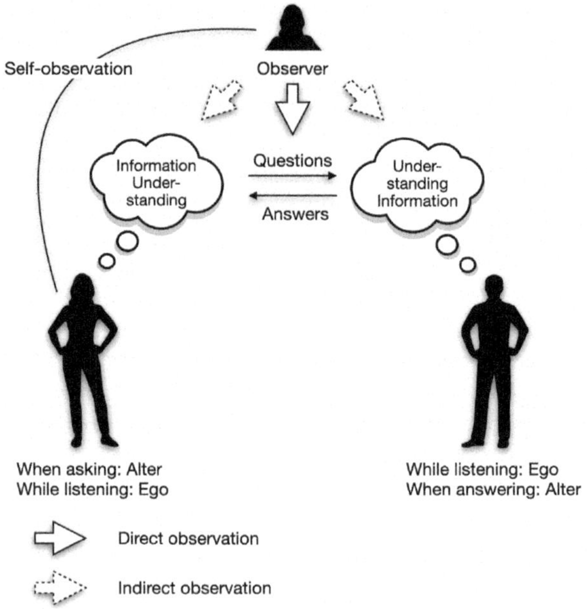

Fig. 4.6 Surveay

In addition, (4) interviews are *asymmetrical* in that the interviewer mainly asks questions and the interviewee mainly answers. However, it is widely known that the one who asks the questions leads. A possible subordination of the interviewee can intimidate them and result in them not having the courage to express themselves openly and honestly. Therefore, one of the skills of conducting an interview is to do so on an equal footing.

Finally, everyday conversations can have *consequences* for the participants, as their contributions are evaluated and possibly sanctioned by the approval or disapproval of the other person. This also applies to other forms of direct, asymmetrical communication such as *interrogation* or *examination*. In contrast to this, it is important in surveys that the questioner adopts a *neutral position* and gives the respondent the assurance that his answers will have no consequences for him. This can be done, for example, by agreeing *confidentiality* or by promising to *anonymize* the statements.

Now that five characteristics of surveys that have a significant influence on their success have been identified, the first three characteristics will be discussed in detail below. A sufficient understanding of the asymmetry and the necessity of the consequences of the respondent's answers should already have been created, since no system-theoretical principles need to be invoked for this.

The success of a survey (1) can generally be described as achieving the pursued goal, but must be described in more detail with regard to the specific application. Now, the goal in general will be to obtain (1a) objective (1b) information about the object of the survey from the respondent.

(1a) *Objectivity* here means the independence of information from the subject, i.e. both from the subjective view of the respondent and from the subjective perception (understanding) of the respondent's statements (messages) by the questioner. However, this is countered by the constructivist insight that people basically have no direct access to real reality, that all perception of reality is subjective. Therefore, objectivity can only be approximated, but never achieved.

In addition, (1b) information is one of the three parts of any communication act, along with the message and understanding, and each of these parts leaves considerable room for misunderstandings or even a lack of understanding. In principle, this cannot be completely prevented. However, the susceptibility to errors can be reduced by what has already been said in connection with the study of documents and what basically applies to any kind of communication: however much an effort to achieve mutual understanding , the lever is not only to be applied to ever more extensive attempts to improve communication, but also to the *fault tolerance* of the systems that depend on communication, and that includes all systems. Furthermore, the *hermeneutic "method"* (Sect. 4.1.4.3) can help to make mutual understanding between communication partners a little more likely. Above all, *patience* and *empathy* are required for this. In addition, all parties involved need to be able to *reflect on themselves* in order to recognize their own self-reference and then to recognize and accept this in the other person as well. Finally, *mutual trust* and *goodwill* are important prerequisites for successful communication.

In this context, the question also arises as to where the respondent gets their *motivation* from to be interviewed (Diekmann 2020, p. 441). This can result from an employment relationship. In such a case, participation can be ordered by a superior and is compensated by remuneration. However, this can result in the imposition of *sanctions*, which, as mentioned above, can stand in the way of a survey's success. One consequence of this can be *socially desirable answers* (Diekmann 2020, p. 447 ff.), which stand in the way of the desired objectivity. Alternatively, the willingness to be surveyed, even for free, can be explained by the self-referentiality of mental systems: people like to talk about themselves. On the other hand, the insight that humans are social beings explains that they like to communicate with others. Among other things, being given a chance to speak and being listened to gives the speaker a feeling of appreciation. With regard to the actual conduct of interviews, this results in the recommendation that participation be *voluntary* and that the interviewer should be *respectful*.

The second characteristic of surveys was identified as the (2) more or less unmediated encounter between the communication partners. If the two meet directly, this is also referred to as an *interview*. In this case, the observer becomes involved, i.e. not only observes the observed object, but also himself at the same time. Although this also applies in principle to all other types of observation, in this case it is particularly difficult to avoid *influencing the object* and thus falsifying the research result (Atteslander 2010, p. 136). Although the survey relies on verbal communication, *non-verbal communication* such as body language and tone of voice is difficult to avoid. However, such influences can be controlled to a limited extent by designing the survey situation. For example, a prepared *interview guide* can contribute to objectification in that the questions are carefully formulated in advance and do not arise spontaneously from the specific interview situation (Mayer 2013, p. 37 ff.). Among other things, *structuring* the questions can also help here. Furthermore, the influence of body language can be excluded by conducting a *telephone survey*, while the influence of the tone of voice remains. Both influences can be avoided by using a *questionnaire*.

Another influence in this context comes from the *situation* in which a survey is conducted (Diekmann 2020, p. 466 ff.). These include the time and place. If, for example, an employee is interviewed in his office during his paid working hours, his answers will be different than in a situation in which he is summoned to his boss's office shortly before the end of his working day. And the presence of third parties also usually has a significant influence on the course of the interview. An employee will express themselves differently in the presence of a representative of the works council than in the presence of a legal representative of their employer. They will also describe a situation differently immediately after it has occurred than they will a few weeks later.

In view of these insights, it is necessary for the interviewer to reflect not only on himself in his role as an observer, but also in his role as a participant (Fig. 4.5) when conducting a survey. This applies in all three phases: in the course of preparing a survey, potentially falsifying influences such as asking leading questions, one's own body language, the choice of space and time, and so on should be eliminated as far as possible. To do this, the interviewer may need to have a relevant qualification (ibid., p. 486). During the interview, the interviewer should continuously observe themselves and control their own behavior (own opinions, emotions, etc.). In the course of the subsequent evaluation of the interview results, it is also important to filter out any distortions that may have occurred. The third of the identified characteristics of a survey is (3) the verbal form of communication. An initial prerequisite for its success is that both communication partners speak the same individual language, for example German or English. In addition, other forms of expression such as sign language, sign languages, as used for example in technical drawing, or formal languages such as those used in mathematics, can be added.

The third of the identified characteristics of a survey is (3) the verbal form of communication. A first prerequisite for its success is that both communication partners speak the same individual language, for example German or English. In addition, other forms of expression such as sign language, sign languages such as those used in technical drawing, or formal languages such as mathematics can be added.

In view of the systemic insights into the functioning of social and psychological and thus socio-technical systems, the deeper layers of communication must also be taken into account. In Section 2.6.1, for example, it was explained that systems distinguish themselves from other systems by the internal use of specific *semantic codes*. So, for example, if an engineer considers a certain technical measure to be necessary, it may occasionally be rejected by the controlling department because it is viewed quite differently from the respective perspectives: For one, it reduces the probability of failure (MTBF) of a machine or plant, for the other, it reduces the money in the till. And different *social milieus* can also have an influence on the possibilities and limits of communication (Diekmann 2020, p. 441 f.). In view of the necessity of maintaining a professional distance between the questioner and the interviewee, which will be discussed below, the aim is not to overcome the boundaries of the respective systems to which the two communication partners belong, but to penetrate them. Metaphorically, the boundary between the social systems represented can be interpreted as a *semipermeable membrane* that allows questions and answers to pass through, but maintains the boundary for everything else.

Another boundary lies between the mental systems involved, to which the same applies as to the boundary between social systems. In contrast to this, however, meaning is processed not through communication but through thoughts (Fig. 2.1), but these too are formed with the help of language, so that thinking can be transformed into communication relatively easily. This only becomes more difficult when deeper layers of consciousness, i.e. the unconscious or subconscious, are to be brought to light, but this is then a matter for psychology and usually no longer the subject of Systems Engineering & Management.

Some of the methods already mentioned will be explained in more detail below. However, this cannot come close to replacing the reading of works such as those by Nina Baur and Jörg Blasius (2019), with more than 1,500 pages, or by Andreas Diekmann (2020), with almost 800 pages, which are dedicated to the field of empirical social research in detail. Therefore, works like these are explicitly recommended as a supplement and for more in-depth study.

Interview

The process of an interview can be structured on the basis of the *problem-solving cycle*, as might be expected. To do this, the interviewees must be informed in good time and in detail about the reason for the interview. This must be prepared in the initiation phase.

As part of the *situation analysis*, the specific questions to be asked during the interview should be formulated in advance. Key questions in this regard could be:

- What don't we know?
- Who can know what we don't know?
- Under what conditions are those who could know what we do not know willing to share their knowledge with us?
- How can we reach them?
- Who should approach them?

In the course of *formulating the objectives*, questions such as those just mentioned are answered as specifically as possible:

- What knowledge should be available after the interviews have been completed?
- Which potential interview partners should be approached when and how by whom. During the initial discussions, it will be necessary, among other things, to at least begin to communicate and discuss the occasion and the initial situation in relation to the intended questions. And the conditions (time, place, remuneration, etc.) can also already be discussed.

Depending on the results of the previous phases, the methods and tools to be used in conducting the interviews must now be defined.

- This includes determining the *degree of structure* of the questions (Atteslander 2010, p. 134 f.). This ranges from an unstructured conversation to a fully structured series of pre-formulated questions and includes a wide range of possible combinations. These may include questions to be answered with a simple affirmation or negation, questions in which started sentences are to be completed (sentence completions), which are known from the program "Was nun, Herr/Frau..." (what now, Mr./Ms. ...) on the Second German Television channel (ZDF), or open-ended questions to which free and detailed answers are to be given. It may also include questions that the respondent should bring up themselves. There are virtually no limits to creativity in this regard.

- Another methodological specification can concern the *type of communication* (ibid., p. 135 ff.; Diekmann 2020, p. 440 f.). This is divided into soft, hard or neutral behavior on the part of the questioner.
  - In a soft interview, the interviewer remains largely in the background and lets the interviewee do most of the talking. The interviewer's interventions serve not only to guide the conversation, but also to encourage the interviewee when needed. The aim is to create a harmonious interview atmosphere based on mutual sympathy.
  - The hard interview, on the other hand, is designed to put the interviewee under pressure. For example, a rapid succession of questions is used to prompt the interviewee to give spontaneous answers, on the assumption that such answers are more authentic than those that emerge from careful consideration and may therefore contain socially desirable elements. In this way, a hard interview is akin to an interrogation.
  - A neutral interview should minimize the personal relationship between the interviewer and the interviewee. Instead, the focus is on a pronounced objectivity. However, this can lead to the interviewee perceiving the situation as rather unpleasantly distant.
- Furthermore, the *interview guide* to be used, if at all, must be formulated specifically in this phase (Mayer 2013, p. 43 ff.). Notes on this are explained in the following Section 4.2.4.2 on formulating questionnaires and should therefore not be anticipated here.
- The *documentation* of the interviews is of particular importance, because the oral form of communication itself does not include any preservation of what has been said. For this purpose, *protocols* (minutes) are used, which, in contrast to *transcripts*, also contain, in addition to the content of what was said, control data such as the occasion and the initiator, the date and time (start, end), the location, the minute-taker, the chairperson, as well as the participants (possibly also absentees) and often also their signatures, the distribution list and the degree of confidentiality, storage location and period, etc. As an alternative or supplement to written minutes, other documentation techniques can also be used. These include, in particular, *audio and video recordings*, whereby the latter can be used in particular when interviews are conducted as video conferences, since corresponding computer programs usually include such functionality. However, the *personal rights* of all participants must be taken into account, which usually requires an explicit declaration of consent. It is important that minutes are complete and accurate in terms of content and that the documented content is relevant. These may also include occurrences such as applause, laughter or interjections, which are often noted with the help of special characters, for example, a dash for a significant pause in the flow of speech. Furthermore, it is important that the documentation does not yet interpret what

has been documented. Rather, it is important that it is as *authentic* as possible (original, genuine).

- If no documentation is to be made during an interview, for example, in order not to disturb the course of the conversation, a *memory protocol* can be prepared as soon as possible afterwards.
- If the exact reproduction of an interview is important, a *verbatim protocol* is prepared in which every word is written down. This is the case, for example, with sessions of the German Bundestag. As a rule, this is only possible with the help of *stenography* or speech recognition software. In a scientific context, the verbatim protocol is also referred to as a *transcription* (Mayring 2016, pp. 89 ff.). If it is not only the spoken word itself that is important, but also dialects and linguistic subtleties, the *International Phonetic Alphabet* (IPA) can be used. However, since this is relatively difficult to learn, in practice it is often helpful to also reproduce linguistic peculiarities in the common alphabet.
- The contents expressed are documented in a *progress protocol* in the order of their expression in a level of detail to be agreed in each case. The present or past tense can be used as the tense; speeches are rendered in indirect speech.
- The *summary of results* only summarizes specific decisions (consequences). The present tense is chosen as the tense for this. A particular difficulty with both the process protocol and the results protocol is to not distort what was said and in particular what was meant due to the associated omission of speech contributions. For this, Philipp Mayring (2016, p. 94 ff.) lists six possibilities for reducing what was said:
  - Omission: Statements that can be inferred from the rest can be omitted.
  - Generalization: Colloquial statements that usually describe a fact using a larger number of words can be described in a generalized (abstracted) way using a technical term.
  - Construction: multiple statements that describe a situation as a whole can be replaced by the name of the situation being described.
  - Integration: A statement that is contained in another statement can be omitted.
  - Selection: of repeated statements concerning the same facts, one example can be selected, while others are omitted.
  - Bundling: Scattered statements can be presented in one place in bundled form.
- A *protocol of information* (selective protocol) represents a hybrid of a process and result protocol in that it assigns essential contents and consequences to the topics discussed (agenda items: TOP) (Mayring 2016, p. 97 f.).
- Finally, the other influencing factors mentioned above are to be designed in this phase. These include the choice of location and time at which the interview is to begin, and its duration. Other influencing factors that can be designed include the dress code, the number of participants, the seating arrangement, the catering, etc.

In the next phase, the interviews are conducted (Dieckmann 2020, p. 486 ff.; Mayer 2013, p. 95 ff.). This can be followed by an introductory explanation of the occasion, the current situation including open questions and, in particular, the questions that are to be clarified in the course of the interview, and also the aim of the interview and the selected methods, techniques and tools (e.g. software).

After an interview has been conducted, it is *evaluated* (Mayer 2013, pp. 47 ff.). This can also include all interviews belonging to a survey. The aim is to obtain generalizable statements. This can be done by *paraphrasing*, in which various statements on specific facts are summarized in one's own words in an interpretive manner (Mayring 2016, pp. 109 ff.). In this context, it can be useful to cite sources in order to ensure the traceability of the insights gained. In the sense of the hermeneutic circle , paraphrasing can be done multiple times.

Questionnaire

Questionnaires can also be kept open-ended, but they are particularly well suited for standardized surveys of large groups of people when it seems both possible and sensible to dispense with direct contact between the questioner and the respondents. This, in turn, requires, among other things, that the questions can be asked in such a way that explanations, discussions, and the like are not necessary.

Closed questions, which at the same time link all possible answers to the question and, if possible, also categorize them, are suitable for this purpose (Atteslander 2010, pp. 147 f.). The following types of questions are distinguished:

- *Identification type* questions require the mention of a person or group of people, a place, a time or a number, in particular by using the W-questions (who, how, what, when, where, how many or which). An example is the question about the place of birth (where), a possible answer: Bremen.
- The *selection type* includes questions with predefined *alternatives* from which the respondent can select one or more answers. An example is the question: What do you prefer, coffee or tea? One possible answer is: coffee.
  - Since only one of the two possible answers is to be allowed in this case, the question type can also be described as an alternative question.
  - If more than two answer categories are allowed, it is also referred to as a *multiple-selection question*. A special form of the multiple-selection question is the *scale question*, which can be used to determine intensities or frequencies. One example is the question about the evaluation of a product or service. A distinction must be made between even-numbered and odd-numbered scales: Even-numbered scales, such as "very good – good – bad – very bad", force the respondent to take a clear position, while odd-numbered scales, such as "good – average – bad", also allow a neutral answer. It is not possible to say in general terms which of the two types of scale is more suitable. It depends on the respective context. However, it is important to make a conscious decision between the two types of scale.
  - *Yes-no* questions are questions that can be answered with a simple yes or no. For example: Would you recommend this book? The desired answer is, of course, yes.

Quantitative and Qualitative Surveys

In empirical social research, a distinction is made between quantitative and qualitative approaches (e.g. Atteslander 2010, p. 12 ff.).

Quantitative methods are based on probability theory (Sect. 4.1.4.2). In this case, a *sample* (partial sample) is drawn from a *population* that is usually too large to be examined in full (full census). Provided that the sample was selected at *random* (other aspects are also possible in principle), it is possible to draw conclusions about the *characteristics* of the elements of the population (observations, statistical units) with a certain probability (confidence interval) with regard to certain characteristics of the elements of the sample. Questionnaires with mainly closed questions are a common basis for this.

Qualitative methods are based on non-standardized data, for example on answers given in the context of interviews to open-ended questions. This is based on the assumption that individuals or small groups can also make generalizable statements. A common form of such methods is the *expert interview*. In this case, the validity of the statement is linked to the status of the respondent as an expert. This eliminates the need to interview other people and, in particular, laypersons. However, this presupposes that the expert is identified as such on a generally recognized basis (qualification, experience, reputation).

The reliability of expert statements can be increased by applying appropriate methods. The *Delphi survey* (Atteslander 2010, p. 143) is an example of this. It is used primarily when individual experts are unable to provide sufficiently reliable statements about a set of facts. This applies especially to statements about the future (forecasts). By including a larger number of experts, such surveys thus at least represent an approximation to quantitative methods, although the typical number of people involved is usually far too small to speak of a hybrid of the methods.

Specifically, a Delphi survey is conducted over several rounds. In the first round, the questions for the selected experts, who are often not made aware of each other in order to avoid possible rivalries or coalition-building, can be about which questions seem relevant at all for the given topic. A moderator summarizes the answers and presents them to the experts in the second round with the question of whether everyone agrees with them. This means that an expert also receives the answers of all the other experts, can reconsider their own answers in this context and comment on the answers of other experts. This can be continued until a consensus has been reached on the relevant questions. In the following round, answers are then obtained for all questions. These are summarized by the moderator and the resulting outcome is presented to the experts again, so that they have the opportunity over several rounds to reflect on their own answers in the light of the answers of other experts, to change them if necessary, and to comment on the answers of other experts. In this way, a moderated discussion takes place, which, however, largely avoids group-dynamic influences. A well-known example of a successfully implemented Delphi method is the one conducted in the 1990s by the Fraunhofer Institute for Systems and Innovation Research (ISI) on behalf of the German Federal Ministry of Education and Research (BMBF) on the global development of science and technology.

4.2.4.3 Observation

In principle, all empirical methods can be understood as forms of observation (Diekmann 2020, p. 548). However, this article is concerned with the systematic perception of activities and events at the time they occur with the aim of gaining knowledge. In this context, it is not so much static as dynamic processes that are observed, i.e. those that are particularly dependent on space and time (Atteslander 2010, 73; Bracht et al. 2011, 86 ff.). The possibility of using all the senses gives observation more comprehensive access to the world than questioning, which is limited to the spoken word.

Observation can be both quantitative and qualitative. *Quantitative observation* is based on critical rationalism. This results in the postulate of freedom from value judgments in the statements to be obtained, the separation into the discovery and justification context in the procedural model for problem solving, which can go so far as to have the evaluations carried out by different people than those who carried out the observations, and the theory test (Atteslander 2010, p. 76 ff.). Reality is regarded as objectively given, and can be grasped with the help of controlled methods, especially through *measurement*.

Qualitative observation draws on hermeneutics (Gadamer 2010) and Phenomenology (Berger and Luckmann 2003) and assumes that reality is constructed by individuals, that meanings are attributed to it that are not inherent in it. This results in specific demands on the observation (Atteslander 2010, pp. 77 f.):

- *Openness*: No theories or hypotheses may be developed in advance that underlie the observation in order to see what one wants to see, but to see what the object of observation reveals. The application of structured and standardized methods is not effective.
- *Process character* of object and research: The aim is to grasp the processes of interpreting reality, whereby observation itself is understood as a process of interpretation and communication.
- *Reflexivity* of the research: theories and hypotheses are only generated, modified and gradually generalized during the observation (hermeneutic spiral). There is no separation between the context of discovery and justification.
- *Explication* of the approach: The process of interpretation must be disclosed to enable intersubjective examination.
- *Communication*: Everyday theories and scientific statements are assumed to be similar because both are based on the everyday rules of communication.
- *Problem orientation*: The aim is not to simply test a theory, but to critically and practically address a specific problem.

Peter Atteslander (2010, p. 80 ff.) distinguishes four components of an observation: the (1) field of observation, the (2) units of observation, the (3) observer and the (4) observed.

- The (1) field of observation includes not only the spatial area in which the observation takes place, but also the time period of the observation and the conditions under which it takes place.
- The (2) observation units are those sub-systems, systems and elements of an observed system that are used to gain insights. For example, the transport processes in a warehouse can be observed on the basis of the trips of forklifts, which are used as representatives

for all other types of transport, so that transports with other means of transport, such as an overhead crane, are neglected to simplify the investigation.

- With regard to (3) the observer, one of the decisions to be made is whether the observer will drive the forklift truck and thus gain their own experience and possibly even be seen as one of the forklift truck drivers by the other forklift truck drivers, thus obtaining information that is often not shared with outsiders, or whether they will merely watch the forklift truck drivers. The former is referred to as participant observation, the latter as non-participant observation.
- With regard to the (4) observed, questions such as whether a selection should be made (e.g. only permanent forklift drivers and no temporary workers) and what information the observed should receive about the observation or need to preserve personal rights and to what extent explicit consent is required.
- A typology of the most important forms of observation can be made on the basis of the degree of
  - their structure
  - their openness and
  - their participation (Atteslander 2010, p. 93 ff.; Diekmann 2020, p. 564 ff.).

Structured observations define the *subject and type of observation* in advance. This requires a good knowledge of the field of observation. Furthermore, a hypothesis is formulated in advance, which is to be verified or falsified by the observation. Categories are also defined, which are used to describe the observed in order to exclude subjective influences of the observer. In logistics, for example, transports are recorded in categories such as duration and time of events, as well as quantities and frequencies, if an underlying hypothesis assumes that these are predominantly mechanistic processes. In this respect, the insights to be gained are limited to mechanistic facts, so that, for example, motives of the executing persons, which may have an influence on the observed categories, remain unconsidered and limit the usefulness of the insights gained.

In an *unstructured observation*, hypotheses are not developed at the beginning, but at the end. Since no specific categories are used to guide the observation, it is often dispensed with during the observation. Instead, the documentation is carried out shortly after the observation is completed, in order to make an initial selection and structuring of what is considered relevant in the reflection of what has been observed. In this respect, an unstructured observation ultimately also provides a structured result. In contrast to structured observation, however, the structuring work is only done retrospectively and is thus influenced by what has been observed.

In terms of the *degree of openness* of observations, *covert observations* represent one of the extremes. In practice, they are rarely used, especially in a business context, because they encounter legal limits (co-determination, data protection), for example. It is also problematic that if findings obtained with the help of covertly gathered information become known, mistrust can arise and thus cause greater harm than good. One reason for covert observations may be the hope that the observer will influence the observed as little as possible. This can be done by spatial camouflage or by taking on a foreign role (so-called V-man). The latter option in particular places very high demands on the observer, since he has to play a dual role and perform two tasks simultaneously.

As a rule, more or less *open observations* are carried out. Although mistrust and behavior adapted to the situation and thus not authentic can be expected in such cases from those being observed, it is often possible to reduce this with increasing observation time.

In the *dimension of participation*, a distinction is made between active and *passive participation* of the observer. In the case of passive participation, which is often used in observations under laboratory conditions, the observer can concentrate fully on the observation and its recording. However, it may be disadvantageous that it is not possible to put oneself in the situation of the observed person, so that the interpretation of the observed is exclusively based on the observer's life world and thus does not do full justice to the observed. If, on the other hand, the observer actively participates in the observed event, the perceived information about the observed is more complex, but there is a risk that the neutrality of the observation will suffer. In addition, the influence of the observation on the observed object can be greater than in the case of a passive observation.

Furthermore, a distinction can be made between third-party and self-observation (Diekmann 2020, p. 564 ff.). However, strictly speaking, this is not entirely correct insofar as every form of observation includes self-observation, since the observer always influences the observed system through his observation. While this is evident in the case of overt observation (see above), the statement requires justification in the case of covert observation and in the case of document study (Sect. 4.2.4.1). This can be found by considering the goal of the observation, because the goal usually consists of influencing the observed system. This can aim at a (re)design or at a control to maintain a balance (regulation). Although the influence may occur later than the observation, so that the observation initially remains unobserved and thus unaffected by the observer, the observed system will usually notice that it was previously observed at the latest when the influence based on it sets in, and it will react to this in some way.

In this context, it is now important to reduce the influence of the observation on the observed system as much as possible in the course of an external observation in order to obtain an unadulterated observation result. A number of measures can be helpful for this. Above all, there needs to be a relaxed relationship between the observer and the observed. A vivid example from everyday life is taking photographs of people by agreement. Experience shows that not everyone likes to be photographed. Good photographers are therefore not only characterized by their technical skills, but also and above all by their ability to put the people they are photographing at ease. A confident manner, empathy and patience are important prerequisites for this. This also applies to observers. For example, training courses can be conducted for this purpose. Tools can also help, such as pre-formulated observation guidelines or the use of cameras (Bokranz and Landau 2012, vol. 1, p. 559). However, such tools can also interfere with the observation. This must be weighed up in each specific case.

In addition, self-observation is also necessary, especially in the course of an external observation. In doing so, it is important to reflect on one's own actions before and during the observation and to correct them if necessary in order to minimize one's own influence on the observed system. Furthermore, it is important to reflect on whether and, if so, to what extent one's own perspectives direct the focus of the observation and distort the interpretation of what is observed. *Supervision* can be helpful here, in which the observer exposes himself to observation by an uninvolved third party.

A practical example of an external observation is the *multi-moment recording* (synonym: multi-moment frequency study (MMR)) (Bokranz and Landau 2012, vol. 1, p. 565 f.). In this process, several short-term observations are carried out at random in order to statistically verify the results of the observations, to observe and categorize events. Examples of this are a machine being productively or unproductively idle, for example due to a lack of material or as a result of technical faults. For two types of incidents, a binomial distribution results; for more than two types of incidents, a multinomial distribution results. If the product of the number n of observations, the proportion p for one type of incident and the remainder $q = 1 - p$ is greater than $50,000$, which can usually be assumed (example: $n = 1,000$ observations, $p = 49$ for one type of occurrence and $q = 1 - 49\% = 51\%$: $n \cdot p \cdot q = 1.5million$), these distributions can be approximated by the normal distribution. The number of observations (sample size) n' required to ensure a defined confidence interval is then calculated as follows:

$$n' = \frac{3.14 \cdot p \cdot (100 - p)}{f'^2} \tag{4.1}$$

In this, f' denotes the desired confidence interval. For an event type with a share value of $p = 49\%$ with a confidence interval of 2.5, $n' = 1,535$ observations are therefore required. A more detailed explanation of the multi-moment recording would go beyond the scope here. Please refer to the source given above, which also contains a detailed description and an illustrative example.

The actual introspection, which goes far beyond the *self-observation* explained above for avoiding errors in the course of an external observation, aims to gain generalizable insights from observing oneself (Diekmann 2020, pp. 568 f.). The difficulty is that in this case the difference between subject (the observer) and object (the observed) is eliminated. Nevertheless, Sigmund Freud's self-analysis showed that intersubjectively verifiable laws can be recognized in this way. However, the risk of self-deception should not be underestimated. Methods of self-observation include keeping *diaries* or *notebooks* and even writing *autobiographies* (hypomnema). In the field of ergonomics, *self-recording* is also common (Bokranz and Landau 2012, vol. 1, p. 560 f.). For example, they are used to log shift sequences and to record times and quantities in production and logistics.

4.2.4.4 Data Analysis

A lot of information is stored in the form of data, mostly in electronic form (EDP). In the course of a situation analysis, it is important to gain information from the data and thereby gain knowledge about the situation (Schneider 2020).

In the context of a situation analysis, elements of data-processing systems are *characters*[12]. These can consist of numbers, letters or symbols. The *syntax* indicates the rules according to which the characters are combined to form a datum. For example, German

[12] The scientific discipline that deals with characters is semiotics. Incidentally, the relevant signs here can be further broken down from other perspectives, such as microelectronics.

syntax requires a date of birth to be encoded using three sequences of digits, each separated by the symbol of a full stop (Unicode: U + 002E), the first two-digit sequence indicating the day of birth, the second two-digit sequence the month of birth and the third, usually four-digit sequence the year of birth. An American syntax, on the other hand, encodes the month of birth in the first group of digits and the day of birth in the second, and uses a short dash as a separator (Unicode: U + 002D). So, to interpret a date correctly, the respective syntax must be known. However, this is only necessary for a correct interpretation, but it is not sufficient on its own.

In addition, *semantics* are needed to indicate the meaning of the date. Without a reference to this, a date of birth in the narrower sense would be an arbitrary sequence of characters. If, in a broader sense, it were at least known that it is a date, then at least readability would be ensured. However, in order to be able to interpret the date as a date of birth, it must be distinguished by semantics as such from other dates. However, that is still not enough to be able to interpret it meaningfully unless it is also indicated whose date of birth it is. This can be formally encoded by relating data, by embedding the date of birth in a coherent data set that consists of several data that together describe a person, including, among other things, the name.

Data is usually stored during operation, so personal data or data that can be linked to a person is stored in a company's HR department, for example. In such cases, the *information* that can be obtained from the data is known. In the context of the IPO model (Fig. 2.6), which is also widely used in computer science, information is defined as a combination of data and commands (Sect. 2.4.1). In this sense, a rule linked to the date of birth can be encoded in a computer system. This rule then generates a list of employees who will retire after a defined period of time, presumably monthly in the application example chosen here, in the human resources department. This is linked to the request (command) to organize the succession (promotion, transfer, new hire, onboarding, etc.). Another rule encoded in the IT system that accesses dates of birth can remind you one week before such a date to congratulate the employee in question and to present them with a bouquet of flowers.

In a social context, information has been defined as one of the three inextricably linked elements of communication (Sect. 2.6.2.2). It is triggered by a message and completed by understanding (interpretation). This understanding differs significantly from the sender-receiver model of information technology (Shannon-Weaver model), which attributes misunderstandings between sender and receiver to purely technical problems such as signal noise or incompatibility of the technology used and assumes error-free transmission of information if the message sent and the message received are identical. For a socio-technical understanding, such an understanding falls short. On the basis of the extended functional model of psycho-technical and socio-technical systems (Sect. 2.6.2), it was explained that messages, information and their understanding are selective acts, that the selections are made by the respective systems at their own discretion and that this can only be observed and interpreted to a limited extent from the outside. For example, information about an employee's upcoming retirement can also spark interest in an early release from the corresponding employee in the human resources department if it is assumed that performance decreases with age and that a higher salary often no longer justifies this with advancing age. In view of the legally, contractually or collectively agreed protection against dismissal, this can, under certain circumstances, lead to the processing of information and to decisions within the human resources department that aim to circumvent rules and, to this extent, will depend on remaining hidden.

Against this background, it seems instructive to aggregate networked information into knowledge. While information as elements of communication acts are too uncertain to ensure existence in the respective social system due to the associated selections that remain hidden, knowledge in pragmatism (Sect. 2.3) is defined by the fact that it can contribute to ensuring this existence through the accumulation and combination of all relevant information. In this context, it is not a matter of completeness or correctness, however these are defined, but only of viability. This understanding, however, excludes the storage or even the processing of knowledge without human participation and thus contradicts the possibility of purely electronic knowledge processing. This applies in particular because knowledge draws on all accessible sources, including, for example, *experience* in the sense of implicit knowledge (Sect. 3.5.4), *feeling* (so-called sixth sense) and morality. The contradiction against the impossibility of purely electronic knowledge processing, which is to be expected above all from computer scientists, is, however, insofar difficult as, in particular, neural networks can be trained under certain circumstances in such a way that they also provide experience-, feeling- and morality-based results by learning to imitate a corresponding human action and possibly even to draw conclusions from it. However, what remains is that machines – at least according to today's legal and moral concepts – do not assume responsibility and thus cannot take responsibility for the consequences of their "actions". For example, there is little point in locking machines up in a prison. Responsibility will – at least in the foreseeable future – presumably continue to be tied to humans.

Now that the differences and relationships between signs, data, information and knowledge have been worked out, the following Section will discuss how data analysis can contribute to situation analysis. This ties in with the observation mentioned at the beginning that data is usually only collected and processed systematically during operation, since this is economical due to its repeated use and use for different purposes. In contrast, a one-time collection of data for just one purpose, as would be required in the course of a design project, is generally uneconomical. However, this is often not necessary either, since the data collected during operation is available and can provide a good basis for a design project.

However, the data for this is usually not available in the required form and is also not always easily accessible. To illustrate this, Fig. 4.7 shows the typical structure of a computer infrastructure in a data flow-oriented representation (Gehring and Gabriel 2022, p. 365). In the simplistic four-layer model, the lowest layer contains different (heterogeneous) data sources, such as external databases, internet sources, unstructured data such as text files, image files or video files, as well as databases of operational application systems, for example for enterprise resource planning (ERP) or Customer Relationship Management (CRM). If companies have not implemented any layers for consolidating the raw data contained in these so-called *data lakes*, a data analysis will have to be set up at this level. To do this, the required data must be identified, extracted, transformed and loaded into a separate data store (ETL) in order to then set up the analyses on it.

Since raw data often also has to be prepared for other purposes and, in addition, is often separated from the data on which the operational business is based, many companies maintain a permanently installed infrastructure for this purpose, which is shown in Fig. 4.6

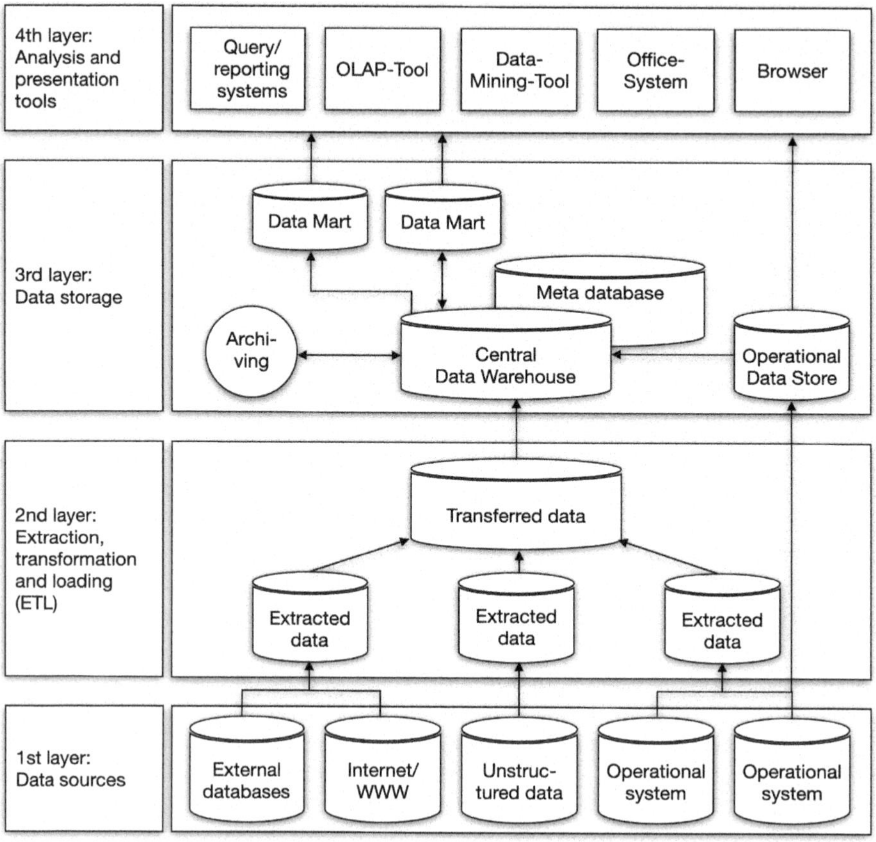

Fig. 4.7 Typical structure of an IT infrastructure (Gehring and Gabriel 2022, p. 365)

in the second and third layers. Fig. 4.8 makes it clear that this is accompanied by a division of labor between Data Engineering and Data Science (Gadatsch and Schreiber 2021, p. 55). While data engineering sets up and maintains the ETL infrastructure and ensures data storage in the third layer, data science carries out specific analyses. To do this, it accesses a central data warehouse (DW) on the one hand and data marts on the other, which contain extracts from the DW for specific purposes.

Technically, database queries are required to obtain data. In relational databases (RDBS), the Structured Query Language (SQL) according to ISO/IEC 9075-1:2016 is used for this purpose. However, such databases are usually only found in the first layer, while DWs are multidimensional databases (MDDBS) in which data is stored, for example, according to the star schema or the snowflake schema. Although such databases can also be queried using SQL, Online Analytical Processing (OLAP) is also available for analyses, which provides a basis for so-called Business Intelligence (BI). However, while BI regularly provides analyses during operation, design projects usually involve one-off analyses that

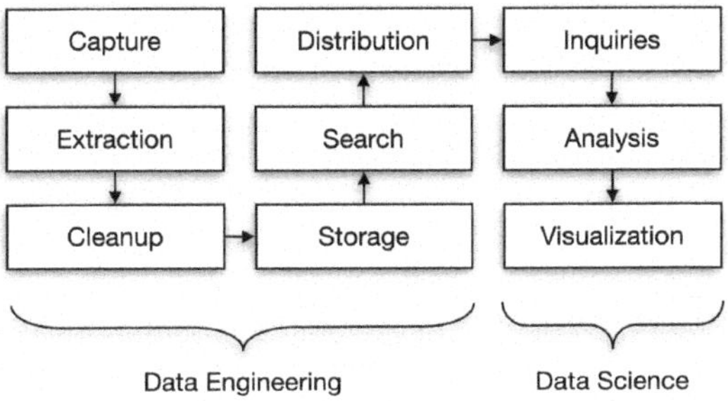

Fig. 4.8 Division of labor between data engineering and data science (Gadatsch and Schreiber 2021, p. 55)

are not part of the BI repertoire. Nevertheless, data science techniques can also be used for one-off queries and analyses.

In addition to OLAP, *data mining* (DM) is another way of analyzing DW data (Jannaschk 2018). However, the original intention of systematically evaluating large data sets has since been extended to include *knowledge discovery in databases* (KDD) (Frawley et al. 1996). The latter begins with the determination of background information for the intended data analysis (situation analysis), continues with the formulation of the analysis objectives, and then includes data selection, data cleansing and data reduction (e.g. through transformation), the selection of a model that describes how the information found is to be represented, and extends to the interpretation of the knowledge gained. Alternatively, the *Cross-Industry Standard Process for Data Mining* (CRISP-DM) follows a comparable approach. This development is currently being continued under buzzwords such as Big Data (Frick et al. 2021). The corresponding methods and tools are not only designed for large amounts of data and high data processing speeds, as the term might suggest at first glance. In principle, both also apply to previous approaches. What is new, however, is the wide range of data types and sources, as well as the question of the authenticity (quality, trustworthiness, validity, etc.) of the data and the added value to be gained from it, the latter especially in view of the fact that the half-life of the relevance of data is decreasing.

A database query is followed by the *evaluation* of the data, i.e., the analysis in the strict sense, as well as the visualization of its results. Essentially, *statistical methods* (Sect. 2.4.5.1) are used for this purpose. In this context, these include, for example, linear or logistic regression, simple classification methods such as decision trees, clustering methods, association analyses, time series analyses, and even machine learning (ML) as a subfield of artificial intelligence (AI). Benjamin M. Abdel-Karim (2022), among others, provides clear practical examples of how the Python programming language is used in this regard.

4.2.4.5 System Analysis

The methods presented so far can be used for all empirical analyses of the initial situation of a design project. The methods explained below draw on these, but now serve to specify the information that is relevant for system design. This primarily concerns the system to be designed itself. In this regard, two situations can be distinguished: (1) In the case of a completely new design, no system exists yet. The deficits (barriers) that are to be identified in the context of the situation analysis then essentially result from the lack of the system to be designed and not from this system itself. Nevertheless, a System Analysis can also be useful in such a situation if it does not refer to an existing system, but to a comparable system, for example to a product of a competitor (*benchmarking*). This may be associated with the goal of reverse engineering, for example. (2) The situation is somewhat different if a system already exists that is to be changed. The associated goal is usually an improvement in general. In detail, this can consist of optimizing existing features (e.g. saving resources, improving performance) or adding additional (new) features (e.g. new technologies).

The analysis of a system is carried out by *modeling* the system, because it is important to distinguish the essential from the non-essential. Therefore, all modeling approaches are used that have already been explained in detail in Chapter 2 (Haberfellner et al. 2018, pp. 205 ff.) and therefore do not need to be described again here. However, this also applies to the development of solutions (system synthesis) P 5. The difference is that System Analysis as part of situation analysis P 2 is used primarily to work out the deficits and potentials of existing systems (2) during the analysis of existing systems (2).

The basic idea of this approach is that innovations should always start from problems (demand-oriented) and not from solutions (supply-induced). The latter follows the so-called *technological imperative* (Lenk and Ropohl 1987, p. 6 f.), according to which a should follows a can. This maxim, deliberately formulated by Edward Teller in reference to Immanuel Kant's categorical imperative, is ultimately not only irresponsible (Jonas 1984), but also uneconomical, because the customer prefers to pay only for what offers him added value. In most cases, they follow the *economic imperative* that there must be a reasonable balance between cost and benefit. Although customers are occasionally happy to receive more than they need, they are usually unwilling to pay for it. So *less is often more*. Therefore, providers of products, services and rights are advised to limit their offerings to what will command a reasonable price. However, in the face of generally high competitive pressure, this must be outstanding. In addition, supply-induced innovations often also cause ecological problems, since anything that exceeds demand carries the risk of wasting resources and unnecessarily polluting the environment.

On the basis of such an attitude, *frugal innovations* (from the Latin frugalis, "from the (field) fruits") can now be observed (Radjou et al. 2012; Weyrauch 2018). With products and services that are limited to the essentials, such innovations are aimed, on the one hand, at price-sensitive segments at the *bottom of the socioeconomic pyramid* (BoP) in order to exploit entrepreneurial opportunities in the service of fighting poverty. On the other hand, however, there is also an increasing awareness among the wealthy population of the need for a sustainable lifestyle, which is closely linked to approaches such as the post-growth economy (Paech 2012), but also combines this with an individual benefit such as a pension from the age of 40 (FIRE: Financial Independence, Retire Early). In the sense of *trickle-up innovations* (Filho et al. 2021, p. 1275 ff.), improvised solutions to problems often serve as a model, with which poor people in developing and emerging countries try to cope with their everyday lives in the face of scarcity. The term *jugaad* (Radjou et al. 2012) comes from India in this context; in more general terms, it is also referred to as *grassroots innovation*. Companies that develop frugal products for emerging markets often find that these products are subsequently also in demand in established markets. This effect is known as *reverse innovation*. One of the companies that has already been successful in this regard is Siemens, particularly in the field of medical technology.

As a result, system analysis essentially provides structural and process models that map the static structure and dynamics of existing systems. The deficits that are identified as a result can take various forms. Some examples may illustrate this:

- Missing, inappropriate or superfluous functions.
- Difficulties and errors when using functions, for example, due to poor ergonomics.
- Lack of quality and reliability.
- The risk of accidents during use.
- Excessive costs for manufacturing, logistics, sales and service.
- Inefficiency of use.
- Technical and organizational problems in production and maintenance.
- Ecological damage during production, operation and disposal.
- Social intolerance.

In addition to the deficits themselves, the barrier (causes) that prevents the deficits from being eliminated immediately and routinely can also be identified, as far as this is possible within the framework of the P 2 situation analysis (Haberfellner et al. 2018, p. 208 ff.). Examples of this are:

- Lack of knowledge and/or skills.
- Third-party intellectual property rights that prevent or hinder the application of solutions in-house.
- Unavailable technology.
- Lack of personnel, materials and/or energy.
- Prohibitions or unfulfillable conditions.
- Lack of space.
- Lack of time.
- Lack of investment capital.
- Risks are too high.

One system analysis that is primarily geared towards the benefits of a product is *Value Analysis* according to DIN EN 1325-1:1996-11 and VDI 2800:2010-08, which in the meantime further developed form as Value Management in DIN EN 12973:2020-05 and leads into Target Costing (Ehrlenspiel et al. 2020, p. 52). Like many other methods, Value Analysis also starts with the situation analysis of the problem solution and is continued in the following phases. In addition, it accompanies the system life cycle: if it is already used in the design of a completely new product, it is initially referred to as value planning and in the further course of the *value design*, in the usage phase of the product then as *value improvement*.

Value is understood to be the ratio of the benefit to the effort required to achieve it. In a first step, the functions of the product to be analyzed are identified. Value Analysis understands these to be the effects that the product should have. The function of the Useless Box sketched in Fig. 2.18 is to close the lid again after opening. Such functions are usually described by a noun and a verb, i.e. "close the lid". In addition, the functions should be quantifiable. In the example, the opening angle in the target state in relation to the vertical axis could be specified as 0 degrees.

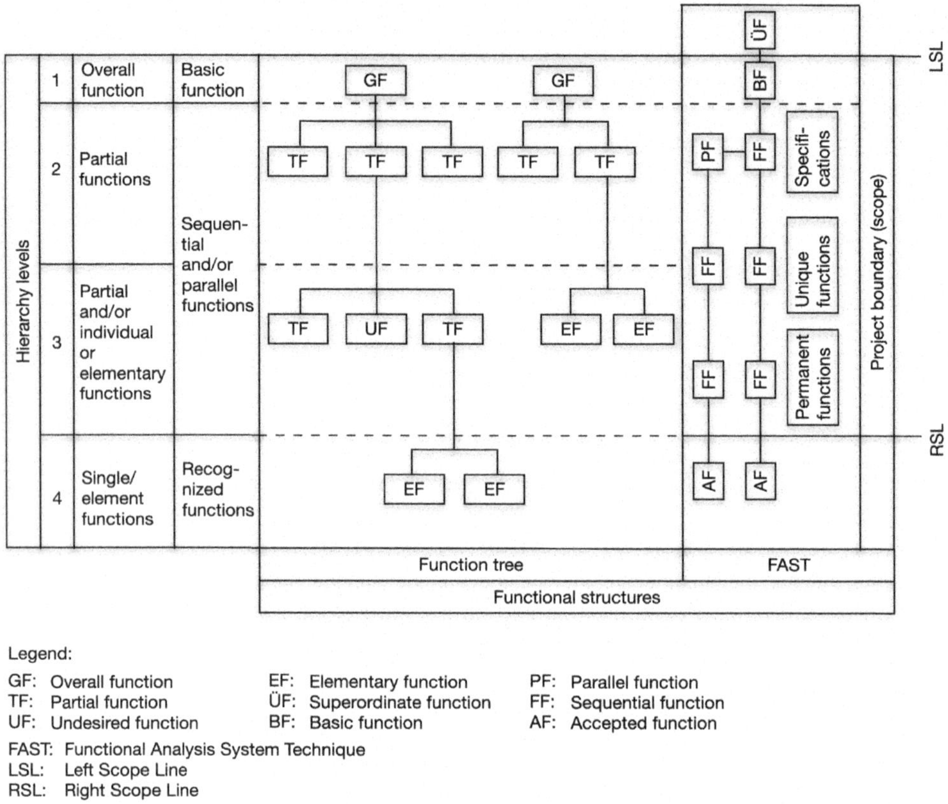

Fig. 4.9 Example of the organization and structuring of a function structure (Naefe 2018, p. 101)

The functions are categorized by their type and by their class:

- A distinction is made between the use function and the validity function as types of function. While the *use function* is relevant for objective use, *validity functions* aim at the subjectively perceived appearance. Although it is debatable whether the Useless Box has any objective use at all, if it is at least partially conceded that it has one, then closing the lid can obviously be classified as a use function, while the color of the box would undoubtedly be counted as an applicability function.
- A distinction is made between main (HF), secondary (NF), basic (GF), partial (TF) and elementary functions (EF) as function classes, whereby each Value Analysis object has at least one HF. What is classified as HF and what is classified as NF depends on the importance assigned to the corresponding functions with regard to customer benefit. The GF corresponds to the overall impact of all functions. The latter are usually

structured hierarchically (Sect. 2.4.3). The GF is broken down into different TFs. For example, a TF of the Useless Box consists of flipping the electrical switch. TFs can in turn be subdivided into TFs at a lower hierarchy level. At the lowest level of the hierarchy, the TFs are referred to as EFs, as elements are what cannot be broken down further from the perspective of the focal system (Sect. 2.4.3). The resulting hierarchical structure is referred to as a function structure in the context of Value Analysis[13] (Naefe 2018, p. 98 ff.). Fig. 4.9 shows examples of classification and structuring from VDI Guideline 2803 Sheet 1:2019-01, as developed using the *Functional Analysis System Technique* (FAST). In the FAST diagram (Fig. 4.10), the functions are entered in such a way that their assignments and dependencies become visible. Slightly different from the function classes mentioned above, the superordinate function (ÜF) indicates the superordinate goal. The basic function (BF) is arranged below the ÜF and describes the overall function. Subsequent functions (FF) are subordinate to the basic function, but are arranged to the right of it in the FAST diagram and result from the basic function. A parallel function (PF) runs parallel to an FF and is directly connected to it; hierarchically, both are at the same level. In addition to the organization and structuring examples shown in Fig. 4.9, Fig. 4.10 shows the basic structure of a FAST diagram as a function structure (ibid., p. 102). In addition, process-oriented representations are also common, which show the interaction of the sub-functions with regard to the overall function (ibid., p. 96).

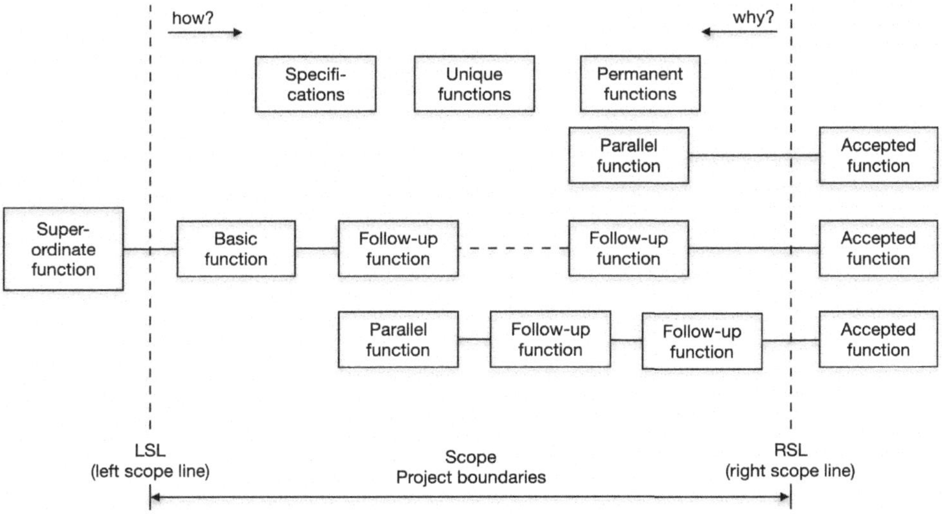

Fig. 4.10 FAST diagram (Naefe 2018, p. 102)

[13] The structure of functions is distinguished from the structure of functions: the former shows the relationships between individual functions, while the latter refers to the active principles required to perform a function (Naefe 2018, p. 95).

On the basis of the function tree, the corresponding costs are determined for each function contained in it. These result from the manufacturing costs of the product's parts and/or assemblies that contribute to several functions, which must be broken down accordingly. The effort for a function quantified by these so-called *function costs* is then compared with its benefit in order to determine the respective utility value (Ehrlenspiel and Meerkamm 2017, p. 678). Low utility values indicate starting points for optimizing the value of the product.

4.2.4.6 Environment Analysis

Now that the focal system itself is known in its current state, the next step is to determine the conditions under which it is to be transferred to the target state and under which it is then to be operated (e.g. DIN EN ISO 9001:2015-11, p. 18; DIN EN ISO 14001:2015-11, p. 45 f.; DIN ISO 21500:2016-02, p. 11; DIN ISO 45001:2018-06, p. 57 f.). The influences that are identified as a result are usually those that cannot be significantly influenced in the course of system design, but conversely must be taken into account in the design. From a design perspective, they are therefore *destiny*.

There are numerous approaches to support environmental analyses (e.g. Thommen and Achleitner 2006, p. 922 ff.), some of which are presented here as examples. In general, the approaches claim to provide reliable insights through a certain system rather than through the use of common sense. Whether and to what extent this claim is justified is a matter for the reader to decide. In any case, the approaches can provide suggestions or be used as checklists.

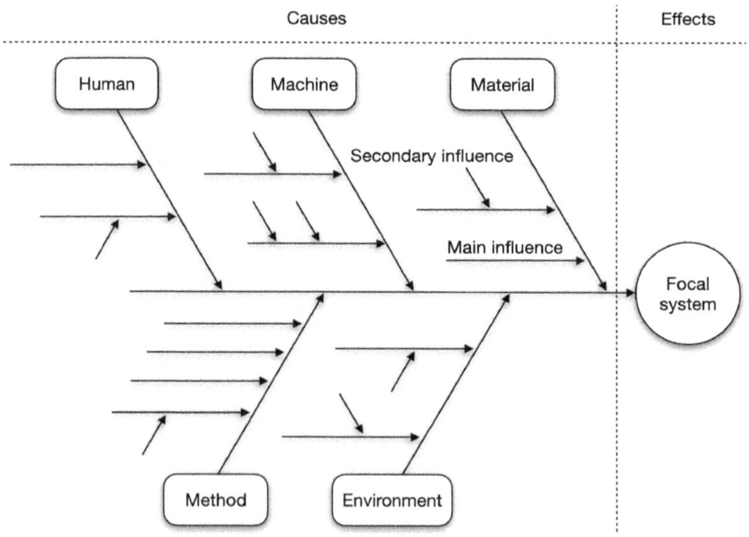

Fig. 4.11 Ishikawa diagram

A relatively general form of recording and structuring influencing variables is provided by the *Ishikawa diagram* (synonym: fishbone diagram, cause-and-effect diagram) shown in Fig. 4.11 (Andler 2015, p. 91 ff.). The influencing variables shown in the figure are people, machines, materials, methods and the environment, because experience shows that a relatively large number of the relevant aspects can often be captured with these. However, it is certainly possible and sometimes useful to consider other aspects as well. The main influences, which can be subsumed under the respective aspects, are represented by horizontal lines, to which secondary influences can be added by diagonal lines. This nesting is not strictly defined either and can be adjusted as needed. This detailing of causes is comparable to the *5W method* used in the concept of Lean Production (Andler 2015, p. 57 ff.; Dombrowski and Mielke 2015, p. 84), which aims to get to the bottom of problems by repeatedly asking about the underlying causes.

Another common approach to analyzing the environment is the so-called *PESTEL analysis* (Andler 2015, p. 311 ff.). The acronym stands for the six factors shown in Fig. 4.12, which are intended to characterize the environment of companies. These factors are compatible with the theories on which this book is based insofar as they show a broad correspondence with the spheres of activity of social systems contained in the AGIL scheme (Fig. 2.4). Deviations, such as the fact that the PESTEL scheme shows technological and legal influences separately, are only superficial. Therefore, a PESTEL analysis, when applied correctly, leads to the same results as an analysis based on the AGIL scheme. In view of this, the AGIL scheme, which is theoretically sound, differentiated and widely recognized (e.g. Münch 1992), can be classified as a generic model, while the PESTEL analysis can be regarded as a highly simplified derivative of it. For practical application, however, this need not be a disadvantage.

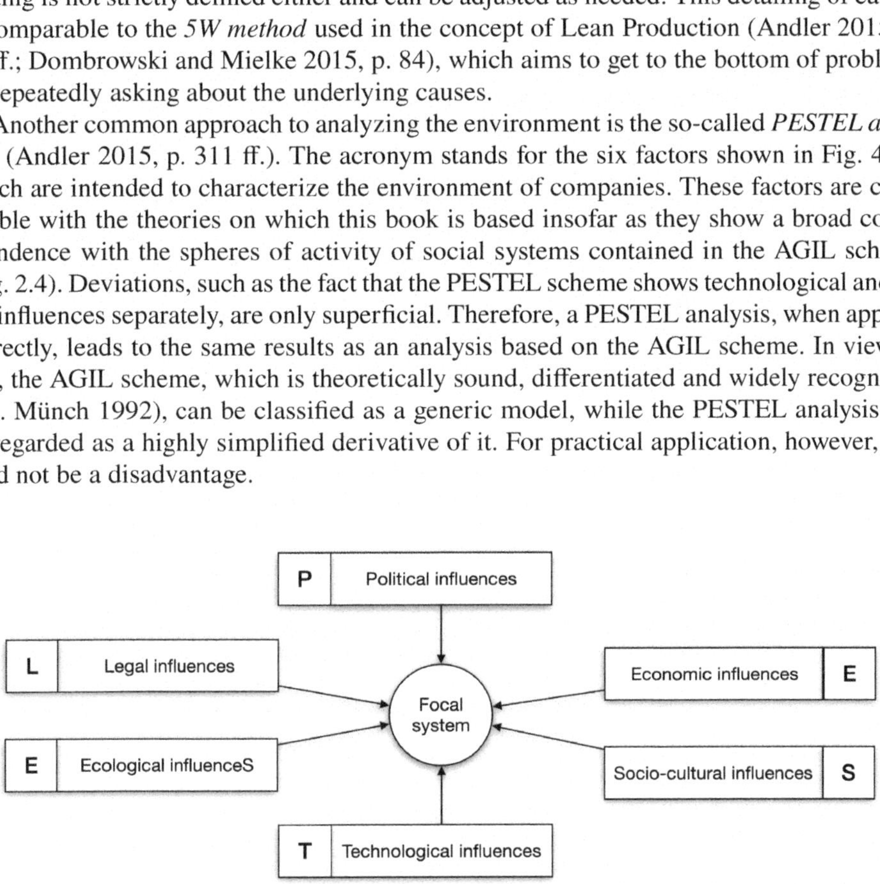

Fig. 4.12 Environmental analysis in the form of a PESTEL analysis (Andler 2015, p. 311 ff.)

The second approach mentioned as an example describes the environment of companies as a *VUCA world* (Fig. 4.13). The factors associated with this acronym describe both influences on a company and success factors (for success factors, see Nicolai and Kieser 2002, p. 579 ff.; Mack et al. 2016) that should enable it to assert itself in such an environment.

As can easily be seen, the factors on both sides are kept so general that anyone, anytime, anywhere, can read almost anything into them. Therefore, it is unnecessary to explain the factors in detail. The claim that the present can be reliably distinguished from the past and/or the future and specifically characterized by it carries the risk of *chronocentrism*: in every generation, there are representatives who believe that their respective present stands out compared to other times, for example, that it is particularly challenging or particularly groundbreaking. This is not particularly helpful, as presumably each generation is more or less right from its own perspective. In this context, an Environment Analysis based on the assumption of a VUCA world can certainly lead to useful insights, but it doesn't have to. However, if an approach is just as likely to be misleading as it is to be useful, it seems obsolete.

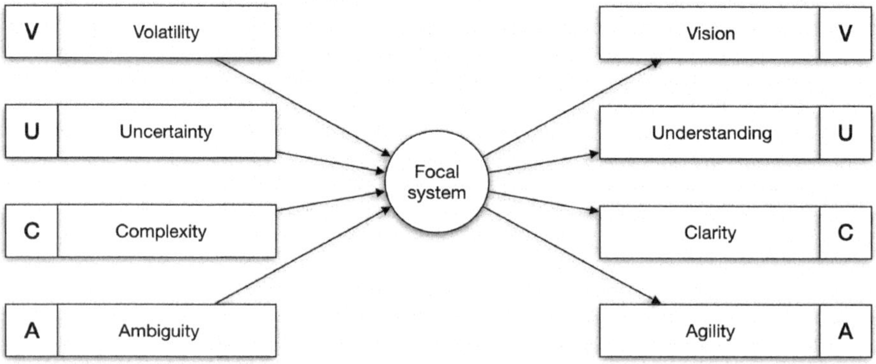

Fig. 4.13 Environmental analysis assuming a VUCA world (Mack et al. 2016)

An *Industry Structure Analysis* can be of interest to companies (Thommen and Achleitner 2006, p. 934 ff.). Approaches to this are at the interface between economics, which characterizes the environment of companies, and business administration, which transforms the aspects of the environment that are relevant for one's own company into operational action. *Industrial economics* (Bester 2016) serves as the main theoretical basis, although it is far too comprehensive to be even roughly explained here. Please refer to the indicated and further literature. One specific manifestation of this general approach, which assumes that the attractiveness of a market is determined by its structure, is the *five forces model* by Michael Eugene Porter (2015; Andler 2015, pp. 294 ff.; Thommen and Achleitner 2006, pp. 934 ff.). The model shown in Fig. 4.14 centers on rivalry with and between existing competitors. This is the central driving force behind market activity. A threat to this given situation comes from potential new competitors who are motivated to enter the market by, among other things, the attractiveness of the market or new opportunities, but who are prevented from doing so by market entry barriers. Factors such as these need to be assessed in order to calculate the extent of this threat. Furthermore, the bargaining power of one's own suppliers and customers must be assessed, which depends, among other things, on the prevailing market forms. For example, many automotive suppliers find themselves in a kind of sandwich situation, in that they are put under equal pressure by powerful customers (oligopoly) and by powerful semi-finished product manufacturers (oligopoly) and cannot counter this with any market power of their own. This often results in a kind of

David-versus-Goliath strategy that relies on ingenuity and agility. Finally, potential substitutes pose a threat. For example, manufacturers of exhaust components such as exhaust systems will find it increasingly difficult to sell their products in the future in the face of the electrification of vehicle fleets for products of this kind. The stronger the five competitive forces are, the less attractive the industry in question is.

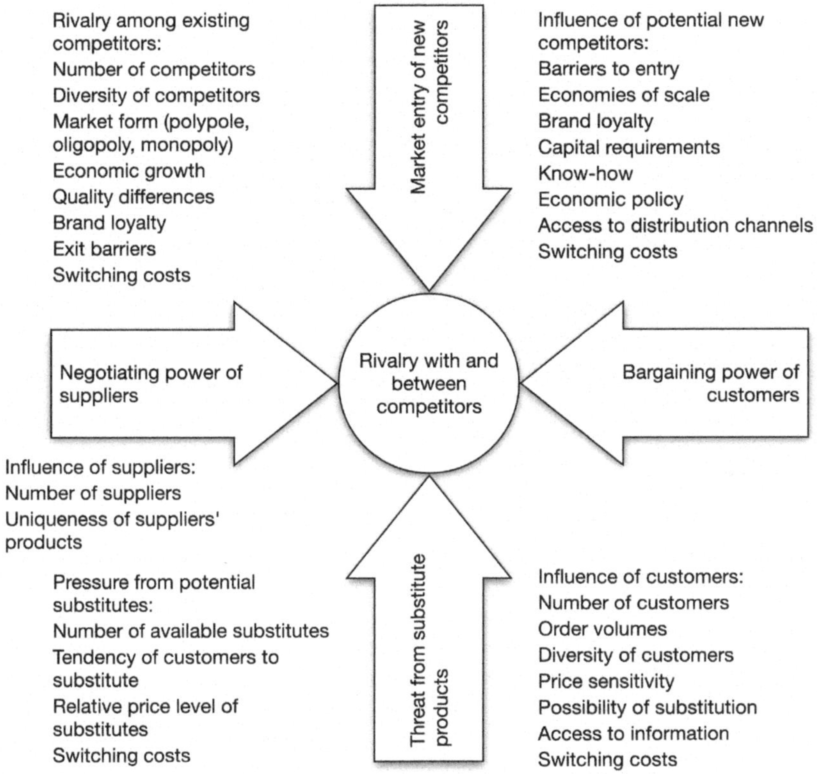

Fig. 4.14 Environment analysis based on the five forces model (Andler 2015, p. 294 ff.)

In summary, it can be said that an Environment Analysis first and foremost requires a solid *general education*. In addition, it requires continuous, attentive and *critical information about current events*. Reliable newspapers and magazines, radio and television, as well as corresponding offers on the internet, are sources for this. In addition, numerous institutions such as market and opinion research institutes, associations and chambers offer more or less specific analyses, such as country and industry reports. Despite or perhaps because of the very extensive and diverse range of information available, an Environmental Analysis always carries the risk of incompleteness and distortion, because the perception of the mostly very general and complex factors is always influenced by personal opinions, individual moral concepts and specific interests.

4.2.4.7 Stakeholder Analysis

Actually, the previous Environment Analysis can also be seen as part of the Stakeholder Analysis introduced here (Göbel 2010, p. 125 ff.), because the can be assigned to stakeholder groups (DIN 69901-2:2009-01, "Process" D.8.2 "Analyze project environment/stakeholders"). These primarily include society, i.e. the social environment, and the natural environment. The fact that it can still make sense to distinguish between these two analyses can result from the insight already mentioned above that many of the requirements that can be attributed to the environment (social and ecological) are not significantly influenced by those who intend to design a system and can therefore be described as fate or boundary conditions, so to speak. Although it will not be possible to dissuade the stakeholders who have more specific demands from the project in every respect, they can be reached by direct communication and can thus be better integrated.

Communication with all participants and stakeholders, or at least with legitimate representatives, is even a mandatory requirement for the success of a design project, because, as a rule, the demands of stakeholder groups are initially formulated in an imprecise manner, are incomplete or over-specified, and, above all, different demands from different stakeholder groups repeatedly come into conflict with each other, which must be resolved by setting priorities and/or making compromises. This, in turn, can only be successfully concluded if all stakeholders give their consent, which is not always easy because it may involve concessions that go as far as completely abandoning demands.

In the following, three categories of claims are distinguished, each of which can be effective on its own specific basis:

- In the first stage, those demands that can be enforced by the stakeholders on a *legal or contractual* basis are to be considered. Examples of such a power base are service contracts, such as employment contracts, contracts for work and services, which are used, for example, to award development contracts to service providers, supply contracts, the German Works Constitution Act (Betriebsverfassungsgesetz, BetrVG), which grants employees participation rights, or framework agreements, such as collective agreements. In the following, such demands will be referred to as *legal demands*, since they can be enforced in court in case of doubt. Documents are often available as sources.
- At a second, more advanced level, claims that are *morally* legitimate are also relevant. For example, there is no law that requires companies to support those in need. However, there may be circumstances in which they can assert claims based on the principle of solidarity, provided that these are reasonable for a company (Ahrens 2001, p. 42). Claims of this kind will be referred to below as *legitimate claims*. One basis for this is *corporate social responsibility* (CSR) (synonym: corporate citizenship), which expects a comparable commitment from companies as legal entities as is the case with citizens of a state as natural persons (Göbel 2020, p. 227). The sources of such requirements are often not as easily accessible as laws or contracts in relation to legal claims. In addition to the culturally anchored values internalized by members of a cultural community, general sources such as textbooks (ibid.) often have to be consulted.
- In addition to these two categories of claims, a third category should be added: *strategic claims* (Ahrens 2001, p. 42; Göbel 2020, p. 144 f.). For example, it is one of the legal and at the same time legitimate rights of employees to have their health and freedom from

injury at work ensured. The situation is different when it comes to income. Although there are enforceable minimum levels for this, whether through the statutory minimum wage or through labor or collective agreements, and the principle of equal treatment is also applicable, Beyond that, however, there is no moral imperative to pay a certain wage. Demands of this kind must be enforced in the course of negotiations. The outcome of the negotiations is determined by the relationship between supply and demand (scarcity). Therefore, incomes are generally not fair, because fairness is a moral category. The same applies to the sufficiency of prices that customers are willing to pay for products. Morally imperative lower limits for prices, as demanded by the Fair Trade initiative, for example, are also subject to market rules and require customers to be willing to base their purchasing decisions on moral considerations in addition to economic ones. Information on strategic demands is provided, among other things, by marketing (e.g. market research) and sales.

With the binding enforcement of strategic and legitimate demands, these are transformed into legal demands. This begins with every design project, in which the relevant stakeholders (those involved and affected) and their requirements are identified in the first step of Stakeholder Value Management (SVM), in order to further develop the latter in the further course into a system of objectives that can be approved by all sides. However, SVM does not end with the completion of a design project P 6, but only with the palliation L 4 at the end of the system life cycle. But here we are now initially concerned with the first steps, which, in the course of the situation analysis of the SVM, begin by identifying the relevant stakeholders and their respective demands for the specific project.

At the beginning of a design project, the client imposes himself. This can and will be relatively common in a company, the entrepreneur or, in a broader sense, a representative of other investors (shareholders, banks, etc.). At the top level, these are the owners themselves in owner-managed companies, otherwise managing directors or board members. At the lower hierarchy levels, they are represented by executives. What they all have in common is that their primary interest is in generating a return on the capital employed. Therefore, projects initiated by this group of stakeholders often also aim to increase profitability or growth.

An example of a perspective dominated by the client can be found in Beate Bender and Kilian Gericke (2021, p. 181), who describe the contents of the handover information and documentation of development orders. On the one hand, these primarily include the interests of entrepreneurs and investors, for example the requirement to minimize costs. In addition, and above all, their own interests are also included. These are referred to as secondary here because they primarily address the customer. However, the primary self-interest of making money through business with customers is hidden behind this. In this respect, it is ultimately also about the concerns of the client. And so the authors mentioned also explicitly explain that the initial phase of determining requirements is determined by the relationship between the client, contractor and user of the product (ibid., P. 182).

However, design projects do not necessarily have to be initiated by companies, investors or their representatives. This can also come from other stakeholders, for example from occupational safety to reduce the risk of accidents or from corporate environmental Protection Management to reduce emissions. What is important here is that the client, by taking the initiative and thus also formulating clear expectations regarding the outcome of the project, is the initial focus. Therefore, those who are entrusted with the project may be inclined to fulfill these expectations exclusively or with priority.

Regardless of who initiates a design project, a whole range of other people and/or groups are always relevant, who must be involved and/or are affected. Since all stakeholders, at least in the narrower sense, are defined in terms of their legal claims in that they have a potential threat on the basis of which they can enforce their claims, it is wise to find out about these claims as early as possible. If stakeholders are involved too late and only then introduce their demands, the changes that are then usually necessary often lead to significant additional costs, delays and conflicts (Fig. 4.15). Stakeholder Value Management at the earliest possible stage makes a significant contribution to avoiding such issues.

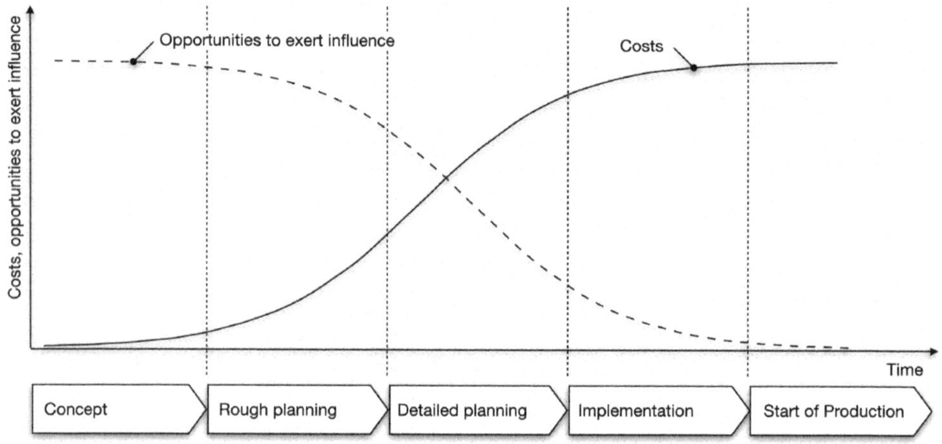

Fig. 4.15 Development of costs and possibilities for influence over time in a design project (Bracke 2022, p. 17)

At this point, it should also be noted that it is not enough just to inform stakeholders and, in the event of resistance, to try to convince them of the correctness of the original plan. Stakeholder Value Management means involving stakeholders in the design (*participation*). The *Change Management* required for this (Sect. 3.5) begins at the same time as every design project.

Once the client has been identified as one of the stakeholder groups with the potential to initiate a design project, the next step is to identify all other stakeholder groups. This is the subject of the situation analysis of Stakeholder Value Management. Fig. 4.16 shows typical stakeholders of a company. However, this selection can only serve as a guide, because depending on the project, other or further stakeholders may be relevant. For example, it may be useful to explicitly name specific neighbors (residents) in addition to society in general when dealing with specific concerns such as avoiding noise pollution. Furthermore, the identification of stakeholders depends on other factors, such as the size of the problem, intervention, solution and impact areas (Fig. 4.4). For example, individual departments within a company may be among the stakeholders.

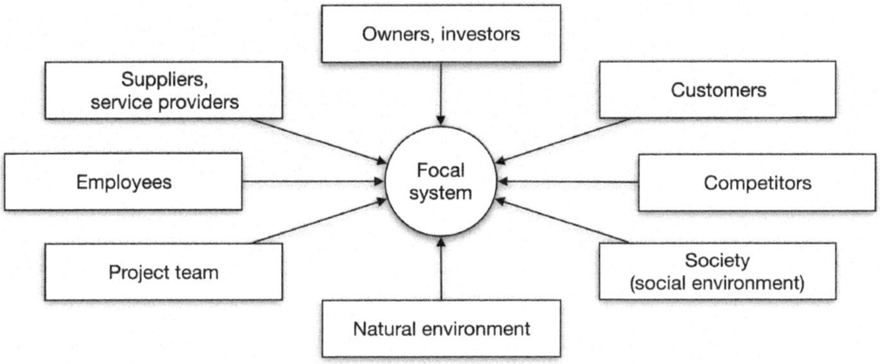

Fig. 4.16 Typical stakeholders

Difficulties often arise when systematically defining stakeholder groups. For example, it can be observed that both customers and Quality Management are identified as stakeholder groups. And employees, the works council, a trade union and occupational safety are also occasionally named separately (ibid., p. 130). As will be explained in more detail in Chapter 6, the demands of stakeholder groups are already represented by institutions in many systems and thus also in companies, so that in such cases it is usually possible to choose whether to name the representing institution or the represented stakeholder group; naming both would be redundant and therefore too laborious. For example, Quality Management represents the legal and legitimate demands of customers within the company (Ahrens and Kleinfeld 1999, p. 64 ff.), while both the works council and the occupational health and safety department enforce certain legal and legitimate demands of the employees. In turn, managers often represent the owners and other investors by being responsible in particular for the profitability of the company. These remarks are intended to make clear that the identification of relevant stakeholders in the concrete implementation is not always as easy as it seems at first glance. A system introduced at an early stage can save a lot of effort in the further course of a design project, because, as already indicated, Stakeholder Value Management accompanies the entire further process.

There are numerous methods available for *more sophisticated analyses* of the identified stakeholders, but explaining them all in detail is beyond the scope of this article. Therefore, in addition to their exemplary mention, reference is made to the relevant literature in each case.

- For example, customers or markets are often segmented (Andler 2015, p. 303 ff.) because they have different requirements depending on the segment.
  - Demographic characteristics such as age, gender and marital status, geographical characteristics such as urban or rural socialization, or socio-economic characteristics such as income such as age, gender and marital status, geographical characteristics such as urban or rural socialization, or socioeconomic characteristics such as income, education or religious affiliation.

- Furthermore, customers can be distinguished, for example, into lead users, expert users, and normal users (Paetzold 2021, p. 137 ff.), with the former being involved in the development of solutions, for example, in the course of co-creation (Deserti et al. 2019), while the latter experiment with beta versions that are about to be launched on the market, for example, in order to be able to offer the broad mass of normal users a product that meets their requirements as closely as possible after the official market launch.
- And customer requirements can also be differentiated, for example, on the basis of the Kano model into basic, performance, and excitement requirements (Bender and Gericke 2021, p. 206).

- In the course of procurement market research (Thommen and Achleitner 2006, p. 306 ff.), suppliers are categorized according to characteristics such as industry affiliation, company size, innovative strength or geographical location. In addition, a comprehensive supplier evaluation is common, especially in Quality Management (Schmieder et al. 2018).
- Employees are sometimes subjected to aptitude testing and categorized according to criteria such as qualifications and motivation (DIN 33430). For example, the Big Five or OCEAN model is widely used to assess personalities, which distinguishes the personality traits of openness, conscientiousness, extraversion, agreeableness, and vulnerability to neuroticism. This results in different need structures and potential for deployment.

Another important activity that begins with the Stakeholder Analysis is *communication* planning. Given the diversity of the stakeholder groups, it often makes sense to plan a specific mode of communication for each group. This concerns, for example, the communication channels, which are shorter within the organization than beyond its borders. Furthermore, the type of communication must be aligned with the communication style of the respective stakeholder groups. This does not necessarily mean being opportunistic, because it is not about telling stakeholder groups what they want to hear, but about promoting mutual understanding with a view to reaching an agreement. For example, the language used to address experts differs from that used to address laypersons, and the language used to address authorities differs from that used to address customers. This also affects the frequency and regularity of communication, which can also change repeatedly over the entire life cycle of the product or service being developed and must therefore be regularly evaluated in terms of its appropriateness and, if necessary, realigned.

Communication with stakeholders should not be confused with project marketing, because it is not about merely selling interim or final results to the stakeholders in the hope of convincing, persuading or enticing them of their usefulness. In fact, the point is to involve the stakeholders in the execution of the project on an equal footing and with sober objectivity (Fig. 3.1). This does not speak against project marketing in principle, but it does speak against it in the context of or as a substitute for Stakeholder Value Management.

The underlying claim of universality includes the reference to reductive approaches, the widespread use of which is not proof of their correctness. These include the so-called *magic triangle of Project Management* (e.g. Alam and Gühl 2020, p. 86; Haberfellner et al. 2018, p. 16 9 f.; Jakoby 2021, p. 81; Madauss 2020, p. 649), which specifies the task, the costs and the deadlines as essential target values of projects. Although this basically corresponds to the stakeholder value approach, it obscures the requirements of all stakeholder groups by classifying only some of them as primarily relevant. Therefore, the application of such approaches in favor of comprehensive Stakeholder Value Management is not recommended. This prevents the overlooking of relevant requirements and at the same time promotes the efficiency and effectiveness of the approach, since only one method has to be applied that encompasses everything, while the application of redundant and at the same time less helpful methods can be avoided.

4.2.4.8 SWOT Analysis

The SWOT analysis establishes relationships between four factors relevant to each project (Andler 2015, p. 282 ff.):

- S: Strengths form the potential to overcome the challenges associated with the design project.
- W: Weaknesses limit the possibilities for overcoming the challenges.
- O: Opportunities refer to favorable conditions that support the project. They should not be confused with the actual objectives of the project.
- T: Threats are, analogous to opportunities, unfavorable conditions that make it more difficult to achieve the objective. They should not be confused with potential problems, which may be solved by the objective of a design project.

In line with the approach of moving from the general to the specific, the SWOT analysis (Fig. 4.17) is carried out in the same way as the previous Stakeholder Analysis at the boundary between the environment of the focal system and the focal system itself, in the sense of the black box view (Fig. 3.19). The black box thus turns into a gray box.

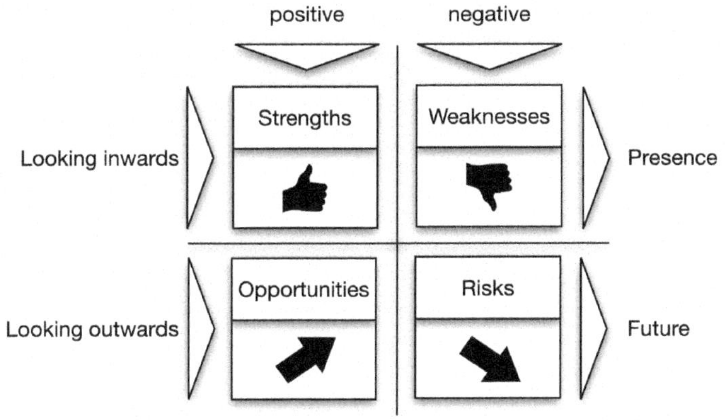

Fig. 4.17 SWOT analysis

The analysis of opportunities and risks ties in with the Environment Analysis (Sect. 4.2.4.4) and focuses primarily on the environment (outside perspective). Furthermore, it is not about the past or present, but about future developments (Sect. 4.2.4.8). By contrast, the analysis of strengths and weaknesses primarily relates to the focal system itself (internal perspective) and is mainly based on the current situation.

As part of the situation analysis P2 of the overarching design project, the SWOT analysis is also initially carried out in the sense of a situation analysis. All further phases of the corresponding problem-solving cycle, which is of course also the basis of SWOT Management, are assigned to the corresponding phases of the overarching problem-solving cycle. And just like Stakeholder Value Management, SWOT Management continues after completion until the end of the system life cycle, i.e. until the completion of palliation.

The previously introduced methods of empirical social research are used to determine the specific strengths and weaknesses as well as the opportunities and risks. The calculation of opportunities and risks based on this is explained in detail below. For the sake of simplicity, this is done using the example of risks; the same applies to opportunities, but with the signs reversed, so to speak.

A risk is calculated using the so-called *insurance formula* by multiplying the probability of a potential loss event occurring by the amount of the loss that can then be expected. Mathematically, this corresponds to the *expected value of a random variable*, which, however, only occurs with an infinitely large number of repetitions (law of large numbers). In practice, however, only a limited amount of data can be used as a basis, statistically speaking, a *sample*. In this case, only the *arithmetic mean* of this sample can be determined, which differs from the expected value, i.e. the true mean (mean of the population). So while different samples can produce different mean values, the expected value is a constant. This shows that in practice, even a calculation of risks is inevitably flawed and therefore risky.

In decision theory, the degree of error leads to further differentiations. For example, they speak of situations of *uncertainty* (Göbel 2018, p. 56; Laux et al. 2018, p. 35) when the amount of damage is known but the probability of occurrence is unknown. However, such a distinction from risk is not recommended here (also Eisenführ et al. 2010, p. 23).

- The first counterargument that is put forward is that any risk calculation is uncertain anyway, as has just been explained. Objective probabilities, as assumed in decision theory (Göbel 2018, p. 56), if these can be calculated statistically, are not objective with regard to statements about future probabilities because, for example, they are based on historical data that may not be representative of the future (Truthahn Illusion[14]). The degree of uncertainty can be estimated, however, even if it cannot be determined with certainty. For example, at the beginning of a design project, the uncertainty will be greater in the face of poor data than after a few years of operation. Such risks of risk calculation are in turn incorporated into this risk calculation. For example, if no specific data from past incidents are known to determine the probability of occurrence, data from comparable situations can be used as a basis or experts can be asked for their assessment. This may not always be satisfactory, but it is better than nothing (uncertainty).
- The second counterargument is that the opposite of uncertainty, i.e. certainty, does not exist, even if decision theory assumes such a thing in a model (Göbel 2018, p. 56; Laux et al. 2018, p. 35). At least a residual risk always remains. Although it is possible to speak of *security* in lay terms or in a model-simplifying way when the limit of the

[14] Gigrenzer 2013, p. 55

residual risk is not exceeded, there is a risk of losing awareness of the remaining risks and of lulling oneself into a false sense of security. Therefore, such use of language is not recommended.

As just mentioned, the probability of occurrence and the extent of damage can be determined in various ways, depending on the options available in the respective situation. Often, reliable data will not be available at the beginning of a design project. If you really have no data to fall back on, you will have to be creative and develop ideas of what can go wrong. System Analysis (Sect. 4.2.4.4) and an analysis of the environment (Sect. 4.2.4.5) can provide a basis for this. In addition, methods of empirical social research such as expert surveys (Sect. 4.2.4.2) can be applied. If there is at least some experience with predecessor models of a successor product or with sufficiently similar products, this can be used, for example, by studying documents (Sect. 4.2.4.1). The longer the design of the object of the risk calculation has progressed or even been in operation, the more accurate the data will be that can be used as a basis for the calculation. This makes it clear that risk calculation, as part of comprehensive Risk Management, is an ongoing process that begins with the start of the design and may even have to be continued after the completion of the palliative phase (Sect. 3.1), if, as in the case of the final disposal of radioactive waste, risks remain even after disposal that need to be managed.

However, the use of specific data is only one way of determining the probability of occurrence and the extent of damage. One disadvantage is that very, very large amounts of data are required for reliable statements (law of large numbers). An alternative is to model risks using *probability distributions*. Such distributions can be described using a few key figures. One of these key figures is the expected value or arithmetic mean, which represents a *location parameter* of a probability distribution by indicating the center of gravity of the area under the probability distribution, which is thought of as a mass. Alternative or supplementary location parameters are the mode, which marks the location of the highest value of a distribution, and the *median*, from which half of the probability can be found to the left and to the right respectively. In the case of symmetrical distributions, the mode and median are therefore identical. The *variance* and the *standard deviation* derived from it describe the degree of dispersion around the expected value, the *skewness* is a measure of the asymmetry of the distribution and the *curvature* characterizes the expression of its peak. If no probabilities are known (poor data situation), a uniform distribution is usually used. If there is an overall risk from a large number of small, independent individual risks, whose probability of fluctuation around the expected value (mean) is equally high in both directions, the normal distribution is selected. If extreme events are to be expected, the Pareto distribution is used. The previous discussions of risks were based on static risks. In reality, however, risks often change over time and with location. How this can be taken into account is explained in Section 4.4.5.3 using the example of the reliability of technical systems.

There are a number of specific methods for calculating static risks, including, for example, the *Failure Mode and Effects Analysis (FMEA)* or the *Fault Tree Analysis (FTA)* according to DIN EN 61025, , which are widely used in Quality Management (Brüggemann and Bremer 2015, p. 38 ff.; Edler et al. 2015; Wälder and Wälder 2017), or the so-called *Risk Assessment in Occupational Health and Safety*. Despite some differences in the details, all risk assessments have in common that they quantify risks as the product of probability of occurrence and extent of damage.

The following steps are to be taken in order to calculate the risk:

- Identifying the causes of risks: For new projects for which no experience is yet available, possible risks and their causes must be thought up using creativity techniques such as brainstorming. The question is simply: what can go wrong? On the other hand, if experience from comparable projects is already available, the problems encountered in those projects can be adopted. However, this should not be done blindly. Rather, it may be necessary to make adjustments to account for any special features of the current project. The result of this step is a list of possible risks and their causes, whereby a risk may have several causes.
- For each cause of risk, possible effects must now be determined in the same way. It should be noted that a cause can have several different effects. Furthermore, the effects should be quantified in terms of their potential for damage. A monetary quantification is useful for this purpose, as it enables a relatively simple cost-benefit analysis in conjunction with a calculation of the costs of measures to reduce a risk. However, in many cases a monetary quantification is not possible in a reliable manner or only with a great deal of effort. In such cases, alternative scales are used as a substitute. For example, a scale from 1 to 10 is common for an FMEA (Brüggemann and Bremer 2015, p. 49).
- In the next step, the probabilities of occurrence of the effects are quantified. Ratios such as percentages are useful for this. However, these are also sometimes replaced by other scales, for example in an FMEA.
- In the final step of the risk assessment, the risks can be sorted in descending order based on the product of probability of occurrence and extent of damage (insurance formula), which makes priorities visible.

Opportunities can be identified in the same way. To do this, it is only necessary to change the sign from minus to plus. The result is a table that looks something like the one shown in Fig. 4.18. The previous explanations assumed that risks are not influenced in a targeted manner. However, this is often not the case when two or more people or groups of people are involved or affected and can influence risk factors in a more or less targeted way. In such cases, one speaks of game situations and these are dealt with on the basis of *game theory* (Göbel 2018, p. 125 ff.; Grimme and Bossek 2018, p. 229 ff.; Holler et al. 2019). In terms of systems theory, this can be categorized as *second-order cybernetics*, i.e., described as an *observation of an observation* (Sect. 2.1). For example, if two players are playing a zero-sum game, i.e. one in which one player wins what the other loses, and the moves are executed simultaneously so that neither player can observe what the other has already done or plans to do, they can only make assumptions (hypotheses) about it. Thinking becomes circular, so to speak: I think that B chooses his strategy 1. But if B thinks that I think that he chooses his strategy 1, then maybe he chooses strategy 2 after all. If B thinks that I think that he chooses strategy 2, then maybe he chooses strategy 1 after all, etc. (Göbel 2018, p. 135).

Game theory knows many different game forms (ibid., p. 129): These differ, for example, in the number of moves per game, in the number of players, in the order of moves (simultaneous or sequential), in the possibility of cooperation or in the degree of information

Opportunity/risk	Causes	Effects (W)	W [€]	Probability (P) [%]	Opportunity/risk W · P [€]
Opportunity 1	Cause 1.1	Effect 1.1.1	100 €	10 %	10 €
	Cause 1.2	Effect 1.2.1	120 €	7 %	8 €
		Effect 1.2.2	90 €	7 %	6 €
Opportunity 2	Cause 2.1	Effect 2.1.1	150 €	2 %	3 €
		Effect 2.1.2	160 €	1 %	2 €

Fig. 4.18 Risk calculation scheme

(symmetry, asymmetry). Since it is not possible to provide a complete introduction here, please refer to the literature cited. However, in order to at least give a rough impression of the approach, it will be explained briefly, based on the example of a two-player zero-sum game with information symmetry, which has already been outlined above (ibid., p. 135 ff.).

For this purpose, Fig. 4.19 shows the so-called payoff matrix. For the sake of simplicity, only the payoffs u_A for player A are given at the respective intersections of the strategies a_i and b_j; the payoffs u_B for player B result from these. If, for example, player A wins with his strategy a_1 7, player B must lose with his strategy b_1 7. The minimax strategy is a good approach: for player A, the respective minima are selected from the rows of the matrix (last column). From these, player A selects the maximum minimum 2, i.e. the payout that he can secure on his own. This is referred to as the lower game value. Player B chooses the column maximum (bottom row), i.e. the minimum of the maximum possible losses (corresponding to the respective winnings of player A). In the example chosen, this is the result of his strategy b_2, i.e. 2. This value is referred to as the upper game value. Between the lower and upper game values lies the indeterminacy interval, in which the finally realized payoff must lie.

		Player B strategies			Min. a_i
		b_1	b_2	b_3	
Strategies player A	a_1	7	1	3	1
	a_2	3	2	6	2
	a_3	-4	0	1	-4
Max. b_j		7	2	6	

Fig. 4.19 Payout matrix (Göbel 2018, p. 136)

In the example chosen, the upper and lower payoffs are identical. This makes the solution unique, in this case the strategy combination a_2/b_2. A game like this is called "deterministic". The equilibrium point is called a *Nash equilibrium* in honor of the mathematician and Nobel Prize winner John Forbes Nash, Jr. No player can deviate from such an equilibrium without risking a smaller win or a higher loss. Player A therefore chooses his best response for each strategy b_j, so for b_2 he chooses a_2. Player B in turn chooses his best response for each strategy a_i, so for a_2 he chooses b_2. If such an equilibrium is missing, mixed strategies are required (Göbel 2018, p. 138), in which pure strategies are considered with certain probabilities. For example, choosing an equal distribution makes one's own strategy unpredictable for the opponent – but also for the player himself. Elisabeth Göbel (2018) cites police traffic controls as an example of this.

Once all opportunities and risks, strengths and weaknesses have been identified, the SWOT analysis can be substantiated in the form of a *Portfolio Analysis* (Thommen and Achleitner 2006, p. 941 ff.). This type of analysis is based on the portfolio selection theory, which was originally developed for capital investment decisions and states that an optimal mix of capital investments can be put together depending on the respective environmental conditions. This also applies to the product portfolio of companies. Based on the system life cycle (Fig. 2.54), the optimum is achieved when products or product groups can be assigned to all product life phases. Against this background, Portfolio Analysis aims to carry out this assignment. This is done in a *market growth-market share matrix* (Fig. 4.20), which compares the relative market share of one's own products in comparison to the market share of the respective strongest competitor with the expected future market growth (e.g. Kirchner 2020, p. 29 ff.). Norm strategies can be derived from the assignment of the products to the four fields of the portfolio. To do this, the sales volume of the respective products is entered into the four-field matrix as a circle with a corresponding radius.

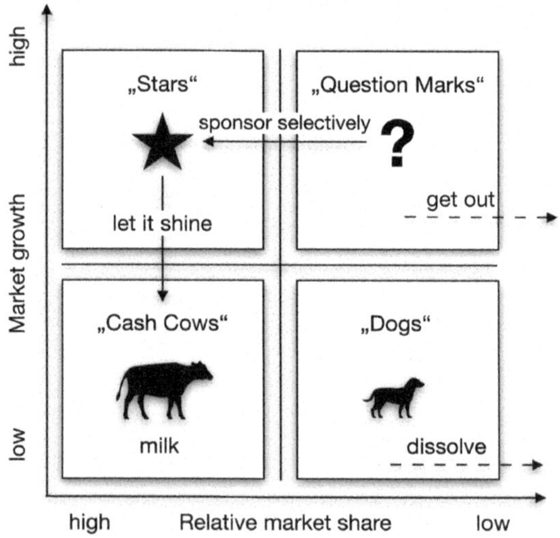

Fig. 4.20 Market growth-market share matrix (Kirchner 2020, p. 29 ff.)

- The products designated as "stars" are offered on a market that is subject to strong growth. However, in order to defend their market position, further heavy investments are needed.
- Cash cows have a high market share, but the corresponding market is expected to grow only slightly in the future. Therefore, the high revenues from the sale of these products should be invested in other business areas.
- "Dogs" are products that are sold with a small market share in a slow-growing market. They tie up resources without making a significant contribution to cash flow. Therefore, they should be removed from the product range.
- Question marks are young products that are difficult to assess in the phase of their life cycle in which they are classified as such. They may become stars through increased use of resources, but may also have to be withdrawn due to their low relative market share.

With a view to the subsequent solution development, four standard strategies can be derived from the SWOT matrix (Fig. 4.21), similar to the market growth-market share matrix (Fig. 4.4):

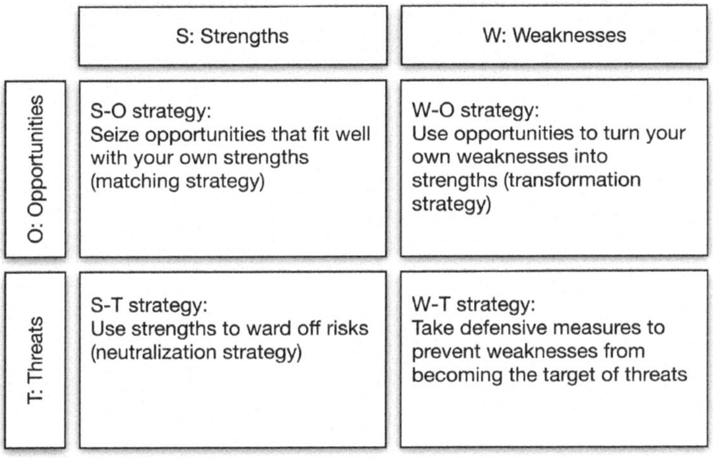

Fig. 4.21 Strategies as a result of a SWOT analysis

- S-O strategy: When your strengths match the opportunities available, you have the optimal combination for making a success of a project. The strategy of combining the two is also referred to as a matching strategy. As an example, consider a company that manufactures semiconductors and benefits when products from other manufacturers in other industries that have not yet been equipped with semiconductors, for example in the course of the development of smart products, are also to be equipped with semiconductors in the future. This offers the semiconductor the opportunity to supply its products and thereby increase its sales volume.
- S-T strategy: In this strategy, internal strengths encounter external risks. For example, a company may have unique selling points in its manufacturing techniques, which have

allowed it to successfully market its products worldwide. Increasing export risks, for example due to an increase in customs duties or significantly rising transport costs, can be neutralized by setting up production facilities in the target regions, among other things.

- W-O strategy: When your own weaknesses coincide with opportunities that arise, investments aimed at converting weaknesses into strengths can pay off. One such opportunity arises, for example, when there is a second chance. Such a scenario is conceivable if car manufacturers who have lost touch with the market leaders in the switch from combustion engines to electric motors can then take the lead in development if it turns out that the current direction of development, to install heavy batteries permanently in vehicles , resulting in long downtimes for charging, is inferior to a technology that relies on replaceable batteries and thus takes up the corresponding principle of stagecoaches, which also did not wait for the horses to rest and eat again and again in between, but simply changed the horses.
- W-T strategy: When your own weaknesses coincide with external risks, only one defense usually helps. One example of this might be the expiration of one's own patent coinciding with competitors entering the market with generic products. To avoid such a situation altogether, defense strategies must be implemented as early as possible, in this case, for example, by developing alternative products in good time, which can then be protected by patent again.

In the course of a situation analysis, it is actually still too early to consider solutions. In the sense of a future-oriented analysis (Haberfellner et al. 2018, p. 211 f.), however, the orientation towards the phases of the problem-solving cycle must not be interpreted too strictly. The cyclical approach also implies previews and returns of this kind to a limited extent. Problem-solving thinking does not follow strict determinism and should not be forced into one by the system of approach recommended here. In fact, the problem-solving cycle offers orientation – no more, but also no less.

4.2.4.9 Requirements Analysis

From all three analyses carried out so far – the Environment Analysis, the Stakeholder Analysis and the SWOT analysis – requirements for the design project and in particular for its result can be derived. These must now be formulated as completely and precisely as possible. This is a critical part of the problem-solving process, in that many projects suffer or even fail because it is not sufficiently clear what is actually to be achieved (Erlach 2014, p. 3 f.).

According to DIN 69901-5:2009-01 (p. 6), a requirement is defined as a characteristic, ability or performance that a product, process or person involved in the process must fulfill or possess in order to fulfill a contract, a standard, a specification or other formally prescribed documents. In principle, requirements should answer the question of who expects what from whom, when, where, why and under what conditions (Fig. 4.22).

Management system standards such as DIN EN ISO 9001:2015-11 for Quality Management, DIN EN ISO 14001:2015-11 for Environmental Management or DIN ISO 45001:2018-06 for Occupational Health and Safety Management indicate that require-

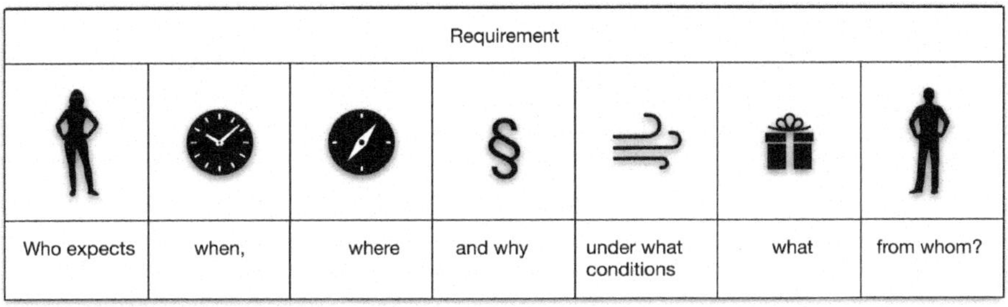

Fig. 4.22 Scheme of a requirements specifications

ments are not limited to formal specifications such as those in contracts or specifications (but are also usually assumed, so can be taken for granted. This opens up gray areas in which conflicts can arise between different assessments of what is and is not self-evident. Therefore, it is important to take appropriate measures in the course of a Requirements Analysis to avoid later conflicts of this kind. A possible, but usually insufficient, measure can be a severability clause. Finally, standards such as those mentioned also refer to mandatory requirements that may arise from laws, regulations, implementing provisions, standards and guidelines. A Requirements Analysis must therefore also include such sources.

Furthermore, it is not always possible to identify all parts of a requirement completely and correctly, and certainly not straight away. For example, if a requirement is derived from the context analysis, it may only be possible to identify the general circumstances from which a requirement arises, for example circumstances associated with an aging society or with a trend towards individualization. In such cases, the relevant information can only be kept relatively general. Nevertheless, the scheme shown in Fig. 4.22 can help to formulate requirements as precisely as possible right from the start. Nevertheless, further specification will usually be necessary in the further course.

The goals associated with a design project must be strictly distinguished from requirements. The terms are often used synonymously (e.g., Bender and Gericke 2021, p. 170), but this can lead to misunderstandings. As already explained in the introduction to Section 4.2.4.5, requirements are often initially formulated incompletely, imprecisely, or over-deterministically. Furthermore, it is not uncommon for them to contradict each other (ibid., p. 171), so that it cannot be the goal to fulfill all requirements equally. Against this background, requirements should be understood as expectations of stakeholders, which in a free society everyone is initially allowed to express, as long as they are within the framework of the values and legal system (except for the demands of blackmailers). The question of which goals are derived from these in the further course of events has yet to be addressed.

At this point, it is worth recalling what has already been explained in Section 3.3.2 on the iterative approach: overcoming a barrier, which by definition is associated with every problem-solving process (Fig. 2.59), usually does not succeed at the first attempt. Instead, what Beate Bender and Kilian Gericke (2021, p. 39), in agreement with other authors, call the *coevolution of problem and solution* occurs. When attempting to solve a problem and thus overcome the corresponding barrier, new problems usually arise that have previously remained undetected. or to which their actual significance was not attributed. This results in a new, usually more detailed problem description, which in turn leads to new solution attempts, which in turn lead to new problems, and so on. It follows inevitably that the phases of the problem-solving cycle must be repeated over and over again until a satisfactory solution is finally found.

In this context, Requirements Management extends over the entire problem-solving cycle: In the course of the situation analysis P2, the requirements of the stakeholders identified in the Stakeholder Analysis are determined. A *requirements specification* often serves as documentation (Rupp and SOPHISTen 2021, p. 48). In the subsequent phase of formulating objectives P3, these are checked and substantiated. The actual goals of the project result from this. This in turn results in a revised and coordinated set of specifications between the client and the contractor. In the method development phase, the team then works out how and with which tools the solution is to be developed. The result of this phase is often documented in a *obligation's specification*. However, since the actual solution is not yet available in this phase, but must be developed in the following phase, the requirement specification can only document intentions and experiences. In the course of the actual solution, new problems often arise that, as explained above, can bring with them new requirements. This then requires a return to the situation analysis, which continues in the manner just explained with the further phases of goal formulation and so on. In view of the resulting complexity, the problem-solving cycle can help to maintain an overview and to always make it transparent where a design project currently stands.

So, for now, the focus is on identifying requirements. The result of the Stakeholder Analysis is recommended as a starting point to ensure that no requirements are overlooked. Each stakeholder group must be addressed specifically. This is shown by the example of customers. In business-to-business (B2B) transactions, those who carry out a design project can often be directly involved. The situation is different in business-to-consumer (B2C) transactions. In such cases, customers are usually anonymous and must be represented internally by marketing or sales (Bender and Gericke 2021, p. 181), who in turn rely on observations, representative surveys and the study of documents. The situation is similar for employees, depending on whether a works council has been set up to represent them or whether they have to be directly involved in the absence of institutionalized representation. Finally, it should be noted that numerous requirements are not always explicitly stated, but are tacitly expected. In the course of the Requirements Analysis, it must be ensured that such requirements are also included.

Another aim of the Requirements Analysis is to achieve a *level of detail* that corresponds to the state of the project. For example, the design phase will not yet be concerned with the requirements for individual screws that can be used to create detachable connections in a product, whereas the actual construction of the product may even involve the design of the thread pitches of individual screws. As a method, a slight modification of a suggestion by Haberfellner et al. (2018, p. 231 f.) is recommended: the *requirement-means thinking* (*teleological system of requirements*). For a requirement formulated at a higher level of detail, the next level of detail is used to search for means by which the requirement can be fulfilled. This usually results in several approaches that appear to be applicable either in addition (AND link) or alternatively (OR link). What were just means can in turn be requirements for the next level of detail, and so on. Fig. 4.23 demonstrates this with an example based on a 2019 project by the Federal Institute for Research on Building, Urban Affairs and Spatial Development (BBSR) in the Federal Office for Building and Regional Planning (BMI) on strategies for inner development, which is offered as a tool for designing lively and mixed-use residential and service locations in smaller towns and communities. Reading this project report is also recommended because it documents an approach that is very similar to the one recommended here.

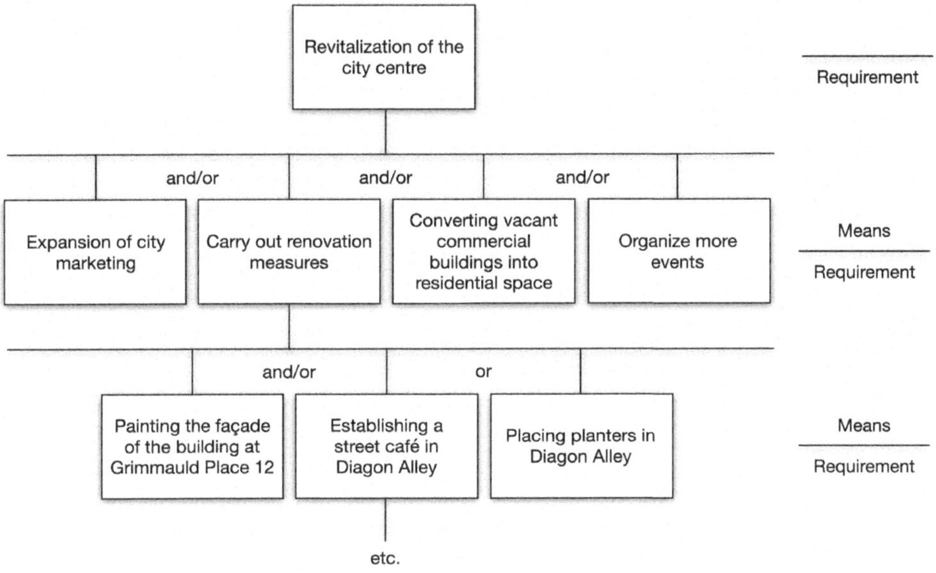

Fig. 4.23 Requirements-means thinking in reference to Reinhard Haberfellner et al. (2018, p. 231)

An alternative to the teleological requirements system, which results from requirements-means thinking, is the *causal requirements system*, which results from cause-and-effect relationships. Such a way of thinking results, for example, in key figure systems such as the one from Du-Pont, which results in the return on investment (ROI) at the highest level (Göbel 2018, p. 68).

In view of the scope and diversity that can result from this, the sources listed below, from which requirements arise, can only be examples (Jenney et al. 2010, p. 85 ff.; Kaffenberger et al. 2012, p. 67):

- Concepts, for example in the form of sketches and/or verbal descriptions, models, prototypes, samples, reference or comparison products [15], and, if the project progresses accordingly, also orders, technical drawings and specifications. In accordance with DIN ISO 21500:2016-02 and DIN 69901-5:2009-01 (p. 12, referred to there as a project order), development orders should also include, in addition to relevant specifications, obligations and assumptions, but also the associated objective, information on the expected

[15] Services and rights are always included when products are mentioned.

results, constraints, responsibilities, planned resources and a concordant declaration of intent by the client and the project manager.
- Minutes, for example of discussions with stakeholders.
- Development contracts (DIN ISO 21500:2016-02; DIN 69901-5:2009-01, p. 12).
- Laws, standards, guidelines, standards, conditions, contracts, specifications, acceptance criteria.
- Proprietary rights (patents, utility models), both one's own and those of third parties, which must not be violated.
- Procedures, for example for submitting applications.
- Requirements for the implementation of design projects, for example, a Project Management Manual in accordance with DIN 69901-1:2009-01 (p. 8).
- Requirements for the participation of third parties, such as project partners.
- Predecessor products or comparable products from competitors.
- Matters of course that are only implicitly present but often remain unspoken, but which lie in a gray area of different assessments and can thus lead to conflicts.

Since requirements are usually diverse and can take on a very large scope, it makes sense to structure them. However, there is no single structuring scheme that experts have agreed on. Therefore, some of the common suggestions are presented below, which can be adapted as they are or in a modified form. Alternatively, it may make sense to develop your own project-specific structure.

Beate Bender and Kilian Gericke (2021, p. 171) suggest a distinction between product-related requirements and framework conditions, whereby the latter include cost- and schedule-related requirements. In the light of the Stakeholder Analysis, however, it can be seen that only the requirements of customers (product-related and deadline-related requirements) and of owners and investors (cost-related requirements) are included, while the requirements of all other stakeholders are left out. Elsewhere (ibid., p. 186), the two authors suggest a categorization into functional and non-functional requirements, with the latter describing the conditions under which the functions are to be fulfilled. One could object to this that conditions under which they are to be fulfilled should be specified for all requirements. For example, a ship will only be able to float if there is a minimum amount of water under the keel.

Christof Ebert (2014, p. 29) uses a similar categorization, but differentiates in detail between functional requirements, quality requirements and constraints. These should be further differentiated into internal and external requirements. In other places, the requirements he refers to as quality requirements are also classified as non-functional requirements. Both categorizations subsume what is abbreviated as RAMSS in the Anglo-Saxon world and includes the aspects of reliability, availability, maintainability, functional safety and security (ibid., p. 31).

Joe Jenney et al. (2010, p. 85) suggest the five categories of functional requirements, performance requirements, design requirements, allocated requirements, and derived requirements. Alexander Kossiakoff et al. (2011, p. 145) recommend a similar categorization into operational requirements, functional requirements, performance requirements, and physical requirements.

It will not be possible to say that any of the categories is not informative. At the same time, it is not possible to decide which of the structures mentioned can claim universal validity. Therefore, the insight formulated by Beate Bender and Kilian Gericke (2021, p. 188) can help, that all categorizations are views of the totality of all requirements, i.e., what was described in Section 2.4.5 as *aspects* of systems (Fig. 2.46). Computer programs that support the use of SysML offer such views, each of which represents excerpts from a model that is stored in its entirety in the database (Fig. 2.48).

While all the categorizations presented so far can be useful for structuring the requirements of different stakeholder groups, one type of categorization is not recommended. For example, there are approaches such as that of Michael Nagel et al. (2020, p. 299 ff.), which divide stakeholders into friends and enemies, so to speak, and pit good (change champions) against evil (saboteurs) in order to turn critics and pessimists into supporters and advocates. If a stakeholder group has different expectations from those driving a project, it should not be condemned for this if and to the extent that the expectations are within the prevailing legal and value system. Nor can it be a matter of expelling divergent or aggravating expectations by force or by persuasion, outwitting, taking by surprise, bribery or the like. As a rule, it can be assumed that stakeholders will present their requirements in a well-considered manner, express legitimate expectations and otherwise be willing to compromise if they are taken seriously and involved in the project on an equal footing (participation). Projects that have to be implemented against resistance cannot be good projects and/or are poorly managed.

Finally, reference is made here to the possibility of categorizing requirements according to the *Kano model* shown in Fig. 4.24 (Kano et al. 1984, p. 38 ff.). Basic features of a product, service or right hardly contribute to the satisfaction of the corresponding stakeholder group, since they are taken for granted. However, they lead to dissatisfaction if they are not present. For performance features, satisfaction is proportional to the level of the corresponding feature. Enthusiasm features do not lead to dissatisfaction, since the corresponding stakeholder group does not expect them to be present. However, they can contribute disproportionately to differentiation from the competition, so that a small effort can achieve a large benefit. Not shown in Fig. 4.24 are irrelevant features that have no influence at all on stakeholder satisfaction, as well as rejection features that lead to dissatisfaction when a feature is present but do not contribute to satisfaction when a feature is present. When applying this model, it should be noted that the assignment of characteristics to the categories mentioned can change over time, for example, due to a habituation effect. In addition, the categorization depends on other factors such as the age or gender of the stakeholder group.

The result of the determination of requirements is often documented in the form of a *requirements specification*. The responsibility for this usually lies with the client and in B2B business they will also have the necessary competence and motivation for this. However, a specification sheet that is created exclusively or predominantly by the client will in many cases only include their primary requirements. Requirements of other stakeholders may be neglected. Therefore, it should be ensured that these are also included in the specification sheet. The contractor's participation can be helpful in this regard. In practice, the creation of a requirements specification often takes place iteratively and is continued in the course of formulating the objectives P 3 in order to further develop the requirements into objectives. Therefore, the explanation of the requirements analysis is continued in Sect. 4.3.

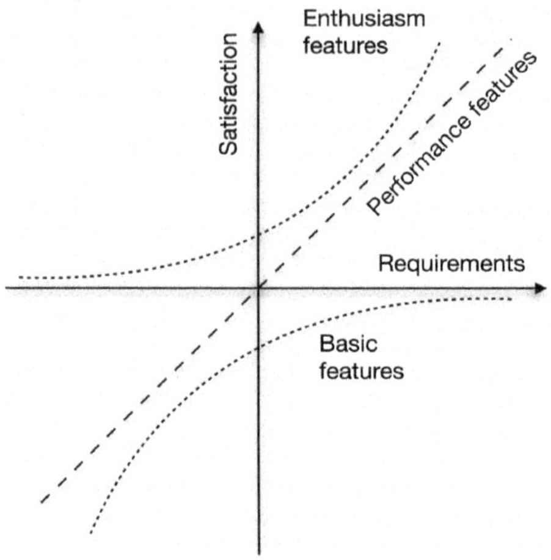

Fig. 4.24 Kano Model (Kano et al. 1984, p. 38 ff.)

4.2.4.10 State of Research and Technology

Even if a situation initially presents itself as a problem, there is a possibility that solutions or at least approaches to solutions have already existed or still exist in another time or place or in another context, which can either simply be applied or at least adapted. A current example is the conversion of vehicle drives from combustion engines to electric motors. So far, vehicles with combustion engines have served as the model, but this has the consequence that the energy absorption of fixed energy storage devices leads to comparatively long downtimes for the vehicles. As a result, a great deal of effort is being put into developing techniques that enable charging times to be shortened. The historical model of the stagecoach, on the other hand, offers an alternative approach. In those days, no one would have thought of putting the horses out to pasture every two hours and waiting to continue the journey until they had rested and eaten their fill. Since horses are continuous feeders, that could take a while. Instead, they simply changed horses – a strategy that would possibly also allow much faster travel today if you simply swap batteries at service stations converted to change stations. This could also be fully automated, making the whole process significantly faster and more convenient than the previous manual refueling. In China, the adaptation of this strategy is possibly more advanced than in the Western world.

Various methods can be used to determine the state of research and technology. For example, the state of the art can be determined by benchmarking, i.e. by systematically comparing one's own products or organization with those of other manufacturers, usually market leaders, or with the organization of other companies. A well-known example of the latter is the widespread attempt by countless companies worldwide to analyze and imitate the so-called *Toyota Production System* (Lean Production) (Dombrowski and Mielke 2015). The fact that this is still going on after more than three decades shows, on the one hand, how great Toyota's competitive advantage is still thought to be compared to other companies, but also, on the other hand, that it is much is much more difficult to analyze the organization of a foreign company than its products, because the latter can usually be bought and then disassembled by a competitor, while a look behind the factory gates of other companies is often only possible to a very limited extent. Therefore, a good corporate culture and a good business organization often include unique selling points and thus offer significant and sustainable competitive advantages (Bokranz and Landau 2012, Vol. 1, p. 8).

For a benchmarking, it is not always necessary to compare systems as a whole or similar systems with each other. What matters is that the sub-systems or elements, the structures or processes or the aspects under which the systems are considered (Chapter 2) are comparable with regard to the relevant characteristics. For example, it is hardly possible for a shipyard with the usual one-off production of ships as a whole to compare its own production with that of an automobile manufacturer like Toyota, which produces vehicles in large numbers. While the building site principle is suitable for the one application, the other will preferably work according to the flow principle in the final assembly (Wiendahl 2014, p. 51). Nevertheless, companies such as the Meyer Werft in Papenburg have also successfully applied the flow principle in their make-to-order production.

Another method for determining the state of research and technology is the *study of documents* (Sect. 4.1.4.1) in a form that is usually referred to in science as a *literature research* (e.g. Kornmeier 2018, p. 94 ff.). While the documents to be studied are usually already available or can be easily obtained, a literature search often involves a more time-consuming search for relevant sources.

- Such sources may include articles in scientific journals (papers), in conference proceedings as written versions of corresponding specialist lectures or in anthologies. Publications that have undergone a review process, i.e. have been reviewed by mostly anonymous referees before publication, are considered to be particularly reliable. So-called gray literature, i.e. working papers of scientific institutes or theses (bachelor's and master's theses as well as dissertations), is considered less reliable. However, innovative approaches can also be found in these. The research is carried out in scientific libraries, which are usually affiliated with universities, and increasingly in electronic databases, which, however, are not always publicly and freely accessible and often have thematic focuses, so that it may be necessary to search in several databases.
- Further reliable sources are patent and utility model specifications (Hahnl 2015). These are provided by patent offices in publicly and freely accessible electronic databases according to a globally valid standard. The targeted search for protective rights is carried out using retrieval languages and corresponding search terms, which are linked together by Boolean operators. In addition, form-based search options are also offered, for which no knowledge of a retrieval language is required.
- Finally, monographs (textbooks, handbooks) such as the one at hand, which cover a specific topic completely and systematically, are relevant.
- Caution is advised with publications that are distributed exclusively on the internet, as they are usually not subject to any review. This also applies to free internet encyclopedias, which are subject to a certain degree of control through the possibility of content-related discussion among members of the corresponding community and through administrators, but are nevertheless susceptible to vandalism and to influence by interest-led institutions. Furthermore, the authors are often anonymous. In science, such sources are not considered citable.

Once the relevant literature has been found in the course of a search, the further literature research process is similar to the study of documents explained in Section 4.1.4.1. In both a scientific and a commercial context, it is important to cite sources. This serves to facilitate retrieval and to protect copyrights (§§ 51, 63 UrhG). There are a number of different citation styles for this (e.g. Balzert et al. 2017). For example, the Harvard citation style is used in this volume. Ultimately, however, it depends on the conventions of a subject area and, within the framework of these conventions, on the personal taste of the author, which citation style is preferred. The only thing that matters is that the cited sources are correctly identified and can be found.

Literature research involves risks in two respects. (1) The first risk is the path dependency of scientific publications (Sect. 2.5). Practices such as the peer review just mentioned favor the maintenance of paradigms that may cloud the view of anything that might be found outside such bubbles. While paradigms offer a certain degree of security within their field of application, they also hinder innovation beyond that field. So anyone looking for the latter will have to stray from the well-trodden path. (2) A second risk arises from language barriers. The widespread use of English in scientific literature offers great advantages because it makes the insights formulated in it internationally accessible. At the same time, however, this obscures the view of publications that are available in less internationally widespread languages. This does not automatically result in lower relevance. (3) Furthermore, more and more sources are being made available in electronic form. This also has significant advantages, but carries the risk that sources that are not available electronically will receive less attention, even though they may be just as relevant. (4) Finally, it is becoming increasingly difficult to distinguish reliable sources from unreliable ones. In particular, the increasing prevalence of artificially intelligent text generators such as ChatGPT makes it more and more difficult to recognize the authorship of texts. In addition, there are people and organizations that deliberately manipulate information, especially on the internet.

4.2.4.11 Forecasts

The methods explained so far are mainly aimed at analyzing the past and the present, although strictly speaking the present has no temporal extent at all, but represents a singularity in which the future melts into the past. However, you don't have to take it that precisely to know that every action is directed towards the future. However, it follows that every action is uncertain, because in principle you cannot know anything about the future at the time of the action. Only the *experience* that the future has often proven to be relatively similar to the past in the past justifies the *hope* that this will remain so in the future. The fact that the remarks in Section 3.5 on the management of exceptional situations assume that this hope will no longer be as well founded in the future as it has been in the past few millennia will be neglected in the following.

Instead, the aim here is to find clues that can be assumed to allow us to at least somewhat reduce the uncertainty about the future and thus the resulting risk, so as not to fall prey to a fatalism (karma, kismet) that, like the story of Oedipus, can have fatal consequences in the literal sense of the word. In this respect, the following remarks are not based on *fatalism*, but nor are they based on its opposite, *voluntarism*, which gives will precedence over reason and, with libertarian manifestations such as radical market economics, appears just as questionable.

Instead, forecasts take a middle course by attempting to recognize and use certain laws as a basis for future developments in the sense of a *determinism*.

If these regularities appear to be very weak, *qualitative forecasts* are used. If the causes of the weakness lie particularly in the fact that predictions are to be made very far into the future, the methodology leads into *futurology* (Gerhold et al. 2015).

Intuition is particularly useful when dealing with uncertainty (Jung 2001b, p. 63 ff.). According to Fig. 3.33, this is based on implicit knowledge, which in turn is composed of experiences and the recognition of an inner logic of the circumstances. Carl Gustav Jung (2001a, p. 7 ff.) also includes what he calls the *collective unconscious*, which makes transpersonal experiences accessible to individuals.

For the concrete implementation of qualitative forecasts, methods of empirical social research are used, in particular *expert surveys*. If and to the extent that quantitative statements are to be made, these are based on *estimates*, which are often also left to experts (Thonemann 2015, p. 37 f.). However, experts in this context are not to be understood as persons who necessarily have an above-average and secure knowledge in the relevant domain, because such knowledge cannot exist about the future. In any case, explicit knowledge and skills in the sense of the SECI model (Fig. 3.33) only explain part of the differences in performance between people (Platz et al. 2014). What is also and above all important when selecting suitable experts, in addition to a certain amount of expertise, which of course cannot hurt, is the ability to recognize large contexts of meaning, to work quickly and accurately, to have a good short and long-term memory, paying attention to structures rather than superficial characteristics, investing a lot of time in qualitative analyses, and correctly assessing their own skills (Chi et al. 1988, p. 108 ff.).

In the following, the focus will be on *quantitative prognoses*, which are used when a sufficiently reliable database is available.

Causal Forecasts

Causal forecasts assume a relationship between a circumstance to be forecasted and a sufficiently known circumstance (Thonemann 2015, p. 41 ff.). If it has rained abundantly in spring, a good harvest can be expected in fall.

A known fact can often be represented as a sequence of data points in a Cartesian coordinate system. Fig. 4.25 shows this using an example. The prediction function is calculated from the points using *Regression Analysis*. To do this, it is first hypothetically assumed that the data points are linearly related to each other. In this case, the prediction function results in a straight line of the form $y = a + bx$. To do this, the regression coefficients a for the axis intercept and b for the slope must be calculated in such a way that the straight line passes through the point cloud as well as possible. This quality is usually measured using the mean square error (MSE) of the known data points from the corresponding forecast value. This is calculated as

n	x_n	y_n
1	1	0,40
2	2	0,55
3	3	0,60
4	5	0,75
5	8	0,90
6	13	1,00

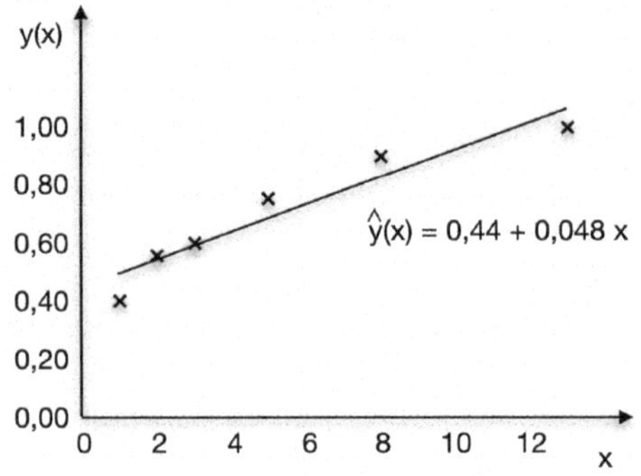

Fig. 4.25 Example of a causal forecast using regression analysis (Thonemann 2015, p. 41 ff.)

$$MSE(a, b) = \frac{1}{N}\sum_{n=1}^{N}(a + bx_n - y_n)^2 \tag{4.2}$$

As is well known, the minimum is calculated by taking the derivative and then setting it equal to zero. In this case, a partial differentiation is required. From this, the two regression coefficients are obtained

$$a = \frac{\sum_{n=1}^{N} x^2 \sum_{n=1}^{N} y_n - \sum_{n=1}^{N} x_n \sum_{n=1}^{N} x_n y_n}{N\sum_{n=1}^{N} x_n^2 - \left(\sum_{n=1}^{N} x_n\right)^2} \tag{4.3}$$

$$b = \frac{N\sum_{n=1}^{N} x_n y_n \sum_{n=1}^{N} x_n - \sum_{n=1}^{N} y_n}{N\sum_{n=1}^{N} x_n^2 - \left(\sum_{n=1}^{N} x_n\right)^2} \tag{4.4}$$

Fig. 4.25 shows $a = 0.44$ and $b = 0.048$ as results, as well as the resulting straight line as a prediction function. However, a visual comparison of the data points with the best-fit straight line suggests that the function might not be linear at all, but rather nonlinear. In such a case, the power function, the exponential function, the logarithm function or the hyperbolic function are often used as approximations, since the two parameters a and b can also be determined for these using linear regression analysis (Thonemann 2015, p. 45 ff.). To do this, however, the nonlinear forecast function must be transformed into a linear function and then transformed back again after the regression. To demonstrate this in detail here would go too far for three reasons: (1) such an explanation would be elaborate and (2) it is well described in detail elsewhere (e.g. ibid.). In addition, (3) market-leading computer programs for spreadsheet analysis and statistical data analysis offer functionalities that calculate functions such as the one shown here almost at the push of a button. Of all the functions that can be determined in this way, the one that best approximates the underlying data is the one in which the MSE takes on the lowest value.

The form of causal prediction explained so far is based on the largest possible amount of data, in which an individual data point is lost even and especially if it shows deviant behavior. If it is identified in the course of data processing, for example, it is usually even eliminated. However, there are also situations in which the individual is important, especially when only a few are involved overall. In such situations, *game theory*, for example, can be used to make predictions about future situations (Grimme and Bossek 2018, p. 229 ff.; Holler et al. 2019). It allows the modeling of conflict situations that are influenced by several participants, each with different prerequisites (actual situation) and intentions (desired situation). To do this, it makes individual assumptions such as those of purely rational behavior (homo economicus), perfect information and perfect memory, in order to use mathematical methods to determine which are the optimal strategies for the respective players and what the outcome of a game will be if these strategies are applied. The basis for this is, for example, the Nash equilibrium, which includes those strategy profiles in which individual players cannot improve by changing their strategy while the strategies of the other players remain the same. To do this, one-time and repeated games are examined.

A well-known example is the *prisoner's dilemma*, which, among other things, underlies the *Braess paradox*. According to this, an additional option for action, such as the construction of a road, can lead to a deterioration of the overall situation despite rational decisions by all parties involved – in the chosen example, this would mean an increase in the driving time for all drivers. Another well-known example is the *tragedy of the commons* (Fig. 2.39), according to which freely available but limited resources are subject to global overuse due to the local optimization of each individual user, which results in damage to all users. This can be used, among other things, to model environmental protection problems. In *Systems Dynamics* (Sect. 2.4.5.2), such phenomena are treated as *positive feedback loops* and modeled, for example, using the so-called *logistic equation*[16]. Findings from such models are applied in practice, for example, to improve environmental protection by introducing a trade in emission certificates.

Time Series Forecasts

Time series forecasting involves extrapolating the past into the future. This is based on the assumption that the future will behave in a similar way to the past, in particular that there will be no breaks in the relevant developments. Fig. 4.26 shows the basic types of progressions: the constant progression, the trend-shaped progression, which, as shown in the figure, can be not only upward but also downward, and the seasonal progression, if a periodicity is recognizable. For each of these basic types, there are specific forecasting methods, which will be explained at least to some extent below. For more detailed explanations, please refer to the further literature (e.g. Thonemann 2015, p. 49 ff.).

A *constant progression* of the initial data (Fig. 4.26, left) can be updated using the moving averages (MA) method or the single exponential smoothing (SES) method. The

[16] The logistic function for continuous fall and the derivation of its solution can be found in Section 2.4.5.2. In addition, the logistic equation can be used to model nonlinear-dynamic (chaotic) behavior and visualize it in the so-called Feigenbaum diagram (bifurcation diagram), for example.

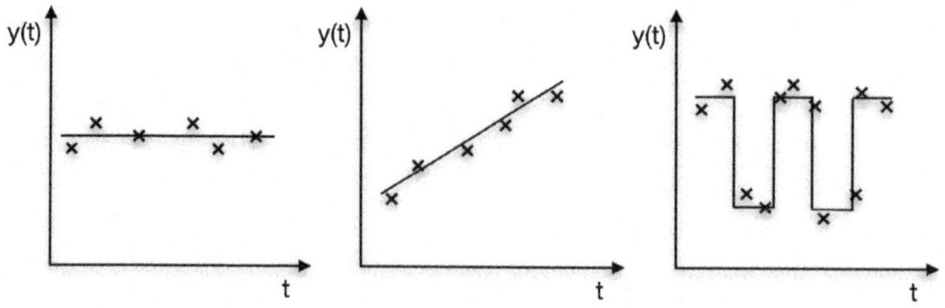

Fig. 4.26 Basic types of time series forecasts: constant, trend-like and seasonal patterns

moving averages method calculates the mean value over the preceding T periods. The forecast value for the period $t + 1$ is then calculated as follows:

$$\hat{y}_{t+1} = \frac{1}{T} \sum_{\tau=t-T+1}^{t} y_\tau \tag{4.5}$$

The number of periods T on which it is based cannot be reliably determined in advance. Instead, this figure is to be determined experimentally by calculating forecast values for different period lengths. The prediction error MSE is then calculated for each of these values. The period length for which the MSE is minimal is the optimal period length.

After each period has expired, it is shifted forward by one period length (sliding). This allows the forecast value to continuously adapt to any changes.

While the moving averages method weights all the periods T equally, namely with $1/T$, simple exponential smoothing weights current periods more heavily than periods further in the past. This is based on the assumption that the more recent past is more informative for the future than values further in the past. However, assigning each period an individual weighting would be cumbersome. Therefore, the method uses a single smoothing factor $\alpha[\%]$ for this purpose. The larger this factor, the higher the current values are weighted. This increase occurs exponentially.

The forecast value for the period $t + 1$ is now calculated as follows:

$$\hat{y}_{t+1} = (1-\alpha)^t y_0 + \sum_{\tau=0}^{t-1} \alpha(1-\alpha)^\tau y_{t-\tau} \tag{4.6}$$

The initial value y_0 contained in it is calculated as the average value of the first N periods.

$$y_0 = \frac{1}{N} \sum_{k=0}^{N} y_k \qquad 1 \le N \le t \tag{4.7}$$

The value that is best suited as the smoothing factor α must be determined experimentally by varying it and then comparing the prediction errors MSE.

For a trend-like course of the initial data (Fig. 4.26 center), it is assumed in the following that it is a linear course. This can then, as already explained in the context of causal forecasts (Parag. 4.2.4.11), be approximated by a straight line function $y(t) = a + bt$ (ibid., p. 57 ff.). In the following, regression analysis and double exponential smoothing (DES) are presented as methods for estimating the intercept a and the slope b.

The regression analysis used here is essentially the same as that explained in Paragraph 4.2.4.11. For the intended purpose, only the point cloud $(x_1, y_1), (x_2, y_2), \ldots, (x_k, y_k), \ldots, (x_K, y_K)$ must be replaced by the point cloud $(1, y_1), (2, y_2), \ldots, (\tau, y_\tau), \ldots, (t, y_t)$. The points must be recorded periodically and must be available for all points in time. Substitution of τ for x_k and y_τ for y_k in the equations given in Equations 4.3 and 4.4 for the regression coefficients a and b yields the corresponding equations for estimating the axis intercept and the slope

$$a_t = \frac{1}{t}\sum_{\tau=1}^{t} y_t - \frac{1}{2} b_t(t+\tau) \tag{4.8}$$

$$b_t = \frac{12\sum_{\tau=1}^{t}\tau y_\tau - 6(t+1)\sum_{\tau=1}^{t} y_\tau}{t(t^2-1)} \tag{4.9}$$

In period t, the forecast value in period $t + \tau$ is calculated as follows:

$$\hat{y}_{t,t+\tau} = a_t + b_t(t+\tau) \tag{4.10}$$

Analogous to the forecast for a constant trend, it is also possible to weight current data more heavily than older data for a trend-like course. However, double exponential smoothing is now required for this. It uses two smoothing factors, the factor α for the axis intercept at and the factor β for the slope b_t. The corresponding estimated values are determined in each period t with the following equations 4.11 and 4.12:

$$a_t = \alpha y_t + (1-\alpha)(a_{t-1}) + (1-\beta)b_{t-1} \tag{4.11}$$

$$b_t = \beta(a_t - a_t - 1) + (1-\beta)b_{t-1} \tag{4.12}$$

The forecast value in period t is calculated for the period $t + \tau$ as follows:

$$\hat{y}_{t,t+\tau} = a_t + b_t\tau \tag{4.13}$$

Additionally, an initial estimate is needed to initialize the procedure (ibid., p. 61). For this purpose, the mean slope of a number of previous periods, determined experimentally or roughly, can be determined. Alternatively, the initial slope can also be calculated using the regression analysis explained above. As with the simple exponential smoothing, it is not possible to optimally determine the smoothing factors in advance. In this case, too, they must be determined experimentally. Again, the MSE can be used as a quality measure. If the trend is not linear, as initially assumed here, the data can be transformed in the same way as explained in the context of the causal forecast, and the double exponential smoothing can then be applied to the transformed data.

As a third possibility for a time course, Fig. 4.26 shows *seasonal fluctuations* on the right. The course of the data repeats itself every P periods. *Triple Exponential Smoothing* (TES) by Peter R. Winters (1960, cited here by Thonemann) is presented below as a way of setting up a forecast based on this. 2015, p. 62 ff.). In period t, the forecast value for period $t + \tau$ is calculated as follows:

$$\hat{y}_{t,t+\tau} = (a_t + b_t\tau) \cdot c_{t+\tau} - P \tag{4.14}$$

This Eq. 4.14 corresponds to the one for double exponential smoothing, but it is extended by the multiplication with a seasonal factor. A value of, for example, $c_3 = 0.5$ means that in the third period, on average, 50% less accrues than on average over all periods. The values for the axis intercept at, for the slope b_t and the seasonal factor c_t are calculated as follows:

$$a_t = \alpha\frac{y_t}{c_{t-P}} + (1-\alpha)(a_{t-1} + b_{t-1}) \tag{4.15}$$

$$b_t = \beta(a_t - a_{t-1}) + (1-\beta)b_{t-1} \tag{4.16}$$

$$c_t = \gamma\frac{y_t}{a_t} + (1-\gamma)c_{t-P} \tag{4.17}$$

However, estimating the initial values for a, b and c is no longer as easy as with double exponential smoothing. Data from $N \geq 2$ seasons is required for this, with each season consisting of P periods. On this basis, it is first estimated how the average data changes from one season to the next. This results in initial estimates for the axis intercept a_{init} and for the slope b_{init}. When estimating the seasonal factors, the data is adjusted for the resulting trend. Only then are the seasonal factors estimated. The most suitable smoothing factors must be determined by trial and error, and the MSE is also used as a benchmark here. Ulrich Thonemann (2015, p. 67) states that, based on experience, good values for α are in the range of 0.1 to 0.3, good values for β are in the range of 0.1 to 0.4, and good values for γ are in the range of 0.2 to 0.6.

In summary, it can be said that, in terms of forecasting, the methods presented last are not mathematically undemanding and always require a certain amount of effort when they are to provide exact values, due to the experiments with different factors that are necessary. However, if such forecasts are to be made more often, it is worthwhile to use commercially available software for spreadsheet or statistical data analysis in order to implement the calculations once and then reuse them, which may only require minor adjustments, for example when customizing row and column references. Incidentally, for those with a deeper interest and for those interested in explained examples, reference is made to the literature cited.

Scenario Technique

Predictions of the kind explained in the previous Section assume that the future will develop in a similar way to the past. This assumption is also referred to as the *time stability hypothesis*. If and to the extent that this cannot be assumed, predictions, in particular due to the accuracy of their mathematically determined results, carry the risk of *spurious accuracy*. The scenario technique is an alternative that is more detached from the past than forecasts and can therefore also take discontinuities (disruptions, wild cards) into account. Scenarios develop possible future states (situations) from a series of causally related hypotheses. They also show different actors possible interventions and their potential consequences.

In doing so, different variants are developed, which are usually illustrated with the help of the funnel outlined in Fig. 4.27 (von Reibnitz 1987, p. 31). The starting point is usually the so-called *trend scenario*, which is based on the time stability hypothesis (ceteris paribus). The edges of the funnel are formed by *extreme scenarios*, which describe, for example, very positive (best case) and very negative (worst case) developments. Further scenarios can be added within these boundaries. Each scenario consists of a bundle of internally consistent assumptions (hypotheses). Furthermore, *interactions* between developments and events are usually analyzed (cross-impact analysis), since the probability of a development or event occurring often depends on the occurrence or non-occurrence of other developments or events.

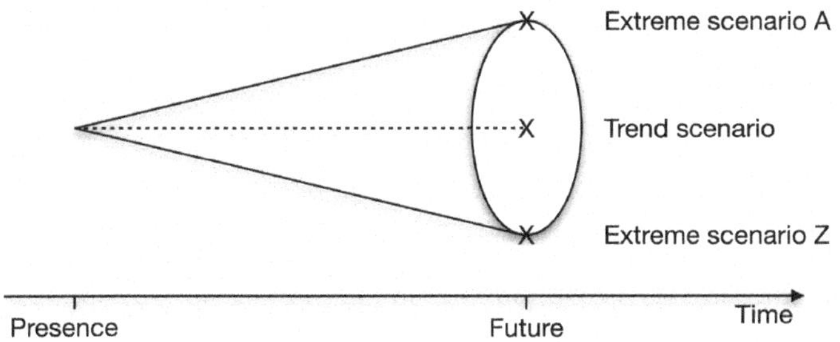

Fig. 4.27 Principle of the scenario technique

From the identified scenarios, which may be a result of the situation analysis, *roadmaps* are derived in the course of the later formulation of objectives (Sect. 4.3). These describe possible ways of implementing strategies or innovating and also include intermediate steps. However, basing specific action on scenarios gives rise to the *paradox of futurology* (Wagenführ 1985, p. 580), in that the occurrence of the expected is prevented or at least altered. As a result, the scenarios may prove to be wrong in retrospect, even if they were correct at the time of their development. For example, if students prepare for a predicted oversupply of teachers by choosing alternative career paths, this may actually lead to a teacher shortage.

4.2.5 P 2.5: Conducting the Situation Analysis

After the detailed presentation of the methods that can be used in the context of situation analysis P2, there is nothing more essential to say about the application of these methods in the course of the concrete implementation of the situation analysis. Therefore, it is only fair to point out once again that the *consistent application of methods* is almost always preferable to an approach based on common sense (intuition), since it significantly increases efficiency and effectiveness. This is not fundamentally opposed to intuition (Weibler and Thielmann 2022, p. 177 ff.). On the contrary, some methods, such as qualitative forecasting (Sect. 4.2.4.10), are based on it, among other things. However, when it comes to the technical implementation of complex projects, intuition should always be embedded in a rational and, to that extent, systematic approach.

Furthermore, attention is drawn once again to the level of detail in which the situation analysis P 2 is to provide insights. This depends on the phase of the system life cycle (Fig. 3.1) in which the corresponding design project is being carried out (Haberfellner et al. 2018, p. 221). The approach *from rough to detailed* (Figs. 3.19 and 3.20) ensures that the situation analysis is always as accurate as necessary and not as accurate as possible.

In practice, the high level of *effort* required for a situation analysis P2 compared to other phases of the problem-solving cycle occasionally leads to uncertainty. In fact, it is not unusual for it to take up to 50% of the total time (strictly speaking, this refers to the resources to be deployed), and occasionally even more. This time is usually well invested, because a comprehensive and thorough understanding of the actual state makes the implementation of the subsequent phases much easier (Sell and Schimweg 2002, p. 71). In *Lean Development*, this insight is referred to as *frontloading*.

In particular, the phase of formulating objectives can be comparatively short after a detailed situation analysis, since the objectives are derived from the requirements determined in the context of the situation analysis. In addition, a good understanding of the problem draws attention to possible solutions at an early stage, sometimes even too early, which can mean that the objectives can no longer be formulated in a solution-neutral way, as required. Conversely, a situation analysis that is not carried out carefully enough can result in more frequent setbacks in subsequent phases. There is nothing fundamentally wrong with this, since setbacks are obligatory anyway, but avoidable setbacks should definitely be avoided, since they always involve a certain amount of additional work, for example, due to the need to re-familiarize oneself with work that has already been completed.

4.2.6 P 2.6: Evaluation

The evaluation of the results of the situation analysis P 2 virtually closes the bracket that was opened with the formulation of the objective P 2.2 by making statements about whether and, if applicable, to what extent the objectives have been achieved. The situation P 2.6.2, which is to be analyzed at the beginning of the evaluation (Haberfellner et al. 2018, p. 273), is given by the results of the situation analysis P 2.5 that has been carried out. These results should be available in documented form. Specifically, the results should include the following content:

- The current situation should be transparently presented. Above all, the problem (syndrome) that is to be solved by the design project, as well as the associated barrier (Sect. 3.2), should be stated so clearly that those who are familiar with the project believe they no longer have to ask any further questions in this regard.
- The relevant structures and processes of the system to be designed should be known in the necessary level of detail (Sect. 4.2.4.4).
- The causes of the effects triggered by the problem in the actual situation should be known.
- The influences from the environment should be known (Sect. 4.2.4.5).
- All stakeholders should be known and already involved (Sect. 3.5.2). In addition, a complete list of all known requirements of all stakeholders in the current phase of the design project should.
- Opportunities and risks, as well as one's own strengths and weaknesses, should be known and related to each other (Sect. 4.2.4.7).
- With regard to possible solutions (therapy), the relevant state of research and technology should be known (Sect. 4.2.4.9).
- Reliable forecasts should be available for relevant developments (Sect. 4.2.4.10).

The goal of evaluation P 2.6.3 is to test the robustness of the insights gained from the situation analysis P 2.5 in order to prepare the decision as to whether the design project should move on to the next phase of goal formulation P 3, go through the situation analysis P 2 again, or be abandoned altogether. Furthermore, the evaluation phase offers the opportunity to characterize the insights gained according to generic criteria. The properties of the identified facts proposed by Robert Sell and Ralf Schimweg (2002, p. 34 ff.) based on Dietrich Dörner (1979) can be used as a basis for this.

In particular, the situation can be characterized by its *lack of clarity*. In such a case, it is particularly important to focus on the essentials. Abstraction and modeling are particularly useful for this purpose. Furthermore, a narrowing down of the object under consideration (system boundaries) or an exemplary processing can be helpful. In particular, visualizations such as those introduced in Chapter 2 can help to provide clarity. Characterizing problems as complex is particularly ubiquitous, perhaps because it brings the problem solver the greatest honor if he dares to tackle it anyway and ultimately even succeeds. However, there are other characteristics that define problems and make their solution no less worthy of recognition.

These include, according to Robert Sell and Ralf Schimweg (ibid., p. 35 f.), the *obviousness* of the solution, which at first glance actually gives the impression that it is far easier to recognize than the solution to a complex problem. In fact, numerous examples show that the level of difficulty is comparable. However, precisely because the view of the obvious solution is initially obscured, it is difficult to classify a problem situation as an obvious problem during the course of the situation analysis. The Bible tells us that when Saul regained his sight, he realized his mistake and let himself be baptized (Acts 9:18). This idiom, which is still in common use today, means that solutions of this kind only become apparent at the moment of realization and that obviousness is often only recognizable as such in retrospect.

In the laboratory, strategies for solving problems of this kind are examined, among other things, on the basis of the problem of the suspension bridge, which is to be crossed by four people at night (ibid., p. 19 ff.). For safety reasons, however, only two people are allowed on the bridge at a time, and they have to carry the only available flashlight for orientation, which, however, only has a remaining lighting time of 60 minutes. The people can walk at different speeds: Alfred needs 5 minutes to cross, Berta 10 minutes, Carl 20 minutes and Dietrich 25 minutes. If two people walk at the same time, the slower one determines the pace. The question is in which order the people have to cross the bridge so that they all reach the other side before the end of the remaining lighting time of the torch. It is usually immediately recognized that two people must always cross the bridge together so that one of them can bring the torch back. It is often assumed, and just as obviously, that the fastest person should do the returning. In fact, the two slowest people have to cross the bridge together, and they have to be the last to do so, so that neither of them has to bring back the flashlight. When dealing with problems involving such properties, it is particularly important to describe the situation very precisely so as not to overlook any details. The question is what exactly is given and which parameters are available for a solution. Robert Sell and Ralf Schimweg (2002, p. 35) describe the corresponding methods as *questioning techniques* and *self-reflection* (introspection). The latter looks for possible thought blocks that obscure the view of the actually obvious solution.

Another generic property of problems can be their *opacity* (ibid.). In the laboratory, strategies suitable for solving such problems are investigated using the candle problem, which goes back to Karl Duncker. A candle, matches and drawing pins are provided in a box. The task is to fix the candle to a corkboard using the aforementioned tools and then light it. The solution is regularly unsuccessful when the tacks are in the box, because then, due to a phenomenon called *functional fixation*, it is not recognized as part of the solution, but only as a means of storage. In most cases, the unsuccessful attempts at solving the problem involve melting the wax of the candle sideways in order to stick it to the wall. However, if the drawing pins and the box are provided separately, it quickly becomes clear that the box, as part of the available tools, can be stuck to the side of the wall and the candle can then be placed in it. As with obviousness problems, the key to solving problems of this kind is to ask questions as precisely as possible, to formulate the initial situation clearly and to reflect on one's own actions.

Another category of problems is characterized by their *dynamics*. This results in time pressure with regard to their solution. Dealing with this requires a certain routine (experience) in solving problems, as well as composure and the willingness to be satisfied with sometimes suboptimal, partially improvised (viable) solutions. Abstraction and modeling, i.e. concentrating on the essentials, can make important contributions here.

Finally, Robert Sell and Ralf Schimweg (2002, pp. 37 f.) recommend the degree of *dependency of the variables* (interactions, networking) as a property for categorizing problems. As a model example, they ask which of the following equations is valid. In it, R stands for ohmic resistance $[\Omega = V/A]$, κ for the reciprocal of ohmic resistance, i.e. for electrical conductivity $[S = A/V]$, ρ for specific resistance $[\Omega \cdot mm^2/m]$, C for capacitance $[F = As/V]$ and ε for the dielectric constant (permittivity of a vacuum) $\varepsilon_0 = 8.854 \cdot 10^{-12} As/Vm$.

$$a)\quad \frac{R}{C} = \frac{\varepsilon}{\rho} \qquad b)\quad R \cdot C = \kappa \cdot \rho \qquad c)\quad R \cdot C = \varepsilon \cdot \rho \tag{4.18}$$

To analyze the dependencies, the authors recommend first breaking down the identifiable facts. This allows us to establish (1) that in all three cases, the dependencies are linear. Furthermore, (2) the two proportionalities apply in a)

$$R \sim \frac{1}{\rho}; \qquad C \sim \frac{1}{\varepsilon} \tag{4.19}$$

(3) In b), the following proportionalities can be determined:

$$R \sim \kappa, \rho \tag{4.20}$$

(4) And in c) the following two proportionalities can also be identified:

$$R \sim \rho; \qquad C \sim \varepsilon \tag{4.21}$$

(5) With recourse to electrical fundamentals, it is also clear that the physical quantities contained in the equations are fundamentally related to each other. This also follows from a check of the units given above. This concludes the situation analysis. At first glance, the exclusion of obviously incorrect equations seems to be a good approach for finding a solution. The result will then have to be verified. On closer inspection, equation b) turns out to be false, since the capacitance C is not matched by any corresponding variable on the other side. This is also shown by a comparison of the units. In equation a), neither R and ρ nor C and ε are proportional to each other, so that this representation also proves to be false on closer inspection. This leaves equation c) as the only possibility for a correct equation. This can be verified by checking the units. Based on this simple example, Robert Sell and Ralf Schimweg (2002, p. 40) derive the following fundamental strategies for solving problems that are particularly characterized by the dependence of variables: This includes determining the degree of dependency, which can be linear, quadratic, exponential, proportional, reciprocal, and so on. Furthermore, the side effects that illustrate the degree of dependency must be analyzed. In the example, this was clarified by asking how, for example, the electrical resistance R behaves when the specific resistance ρ or the electrical conductivity κ increases or decreases.

By checking whether all the facts that can be identified in a situation analysis are present, and by categorizing the problem more clearly than before on the basis of its essential characteristics, the conditions are created for the following formulation of objectives. In the course of initiating the solution to problem P1, the objectives of the project are initially formulated in a rather vague manner so S.M.A.R.T (Doran 1981), i.e. so clearly (specific), measurable, acceptable (accepted), realizable (realistic) and within the given time achievable (timely) that it can offer the subsequent problem solution a promising basis. In the context of Project Management, this is usually formally determined before the transition to the following phase is released (DIN 69901-5:2009-01, p. 11).

4.3 P 3: Target Definition

The recommendation here to also run through the problem-solving cycle within its individual phases may seem excessive for straightforward projects. However, since the recommendations apply to all types and sizes of design projects, it may help to imagine that a project is being carried out that serves only to develop a consistent set of objectives for a larger design project.

4.3.1 P 3.1: Occasion

One reason for carrying out a project to develop a target system could be, for example, a plan that is judged very differently by different stakeholders and sometimes rejected altogether. In such a case, Stakeholder Value Management (SVM) will be a key focus of the work. A historical example was the arbitration proceedings that preceded the conversion and expansion of Stuttgart's main train station (*Stuttgart 21*) from November 2010 and ended with a compromise proposal that was recognized as legally binding by the local administrative court. Since this project took place in public and is well documented, it provides very good insights into the approach and the possibilities and limitations of finding a compromise with regard to other projects of this kind.

Another example of an occasion for implementing the target formulation P 3 may be that, after the occurrence of conflicts in the further course of the project, there is a return from a subsequent phase with the aim of resolving the conflict by adapting the previous target system. In such a case, it will be useful to additionally analyze how the conflict could arise despite the previously formulated objectives, which involved all stakeholders, and how this can be avoided in the future if possible.

4.3.2 P 3.2: Situation Analysis

In any case, a situation analysis must precede the formulation of objectives. The subject and scope of this analysis P 3.2 in the context of the goal formulation phase P 3 explained here depend on whether the goal formulation P 3 directly follows a previous situation analysis P 2 and can draw on the results obtained there, or whether and, if so, to what extent different or additional information on the current situation has to be obtained for the goal formulation P 3. The latter was the case in the example "Stuttgart 21" introduced above. In particular, the opponents of this project felt that the analyses presented by the project sponsor, Deutsche Bahn (DB), were insufficient, so that in the course of the conciliation process, further experts such as geologists, engineers and architects were consulted, among others, in order to substantiate the arguments of the various stakeholders with facts. In such situations, everything that was said in Section 4.2 about conducting situation analyses in general applies to the implementation of analysis P 3.2.

If, on the other hand, the analysis of the actual situation P 2 is complete, the situation analysis P 3.2 can be limited to the analysis of those aspects that must be considered immediately for the implementation of the target formulation P 3. To this end, the second example mentioned in the previous section can be continued. In such a case, the analysis of the causes that led to conflicts in the further course, despite a complete, competent and conscientious situation analysis P2 and a consistently structured target formulation P3, will have to be carried out in this phase.

Generally speaking, the question at this point could be: What information about the situation in which the formulation of objectives P 3 is to be carried out is not yet available and does not justify a return to the previous phase P 2. Obtaining this information is then the subject of the situation analysis P 3.2. Furthermore, it may be necessary to process the information taken from the previous situation analysis P 2 again in a way that is necessary, useful or desirable for formulating the objective. This may require, for example, a condensation of data or its visualization.

4.3.3 P 3.3: Target Definition

Goals describe the desired state of a solution to a problem. The purpose justifies the meaningfulness (motivation) of a goal. In phase P 3.3, the focus is initially on the goals of goal formulation P 3 and not yet on the goals that are to be developed in the course of the goal formulation phase P 3. This is only done in phase P 3.5. This once again highlights the *self-similarity* of the problem-solving cycle, similar to the situation analysis P 2.2 of the situation analysis P 2. Experience has shown that the application of the associated *recursion principle* can initially be perceived as not very easy. However, once you have got used to the strict logic of *scaling* at the various levels of the procedure, experience shows that it is difficult to do without recursion.

The formulation of objectives not only serves as a guideline for the subsequent phases, but also as a basis for evaluating the results achieved, by providing the criteria for this. In this respect, goal formulation and evaluation form a kind of bracket around the intermediate phases of method selection and solution development. This can also be embedded[17] in *decision theory*, which, although it can lead to misunderstandings due to its significantly different understanding of the central concept of decision (Sect. 2.6.2.3), apart from that, however, makes significant contributions to Systems Engineering, both in the context of formulating objectives and, even more so, in the context of the final evaluation, which, after all, actually has something to do with decisions to the extent that it prepares them, while the decisions themselves are no longer the subject of SE.

The fact that decision theory refers to the approach to decision-making as a process and thus also differs from the distinction emphasized here between cycle (problem solving) and process (task completion) may be overlooked, because the approach is not part of the core theory there. However, it is precisely for this reason that it seems remarkable that the rather rudimentary understanding of the procedure leads to a result that is almost the same as the research on SE, which is based, among other things, on cognitive science. Helmut Laux et al. (2018, p. 12 ff.), for example, outline the decision-making process in terms of the steps of problem formulation (situation analysis), the specification of possible alternative actions (solution development), and the decisions in the realization phase (evaluation). The designations of the phases of the problem-solving cycle added in parentheses have already made clear the similarities with the process model previously used in SE (Haberfellner et al. 2018, p. 70). The only thing missing is the added phase of method selection (Ahrens 2020, p. 74). Against this background, the strengths of decision theory can be found in the formulation of objectives and in the evaluation based on them, while SE contributes its strengths in system analysis and solution development. Together, you could say that these two theories make a good team. This in turn makes it clear that the critical examination of decision theory in Section 2.6.2.3 was not intended to refute this theory, but to harmonize it with SE and to bring its strengths to bear.

However, the large scope of decision theory precludes an even approximately complete integration of its findings into this book. Therefore, reference is made at this point to the relevant literature (e.g. Göbel 2018; Laux et al. 2018). Here and in the following, only some of the results that appear to be essential can be included, in particular those that should find practical application in SE. These include, in the phase of formulating goals P3, the *requirements* for goals (Laux 2018, p. 43 ff.), which in decision theory are formulated, among other things, in the form of axioms. These requirements are explained below.

4.3.3.1 Order and Transitivity Axiom

One problem with extensive target systems is that the target sub- and part-systems they contain, as well as the target elements, are linked by a usually large number of relations, i.e. they are interdependent and sometimes even contradictory. In order to cope with the resulting complexity, simplifications must be made. While it would actually be necessary to keep all the objectives and their interactions in mind at the same time, they are subjected

[17] This embedding can be clearly seen in Elisabeth Göbel (2018, p. 163). She devotes only one paragraph to what she calls alternative search in her Section 7.3.2, but in it she does explain what is at the core of the concern in Systems Engineering, namely solution development P 5 and the necessarily preceding development of the necessary methodology P 4.

to a *successive comparison*; the whole problem is thus broken down into sub-problems (dissected). Formally, the *axiom of ordering* states that for any pairing of two targeted outcomes (goals), one of the following relationships can be stated (ibid., p. 44):

- $e_i \succ e_j$ $\hat{=}$ the result e_i is preferred to the result e_j (preference)
- $e_i \sim e_j$ $\hat{=}$ the results e_i and e_j are equal (indifference)
- $e_i \prec e_j$ $\hat{=}$ the result e_j is preferable to the result e_i (preference).

The transitivity axiom states that the preferences or indifference with respect to three arbitrary outcomes ei, ej and ek are consistent in the following way:

- if $e_i \prec e_j$ and $e_j \prec e_k$, then $e_i \prec e_k$
- if $e_i \sim e_j$ and $e_j \sim e_k$, then $e_i \sim e_k$
- if $e_i \succ e_j$ and $e_j \sim e_k$, then $e_i \succ e_k$.

A violation of the transitivity axiom can have undesirable consequences. Howard Raiffa (1973, pp. 99 f.) demonstrates this using the example of a house purchase based on the intransitive preference order of the buyer A$\succ$ B, B $\succ$ C and C $\succ$ A. Once the buyer has purchased house A, the real estate agent offers him a small sum of money to swap house A for house C. If the buyer follows his preference order, he will be willing to do so. This also applies if he then receives an offer from the real estate agent to swap house C for house B in return for a small sum of money, and so on. In this way, the buyer would eventually lose all his money. Empirical studies have shown, however, that the axiom of transitivity is not always fulfilled in practice (Schauenberg 1978, pp. 166 ff.). For example, the preference order $e_i \sim e_j$ and $e_j \sim e_k$ and $e_i \succ e_k$ can be observed (Laux et al. 2018, p. 45). Hans Schneeweiß (1966, p. 131) explains this by the fact that in practice, slight differences in perception often lie below the threshold of perception. Only on closer inspection can they be recognized as such. In a professional context, there is no other way to fulfill the transitivity axiom than to take such a closer look.

4.3.3.2 Preference Relations

The aim of combining Systems Engineering, with its strengths in solution development, with the strengths of decision theory in the choice between different solution options (alternatives, more generally: variants) requires some conceptual coordination. In this context, preference in the sense of decision theory is understood as the attitude of a participant towards different *options* (possible actions, strategies, actions, solution variants) (Göbel 2018, p. 63). *Target values* (attributes, criteria, cues) are those values by which the quality of an option is to be measured. In this context, *quality* is understood in the sense of the international standard DIN EN ISO 9000:2015-11 (p. 39) as the degree to which a set of characteristics fulfills requirements. However, two restrictions are lifted here that deviate from the standard: firstly, the restriction to objects and thus the extension to services and rights, and secondly, the restriction to inherent characteristics, i.e. those that are inherent to the object, service or right (e.g. weight or length of an object). In addition, associated characteristics such as the price of a product are also included. Objectives are reflected in their results e_{ij} (consequences), which result from the combination of options ai and different *environmental states* U_j (Göbel 2018, p. 49).

In addition to the target values, five preference relations must be taken into account:

- The *height preference ratio* indicates the desired extent of a target size. It can take three forms:
  - extreme values, i.e.
    - the maximum or
    - the minimum,
  - fixed values, which signal the achievement of a precisely defined result, and
  - satisfiable values, which indicate whether results are above a certain level.
- The *time preference relation* indicates when the participant wants to achieve results. For example, quick wins that are achievable in the short term but may be less sustainable can be preferred to long-term results (Fig. 2.33). Methodically, this is done, for example, by *discounting* a future payment to the present. This is logical insofar as a sum of money received in the present can be invested at interest and will thus have gained in value in the future.
- While decision theory only takes the time dimension into account, the spatial dimension arises from the systems theory perspective as a complement to it, i.e. the *spatial preference relation* (Haberfellner et al. 2018, p. 228). Colloquially, this complement is often referred to as the here and now. For example, a participant may prefer a solution that has an impact in their immediate environment to another solution that is less effective regionally but also brings advantages beyond the region.
- The *risk preference relation* has already been explained in Section 2.6.2.3 in the context of decisions. In principle, it is divided into the categories of risk appetite, indifference and risk aversion. However, it should be noted that a participant would always choose a lower risk under *ceteris paribus conditions*. Therefore, the risk preference relation is usually related to the preference for height, since, for example, a higher return is often accompanied by a lower probability of occurrence. In such cases, magnitude and probability must be weighed against each other. For example, participation in a lottery may depend on the size of the jackpot.
- The *species preference relation* comes into effect when goals are in conflict with each other. These can be resolved by *priorities* (Haberfellner et al. 2018, p. 236 f.) or by *compromises*. Setting priorities means that one goal is pursued at the expense of the other. In this case, one side loses everything and the other wins everything. A compromise, on the other hand, means that both goals are adapted in such a way that the conflict is resolved. In this case, concessions are required from both sides. You meet in the middle, so to speak.

4.3.3.3 Transfer of Requirements into Targets

Each objective formulation takes up the *requirements* of the various stakeholders identified in the context of the respective previous situation analysis. In this respect, requirements and objectives are related, but must be strictly distinguished from one another. For example, the requirement mentioned by Beate Bender and Kilian Gericke (2021, p. 169 ff.) with reference to Klaus Ehrlenspiel that a dishwasher should be quieter than all other dishwashers on the market is a requirement, not an objective. On the one hand, it differs from an objective in that the formulation is still too vague to be a could describe a target state with sufficient precision. Furthermore, it has not yet been clarified whether this requirement can and should actually be incorporated into the design project's target system without restriction. When it comes to finding a compromise between the possibly conflicting requirements of different stakeholders, it cannot be ruled out that this one requirement cannot be fully realized. This is also evident from standards for management systems, for example when DIN ISO 14001:2015-11 for Environmental Management in Section 4.2, point c) requires that in the course of Stakeholder Value Management, it must be determined which of the requirements (needs, expectations) for the project should become binding obligations (here: goals).

In the course of formulating the objectives, the first step is to *specify* requirements to such an extent that their fulfillment can be measured and verified if possible (Haberfellner et al. 2018, p. 228 f.; Göbel 2018, p. 66). A suitable specification that meets the above requirement could, for example, be that the noise of the machine in a normal room at a distance of 2 m must not exceed 70 dB.

Secondly, it must be ensured that the objectives reflect the requirements of the stakeholders as *completely*as possible (Haberfellner et al. 2018, p. 234; Göbel 2018, p. 66).

Thirdly, however, this conflicts with the requirement for a *simple* and clear (transparent) target system (Haberfellner et al. 2018, pp. 239 f.; Göbel 2018, pp. 66 f.). In this respect, a compromise must be found between these two requirements for target systems.

Fourthly, the requirements must be *examined* to determine whether they are achievable and should actually be included in the target system, because in a free society everyone is allowed to express wishes, but not all wishes will always be fulfilled.

Fifthly, requirements are associated with *opportunities and risks*. If and to the extent that these have not already been incorporated into the SWOT analysis in the course of the situation analysis P2, they must now be added or further specified.

Sixth, the requirements must be *related to each other* in order to make it transparent whether and, if so, to what extent they contradict or even exclude each other. Such a relation also shows which goals may mutually support each other. The requirement mentioned above, for example, could come into conflict with another requirement that it should be the dishwasher with the lowest price of all those offered on the market. In this case, elaborate soundproofing measures would probably be too expensive. A particular challenge when dealing with interrelated objectives is that redundancies should be avoided (Göbel 2018, p. 66). Redundant and interrelated objectives can lead to their overemphasis. An example would be if both the minimization of capital commitment costs and the reduction of inventories were required. Avoiding this can be helped by using a requirements-means approach (Fig. 4.23). Furthermore, it should be possible to evaluate objectives independently of other objectives, because only then can the individual evaluations be added up (Göbel 2018, p. 67).

Seventh, test cases must be defined that enable the measurement and testing of the objectives (Ebert 2014, p. 192 ff.). In this context, agile approaches speak of *Test-driven Development* (TDD), which is particularly emphasized in the so-called *V-model* (Fig. 4.28) (ibid., p. 193; Haberfellner 2018, pp. 85 ff.; Informationstechnikzentrum Bund 201 9, p. 38; Kaffenberger et al. 2012, p. 27; Weilkiens 2014, p. 34 f.; Walden et al. 2023, p. 36; Winzer 2016, p. 42 f.). While the determination and realization of requirements takes place on the left, descending branch of the V-shaped process model, the corresponding tests are carried out on the right, ascending branch at the respective levels of detail of the integration (VDI 2206:2021-11). The latter must, however, be developed on the left, descending branch, so that the yardstick for judging the success of the development is already known in the course of the development. Incidentally, the V-model corresponds (essentially) to the problem-solving cycle preferred here (Fig. 3.23): the determination of the requirements (specifications) corresponds to the situation analysis, the system specification (functional specifications) to the formulation of objectives, the specification of the elements and their realization to the development of solutions, and the tests to the evaluation. What is missing from the V-model as well as from the original problem-solving cycle (Haberfellner et al. 2018, p. 70) is the added phase of method definition P 4. Furthermore, the boundary between analysis as part of solution development (synthesis) P 5 and evaluation P 6 is to be defined for each specific application. In the lower part of the V-model, the analysis tends to deal more with verification and the evaluation P 6 more with validation in the upper part of the V-model, i.e. at the end of the problem-solving cycle.

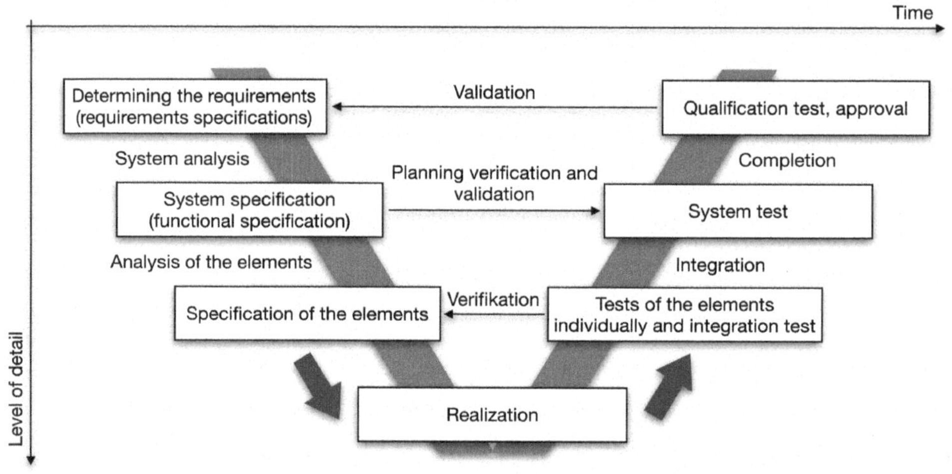

Fig. 4.28 V-Modell XT

Eighth, one goal of the goal definition should be to formulate the goals as *solution-neutral* as possible (Haberfellner et al. 2018, p. 234). This is not that easy, because the problem-solution cycle is often run through in the course of a problem solution from the rough to the detailed, so that possible solutions have always been identified in previous runs. Beate Bender and Kilian Gericke (2021, p. 39), in agreement with other authors, refer to this as the *coevolution of problem and solution*. However, if and to the extent that a problem solution has not yet been completed, this is particularly due to the fact that no satisfactory solution has yet been developed. Therefore, further work on it should be open to solutions that have not yet been included; the related *innovation potential* should be preserved as far as possible. This would be prevented by a pre-commitment to solutions that had already been considered. Against this background, the commitment to solutions should be limited to those solutions that have already been explicitly selected in previous development phases at a lower level of detail. Since it is assumed here that this has already been done on the basis of Systems Engineering, i.e. in a systematic and methodical way, such solutions can be classified as further-reaching. All other solution components that will need to be developed in the course of further detailing should not be included in the objectives of the relevant problem-solving phases.

A simplified formulation of the requirements for target systems is expressed using the acronym *SMART* (Doran 1981). According to this, goals should be

- specific, i.e. clearly defined,
- measurable, i.e. measurable and verifiable,
- accepted,
- realistic, and
- tme-bound, i.e. achievable within a specified time.

4.3.4 P 3.4: Method Planning

The following Section explains methods for the steps in goal formulation mentioned in the previous Section. In addition, numerous other methods can be found in the literature cited in each case.

4.3.4.1 Clarification of Requirements

In general, it is desirable to formulate requirements as precisely as possible as early as the requirements identification phase in the context of the situation analysis. For this reason, a scheme has already been provided in connection with Fig. 4.22 that can be used for this purpose. However, there are a number of other, alternative suggestions for supporting the most precise possible formulation of requirements. A selection from these should be added here.

For example, Model-based Systems Engineering (MBSE) with SysML offers the *requirements diagram* (req) (Alt 2012, pp. 51 ff.; Weilkiens 2014, pp. 314 ff.). In this diagram, a requirement element consists of a unique identifier and an explanatory text. Computer programs that support modeling with SysML usually offer the option of referencing external sources such as laws, standards, or contract documents, for example, by using links. Furthermore, functional requirements can be described with other model elements, such as use cases or interfaces. Fig. 4.29 shows an example of the use case of a useless box used above. For the sake of simplicity, it contains only one requirement that is addressed to the process. The process is referenced as a whole. The details can be found in the corresponding diagrams (Figs. 2.45 and 2.46).

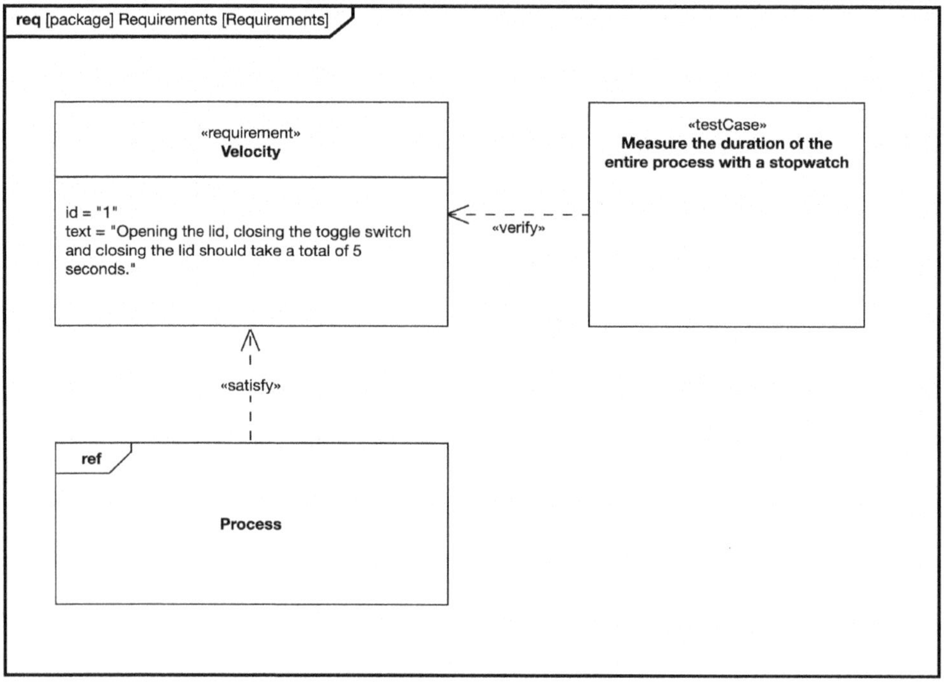

Fig. 4.29 Requirements diagram in SysML notation, created with Enterprise Architecture software

Despite the clear structuring that SysML supports, there is usually no way around formulating requirements in text form. At least it has already been demonstrated on the basis of Fig. 4.22 that texts can also reveal at least some structure, and how. In addition to this, another *formulation template* is offered here on the basis of Fig. 4.30 (Alt 2012, p. 13; Ebert 2014, p. 107; Rupp and SOPHISTen 2021, p. 358 ff.). It begins with a temporal or logical condition that can be omitted if the corresponding requirement always applies without exception. The second item is the indication of the binding character (priority). The terms given in the template can be replaced by others, for example "will" by "can". In any case, it is important to clearly define how the terms used are to be understood.

For example, it can be specified that a must-have requirement leaves no room for discretion, while a should-have requirement allows for intended discretion, where the usual procedure is to proceed in the manner corresponding to the requirement, while exceptions are only permitted in justified individual cases. The *MuSCCoW prioritization* adds the Won't-have requirements to the Must-have, Should-have and Could-have requirements, which should provide the option of including requirements at the given time that are fundamentally considered important but can or should only be fulfilled at a later point in time. This can be advantageous if these requirements are not time-critical and are actually taken up again later. Furthermore, such requirements can already be taken into account in the current solution development. However, classifying requirements as Won't have carries the risk that the realization of so-called *Quick Wins* will remain, which can lead to a lack of sustainability.

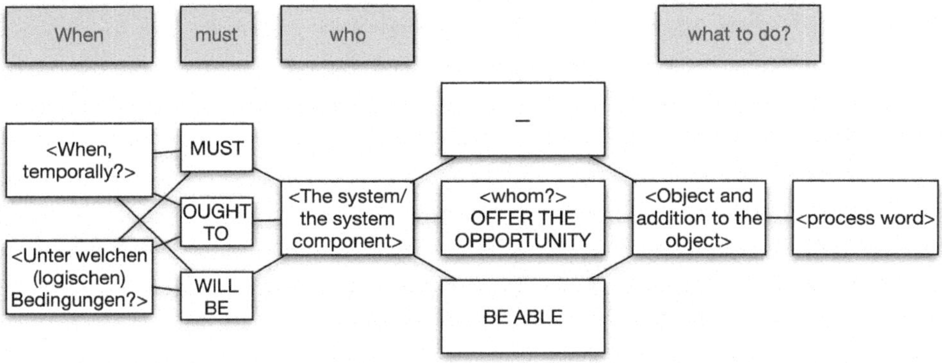

Fig. 4.30 Template for formulating requirements (Alt 2012, p. 13; Ebert 2014, p. 107)

The third part is the specification of the system to which the requirement is addressed. The remaining parts then describe what the system must, should or will (can) do. An example of a requirement formulated in this way, with reference to the example of a dishwasher already cited above, could be: In a normal room at a distance of 2 m, the dishwasher must be able to not exceed a noise level of 70 dBA.

4.3.4.2 Review of Requirements

As a rule, the process of defining the requirements more precisely is accompanied by an initial review, because this helps to filter out those requirements that cannot be formulated with sufficient precision. This can often be interpreted as an indication that the corresponding demand is not constructive. The actual examination now focuses on two key questions: firstly, whether and, if so, on what basis the demands are justified, and secondly, whether the justified demands can be met.

Regarding (1), it is generally true that in liberal societies everyone has the freedom to make demands within the prevailing values. But not every demand is justified (founded). Fig. 4.31 shows in the notation of event-oriented process chains (Sect. 2.4.5.2) the steps in which it can be checked whether and, if so, on what basis a demand actually has to be considered. The basis for this are the categories of legal, legitimate and strategic demands introduced and explained in Sect. 4.2.4.6. However, it is not always immediately apparent to which category a specific demand can be assigned. For example, a strategic demand such as a demand for a higher salary may be morally justified, which is possible in principle with reference to the ethical dimension of justice, among other things, but not always appropriate, since prices and thus also wages are based, among other things, on market rules.

Fig. 4.31 Requirements review process

If a requirement is identified, the first step (Fig. 4.31) is to check whether it is legal, i.e. whether it is required by law. Illegal demands are rejected. Demands that can be enforced in court in the event of a dispute must be taken into account. Demands that are not illegal but have no legal basis on which they could be enforced in court could be rejected if no further standards need to and/or are desired.

If and to the extent that this is not the case, the second step is to check whether the corresponding claim is legitimate. Immoral demands are rejected, as are illegal demands in the previous step. Legitimate demands are to be considered if and to the extent that the examining instance embraces the underlying values and norms (morality). In practice, it can occasionally be observed at this point that demands that appear immoral (illegitimate) are rejected, and rightly so. ethical justification, while demands that are ethically well founded but stand in the way of the interests of the inspecting authority are also rejected. Such inconsistency can lead to conflicts.

Demands that explicitly or implicitly refer to a market situation, i.e. to the scarcity of a service, and require negotiation on the basis of supply and demand, can be immediately recognized as *strategic*. In companies, such demands are channeled into the area of business administration and there in particular to purchasing or to sales or to Human Resource Management. If such demands can be implemented on the basis of a cost-benefit calculation, they are fulfilled. If this is not successful, demands may under certain circumstances be classified as legal or legitimate and forwarded to the Stakeholder Value Management.

Claims that are based on their legality or legitimacy from the outset are taken up immediately by the SVM and, via the stages of the legality and legitimacy check already explained above, to the branch (Fig. 4.31) where it is recognized that, despite initially different assessments, no ethically justifiable basis is recognizable. From this branch onwards, the corresponding requirement is treated as a strategic requirement and analyzed for its economic efficiency. If it is associated with an economic benefit, it is fulfilled; otherwise, it is rejected.

Methods of profitability calculation can be found in the context of business administration and will therefore not be discussed in more detail here. For more information, please refer to the relevant literature (e.g. Becker and Peppmeier 2022; Müller 2020; Plinke and Utzig 2020). The following Section explains the legitimacy test, which is methodically carried out in the form of a *discourse* (Sect. 4.1.4.4 4.1.4.4), i.e., through an ideally impartial clarification of clarification of whether the fulfillment of a specific demand is morally required in a particular situation (Ahrens 2001, p. 43; Göbel 2020, p. 145 ff.). This is done by a fictitious role reversal of the discourse partners, with each of them putting themselves in the other's shoes. If they come to the same conclusion from this perspective as from their original perspective, the claim under discussion can be considered resolved. However, this requires that all participants in the discourse have the ability and the recognizably good will to make the maintenance of their respective claims dependent on whether they pass the legitimacy test. A lack of such willingness often leads to the failure of discourses in practice.

How such discourses are conducted in concrete terms has not yet been conclusively clarified. According to Jürgen Habermas (Seiler 2015, p. 31 ff.), an individual can also conduct a discourse if he applies the rules of discourse and, in this case, in particular, the exchange of roles. Others, such as Seyla Benhabib (1995), object that this is not possible because, for example, a man cannot, in principle, put himself in a woman's shoes and vice versa, so that for this reason alone, people of different genders must always be involved in a discourse if the issue at hand affects more than one gender. In this context, it is recommended that the discourse be conducted with all participants and stakeholders wherever possible. If the group becomes too large, representatives can be elected. However, their acceptance is enhanced if they are elected by the stakeholders themselves according to a generally accepted procedure, for example democratically.

In states where most moral imperatives and prohibitions are also enshrined in law and violations are sanctioned, legitimacy tests are rarely necessary, and if they do take place, then usually at a relatively high level. The situation is different in parts of the world where even human rights are not respected. If, for example, there is no law prohibiting child labor in a country, companies whose home countries have such laws are obliged to apply this morally based prohibition elsewhere as well. However, given the international competition and the widespread pursuit of profit maximization, it is clear how difficult it seems to be to apply even such simple and self-evident principles, as the German legislator saw fit to create a legally binding basis for this from the home location with the *Supply Chain Due Diligence Act* (LkSG). In addition to this approach of *external commitment*, the international standard DIN EN ISO 26000:2021-04 on social responsibility attempts to provide a basis for companies to *voluntarily commit themselves* to moral values and standards.

In the following, typical legal and legitimate demands of typical stakeholders are listed as examples to provide starting points for further analysis:

- In a market economy, *owners* and other *capital providers*, such as banks, have the right to an *appropriate return on their capital employed*. What is considered appropriate depends on the values of the respective society. In addition, in many societies, the return on capital is subject to the proviso of *social responsibility* (Art. 14 GG) in the sense of *Corporate Social Responsibility* (CSR): ownership entails obligations. Furthermore, owners are entitled to make profits that exceed an appropriate return if and to the extent that they succeed in enforcing corresponding prices in accordance with market rules in the course of negotiations. This can be achieved, for example, on the basis of unique selling points. However, there are also moral and, consequently, legal limits. Usury, for example, exploiting the weaknesses of a business partner to enforce significantly excessive prices, is, for example, in Germany, due to the *immorality* (§ 138 para. 1 BGB) of such action, civilly (§ 138 para. 2 BGB) and criminally (§ 291 StGB) pursued. The same applies to the conclusion of so-called *adhesion contracts* or the formation of *cartels*, as has been legally established and prohibited, for example, to limit the useful life of incandescent lamps (planned obsolescence) (Phoebus cartel). In other societies, however, different principles and rules derived from them may apply. For example, haggling in bazaars is common in oriental cultures. The initial prices there, also known as moon prices, would not infrequently constitute the offence of usury in Germany.
- Employees have a dual role (Göbel 2020, p. 134 f.): on the one hand, they themselves are a stakeholder group vis-à-vis their employer and, in particular, have the right to physical integrity, i.e. to the avoidance of injuries and illnesses, as is to be ensured by occupational safety and health (DIN ISO 45001:2018-06). Furthermore, their dignity must be preserved, which, for example, prohibits discrimination and bullying. At the same time, however, employees also work for their employer and are therefore among those who are called upon to fulfill the legitimate claims of other stakeholders, such as those of customers. Managers are employees themselves, but at the same time they are also superiors of other employees, so in this role they are responsible for ensuring that the legitimate claims of these employees are met. Furthermore, this dual role comes into play

in relation to society and the natural environment. Finally, employees also have strategic interests vis-à-vis their employers, particularly with regard to their remuneration.

- Customers have the right to products of impeccable quality (DIN EN ISO 9001: 2015-11). In addition, price and performance must be in reasonable proportion to one another (§ 138 para. 2 BGB; § 291 StGB). However, the definition of the term quality according to DIN EN ISO 9000:2015-11 explicitly distinguishes between inherent and assigned characteristics and explains that only inherent characteristics are relevant to quality.
- Suppliers have a right to fair and respectful treatment. A customer's possible market power must not be abused (Sect. 138 (1) of the German Civil Code (BGB)). This includes, among other things, the protection of property rights (patents, utility models) and trade secrets.
- The company expects corporate citizenship (CC) from companies. Companies should behave in society in a manner comparable to that expected of good citizens (DIN EN ISO 26000:2021-04). This is particularly aimed at civil society engagement.
- The natural environment is to be protected (DIN EN ISO 14001:2015-11). In particular, natural resources are to be used sparingly, waste is to be recycled (Sect. 1 of the Closed Substance Cycle and Waste Management Act (KrWG)) and emissions (Sect. 3 of the Federal Immission Control Act (BImSchG)) are to be minimized.
- Those involved in or affected by the implementation of the design project expect to be challenged and supported. At first glance, this statement may seem confusing in that challenge does not have positive connotations in every respect. What is meant here is that the participants should be deployed in line with their respective qualifications and motivation and, in this respect, should not be overburdened or underchallenged. Only when the participants approach their work with self-confidence and enjoyment can design projects achieve the success that is usually expected. This applies all the more since many design projects go through phases that demand everything from the participants, not infrequently also to surpass themselves. This is sensible and possible as long as it is not exaggerated and is always balanced out by quieter phases. In addition, there is the necessity of support, which includes not only adequate material support but also respect, fairness and recognition of achievements.

The second of the questions raised at the beginning (2) aims to examine the *content* of the requirements. This refers to the following sub-tests (Haberfellner et al. 2018, p. 234 f.; Bender and Gericke 2021, p. 185):

- The requirements must be checked for *completeness*. However, there can only be an absolute benchmark for this when the entire design project has been completed, since requirements can change continuously during the course of the project. The basis for the completeness check is therefore always the level of knowledge available in the respective project phase. The specific question is therefore whether the project team believes that it can achieve the next development goal on the basis of the known requirements or whether essential information is still missing. However, the criterion of completeness of the requirements catalog is in conflict with the equally important criterion of *clarity* (Göbel 2018, p. 47; Haberfellner et al. 2018, p. 239 f.). Particularly in the case of very extensive design projects, the use of software to manage requirements (databases), the structuring and, in this respect, in particular the categorization of requirements, as

well as the consolidation by combining similar requirements and by dispensing with subordinate, dispensable or even irrelevant requirements, can help in this regard.
- Furthermore, the requirements known at the respective stage of the project progress must be checked for *consistency* (Laux 2018, p. 47 f.). However, such consistency is hardly to be expected. Therefore, the specific aim is to make contradictions transparent. These contradictions are then resolved only in the course of formulating the objective P 3, but this requires an iterative approach that includes the situation analysis P 2 and here in particular the requirements analysis, but also the solution development P 5 (Bender and Gericke 2021, p. 174; Haberfellner et al. 2018, p. 237 ff.).

4.3.4.3 Relating the Requirements

The consistency check is so important that it is given its own Section, because contradictions harbor potential for conflict between the stakeholders and thus represent a significant risk to the success of a solution. In particular, Reinhard Haberfellner et al. (2018, pp. 237[18]) distinguish the following types of relationships between requirements:

- Mutual support: Fulfilling one requirement simultaneously supports the fulfillment of the other requirement.
- Independence: One requirement can be fulfilled independently of the other requirement and vice versa.
- Contradiction: Reinhard Haberfellner et al. (ibid.) distinguish between antagonism (competition) and contradiction per se (conflict). While they understand antagonism to mean a gradual contradiction in which one requirement can be fulfilled to a lesser extent the more another requirement is fulfilled, they use a different term for the case in which one requirement cannot be fulfilled at all when another requirement is fulfilled. Although such a differentiation seems possible in principle, practice has shown that it often leads to confusion. Therefore, it will not be used here. A contradiction therefore always exists when the fulfillment of one requirement is at the expense of the fulfillment of another requirement, regardless of the extent. The latter is to be used as a basis for finding compromises or setting priorities.

To determine these relationships, a pairwise comparison of the requirements in the requirements-relationships matrix shown in Fig. 4.32 is useful, using the example of a cargo bike supported by an electric drive. In it, the conflicting requirements are marked with a "w" at the respective intersections, and the mutually supportive requirements are marked with a "u". The independent requirements are not marked with any symbol, which aids clarity in this case. The matrix reveals a relatively large number of conflicting goals. For example, there is often a conflict between the desire to pay a low price and the simultaneous formulation of requirements that incur costs, which is the case, for example, with a large range that requires a battery with a correspondingly large capacity. A steering damper also stands in the way of a large range due to its own weight, but supports the requirement for a high level of road safety. It is not entirely easy to clarify whether high speed also contributes

[18] A similar scheme is used in decision theory (e.g. Göbel 2018, p. 69).

to road safety: on the one hand, it makes it possible to keep up with other cyclists, but on the other hand, greater damage often occurs in an accident at higher speeds. This example shows the limitations of the application of the requirements relations matrix. Despite these limitations, the method is generally well suited to providing a systematic basis for resolving conflicting goals.

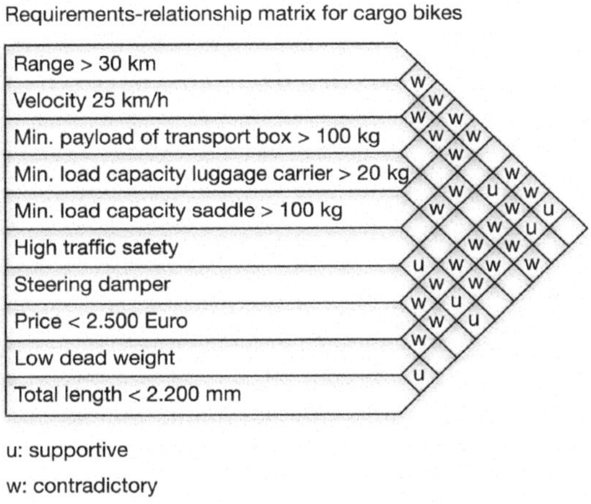

Fig. 4.32 Requirements relation matrix (Böhm 2005, p. 176; Haberfellner et al. 2018, p. 237)

Furthermore, the transparency of conflicting goals can also promote innovation (Haberfellner et al. 2018, p. 237 f.). One example of this is assembly line production, which originally seemed suitable only for mass production where each product is the same as the next. Today, the automotive industry, for example, is able to assemble vehicles on production lines where no two vehicles are the same.

4.3.4.4 Transferring Requirements into Targets

While the requirements express the expectations of the stakeholders and thus result from the situation analysis, they must now be transformed into goals. To do this, the requirements that are to be included in the catalog of goals must first be selected. This means that there may always be requirements that are explicitly not met. However, the selection should not be arbitrary. Therefore, criteria must be defined on the basis of which the requirements are included in the catalog of goals. These criteria should, in turn, be defined together with all stakeholders in order to avoid later conflicts. The selected requirements may also need to be modified in order to produce an overall consistent (non-contradictory) set of objectives.

The previous remarks have already shown that objectives are understood here as those requirements whose fulfillment is intended. This should be explicitly reflected upon again at this point, because the term "objective" can be understood somewhat differently in different areas. In particular, the degree of detail and concretization varies. For example, the corporate objectives discussed in business administration may include achieving a certain annual turnover. This is quantifiable and specific in that it is expressed as a monetary amount, but it lacks detail. The contribution that the pitch of a screw thread makes to achieving the sales target is also quantifiable and specific, but it is much more detailed. The goal of leading a fulfilled life, which is addressed in philosophy, is neither quantifiable nor detailed, but it is specific. Despite such differences in the way the term is used, it is common in all cases to speak of goals when the associated action is directed towards consciously achieving a desired state or event. Therefore, it seems sensible and possible to also describe the intention to fulfill a requirement as a goal here. However, this does not change the characteristics of the requirement, so that it remains possible to continue to speak of requirements as synonyms for goals. Although some of the clarity and distinctiveness intended here is lost, the benefit of a language that corresponds to common usage is gained. However, it should be recalled that there is a difference between the requirements formulated by the stakeholders and the requirements that a design project aims to fulfill.

A distinction is often made between goals and *constraints* (synonym: framework conditions, secondary conditions) (e.g. Bender and Gericke 2021, p. 172). Whether and, if necessary, on the basis of which criteria this should be done in the course of formulating the goal depends on the application and on what is to be expressed by such a distinction. In mathematics, for example, boundary conditions may be necessary if a calculation procedure delivers more solutions than are practically (e.g. physically) possible. In operations research, for example, in linear optimization (synonym: linear programming), a non-negativity condition is often set up to restrict the search space for a solution to positive values , because when calculating optimal production quantities, it is not possible, for example, to produce a negative number of products (Domschke et al. 2015, p. 17 ff.). In other cases, objectives are referred to as constraints if these conditions cannot be avoided at all or only with an unjustifiably high level of effort. In the chosen example of the cargo bike, roadworthiness, which, among other things, requires lighting, can be understood as a constraint because it is hardly possible for a bicycle manufacturer or a cyclist to change the general traffic rules. In the light of Stakeholder Value Management, it can be said that the owner's goal is to make money (shareholder value), while all other stakeholders set the (constrained) conditions to which this is subject: protecting the natural , protecting employees from work-related illnesses and injuries, paying employees, service providers and suppliers appropriately, complying with laws and regulations as well as moral standards and values, delivering high-quality products and services to customers, competing fairly with other companies, and so on.

4.3.4.5 Development of a Common Set of Targets

The considerations regarding discourse in Section 4.3.4.1 can be assigned to a conflict resolution strategy that James Gardner March and Herbert Alexander Simon (1993, p. 149 ff.) describe as *problem-solving orientation* (Göbel 2018, p. 160 f.). The groups involved share common requirements and work together to find the best solution. The basis of this strategy is therefore a fundamental consensus. However, if this goes too far, there is a risk that group cohesion will have a stronger effect than the pursuit of an optimal solution (Janis 1972).

Beyond that, however, the authors mentioned above identify three further situations. One of these is when a group agrees on fundamental requirements, but not on how to achieve them, i.e. on instrumental requirements. In conflict situations, it can be helpful to consciously make such a distinction between fundamental and instrumental requirements, as it often makes it easier to reach an agreement (Göbel 2018, p. 162 f.). In most cases, the differences can be overcome through argumentation. The basis for resolving such conflict situations is therefore the *power of argumentative persuasion*.

A third possibility is that conflicts of interest cannot be resolved by persuasion. This quickly leads to threats, false allegations or horse-trading. In such a situation, the only way forward is through *negotiations* aimed at *compromise*. These are characterized by all sides making concessions on their own ideas, without being convinced that these will contribute to an optimal overall solution. The overall solution is ultimately not considered to be the best solution in objective terms, but rather the best solution that can be enforced. Negotiations follow defined rules that are intended to avoid possible, but undesirable, negotiation outcomes from a legal and moral perspective, such as overreaching (e.g. § 138 BGB on immoral legal transactions). In this context, so-called *win-win solutions* are desired. This does not mean that there are no situations in which one side loses what the other wins (see below). However, this is subject to different rules. The basis of negotiations is *influence*: the negotiating partners can influence the outcome of the negotiations, but cannot determine it unilaterally (dominantly).

Political solutions are ultimately also based on transparent conflicts of interest, but they do not aim for compromise, but for one party to prevail against another; one side wins what the other loses. In order to protect the losers from extreme disadvantages, however, such constellations provide for the protection of minorities. In particular, no inalienable rights (ius cogens) may be at stake, only rights that can be waived (ius dispositivum). As a rule, a higher instance watches over this, for example the constitutional court at the state level. The basis of political decisions is *power*. Therefore, it can be assumed that those who have power do not need a factually sound assessment, since they can override it because of their power. However, power is also based on success, at least in the long term, and for that, problem solutions that work are a necessary, if not sufficient, condition.

While the first two conflict resolution strategies can be used in the context of a design project with the help of systems engineering methods, the last two strategies in this context reach the limits of this subject-oriented type of problem solving and in the last-mentioned case even go well beyond it. In both situations, *decisions* (Sect. 2.6.2.3) must ultimately be made that are based on a different, in particular political, form of rationality. Therefore, negotiations and political decisions must be delegated to the responsible instances. Methodically, they are supported, among other things, by game theory (Göbel 2018, p. 125 ff.; Grimme and Bossek 2018, p. 229 ff.; Holler et al. 2019).

Despite the fundamental willingness to cooperate that underlies the first two conflict resolution strategies, conflicts between different stakeholders remain, as does the interest in incorporating one's own requirements as far as possible into a solution. Decision theory knows two basic approaches to how stakeholders can arrive at a common system of objectives (Göbel 2018, p. 161 ff.).

(1) All groups or their representatives deliberate *simultaneously*, i.e. together, and discuss differences until a consensus or a compromise is found or a priority is set. To a certain

extent, there can also be negotiations in a fundamentally cooperative constellation that were actually attributed to the conflict-oriented strategies above. However, it is crucial that the will to achieve consensus is clearly recognizable on all sides.

(2) Alternatively, the groups or their representatives can *sequentially* develop a position alone or among themselves in a first step, in order to then incorporate this into the collective development of the target system in a second step.

Both approaches can also be combined, for example by first agreeing on a procedure together and then working out one's own proposals for weighting the requirements, which are then introduced into a discussion or negotiation in plenary.

Once all the requirements are available, the next step is to *weight* them according to their importance, because usually not everything is equally important. For this purpose, a *value function* is set up (ibid., p. 163 ff.). This assigns values to the requirements, which can then be added up in order to be able to determine an overall evaluation in the course of the final evaluation (Sect. 4.6). To do this, different values, such as the range measured in kilometers and the speed measured in kilometers per hour in the example chosen above, must be converted into a common variable. This variable is often referred to as *utility*. In a group, each person can then indicate how great they consider the benefit to be for each requirement. The average of all the values can then be determined, which can be understood as a compromise.

A simple option is for each participant to enter a specific utility value. For example, participant B_1 can rate the utility of the range of variant V_1 on a scale from 0 to 10 as 5 and participant B_2 can rate it as 4. The arithmetic mean of this is a utility value of 4.5. However, it is not always easy to specify an exact value. In such cases, you can use intervals, as is the case, for example, when determining school grades based on the number of points achieved in an exam. Here, too, a grade usually results from a range of points. For example, B_1 can specify the range of the cargo bike of variant V_1 in a range between 4.5 and 5.5, B_2 between 5.5 and 6.5. In this case, both could agree on a utility value of 5.5. Formally, the minimum of the individual upper value limits (here: 5.5) must be greater than or equal to the maximum of the individual lower value limits (here: 0.55) (ibid., p. 166). However, such mean values are only meaningful if the individual values are sufficiently close to each other. In such cases, the value ranges are retained and scenarios are derived from them, one based on the upper value range and the other on the lower value range.

Voting in groups depends significantly on the *voting rules* (Göbel 2018, p. 168 ff.). In a fundamentally cooperative group, it is usually assumed that each group member has a complete and transient preference order (Sect. 4.3.3.1) with respect to all variants and that the true preferences are revealed. If a group member secretly had different preferences (hidden agenda) and voted strategically, this would be a case for game theory. Under the stated conditions, the following voting rules are possible (ibid.):

1. Approval voting
2. Simple majority
3. Absolute majority
4. Rule of the majority of paired comparisons
5. Rule of successive paired comparisons
6. Borda rule

Elisabeth Göbel (ibid.) uses the example outlined in Fig. 4.33 to explain these rules. A committee of nine people is to choose between five variants A to E. Three factions have formed, each of which agrees on the order in which they rank the variants. However, there are different preference orders between the factions. Together, the three *preference orders* form the *preference order profile*.

Rank (R)	Group 1 4 members	Group 2 3 members	Group 3 2 members
1	A	B	C
2	E	C	D
3	D	E	E
4	C	D	B
5	B	A	A

Fig. 4.33 Voting rules (Göbel 2018, p. 168)

In an *approval voting* (1), each member of the voting group receives more than one vote, but no more than the number of variants. Each vote can be used to select one variant. The variant that receives the most votes is accepted. In the example shown (Fig. 4.32), let us assume that the members of faction 1 give one vote each to the three variants A, E and D that they consider to be the best, so that each of these variants receives 4 votes. The members of the second and third factions each give only the first two of their best variants a vote, so that B and C now each receive two votes from the second faction and C and D receive two votes from the third faction. This means that variant D is selected; the collective preference order is D > C > A, E > B.

The election by *simple majority* (2) is a borderline case of approval voting (1) in the event that each group member can only cast one vote. In the example, option A is chosen because faction 1 with four members decides in favor of this option, while option B receives three votes from faction 2 and option C receives two votes from faction 3. The collective preference order is A > B > C > D = E.

If an *absolute majority* (3) is required, one of the variants must receive more than 50% of the votes, in the example chosen (Fig. 4.33) at least five votes. This often requires several rounds of voting. In the example chosen, a *run-off vote* is held for the two best variants, while all other variants are eliminated in the first round. This corresponds to the *Hare rule* named after Thomas Hare. In the run-off vote, variant A again receives four votes from faction 1. Since the other two factions consider this variant to be the worst, they will vote for variant B, which thus receives the absolute majority with five votes. In this case, however, there is no complete preference order for all variants.

To determine the *majority of the pair comparisons* (4), all variants are compared in pairs (Fig. 4.34). The most frequently preferred variant is selected, in the example shown, this is variant C. The collective preference order is now C > E > D > B > A.

Successive pairwise comparisons (5) start with any juxtaposition of two variants and then continue the comparison with the superior variant and another randomly selected

Pair comparisons	Number of votes	Preferred variant
A : B	4 : 5	B
A : C	4 : 5	C
A : D	4 : 5	D
A : E	4 : 5	E
B : C	3 : 6	C
B : D	3 : 6	D
B : E	3 : 6	E
C : D	5 : 4	C
C : E	5 : 4	C
D : E	2 : 7	E

Fig. 4.34 comparison of couples (Göbel 2018, p. 169)

variant, until all comparisons have been made. The last superior variant is chosen. In her example (Fig. 4.34), Elisabeth Göbel (ibid., p. 170) starts with the comparison A: B. Since five members of the group rank B ahead of A, A is eliminated. B is then compared with C, with C emerging as the winner. C also emerges as the winner each time it is compared with D and E. C is the collective preference order. There can be no complete preference order, as only one variant can win.

In the *Borda rule* (6), named after the French engineer Jean Charles Borda, each group member expresses their preferences by awarding points. With k variants, the best-ranked variant receives k – 1 points, the second-best k – 2 points, and finally the worst receives 0 points. For each variant, the points are then added up and the variant with the highest number of points wins.

This rule is referred to as positional because the positions of the variants in the ranking are included in the group preference (Fig. 4.35): A receives 16 points, namely 4 points from each member of faction 1. B receives 14 points, namely 3 points from each member of faction 1 and one point each from the two members of faction 3. C receives 21 points: 4 each from faction 3 and 3 each from faction 2, as well as one each from faction 1, and so on. In this way, variant E is finally chosen; the preference order is E > C > D > A > B.

A comparison of the voting rules shows that each of them leads to a different result despite the identical starting point. And indeed, Kenneth Joseph Arrow (1951) proved that with more than two variants, there is no voting rule that fulfills all the conditions that must be met in order to be able to choose a voting rule that leads to a transitive collective preference order. This is referred to as the *impossibility theorem* (Bamberg et al. 2008, p. 224 f.). In addition, strategic behavior with the aim of manipulation can occur, especially in small groups (Göbel 2018, p. 174). This can be counteracted, among other things, by open discussions, since these lead to group members presenting and justifying their respective points of view. However, especially in harmonious groups (social systems), there may be *pressure to conform* (group-think) (Janis 1972), which prevents controversial

Rank (R)	Group 1 4 members	Group 2 3 members	Group 3 2 members
1	A = 16	B = 12	C = 8
2	E = 12	C = 9	D = 6
3	D = 8	E = 6	E = 4
4	C = 4	D = 3	B = 2
5	B = 0	A = 0	A = 0

Fig. 4.35 Bora Rules (Göbel 2018, p. 170)

discussions. It can also lead to the formation of camps, which can result in the desire to defeat the opposing camp being stronger than the desire to make the best collective choice. Risks such as these can be reduced, for example, by means of suitable *context design* (Eisenführ et al. 2010, p. 365 f.). In this context, higher-ranking group members should avoid revealing their preferences at the beginning of a discussion so as not to pre-determine a general direction. Furthermore, Elisabeth Göbel (2018, p. 176) points out the possibility of *consulting advisors* who can provide an unbiased view. In addition, a skilled *moderator* can positively influence group dynamics. And asking a group member to take on the role of *devil's advocate* and present a counter-position to each position can also be helpful. Finally, Elisabeth Göbel (ibid.) points out software that can be used to support group decision-making under names such as Group Decision Support Systems (GDSS).

Ultimately, this results in a catalog of goals and constraints (Haberfellner et al. 2018, p. 230), to which the same applies as to the underlying catalog of requirements:

- The catalog of objectives should be structured for clarity. Structuring criteria can arise from Stakeholder Value Management by assigning the objectives and constraints to the respective stakeholders. This can result, for example, in quality objectives with regard to customers, objectives regarding occupational safety with regard to employees, objectives regarding environmental protection and so on. Furthermore, goals can be structured according to whether they relate to the result of the design project (product, service, etc.) or to the procedure for designing the result (Project Management) (ibid., p. 229 f.).
- Furthermore, the goals (including the constraints) should be solution-neutral, i.e. they should not anticipate or unnecessarily restrict what is yet to be developed. And the application of certain methods is usually not one of the goals either. These methods are not selected until the difference (barrier) between the actual state and the desired state is sufficiently known.
- The objectives should be formulated clearly and comprehensibly. For this purpose, the templates already used in the course of the requirements determination (Figs. 4.22 and 4.30) can be adopted. In the next Section, however, further aspects will be added with the test cases, which will then have to be added to the catalog of objectives.
- It is not easy to achieve that the target catalog should be as complete as possible, but at the same time as manageable as possible. Especially in the case of large-scale design projects, hundreds or even thousands of objectives can arise. In such cases, only

databases or specific programs based on them that offer search, selection, sorting and linking functions can usually help.
- The requirements finally adopted in the target catalog should be as consistent as possible (Haberfellner et al. 2018, p. 237 ff.), i.e. contain appropriate compromises and priorities.
- Finally, the calculation of opportunities and risks, which was already started in the course of the situation analysis, must be revised and, if necessary, supplemented, because the targets that have finally been formulated in concrete terms put some of the previously available ones in a new light.

4.3.4.6 Definition of Test Cases

Goals are designed to be achieved. Whether and, if applicable, to what extent they have ultimately been achieved can be seen at the latest when what has been achieved has to prove itself in its practical application in the intended context. However, by then it is usually too late, because correcting mistakes usually becomes more complicated the further a project has progressed (Fig. 4.15) (VDA Volume 3, p. 157 f.). Therefore, verification and validation are usually carried out during the design process to determine whether the characteristics and properties of the design objects are target-oriented (in accordance with the specifications) and suitable for the intended purpose. To ensure objectivity, the definition, development and execution of tests are carried out *independently* of the development and realization of the design objects.

The necessity of *verification* and *validation* during development and realization, as well as their integration into the system life cycle (Fig. 3.22), is particularly emphasized by the V model (Fig. 4.28). On the descending branch of the V, the design object is developed from the rough to the detailed (Sect. 3.3.1), i.e. starting with the overall system, via its sub- and part-systems, down to the level of the elements. At each level of detail, the corresponding tests are defined, which serve the later verification and validation of the development results achieved at this level (sub-/partial systems, elements). On the ascending branch of the V, the design object is then realized from the detailed to the rough, i.e., starting with the elements, via the sub- and part-systems, to the overall system. In these phases, the previously defined tests are now carried out at each level of detail (Informationstechnikzentrum Bund 2019, p. 34).

However, the V-model is just another way of presenting the design phase of the system lifecycle and is therefore located at a meta-level from the perspective of the actual problem-solving. Here, on the other hand, we are dealing with the problem-solving itself, the procedure of which is represented by the problem-solving cycle (Fig. 3.22). At this level, tests are defined for the purpose of verification and validation in the course of defining the objective P 3, because it is important to prove the fulfillment of each objective in detail, so that the direct reference of a test to the corresponding objective is of particular importance. Verification and validation then accompany phase P 5 of the actual problem-solving process and are concluded in the evaluation phase P 6. The success or failure of verification and validation is crucial to determining whether a return is necessary in the sense of a cyclical repetition of the problem-solving cycle or whether the problem-solving process must be terminated altogether due to a lack of prospects for success.

Tests are used for verification and validation (Informationstechnikzentrum Bund 2019, p. 34; Wenzel et al. 2008, p. 33 ff.).

- Verification answers the question of whether a design object meets its specification, i.e. whether the requirements adopted in the target catalog in the course of the target formulation are fulfilled. The question is often formulated in a simplified way: Is the product being developed correctly?
- In the course of validation, the question is answered as to whether the design object is suitable for the intended use. A simplifying question in this regard is: Is the right product being developed?

The *tests* required for verification and validation are understood to be methodically conducted experiments, the results of which reveal whether the tested characteristics or behaviors of the object meet the corresponding requirements (DIN 1319-1:1995-01). At the same time, however, an examination is also a procedure in which competencies (Fig. 1.3) are determined and evaluated, for example in the course of a degree program in an exam.

Tests are to be distinguished from *experiments*, which are open-ended or where the outcome can only be guessed at. So while experiments are mainly used in science, tests are mainly carried out in practice. Finally, the term *trial* is also occasionally used (e.g. Haberfellner et al. 2018, p. 67; VDA Vol. 3:2008, p. 147). However, this term is closely related to trying out, which in turn refers to the application of the heuristic method of trial and error and thus does not belong in this context. Therefore, the term testing seems less suitable and should be avoided if possible in view of the alternatives mentioned above.

According to DIN 1319-1:1995-01, the results of tests can be determined *objectively* by *measuring* or *gaging*. When measuring, a physical quantity is recorded with a measuring device that displays the result as a numerical value. When gauging, a gauge (reference standard) is used to determine whether a characteristic or property of the object to be tested lies within specified limits. This only allows a yes-no statement. In addition, it is usually possible to determine in which direction the limits are approached or exceeded. In addition to objective testing, subjective testing is also possible if only the *sense organs* are used and if these are only supported by aids whose purpose is to compensate for weaknesses of the sense organs. According to this understanding, for example, glasses are an aid, while a microscope is to be classified as a testing device.

The requirements for measurements and tests are defined in countless norms, guidelines and standards (quasi-standards) (e.g. DIN 863-2:1999-04). However, the individual test conditions (test severity) must be agreed in each individual case (VDA Volume 3, p. 206). Since organizations such as industrial companies usually manufacture products of a certain type, for example only aircraft, only dishwashers or only fasteners, comparable tests will be required time and again. It is therefore recommended (VDA Volume 3) to create company-specific catalogs in which the relevant measurements and tests are recorded and explained. In the interest of Knowledge Management, experiences from previous measurements and tests should also be added to support continuous improvement (Kaizen).

4.3.4.7 Documentation of Targets

The documentation of the objectives is linked to the documentation of the requirements that have already been determined in the context of the situation analysis and in particular in the course of the stakeholder analysis and should have already been documented in this phase. As explained in Section 4.2.4.8, such documentation usually takes the form of a *requirements specification*. However, with reference to the so-called *co-evolution of problem and solution* as described by Beate Bender and Kilian Gericke (2021, p. 39), it has already been pointed out that this is not the end of the story. In fact, documents such as the requirements specification and, in the further course, the resulting specifications (in the course of a cyclical procedure, these must be continually further specified, completed and usually also repeatedly varied (Sect. 4.3.4.7) must be continually further specified, completed and usually also repeatedly varied (Sect. 4.3.4.7). Against this background, the documentation of objectives discussed here is an essential step, but it is one that follows on from other essential steps that have already been taken and will follow.

In this phase of a design project, the documentation of the objectives is based on an already existing requirements specification (Sect. 4.2.4.8), which, however, so far contains the requirements that the stakeholders place on the project. This document now forms the basis for documenting the work carried out in the previous Sections on formulating the objectives. To do this, a new, extended version of the requirements specification is usually created, which should, however, retain the references to previous versions to ensure traceability, because the new version of the requirements specification is still an important basis for communicating with the stakeholders and for maintaining an understanding with all participants and affected parties.

The way the objectives are documented is based on the way the requirements are documented and will therefore not be explained again here. In view of a further concretization and the changes that go hand in hand with a transformation of requirements into objectives (Sect. 4.3.4.4), a possibility for a more extensive structuring of a requirements specification is shown below, which does justice to the difference between requirements and objectives. This possibility is based on the structuring criteria recommended in the international standard ISO/IEC/IEEE 20148:2011.

- *Reference number*: The requirements identified in the course of the situation analysis should already have been made accessible for unambiguous identification with the help of a suitable numbering system. This numbering system should be flexible enough to be retained during the formulation of objectives, but also to be varied and differentiated. Examples of suitable numbering systems can be found in the context of parts lists (Wiendahl 2014, p. 168 ff.).
- *Designation*: Each goal should be given a short, meaningful and, as far as possible, unambiguous designation to support quick recognition and classification.
- *Time and place*: The time by which or the period within which the goal should be achieved should be stated, along with the expected location. A point in time does not have to be specified in a time dimension. Alternatively, it can also be linked to specific events, for example, the completion of a trade. Similarly, the location does not have to be given as a geographical indication. Instead, it can also refer, for example, to the installation location of a component in another component. Finally, intermediate

times (milestones) or locations where a temporary stay (camp) is intended can also be specified.

- *Status*: This can be used to indicate the state of realization. Under certain circumstances, this can also be part of the numbering system (see above).
- *Description*: On the basis of a suitable template, as already introduced here in Figures 4.22 and 4.30, the objectives should continue to be described as clearly as possible. However, since the transformation of the requirements into objectives has now added further aspects such as testing, the templates used so far are no longer sufficient to describe an objective with the completeness and accuracy that is now possible.
- *Source*: Even after a transformation of requirements into goals, it should remain comprehensible to which stakeholder group a goal can be traced back in order to maintain the possibility of accurate communication.
- *References*: These are, for example, laws, standards, guidelines and contractual documents to which objectives are linked.
- *Explanation*: Additional information that helps to better understand an objective is often helpful. This can also include references to other documents such as meeting minutes, correspondence or drawings.
- *Constraints*: These are usually understood to mean circumstances that can only be influenced with great effort or not at all in the course of implementing a design project and are therefore to be regarded as given.
- *Purpose*: To promote acceptance of objectives, it makes sense to state the reason (benefit, meaningfulness, advantage) that motivates the relevant stakeholder group to pursue the objective.
- *Priority*: This is usually already part of a formulation template (see above) and therefore only needs to be mentioned separately if no template is used. The priority can be indicated as an extract of the description, for example as "must", "should" or "can", but it can also be given its own code, for example A, B or C, 1, 2 or 3.
- *Cross-references*: These are other objectives that are affected by the documented objective or affect others (restrict, support, change, etc.).
- *Influences*: These are aspects that could influence the realization of the goal.
- Effort: This is a measure expressed in time or monetary units to quantify the effort calculated and/or already expended in realizing the goal.
- *Responsibilities*: These include responsibilities such as ensuring that the goal is achieved, for approvals and releases, for communication with contact persons or groups of people or organizations (authorities, suppliers, etc.).
- *Acceptance criteria*: These include, in particular, test cases including corresponding target values.
- *Comments*: These are aspects that are not mentioned in the other Sections but are nevertheless relevant, for example notes on the CV of the target.

4.3.4.8 Management of Changes to Objectives

Due to the learning process that accompanies every solution, the requirements inevitably change (Andler 2015, p. 30 f.; Erlach 2014, p. 8 f.). The problem-solving cycle is set up for this. Since requirements represent a significant part of the initial situation, any significant change requires a *return to situation analysis* P 2 in order to compare the changed requirements with the requirements available up to that point (requirements determination), to check the changes in a second step and to document the entire process. To do this, the requirements are usually *versioned*. The numbering system recommended in the previous Section is suitable for this purpose, and can include a *change index*, among other things. The changed requirements must then be *distributed* to all parties involved and affected. At the same time, requirements that are no longer valid are withdrawn, and all of this must also be *documented* and checked. Finally, requirements that have been withdrawn from circulation must be *archived* in such a way that they can be retrieved again if necessary.

The procedure explained here is sometimes referred to as the *waterfall model*. Although there is nothing to be said against proceeding quickly and flexibly, experience has shown that this does not always work in practice. This has led to the spread of *agile approaches* such as *Scrum*: 'Scrum dispels the myth that all requirements can be precisely defined at the beginning of a project. Instead, Scrum focuses on a few, but important requirements and their implementation in the context of a sprint. The sprint creates certainty for a limited time. While requirements are difficult to predict in the medium term, they are fixed for the duration of a sprint, and it is possible to predict relatively accurately which features will be delivered at the end of a sprint. [...] The requesters leave the team alone for the duration of a sprint and in return get a chance to correct things every 30 days' (Wirdemann, 2017, p. 26).

The fact that the insight into the limited plannability of problem solutions and thus also into the limited formulability of the requirements to be placed on them is cognitively justified was already explained in Sect. 3.2 (Dörner 1979, p. 19; Duncker 1974, p. 1; Ehrlenspiel and Meerkamm 2013, pp. 56 f.; Funcke 2003, p. 25; Lüer and Spada 1990, p. 256). In this respect, it also underlies Systems Engineering and is reflected in the cyclic approach. Agile approaches do not differ significantly from this, but they emphasize the necessity of *speed* and *flexibility* as well as the necessity of the *cyclic approach* that can be derived from this (Goll and Hommel 2015, p. 73) and express this, among other things, by calling the so-called problem-solving phase P 5 in the problem-solving cycle (Fig. 3.23) a *sprint*.

However, the usual visualizations of models of an agile approach, such as the model by Roman Pichler (2008, p. 7) shown in Fig. 4.36, do not explicitly reflect the change of requirements in the course of a project. Rather, such representations give the impression that the catalog of requirements (specifications) known as the product backlog is defined at the beginning and not changed again. This also applies to the *sprint backlogs*. These contain selected items (selected backlog) from the requirements catalog (*product backlog*) and are broken down into individual *tasks* during a *sprint planning meeting* (Wirdemann 2017, p. 29). In this respect, this step can be compared to an at least rudimentary creation of a requirements specification. On this basis, sub-systems or partial systems or elements (increments) of the overall system are then developed. In software development, for example, these can be individual subroutines that each contribute certain functionalities to the overall range of functions, for example by enabling a file to be saved.

The necessity of continuously taking changed requirements into account must be explained in separate illustrations and/or in text form in addition to the visualizations of the

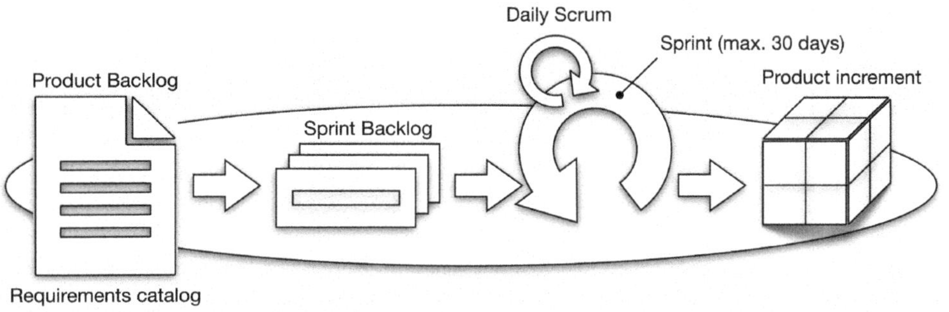

Fig. 4.36 Agile Project Management (Pichler 2008, p. 7)

process models on which agile approaches are based (e.g. Wirdemann 2017, p. 26). In contrast to this, the cyclical integration of the situation analysis P2 and, with it, the integration of the consideration of changed requirements in the model of the problem-solving cycle is explicitly shown. Therefore, this model appears to be superior to the process models of agile approaches and thus also fulfills the requirement of being generalizable as a generic model in this respect. At the same time, however, a detailed comparison of agile approaches with the Systems Engineering approach shows no significant differences. Systems Engineering is no more of an obstacle to rapid and flexible problem solving than, for example, Scrum.

In agile approaches, the concrete formulation of requirements takes the form of so-called *user stories*. However, these only describe customer requirements, formulated in the customer's language, for example in the form "As a customer, I want to transfer money to other accounts" (Wirdemann 2017, p. 55). So-called *epics* are written in an even more general way, for example in the form "Invoicing" (ibid.). No technical aspects should be considered (ibid., p. 61).

Non-functional requirements, such as those for performance or reliability (Sect. 4.3.4.5), undergo a two-stage process in agile approaches (ibid., p. 115 ff.). In the first step, an attempt is made to reformulate such requirements into user stories. For example, a performance requirement can be reformulated from a customer's perspective as a wish for a transfer process to be completed as quickly as possible. Non-functional requirements that cannot be reformulated in this way are treated as constraints and bugs by agile approaches such as Scrum (ibid.). Such a constraint may be that an availability of 99.9% is required. Constraints and bugs are also included in the product backlog (requirements catalog, specifications).

In addition, agile approaches take into account purely technical and organizational requirements, such as refactoring, which is widely used in software development and involves reworking the source code of a computer program to improve clarity or complete the documentation without changing the functionality. Such requirements are managed in a separate technical backlog (Wirdemann 2017). The crucial difference compared to the conventional approach is that the translation from the customer's language into the engineer's language (obligations specifications) does not take place until the problem-solving phase P5 (sprint). This is intended to increase flexibility when taking into account continuously changing requirements.

4.3.4.9 Quality Function Deployment (QFD)

However, speed and flexibility are not an end in themselves. In fact, there is also justification for design projects that place particular emphasis on *thoroughness*, for example when the *safety* and *reliability* of solutions are important.[19] The quality function deployment explained below takes this into account by combining many of the previously explained subtasks of requirements determination and their transformation into goals in a self-contained method.

A particular challenge for Requirements Management is that stakeholders often formulate their requirements in an amateurish way. For example, a customer of a bicycle bag for a carrier may expect it to look good. However, good looks is not a characteristic that engineers can specifically design. QFD can help to translate the language of the customer and thus the language of all stakeholders into the language of the engineer (Brüggemann and Bremer 2015, p. 30 ff.; Erlach 2014, p. 222 ff.; Geiger and Kotte 2008, p. 164; Hering and Schloske 2022; Jenney 2010, pp. 133 ff.; Kaffenberger et al. 2012, pp. 78 f.; Kossiakoff et al. 2011, pp. 306 f.). Fig. 4.37 shows the *House of Quality* (HoQ), which, among other things, represents a comprehensive interface between the requirements of the stakeholders and the characteristics of the product. The HoQ presented here consists of eight sub-documents, also referred to as rooms. Other variants may differ from this number and also from the arrangement of the rooms. This shows that the method can be used flexibly. In this respect, the aim here is merely to present the principle.

- Room 1 contains the "voice of the customer", i.e. their requirements, which are usually unspecific and expressed in an amateurish way. The fact that the method in its original form only includes customers can be explained by the fact that it originates from Quality Management. Here, on the other hand, it is applied universally, so that the requirements of all stakeholders are included. When the requirements are recorded, their weightings are also queried and entered in the last column of this room. For this purpose, a scale can be used that can be freely selected. It should only be noted that an even number of weighting factors forces a tendency, while an odd number of factors also allows the choice of the middle value, which allows a neutral weighting.
- Room 2 lists essential features of the product to be designed. As a rule, listing all features is beyond the scope. Therefore, it is important to select features that are representative of all other features. An essential criterion will be the contribution of a feature to meeting the requirements of the stakeholders.
- This degree of contribution of a product attribute to the fulfillment of each listed requirement is quantified in room 3. For this purpose, weightings are also to be carried out on a scale to be defined. In order to enable the clearest possible differentiation, the weights on the chosen scale are often not distributed linearly, but asymmetrically. For example, on a scale of 0 to 10, 0, 1, 3 and 9 can be selected. The products of the weightings of the requirements from area 1 and the sum of the influence values of the product features on the fulfillment of the requirements from area 3 are noted in the

[19] For such applications, Joachim Goll and Daniel Hommel (2015, p. 79) explicitly exclude agile approaches, while Systems Engineering has precisely such design projects in mind.

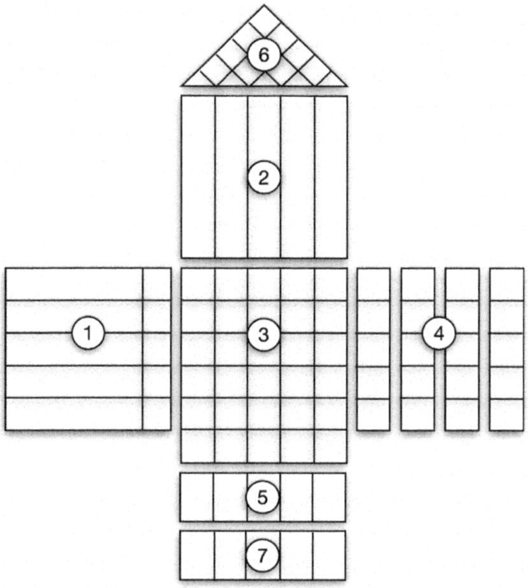

1 Requirements of the stakeholder groups incl. weighting
2 Characteristics of the product or service
3 Influences of the characteristics on the fulfillment of the requirements
4 Competitive comparison from the stakeholders' perspective
5 Competitive comparison with regard to product features
6 Pairwise comparison of characteristics to identify conflicting objectives
7 Quantified objectives

Fig. 4.37 Schema des Quality Function Deployment (QFD)

bottom line of area 3. This results in a prioritization of the features in relation to their respective influences on the fulfillment of the requirements.

- Room 4 is used for competitive comparison. Here, the perception of the stakeholders serves as a benchmark. A simple way to do this is to compare only with the competitor that is considered the strongest. Alternatively, but somewhat more complex, a comparison with several competing products is possible. From the perspective of one's own product, a distinction must be made as to whether it is rated better or worse by the stakeholders than the competing product in terms of the respective requirements, whereby the same applies with regard to the scale to be used as was explained above. This results in a strengths-weaknesses profile from which priorities can be derived for one's own developments.
- Room 5 also serves to compare with the competition. However, this is about technical strengths and weaknesses. Otherwise, what has been explained in relation to room 4 applies. Room 6 is used to compare product features in pairs with regard to whether their respective realization can lead to conflicting goals. For example, reinforcing the bottom

of a bicycle bag to increase the permissible load capacity can conflict with minimizing weight. Priorities for development can also be derived from this consideration.

- Finally, room 7 provides an opportunity to define goals as a result of the analyses, which should also be quantified if possible. In this respect, the method goes beyond the situation analysis P 2 and anticipates the subsequent phase of goal formulation P 3. From this, the recommendation is derived here to only fill in room 7 in the course of formulating the goals in P 3.

The methods used to collect the data include the usual methods of empirical social research, in particular the study of documents and surveys. Fig. 4.38 demonstrates QFD using a simple example.

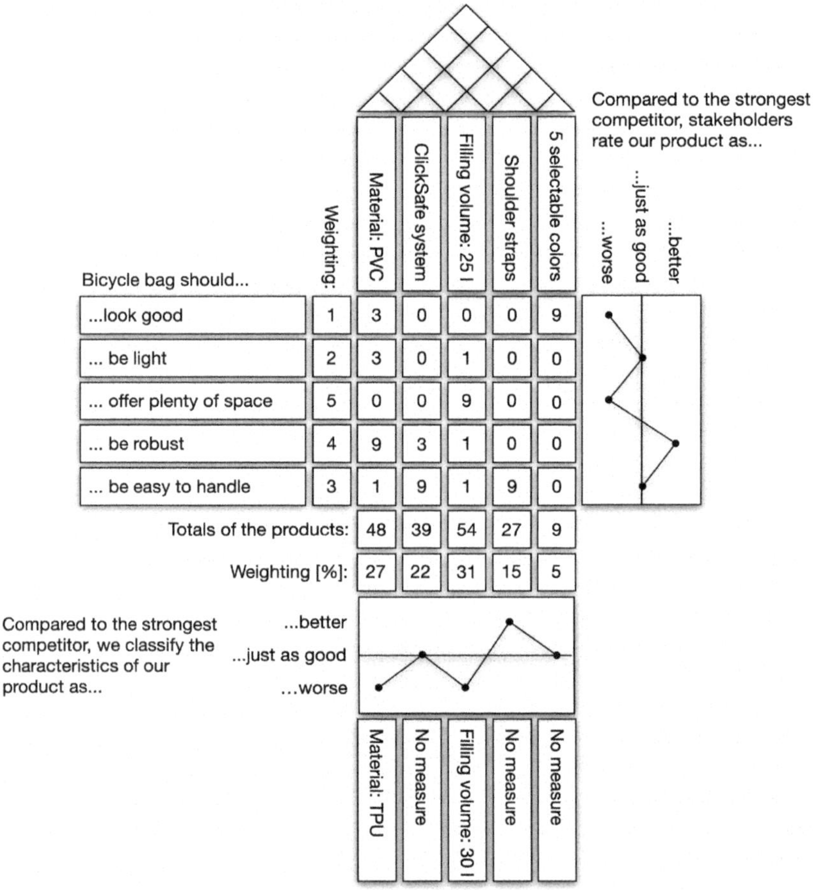

Fig. 4.38 Example of a QFD

4.3.5 P 4.5: Target Definition

Now that the methods for formulating goals have been explained in detail, it is not necessary to repeat their application to create a specific goal system. Instead, the limits of the approach recommended here will be discussed below, because, as is well known, the dose makes the poison – in other words, too much of a good thing can become just as much of a problem as too little.

The ways in which a goal orientation can go too far have been researched, for example, by Lisa D. Ordóñez et al. (2009, pp. 6 ff.). They found that goals can cause a strong, and in some circumstances, an overly strong *focus* and thereby distort the view of aspects that may also be relevant or become relevant in the course of a problem solution. In the context of formulating objectives, this effect requires particular attention because requirements can change over time and with increasing knowledge. An excessive focus can cause the original goals to be adhered to too rigidly and the willingness to learn to suffer.

Furthermore, the authors mentioned point out that a very committed focus on achieving goals can distort *risk preferences*. This is especially true when the reward for success is of great importance. Above all, it can also provoke *immoral behavior*. However, achieving a goal at any cost is usually not desirable.

After all, an excessive focus on goals and the associated structure of the difference between the actual and target state can limit the *creativity* that is central to solving problems in the further course. Yeun Joon Kim and Chen-Bo Zhong (2017, p. 15 ff.) have demonstrated this effect in their studies, among other things, using LEGO bricks: subjects who built a fantasy figure from a disordered pile of bricks came up with more creative results than those who were given the bricks in an organized form.

4.3.6 P 3.6: Evaluation

If and to the extent that the objectives have already been reviewed in the course of the previously explained steps of phase P 3 of the goal formulation, this is no longer part of the final evaluation. Only if questions remain unanswered in this regard do they have to be dealt with at this point. What is important is that the essential requirements for the catalog of objectives are met (Haberfellner et al. 2018, p. 225 ff.).

- Each target should be given a unique reference number, making it unambiguously addressable.
- Each goal should be given a short, meaningful and, if possible, unambiguous name to support quick recognition and classification.
- For each goal, it should be stated by when or within what period the goal is to be achieved and at what location this is expected.
- It should be possible to indicate the status of each objective in terms of its degree of achievement.
- The objectives should be described as precisely as possible using an appropriate template (Figs. 4.22 and 4.30).

- Even after the transformation of the requirements into objectives, it should be possible to trace back to which stakeholder group an objective can be attributed.
- References such as laws, standards, guidelines and contractual documents to which objectives are linked should be fully stated.
- Where necessary, additional explanations should be provided, for example in the form of minutes of meetings, correspondence or drawings.
- Boundary conditions should be indicated as such.
- The purpose (benefit, reasonableness, advantage) should be stated for each objective.
- The objectives should be prioritized.
- Cross-references to other objectives should be producible.
- Influences on the achievability of the objectives should be indicated.
- The effort required to achieve a goal should be indicated.
- Responsibilities for ensuring that the objectives are achieved, for approvals and releases, for communication with contact persons or groups of persons or organizations (authorities, suppliers, etc.) should be specified for each objective.
- The catalog of objectives should contain acceptance criteria for each objective, in particular test cases and corresponding target values.
- The target catalog should be structured according to suitable criteria to ensure clarity.
- Conflicting objectives should be resolved through prioritization or compromise. At the very least, it should be possible to resolve them in the further course of the project.
- The objectives should be formulated in a solution-neutral way and should not already provide for the application of certain methods.
- The catalog of objectives should be complete and factually correct.
- Access rights to the catalog of objectives should be regulated, especially if it is stored electronically. In particular, the requirements for data protection and data security should be met.

In addition to these mainly formal aspects, it is important in phase P 3.6 of the evaluation that all stakeholders agree with the system of objectives. This agreement should be explicitly sought and formally confirmed (documented) if possible.

If the evaluation of the catalog of objectives is unsuccessful, it is necessary to go back to the phase in which the cause of the failure can be eliminated. From there, all further steps must be gone through again. While it is possible to consider the option of moving on to the next phase of defining methods for problem solving and their final evaluation at the same time, while still clarifying aspects of the formulation of objectives, there is a not insignificant risk of losing sight of the big picture or of leaving stakeholders behind. The latter can lead to conflicts in the further course and should therefore be avoided.

4.4 Method Planning

The methodology to be developed in the now beginning phase P4 refers to the subsequent problem solving P5 as such as well as to the subsequent evaluation P6. While the steps of the method planning described in the phases explained above the next lower recursion level (this is made clear, among other things, by the third outline level of the table of contents), the method planning that is now at issue is to be found at the higher recursion level (recognizable in the table of contents by the second outline level).

However, such a consistent application of the recursion principle results in the apparent difficulty that the phases of the actual problem solving P 5 and the evaluation of the results P 6 again contain their own phases of the method planning (P 5.4 and P 6.4). On closer inspection, however, this characteristic of the self-similar problem-solving cycle (Fig. 4.1) turns out to be one of its strengths if you concentrate on planning the methods as such in phase P 4 in order to then *configure* them in the following phases P 5.4 and P 6.4. For a better understanding, this will be explained using specific examples.

At this level, P 4, the survey can be defined as a suitable method. To do this, it is necessary, among other things, to weigh up its advantages and disadvantages in comparison to other possible methods, such as observation or document study, in order to justify that and why the survey is the most suitable method in the specific application.

However, with many methods, and this is also the case with surveys, it is not enough just to select the method and justify your selection. After the survey has been chosen, it must be configured (Sect. 4.2.4.2). For example, it must be determined in detail which questions are to be asked, how the questions are to be formulated (e.g., open or closed questions), in what form answers are expected, who is to be surveyed when and where by whom, and so on.

The situation is similar when, at this level, P4 has to weigh up between an observation in the field and an experiment, for example. Here, too, some consideration is needed first from a higher-level perspective, in particular a comparison in the light of the respective advantages and disadvantages, opportunities and risks, own strengths (possibilities) and weaknesses (limits), etc. Once the choice has been made, the selected method also requires configuration in the further course of the process, and this too is the subject of the methodological considerations in due course, which arise in the course of the concrete problem solution (P 5.4) and the subsequent evaluation (P 6.4) of the developed results.

For a better understanding, it may also help to imagine that the application of the methods to be chosen in this phase P 4 involves third parties, for example, experts in-house or service providers who have specialized in certain methods. Computer-aided simulation (Sect. 2.4.5.3) can be an example of this. A typical application of this experimental technique can be found in factory planning (Bracht et al. 2018, p. 27 ff.). Such extensive design projects are not usually carried out very often in a company, but they have a major impact in the situation in which they take place, so a systematic approach and the consistent application of suitable methods are required. However, it is often not worthwhile to develop internal expertise for specialized methods or to purchase the necessary software. Therefore, an alternative or supplement is to involve service providers, who then also participate in the detailed configuration of the corresponding method, in particular in the creation of the relevant specifications and requirements (VDI 3633, Sheet 2). This then takes place in phase P 5.4.

This makes it clear that the approach *from the rough to the detail* (Sect. 3.3.1) is also applied in this context. This also explains the structure of the further explanations, because these also initially refer to general statements here (P 4), while the respective details are dealt with in more depth later in the problem-solving (P 5.4) and evaluation (P 6.4) phases.

In the phase of method planning, the aim is not only to select methods, but also to define the *tools* and *environments* (operating conditions) that are required for the application of the respective methods. Fig. 4.39 clarifies what is to be understood as a tool (synonym from Latin: instrument). For this purpose, a distinction is made on the basis of a *functional modeling* (Sect. 2.4.1) between those functions that humans[20] can perform with the help of their physical, mental and emotional abilities and those that can be transferred to a technical system. Those functions that cannot be supported by a technical system remain the preserve of humans. This is the area of *reservation*. It includes, for example, evaluating, feeling or dreaming. Although humans can perform other functions, these can be improved with the help of tools. For example, humans are often not very good at mental arithmetic; a pocket calculator is better at this. And by using a bicycle, people can also increase the speed of their locomotion compared to walking. This is the area in which tools are used that *substitute* human functions. Finally, there are functions such as flying that humans cannot perform at all without technical support. In this area, tools serve to *complement* human functions (Ropohl 2009, p. 184).

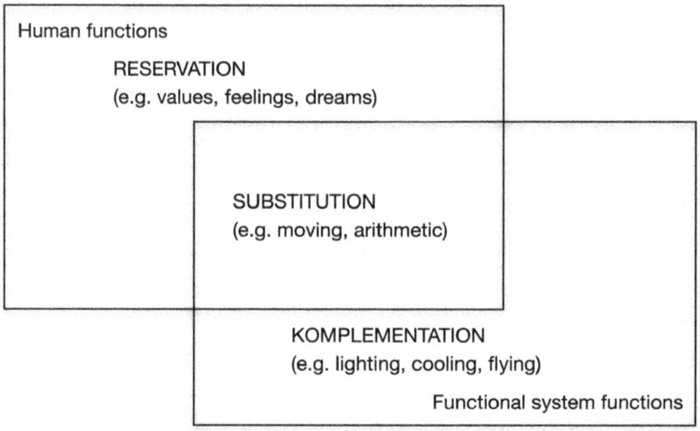

Fig. 4.39 Principles of mechanization (Ropohl 2009, p. 182)

A *hierarchical modeling* (Sect. 2.4.3) leads to the insight that tools can also use tools, with one tool substituting or complementing the functions of the other. For example, you can easily cover many distances by bicycle within a city, but it is often more convenient to take the train with your bike from one city to another. In such a case, you can substitute the function of the bicycle with that of the train when covering longer distances. Complementation can be illustrated using the example of a machine tool. While the machine provides functions such as those of the drives, the actual shaping is done with the help of tools such as the lathe tool. In the light of the previous considerations, however, it remains the case that the machine is also a tool, albeit one at a higher hierarchical level.

Such an understanding has far-reaching consequences. It makes it clear, for example, that widespread definitions such as those according to which tools are objects that do not belong

[20] The following also applies to animals, but that is not relevant here.

to the body of living or artificial objects, but that are used to enhance the body's functions in order to achieve a specific goal (Lawick-Goodall 1970, p. 195 ff.), are not helpful in all respects. For example, they only refer to bodily functions and ignore mental and emotional functions. In the light of increasing *digitalization*, however, support for mental functions is becoming more important. And *emotional intelligence* (Goleman 1996) is also relevant for understanding psycho- and socio-technical systems (Sect. 2.6.2). The focus on the physical is also emphasized by the characterization of a tool as an object, since this term is usually used to refer to physical objects. In the course of digitalization, however, so-called software tools are now also being used that are not usually subsumed under the term object. Software tools only acquire physicality through their connection to computer hardware.

Furthermore, the demarcation of the functional area of reservation from the two functional areas that can be supported by the use of tools is important for understanding the possibilities, but especially the limits, of *artificial intelligence* (AI), because it makes it clear that tools cannot be autonomous in the same sense as humans, because their intelligence, if one wants to use this term at all in this context, is limited to substitution and complementation, while for an autonomy that would be comparable to human autonomy, the functional area will always be lacking, since this remains reserved for humans.

Finally, the definition of tools cited as an example includes the necessity of goal-oriented use, which, however, cannot be derived from the function of tools, but is derived from what is understood in the cognitive sciences in contrast to behavior under action (Fig. 3.15). A tool use is therefore only aligned with an objective by being carried out in the course of an action, which does not exclude the possibility that it can also support a behavior, but in that case it is not tied to an intention.

As already mentioned above, the tools and also the *environments* in which the tools are to be used are to be planned in the course of method planning. This applies above all if the results are to be reproducible, as is the case, for example, with experiments (Sect. 4.4.5.1) and tests (Sect. 4.4.5.3). One example is the measuring system (Prorok 2022, p. 165 ff.), which consists not only of the tool, in this case the measuring equipment. In fact, the environment and the situation of the measuring process also have a significant influence on the measurement result. Above all, tests during production are affected by this, since they are characterized by frequent and often significant changes in the influencing factors: different people with different qualifications carry out measurements, each with their own measuring equipment, under different environmental influences such as temperature, humidity, time of day (consequently, lighting conditions, fatigue of the people involved, etc.), posture during measurement, etc. When designing the work system (Bokranz and Landau 2012; Schlick et al. 2018), care must be taken to ensure that influencing factors such as those mentioned above enable reliable measurement. The remaining influencing factors must then be quantified in a *measurement system analysis* (MSA) in order to be able to quantify statistical key figures such as accuracy, correctness, repeatability, traceability (comparative precision), stability and linearity, as well as ability indices such as the C_g or C_{gk} value (method 1 of the MSR) and the %GRR value (method 2 of the MSR), which can be used to make statements about the accuracy of the measurement results.

4.4.1 P 4.1: Occasion

If a method P4 is directly integrated into a problem-solving cycle, as in the present case, it is directly linked to a goal formulation P3 and also directly transitions into the problem-solving phase P5. The explicit explanation of the occasion can often be omitted because it is immediately apparent. However, the underlying claim of general validity means that solutions must also be provided for cases in which not everything is clear from the outset. In terms of method planning, this may apply, for example, when a method for solving a specific problem has yet to be developed. Such projects are often found in science.

In the course of a corresponding research project, the reason for this is explained at the superordinate level in the corresponding phase P 1 (Ahrens 2020, p. 85 ff.). However, if and to the extent that such a project is part of an overarching project, the integration of a research project into this overarching project can be done at this level P 4.1 and then also with a correspondingly detailed explanation of the reason for doing so. In this way, an overall project and several associated sub-projects can be very systematically interlinked.

4.4.2 P 4.2: Situation Analysis

For the situation analysis to be carried out as part of method planning (P 4), everything that has already been said about this phase of a design project (Sect.s 4.2, 4.2.2, 4.3.2) initially applies in a similar way. These will now be supplemented only by aspects that are specific to method planning.

In a first step, in the light of the actual situation P 2 and the objective P 3, as well as the barrier derived from them, which needs to be overcome in the course of solution development (Sect. 3.2), possible methods must be identified that can be used to overcome the barrier. Often, several methods are available for this purpose, from which the most suitable methods for the specific application will have to be selected in the further course (Sect. 4.4.5). In addition to what is offered in this book, numerous and mostly highly recommended works are available as sources for known methods (e.g. Andler 2015; Atteslander 2010; Balzert et al. 2017; Baur and Blasius 2017; Diekmann 2020; Domschke et al. 20 15; Dörner 1979; Funke 2003; Lödding 2016; Schönwandt et al. 2013; Schurz 2014; Sell and Schimweg 2002; Stachowiak 1973; Weilkiens 2014, and many more).

However, as already mentioned in the previous section, not all methods are always available and accessible to solve a specific problem. Often, methods still need to be researched and developed or modified. In this case, the *state of research and development* (R&D) or the *state of science and technology* must be determined as part of the situation analysis. This also includes estimating the potential of possible solutions with regard to the timely readiness for use of the methods that still need to be researched and developed as part of the design project (feasibility analysis).

The requirements determined in the context of situation analysis P2 relate to the overall project. Although they usually also include requirements for the approach to solving the problem, the requirements will essentially express the expectations of the stakeholders regarding the outcome of the overall project (specifications). Therefore, it may now be necessary in this phase P 4.2 to determine additional requirements for the methodology, tools, and conditions of use (environment) (specifications). This is also done by means of a *Stakeholder Analysis* (Sect. 4.2.4.6).

For example, a method may be desirable from the perspective of those carrying out the project in terms of the knowledge gained from its application; however, from the perspective of those providing the funds, it may prove to be too costly. In such a case, the costs and benefits must be weighed against each other. A method may also be rejected on moral grounds, for example if it involves animal testing. The application of certain methods can also result in risks to the natural environment, as may be the case with experiments involving genetically modified organisms that are released and could spread indefinitely. Whether or not it is justified, the mere suspicion that the SARS-CoV-2 virus, which caused a pandemic in 2019, was released as a result of a laboratory accident highlights the potential for disaster that can arise from the use of certain methods, procedures and tools. In this context, the consideration of a method's use can range up to a comprehensive technology impact assessment.

4.4.3 P 4.3: Target Definition

A general goal of the use of methods should be to achieve the supported knowledge progress as well as necessary, but not as well as possible. This approach alone will contribute significantly to minimizing conflicts between the requirements of different stakeholders. At first glance, it may appear that this primarily serves those who, for economic reasons, are interested in a faded relationship between effort and benefit. However, on closer inspection, it can be seen that other interests are also taken into account. For example, the natural environment is generally protected when the use of resources is limited to what is necessary. Furthermore, the social environment can benefit if, for example, neighbors are only affected by the effects of the application of certain methods for a short time or with a limited intensity, whether through noise, increased traffic, etc. And parsimony in the collection and processing of data can also contribute to social peace and mental health, especially when personal data or data that can be linked to individuals is involved.

Beyond a commitment to this general goal, the aim in this phase of method planning is to derive goals from the requirements identified in the course of the situation analysis P 4.3.2, which are linked to the use of the methods to be planned. In this context, a distinction can generally be made between *therapeutic* and ingenious goals. While therapy aims to restore an earlier state (structure, process), ingenious projects aim to create something new, something that has not existed before. The methods used for this are partly the same or at least comparable when, for example, it comes to promoting and optimizing creativity and finally subjecting the result to a test for success. However, the methods also differ in some cases.

For example, therapies often make use of the *self-healing powers* of the system to be treated, so that a corresponding solution to the problem primarily serves to stimulate and support this and to give it the opportunity (time, space) to develop its effect. By contrast, the creation of something new often also and primarily requires creative activities. In addition, the desired state in a therapeutic endeavor is already known, so that only the obstacles (barriers) on the way there have to be overcome, while the desired state in an ingenious endeavor is often subject to uncertainties in addition to the path to get there. In the case of therapeutic projects, the solution strategy proposed by Pappus of Alexandria (c. 300 AD), which starts from the expected result and then searches for the solution path by stepping backwards (analysis) and finally proves its success by stepping forwards (synthesis), is more suitable than for ingenious projects. These preliminary considerations make it clear that, with regard to the selection of suitable methods, it is more useful to state the objectives of the solution search as clearly as possible.

4.4.4 P 4.4: Method Planning

The considerations regarding the methodological support of solution development (P 5) already start with the preceding situation analysis and the subsequent formulation of objectives, because a detailed examination of the actual and target state often already gives rise to the idea for overcoming the barrier in between. As an example of this, Robert Sell and Ralf Schimweg (2002, p. 71 ff.) cite the *explication* of the problem, originally recommended by Friedrich Schleiermacher, which can bring to light aspects that would be overlooked or insufficiently complete, detailed and clear, as well as taking into account their interactions, in an initially only cursory, superficial examination, as is usual when initiating a problem-solving project P1, overlooked or not worked out sufficiently completely, in detail and clearly, and taking into account their interactions. Explicitly writing down what is given and what is sought, abstraction and modeling, and the associated separation of the essential from the unessential, the structuring and visualization (Sect. 4.2) of contexts, but also of the unknown or unrelated (fractures, conflicts), and finally a concentrated, systematic and methodically supported way of working using suitable tools. All of this can reveal the solution in the course of the situation analysis and the subsequent formulation of objectives or an iteration between the two. It is therefore worth taking a lot of time and care with this.

The selection of the methods explained below is based on two essential aspects: *creativity* (innovation) and *optimization*. To this end, a basic understanding is first provided before specific methods are explained in detail.

Haberfellner et al. (2018, pp. 245 ff.) assign the mentioned aspects to the *synthesis* of solutions and emphasize the necessity to strictly distinguish this phase of the actual problem solving P 5 from the subsequent *analysis* (ibid., pp. 257 ff.) of the found or developed solutions. The separation between creative and critically reflective phases corresponds to the *principle of deferred judgment* (ibid., p. 75; Sell and Schimweg 2002, p. 77; Traut-Mattausch and Kerschreiter 2018, p. 274). This is based on the insight that with almost every idea of how something could work, the idea is also immediately born of why it probably won't work after all. As a result, many ideas are lost that could prove valuable on closer inspection. Since such an analysis is usually done intuitively, it can neither be avoided nor controlled intuitively. This requires a conscious and methodically supported approach. As will be shown, creativity techniques offer concrete solutions for this.

In the sense of Test Driven Development (TDD), the analysis includes, above all, the execution of *tests*, so that these form the third focus of the methodical considerations, alongside creativity and optimization. However, since tests were explained in general in Section 4.3.4.5, no further general introductory section is devoted to them below. In the course of the presentation of specific methods (Sect. 4.4.5), however, methods are also included (Sect. 4.4.5.3).

4.4.4.1 Innovation

The barrier between the actual and the desired state (Fig. 3.15) is characterized by a lack of knowledge and/or means to overcome it. To overcome this deficiency, either an *concoction* or a *discovery* is needed. While an concoction produces something new, a discovery draws on what already exists. If something known is changed, the change can in turn be categorized as an invention or as a discovery according to the criteria just given, so that combinations of both are also possible.

As long as an concoction merely produces an idea, a concept in the course of further concretization, and finally prototypes, it is referred to as an *invention*. It only becomes an innovation when it proves its usefulness, which in an economic context is demonstrated in particular by successful sales on the market (product, service, right of disposal[21]) or by successful use in the production of marketable products (means of production, business process, etc.). Therefore, Systems Engineering is part of *Innovation Management* (e.g., Hauschildt et al. 2023; Wördenweber et al. 2020), which not only includes development and realization, but also marketing and innovation controlling.

Furthermore, both discoveries and concoctions can occur spontaneously or intentionally. In Systems Engineering, however, the focus is primarily on the intentional, since spontaneity is influenced by chance, while design projects are carried out in a targeted and systematic manner (planned, guided, methodically supported). In addition, only a limited amount of material resources is usually available for design projects, which does not allow for waiting for random inspiration. However, a spontaneous invention can be the cause of a solution to a problem (Sect. 4.1). Inventions and discoveries must be supported by appropriate measures so that they can be carried out efficiently and effectively, which requires at least a rudimentary understanding of the underlying principles. Therefore, these should be mentioned at the beginning of the explanations, which will become increasingly more specific in the further course.

In technology, inventions (inventions and especially innovations) usually take precedence over discoveries, since they can be legally protected under certain conditions by patents and utility models (PatG) or by invoking copyright (UrhG) and thus create *Unique Selling Points* (USP) that in turn can lead to *competitive advantages*. In contrast, discoveries cannot be legally protected and may even have to be paid for, for example in the form of *license fees*. For example, anyone who discovers a new animal or plant species does not acquire ownership of it or any exclusive right of use. Only if a new technical process has been developed that can be used to produce or isolate biological material does an industrial property right also extend to something that already exists in nature (Sect. 1 PatG).

[21] What is meant in particular is a license as permission for commercial use.

Despite the greater attractiveness of an invention, a discovery can also be very useful for solving a problem. This applies, on the one hand, because many inventions cannot be legally protected at all, such as scientific theories and mathematical methods, aesthetic creations, schemes, rules or methods for performing mental or business activities, programs for electronic data processing[22] (however, restrictions on free use may arise from copyright law in this regard (Sect.s 66a-g of the German Copyright Act (UrhG)) or the reproduction of information (Sect. 1 of the German Patent Act (PatG)). The discovery of such research and development results can open up the possibility of using them for one's own problem solving without much effort. For example, the Berlin-based investment company Rocket Internet pursues the copying of existing business models as a business model and is quite successful at it.

On the other hand, a discovery can be interesting if it provides suggestions for new ideas or forms the basis for new ideas through changes (further developments, transfer innovations) (Kaschny et al. 2015, p. 26). Fig. 4.40 shows the possible stimuli between science and practice. Among the examples given, Systems Engineering stands out, which has its origins in practice (Walden et al. 2023, p. 3 ff.) and is now also applied in science.

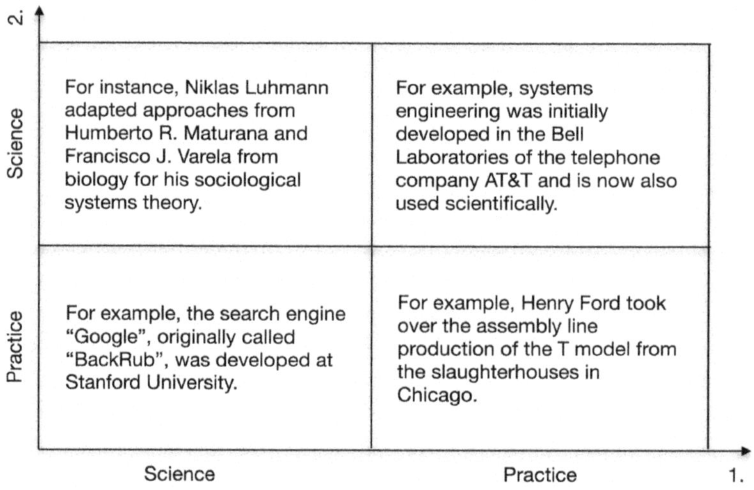

Fig. 4.40 Suggestions for interaction between theory and practice

However, the measures to support invention on the one hand and discovery on the other are very different. The method of discovery is, simply put, searching. In most cases, this requires one thing above all: *diligence*[23]. Concrete methods available for discovery

[22] This is one reason why software manufacturers themselves put a lot of technical effort into copy protection and a lot of organizational effort into keeping the source code secret.

[23] The moral of the legend of the Seven Lazy Brothers (Wagenfeld) is that you can be lazy, but only if you know how to help yourself. However, the seven sons of the poor farmer were not really lazy, but rather industrious in the sense that they spent many years in foreign countries and discovered there how to make life easier for themselves. They still had to implement this after their return before they could sit back

include the study of documents (Sect. 4.2.4.1), questioning (Sect. 4.2.4.2) or observation (Sect. 4.2.4.3). Therefore, the discovery will not be discussed further in the following. Instead, the further remarks will concentrate on inventing.

4.4.4.2 Creativity

The basis of inventing is creativity. While this term allows for a good understanding of what is meant by it in everyday life, there is still no consensus-capable definition in science and professional practice. This may be an indication that creativity describes a very complex set of facts. And so those who attempt a definition (probably partly without knowing it) draw from the repertoire of systems thinking (Chapter 2), since it helps to manage complexity.

One example of this is the well-known four Ps, which describe the person, the process, the product and the environment (Rhodes 1961, p. 305 ff.) as the main factors influencing creativity. This approach could be visualized, for example, using the cause-and-effect diagram shown in Fig. 4.41 (Ishikawa diagram, fishbone diagram). Alternatively or in addition, the *four P* can also be understood as aspects of a system (Sect. 2.4.6). The individual P can in turn be examined using other models, for example the process using the *IPO model* explained in Section 2.4.1 (Fig. 2.6), the person using the functional model of psycho-technical systems explained in Section 2.6.2.1 (Fig. 2.54) and the product using the system analysis explained in Section 4.2.4.4.

One of the widespread definitions of this type of thinking is the so-called *divergent thinking* by Joy Paul Guilford (Guilford and Hoepfner 1971), which is contrasted with converting (focused, goal-oriented) thinking. While the former promotes creativity, the latter is considered to be a hindrance in this respect. Edward de Bono has chosen an equally popular but comparable approach with the concept of *lateral thinking* in contrast to vertical thinking (Dudgeon 2001). Both definitions touch on the very heart of systems theory, namely, *interdisciplinary thinking* that allows for the recognition of analogies and *isomorphisms* that support the transfer of knowledge from one domain to another. Niklas Luhmann's sociological systems theory (1996) is an example of this, because he received significant input from the biologists Humberto Rosmesín Maturana and Francisco Javier Varela García.

By now, it should be clear why it is not possible to define creativity in a simple sentence and then move on to the agenda. This would do justice neither to its complexity nor to its central significance for Systems Engineering. Anyone who wants to successfully carry out or manage a design project cannot avoid a more thorough examination of the question of what creativity is and how it can be developed in the sense of Innovation Management.

This becomes clear, for example, from the widespread view that a targeted (effective), efficient and, in particular, methodically supported approach hinders creativity (Steiner and Landes 2022, p. 95 ff.). Since such an approach is recommended here, it seems necessary to resolve such apparent contradictions. However, as will be shown, it is entirely possible to reconcile creative thinking with a systematic way of working.

---

and be lazy, to the annoyance of their hard-working neighbors. They would have been truly lazy if they had developed the solutions to their problems in a creative way. The moral of this consideration is that searching actually requires diligence, while creativity requires leisure. The latter will be derived later.

To do this, the systems thinking developed in Chapter 2 is used and examined at and between the levels of psychological and social systems[24]. As explained on the basis of Fig. 2.1, however, the two levels are only structurally linked. Their interaction is characterized by emergent phenomena, so that a conclusion from the behavior or actions of individuals to the collective cannot be causally explained in every respect and vice versa. Nevertheless, social psychology attempts to establish the connections between the levels, on the one hand from the perspective of psychology (bottom up) and on the other hand from the perspective of sociology (top down). Using the distinction between human and person introduced in Parag.. 2.6.2.2, the scientific domains of psychology at the micro level, social psychology at the meso level, and sociology at the macro level, which are used in the following, can be level, in that psychology deals with the human being, while social psychology is interested in the person, who is understood as what is relevant about a human being in a collective (Luhmann 2011, p. 89). Finally, sociology is concerned with the collective. The following is an examination from these three perspectives of the contributions made by the respective levels to the development of creativity and the hurdles placed in their way, which System Engineering is designed to support.

Psychological Perspective

Creativity is generally understood as the ability of a person to produce something new (synonym: original) that is considered valuable. This understanding modifies widespread definitions of this kind (e.g. Sell and Schimweg 2002, p. 76) to the extent that it does not declare the usefulness of the new as the yardstick, but rather, more generally, a value that is ascribed however and whenever. On the one hand, this definition takes into account the widespread view that dispensing with the condition of recoverability or usefulness would also classify creations that are completely worthless or even harmful as a result of creative work, which is possible in principle but usually not desired[25] (Feist 1998, pp. 209 ff.). On the other hand, the understanding on which this is based opens up creative work to everything that has value, because the Useless Box, which is occasionally used as an example here, is an example of how even obviously useless creations can find interested parties who are willing to invest time and money in them. Furthermore, the understanding on which this is based avoids conflicts that can arise from the fact that the usefulness of a creation can be classified differently by different judges. For example, some people consider nuclear power plants to be useful and emphasize the corresponding technology as the result of particularly creative achievements. Others, on the other hand, see it as harmful, although some of them may still admit that the research into the scientific fundamentals and the development of the technology based on them are indeed creative achievements. The latter would be excluded if creativity were tied to both the creation of something new and its usefulness.

[24] In creativity research, the neurobiological level (Fig. 2.1) is also examined. However, this will not be discussed in detail here, but rather with reference to the relevant literature (e.g. Vartanian et al. 2016).

[25] In this way, a particularly perfidious method of breaking into a building could be described as creative, but this is not the intention here.

This brief discussion of the concept of creativity may have started with the corresponding human ability, but it has not yet led very far in this respect. This is now to change with the not only historically relevant reference to Joseph Schumpeter (1997, p. 138), who attributes creative achievements not only to the creator's *economic considerations*, but also to his joy in designing, in a more general sense, to his *intrinsic motivation*.

Motivation is a *cognitive ability*, since the activation of definable types of motives can be associated with the release of corresponding neurotransmitters in the brain. In the further course, a second category is added with the personality[26] traits. Both categories are used to determine what creative people differ from others. First, the motivation already introduced and then the intelligence will be used to analyze the cognitive abilities in more detail.

David Clarence McClelland (1987) distinguishes between the power motive, the affiliation motive, and the achievement motive (Fig. 4.41). The power motive is associated with the release of adrenaline and noradrenaline (McClelland et al. 1980, p. 93 ff.), the affiliation motive with the release of dopamine (McClelland et al., 1987, p. 51), and the achievement motive with the release of vasopressin and adrenaline (McClelland, 1995, p. 59). In the first case, attention to and integration of the contributions of creative people serve as measures to promote motivation, in the second case, integration into decisions (Sect. 3.5) and in the last case, recognition of performance.

Motives, wishes, hopes	Fears, anxieties
Belonging (safety, affection, security, friendship) Promotion through consideration and integration of contributions	... unpopular, rejected, isolated, lonely Feeling: Worthlessness
Power (control, dominance, importance, status, influence, struggle, competition) Promotion through involvement in decisions	... loss of control, unimportant, dependent, disregarded Feeling: Powerlessness
Achievement (success, progress, variety, curiosity, imagination) Promotion through recognition	... incapable, weak, humiliated, useless, stupid, failure, loser Feeling: Failure

Fig. 4.41 Basic motives (McClelland 1987)

[26] This term is used here only because of its widespread use. In fact, its use in a psychological context contradicts the previously agreed-upon convention, according to which psychology deals with people, while persons are the subject of social psychological research.

Based on these and other approaches to a differentiated explanation of motivation, John E. Barbuto and Richard W. Scholl (1998, pp. 1011 ff.) distinguish between two intrinsic and three extrinsic sources:

- According to this, there are the following intrinsic sources of motivation:
  - *Process motivation*, which moves a creative person to tackle a task or solve a problem for its own sake. Writing this book is an example of this.
  - *Internal self-concept*, which underlies one's own actions with certain values underlying one's own actions, which have often entered the subconscious and are no longer questioned.
- Among the extrinsic motives, which would actually have to be considered from a socio-psychological perspective (Sect. 4.4.4.2) and are only discussed here to avoid interrupting the context arising from the quoted source, John E. Barbuto and Richard W. Scholl (ibid.) include the following:
  - *External self-image*, which is fed by the collective's expectations of the person and the roles assigned to them and can be attributed to the affiliation motive identified by David Clarence McClelland (McClelland et al. 1987, p. 51).
  - *Instrumental motivation*, which is geared towards a reward and can be attributed to the power motive identified by David Clarence McClelland (McClelland et al. 1980, p. 93 ff.).
  - *Internalization of goals*, which no longer reveals that the ostensibly own goals of action actually correspond to externally prescribed goals. This source of extrinsic motivation can be attributed to both the affiliation motive identified by David Clarence McClelland (McClelland et al., 1987, p. 51) and the achievement motive (McClelland, 1995, p. 59).

Only intrinsic motivation promotes creativity (Amabile 1984, p. 223 ff.; Harrington and Chin-Newman 2017, p. 442 ff.), while extrinsic motivation hinders creativity (Sprenger 2000, p. 120 ff.). However, this applies to every form of motivation, regardless of the purpose it is supposed to fulfill: *All motivation is demotivating* (ibid., p. 12).

Another mental ability that has a significant influence on creativity is *intelligence* (Jauk et al. 2018, p. 212 ff.). The *threshold hypothesis* says that the intelligence quotient must be above a certain minimum value in order to enable creativity. Above a higher quotient, for example at the borderline to giftedness, particularly pronounced creativity can be expected. However, the influence of personality traits is considered to be greater than the influence of intelligence in this area. Furthermore, the only benchmark used is the number of ideas produced, so that a transfer to other aspects of creativity is not verifiable.

While motivation and intelligence are very general prerequisites for achieving any kind of performance, Gregory John Feist (1998, p. 299) has identified a number of factors that are specific to creative people, particularly in their combinations, on the basis of a meta-study. These include (1) autonomy, (2) introversion, (3) openness to new experiences, (4) doubting social norms (unconventionality), (5) self-confidence, (6) self-acceptance, (7) commitment, (8) ambition, (9) dominance, (10) hostility and (11) impulsivity. However, only the cognitive abilities (3) openness, flexibility and imagination, as well as the personality traits (7) commitment and (8) ambition, are psychologically relevant. All other characteristics belong to the field of social psychology and will be discussed in more detail in Section 4.4.4.2.

Openness refers to the willingness to engage with new ideas, strangers or unfamiliar situations (*ambiguities*). Open-minded people, as the saying goes, can think outside the box. This is associated with an *imagination* that is not limited to the familiar, but also includes imaginative possibilities and limits. *Commitment* refers to motivation and endurance (perseverance). *Ambition*, in turn, can be understood as an interest in realizing ideas. This also includes the ability to present ideas in such a way that third parties, even skeptics, can be convinced of them. This in turn requires a certain *tolerance for frustration*, because persuasion can be exhausting, takes time and usually does not lead to complete success.

In addition to the creativity-promoting abilities and characteristics of psychological systems, however, inhibiting behaviors and actions must also be taken into account. The basis for this is the *Prospect Theory* (Kahneman and Tversky 1979, p. 263 ff.), which replaces the model of *Homo Economicus*, on which many problem-solving methods are based and which assumes that people act predominantly rationally (Commander Spock), with a model that also and especially takes into account irrational behavior and action (Dr. McCoy). Nevertheless, such behavior can also be modeled mathematically, but this is not done here in favor of a more descriptive explanation.

Instead, some examples of the causal *cognitive biases* will be used, of which there are now so many known that a complete list is not even possible here. The starting point of the research for which Daniel Kahneman received the Nobel Prize was the realization that the individual's approach to *risk* does not depend on economic self-interest, but varies depending on certain observations and expectations. For example, people behave risk-averse when positive events occur and risk-seeking when negative events occur. If, for example, they are given the choice between receiving a certain amount of money or doubling the amount with a 50% probability, they will mostly opt for the fixed amount. In a negative situation, for example when stock market prices are falling, they are more likely to choose an uncertain but high loss than a small but certain loss. Mathematically, this results in an S-shaped value function that is concave in the positive range and convex in the negative range (disposition effect).

The *availability heuristic* is mentioned as an example of further cognitive biases that can impair creativity (Funke 2003, p. 101). In this case, uncertainty is not reduced by collecting, processing and interpreting reliable information, which can be explained by a lack of time, opportunities or willingness to put in the effort. Instead, a simple path is chosen, which may consist of using suitable examples or memories (*anecdotal evidence*). One example of this is the visualization of data, which makes the consumption of the associated statements easier than the comparatively laborious and dry evaluation of data sets. However, ease of availability is not an indication of the reliability of the statements. A suggestion evokes matching associations, which then influence behavior and action. This is also referred to as priming, in which the processing of a stimulus depends on what memory content a previous stimulus has activated. In practice, event organizers, for example, use this effect to give an event the desired direction from the outset. However, if events are supposed to produce creative results, such anchoring is counterproductive. In fact, certain creativity techniques actively counteract the anchoring heuristic by, for example, randomly displaying images at the beginning of a corresponding event that are intended to evoke a very broad spectrum of associations.

In any case, *associations* are not an optimal basis for creative work, since they promote unconscious associations that can stand in the way of making new associations. One example is *Pavlov's dog* with regard to conditioning, the unconditional reactions that are triggered by both an unconditional and a neutral stimulus can be observed after a certain time even if only the neutral stimulus precedes it. Therefore, in the further course of the concept of *bisociation* derived from the concept of association is used (Sect. 4.4.5.1), which, in the context of creativity, better describes how it can be promoted. Closely related to the anchoring heuristic is the so-called *framing*, which, however, is to be classified as a social heuristic and is therefore explained in Sect. 4.4.4.2.

The last example is the well-known *selective perception*. It not only ensures that perceptions of physical needs (if you are thirsty, you pay more attention to the coffee pot at the other end of the meeting table than to the topic under discussion) or of immediate sensory impressions (if the sun is shining in your face, you are more likely to look for shade than for the conversation) are directed. It also reinforces existing perceptions, for example by seeking further evidence of a tense atmosphere in the room in an attempt to confirm the initial impression. And anyone reading this book who does not consciously and successfully suppress their selective perception will find themselves confirmed in what they have always believed and more likely to question what is not compatible with this framework.

Social-Psychological Perspective

Social psychology investigates the interactions between the individual and the collective and derives principles from them. Of the mental abilities and personality traits of creative people identified by Gregory John Feist (1998), their interactions with other people are influenced by (1) their autonomy, their (2) introversion, their (4) doubting social norms, (5) self-confidence, (6) self-acceptance and (9) dominance, (10) hostility and (11) impulsivity (spontaneity).

To the extent that it is about the behavior and actions of a person (psychological system) in a social system (collective), for example in a department of a company or in a project team, the person is only relevant as a person in the sense explained above. The mental abilities and personality traits are no longer tied to the person, whose true face remains hidden behind the mask from third parties. What defines the person are abilities and *characteristics* that are constructed through communication and are thus largely *fictitious*, attributed to him by the other members of the collective (Luhmann 2011, p. 89). These characteristics can result from expectations, as expressed in practice, for example, in employment contracts and job advertisements, or from observations, the insights of which are expressed and discussed, for example, in performance reviews.

(4) *Acceptance of social norms*, for example, describes a person's relationship to the social system in which the norms apply. One of the characteristics of creative individuals is that they tend to question such norms and do not consistently adhere to them. As a result, they are often what is commonly referred to as *outsiders*, i.e., as those who, although they

belong to the social system in a certain way, are not fully integrated. This results in a field of tension that often leads to conflicts. This correlates with the personality trait of *introversion*, which is often observed in creative people. This refers to the orientation of attention to one's inner life and a certain passivity within the group (Jung 2001b, p. 71 ff.). In this respect, Gregory John Feist (1998, p. 300) summarizes that creative people have a relatively asocial or even anti-social attitude and explains this by the fact that they need to be able to spend time alone. For example, you cannot write a novel, compose a symphony or paint a picture when you are in company. Although inspiration may come from the social environment, the creative process itself usually takes place in solitude – one reason why, for example, university laws provide for sabbaticals during which professors are released from teaching and administrative duties for the purpose of research. This is also common in companies under the term "sabbaticals".

These circumstances also give rise to the (10) *hostility* often attributed to creative individuals. It follows from the experience that new ideas usually have to be defended against others, because novelty also means otherness and thus a deviation from the norm, so that rejection by the collective is almost inevitably associated with the maintenance of the norm. This corresponds to the behavior of a control loop (Sect. 3.4).

Sticking to one's own ideas (commitment) requires a pronounced (5) *self-confidence* in order to compensate for the collective, not always benevolent, external perception. This also means that creative people have to accept themselves as they are, despite the personal criticism to which they are often exposed. However, this being as one is is in the field of tension between human and person, because while the individual perceives himself both as a human and as a person, the collective's perception is limited to the person, whereby the latter always also includes the roles that are assigned to the person in the collective. Therefore, creative people must be able to accept that they are reduced to one person and to the roles assigned to them in the collective, while their potential and thus, in this context, their creative potential in particular, goes significantly further.

This can result in a certain (9) *dominance* of creative individuals within the group, because, as is well known, exercising power means being able to persuade others to do or refrain from doing something that they would not otherwise do or refrain from doing voluntarily. This makes it easier to implement one's own ideas where objective persuasiveness is not enough: Remaining contingency can be absorbed by decisions (Sect. 2.6.2.3).

Finally, creative individuals often act (11) *impulsively* (spontaneously) by deviating from plans, not always adhering to principles or disregarding rules. Like many of the personality traits mentioned earlier, this is not always socially acceptable either, since such behavior and actions are unpredictable for others, thus generating uncertainty (contingency) that must be absorbed.

In summary, from a social-psychological point of view, it follows from the sum of all personality traits and mental abilities associated with creativity that the integration of creativity into problem solving must, above all, build bridges between the behavior and actions of creative (unconventional) individuals (Feist 1998, p. 300) and the conventional collective, which are perceived by the collective as asocial or even anti-social in the value-free sense of this word. The methods (creativity techniques) explained below contribute to this. Above all, however, it is also important that all those involved and affected, including in particular those with personnel responsibility, are aware of the tension between creativity and convention and the resulting potential for conflict between those who represent one and those who represent the other, and are willing and able to deal with it. In particular, this requires tolerance and mutual *respect*.

Furthermore, the relationship between creative individuals and the collective to which they belong makes it clear that the often-touted *teamwork* is not helpful in every respect when it comes to developing and implementing creative ideas. It can succeed when the team members are familiar with each other, because this opens up the chance for introverted individuals to become active in the group. In a less familiar group, which in practice can happen, for example, when teams are put together on a case-by-case basis or superiors are involved, creative individuals are unlikely to express themselves. Creative work in a familiar group, on the other hand, can be hampered by the fact that such teams are particularly concerned with conformity, so that anything that jeopardizes this is absorbed. This can especially include new ideas if they affect the team itself. Against this background, it is necessary to also enable, recognize and promote *individual achievements* (Ehrlenspiel and Meerkamm 2013, p. 433 f.). As will be shown, creativity techniques take this insight into account in different ways. In addition, however, it also depends in this respect on the corresponding social competence of all those involved and affected, and here in particular those with personnel responsibility.

Sociological Perspective

Due to the emergence between the micro and macro levels (Fig. 2.1), the principles that apply at the macro level under consideration here and the recommendations for action that can be derived from them do not arise directly from what has been said so far about creative people or individuals. For the organization of a social system, this means that only conditions under which creativity can flourish can be created. Whether and how it actually develops remains to be seen. At least some factors that have proven effective in promoting creativity at the macroscopic level are now known. For example, it has been shown that high *pressure to succeed* (stress, pressure to perform), as caused, for example, by the incentive systems (bonuses, bonus payments, etc.) that are widespread in many companies and increasingly also in science (Kremer 2022), can inhibit creativity (Poriger and Brücklmeier 2022, p. 80). However, the associated *performance evaluation* does not have consistently negative effects. In fact, it has been found that fear of negative evaluation at a low level can be quite beneficial and only has a counterproductive effect at a high level (Byron et al. 2010, p. 201 ff.). Furthermore, a high degree of *autonomy* promotes creativity (Hansen 1999, p. 82 ff.). One example of this is the so-called *start-up companies*, which, in addition to their extensive independence, are often small and can therefore act quickly and flexibly (unconventionally, unbureaucratically). Furthermore, they focus on the realization of an idea. Due to the high risk associated with this, they do not finance themselves through the conventional capital market. Instead, they are supported by venture capitalists, business angels or crowdfunding. They are often located in business incubators that are subsidized by the state or by companies. Even established companies sometimes found start-ups when they want to bring certain ideas to market quickly while limiting the associated risk (time to market). Well-known examples of highly successful start-ups are companies like Hewlett-Packard, Apple and Google, which were founded in garages and have since grown into global corporations.

The *environment* in which creative people work also has an influence, and *materials* can be used to support them. From a sociological perspective, however, this insight is not entirely self-evident, because objects are rarely included in the social sciences and humanities (Sect. 3.3.1). Instead, their explanation is the subject of the natural and engineering sciences, while the humanities and social sciences try to understand subjects. However, the subject-object dichotomy can be mitigated if one considers that the elements of social systems do not, as is often assumed, consist of people or persons (subjects), but of acts of communication (Sect. 2.6.2.2). This leads to the insight already explained in Section 3.5.5 that objects also communicate (Latour 2005, p. 37 ff.). James Jerome Gibson (1973) has also found that objects have an affording character: they can encourage behavior and action. Donald Arthur Norman (2013) has applied this insight to everyday objects, among other things. This also includes spaces and their furnishings, as well as materials that are provided in the spaces for viewing (catalysts) or for handling (converters) (Sect. 3.5.5). Findings such as these provide evidence that creativity is also promoted by a stimulating spatial environment (Fig. 4.42) and by the provision of materials (Kohlert and Cooper 2018).

Fig. 4.42 Creative space at the ZAL Center for Applied Aviation Research in Hamburg, Copyright: ZAL DReinhardt

As a rule, companies' strategies provide for innovation within their own organization (*closed innovation*) in order to gain an advantage over competitors by developing *Unique*

Selling Points. To do this, it is necessary to recruit employees from the labor market who have a high level of creative potential (Sect. 4.4.4.2). Furthermore, intensive *market research* must be carried out (e.g. Olbrich 2022, p. 19 ff.). The corresponding capital must be available for the pre-financing of the innovation, and the associated risk must also be borne by the company alone. Finally, innovations must be protected. For this purpose, *industrial property* rights such as patent law (Hahnl 2015), trademark and design law (DesignG), semiconductor law (HalblSchG), plant variety protection law (SortSchG) and unfair competition law are available, whereby the latter is also part of competition law and in particular includes the *law against unfair competition* (UWG). And copyright law (UrhG) also offers a certain degree of protection, but it applies more to artistic than to commercial creations (Dietrich and Meitinger 2021, p. 111 ff.).

Alternatively or in addition, an open innovation strategy (*open innovation*) is also possible (Chesbrough 2006). Instead of relying only on their own employees, companies also involve people or companies from outside. One example of the integration of external people is the *lead-user method* (Hippel 1986, p. 791 ff.). This involves trend-setting users in creative processes. These users often expect to derive great benefit from the targeted innovations themselves, and they contribute their own ideas and possibly their own expertise. These users can often be found in Internet forums (communities). Closely related to this are so-called *prosumers* (Toffler 1997). This portmanteau word, composed of the terms producer and consumer, is used to describe consumers who also produce a consumed product. A vivid example is the owner of a house who produces electricity for his own needs with a photovoltaic system and feeds surplus electricity into the public grid. Interested operators of such systems can also be involved in research projects, for example, that aim to advance the networking between decentralized energy sources and energy sinks (Holstenkamp and Radtke 2018). A comparable approach is called *co-creation* (Kurzhals et al. 2022). *Open-source software* is often mentioned as a practical example in this context.

In addition to individuals, organizations can also be involved in their own innovation processes. This includes, for example, cooperation with a university or college or with research and development service providers such as the Fraunhofer Society for the Promotion of Applied Research, the Max Planck Society for the Advancement of Science (MPG), the Helmholtz Association of German Research Centers or the Gottfried Wilhelm Leibniz Science Association. In view of the low level of in-house production in many companies and the large proportion of value added by suppliers, innovative suppliers are also involved in research and development projects.

While the examples of an open innovation strategy just mentioned can be categorized as *outside-in concepts*, *inside-out concepts* involve third parties in innovations that were initially developed in-house. This can be useful, for example, if the developed products or services do not fit into the company's own portfolio or if the company's own marketing opportunities, for example in foreign or distant markets, or its own production capacities are limited. The granting of licenses is particularly suitable for this purpose.

The advantages of closed innovation are usually the disadvantages of open innovation and vice versa. For example, organizing open innovation processes is often much more expensive due to the need to involve external individuals or organizations. The risk of partners dropping out is greater due to the weaker ties to the innovation community, and, above all, it is more difficult to legally protect innovations. For example, the authorship of contributions cannot always be clearly attributed. Furthermore, the weaker ties between external partners make it more difficult to ensure confidentiality vis-à-vis uninvolved parties. Accordingly, legally secure protection of innovations is more difficult. If, for example, publicly funded universities or colleges are involved, or if research projects are funded by public institutions such as the German Research Foundation (DFG), research results must be published. This runs counter to the desire to gain unique selling points. This stands in the way of the opportunity for open innovation processes to tap significantly more creative potential under certain circumstances and to market the results more successfully.

Just as on the psychological level, there are also heuristics on the social level that stand in the way of creativity. These include, for example, *framing*, which is related to the psychologically relevant anchoring heuristic (Sect. 4.4.4.2), and which aims to anchor *narratives* in social systems. This establishes certain collective patterns of thought and interpretation, which can serve to stabilize a social system, but which hinder creative breaking out of the usual framework. Closely related to this is *agenda setting*, which deals with the question of why and how certain topics (problems) enter the collective consciousness and not others that may be just as important or even more important. This is opposed by the *spiral of silence*, which causes people to preferentially comment on topics that are generally desired, and to do so in a generally expected way (opportunism).

The basic considerations of creativity from a psychological, a socio-psychological and a sociological perspective form the starting point for the specific methods (often also referred to as creativity techniques), which take up the aspects explained so far in different ways and translate them into specific measures for their application. The specific methods are explained in Section 4.4.5.

4.4.4.3 Optimization

Those who expect to reap the benefits of applying Systems Engineering usually face stiff competition. There are two mottos for surviving in the competitive arena, both of which claim to be the *motto of the Olympic Games*. Officially, they are: citius, altius, fortius – *faster, higher, stronger*. The best is just good enough. The winner takes it all, you could also say. But does the winner really get everything and everyone else gets nothing? Even in sports competitions like the Olympics, medals are awarded not only to the best but also to the second- and third-best. And just about everyone who has managed to make it to the Games, as well as everyone who has tried and failed to make it, will have experienced the unofficial motto more accurately: *being there is everything*. This corresponds to the principle of *frugal innovation*(Sect. 4.2.4.4), which states that it is sufficient for something to be *good enough* (Weyrauch 2018, p. 13). When it comes to products and production processes, being good enough is often not enough to survive the competition, and it is often too expensive and not infrequently harmful to the natural environment. And organizations usually do not choose the best possible organizational form (structural, process, etc.), but are satisfied with workable solutions (Simon 2015, p. 32).

These considerations should at least somewhat relativize the goal of design projects to always achieve an optimum, because in fact, if one chooses an etymological derivation, the best achievable result is meant. After all, an optimum is not an ideal, i.e. not the best conceivable solution, due to the condition of achievability. But even achievability appears to be an unsuitable yardstick when it comes to not being better than good enough. Therefore, optimization in business and technology is usually based on a specific *goal* (Sect. 4.3) and is subject to defined (auxiliary/marginal) *conditions*. So the goal (requirements) is the yardstick by which the optimality of a solution is to be evaluated. Achieving more than the set goal is not required – but neither is achieving less.

Mathematics offers a particularly reliable way of determining an optimum, and in relation to typical technical and economic problems, in particular *Operations Research* (OR) (Dittes 2015; Domschke et al. 2015; Grimme and Bossek 2018; Nickel et al. 2022; Papageorgiou et al. 2015; Werners 2008). *Quantitative methods* allow an *exact formulation* of the objective function and the constraints. If the function has an extreme value, i.e. a maximum or a minimum, it is possible to *calculate* this value and thus an *absolute optimum* using appropriate methods (e.g. differentiating the function and setting it to zero). This means that the results can be *reproduced* at any time and in any place by anyone who is familiar with the problem and has mastered the calculation method. Furthermore, the validity of statements can be *proven* in mathematics.

However, the practical applicability of mathematical methods often reaches its limits. For example, numerous optimization problems cannot be solved efficiently at all. This means that on real computers they require such long calculation times that no results can be expected within a person's lifetime. Nevertheless, some problems of this kind can be solved with the help of *approximations*. For other problems, there are substitute *heuristics* such as trial and error, *exclusion methods* or *statistical evaluations* of randomly selected samples. Neither approximations nor heuristics provide exact solutions, but they are usually sufficient for practical applications.

Another problem with mathematical optimization methods is that they can often only take a few parameters into account. For example, methods for calculating the shortest route, such as the *Dijkstra algorithm* (Dijkstra 1959, p. 269 ff.), which is implemented in navigation systems and Internet routers, among other things, only allow for one parameter, which can be interpreted either as distance, as utilization or the like (Werners 2008, p. 197 ff.)[27].

If many parameters are to be included in the solution, *qualitative optimization* methods can be used. With their help, you will find not an absolute, but at least a *relative optimum*. In this context, relative means a solution that is the best in comparison (in relation) to other solutions, but this does not mean that there are no better solutions. In this case, the optimization problems are not formulated in the mathematical sense in the form of objective functions and constraints and can therefore neither be calculated nor proven. If calculations are carried out, they are, for example, comparative calculations as they occur in a utility analysis (Kühnapfel 2021). This in turn means that the optimality of a solution is also not reproducible. These are disadvantages that may be important primarily in science. In practice, however, a major advantage of qualitative optimization methods is that their

[27] In his famous novel "The Hitchhiker's Guide to the Galaxy", Douglas Noël Adams (1979) parodied the limits of mathematical optimization with the answer "42", which the supercomputer "Deep Thought" calculated after several million years in response to the question "about life, the universe and everything else".

solutions are usually much closer to reality than many solutions obtained using quantitative methods.

In Section 4.4.5, typical optimization problems are explained by way of example in order to demonstrate their basic functioning as well as their respective possibilities and limitations. In fact, the number of available optimization methods is very large, so that completeness cannot be guaranteed at this point. For more information, please refer to the relevant literature (e.g., Dittes 2015; Domschke et al. 2015; Grimme and Bossek 2018; Papageorgiou et al. 2015; Werners 2008).

4.4.5 P 4.5: Concrete Methods

The core of the methodical phase P 4 of the problem-solving cycle (Fig. 3.23) is the concrete selection of methods and corresponding tools that appear suitable in the light of the situation analysis P 2 and, in particular, the objective P 3 to support the work in the subsequent phase P 5 of the actual problem solving as well as the final phase P 6 of the solution evaluation. The cyclical approach makes it clear, however, that it is not possible to ensure in the methodical phase P4 that the selected methods and tools will actually prove their worth in the further course. If and to the extent that this should turn out to be the case, a return to the methodical phase P4 is necessary. However, experience from comparable projects is often available, which can provide a certain degree of planning security.

Before explaining the selected methods, it should be noted that, given the great variety, no completeness can be offered, whether with regard to the methods and tools themselves or to the description of the associated procedures. In this regard, the relevant literature should be referred to once more, for which numerous, but by no means all, sources are given, because there is also a great variety here.

4.4.5.1 Creativity Techniques

The term "creativity techniques" is so widely used to describe methods for supporting creative work that it should be retained here, although the term "method" is actually preferred, because a system that is followed through to its logical conclusion carries the risk of resulting in methodism. However, a brief discussion of the term in the context of what is striven for in Systems Engineering seems necessary, because it is not just creativity as such in the sense of a person's ability to produce something new (synonymous: original) to which a value is attached (Sect. 4.4.4.2). In fact, it is about more than that, namely *innovations*. These require a targeted generation of something new, almost on command, that is, in the course of a specific design project at a specific time and place with a clearly defined goal (Sect. 4.3). In this context, creativity techniques must achieve more than just support creative work; above all, they must do so under the circumstances just mentioned. Therefore, a designation as innovation techniques would be more appropriate under certain circumstances. Nevertheless, in order to maintain compatibility with current practice, the term "creativity techniques" will be retained here.

Many creativity techniques have in common that they are intended to support *groups*[28], even though some of them can also be used by individuals. The creativity blockades that can arise from a collective have already been addressed in Section 4.4.4.2. However, the advantage of creative work in groups is often countered by the fact that the more people involved in the idea generation, the greater the number and diversity of ideas. To maximize this opportunity, many creativity techniques include measures to limit the risks associated with working in a collective.

- These include, for example, the often-used limitation on *group size*. Typical information on this ranges from 5 to 12 participants (Böhm 2005, p. 185; Ehrlenspiel and Meerkamm 2013, p. 433).
- Furthermore, it is usually recommended that the groups be composed of members *from different fields* (interdisciplinary) and with a *homogeneous hierarchy* (Ehrlenspiel and Meerkamm 2013, p. 433). This means that the members of the group should have different qualifications and experiences so that they can mutually inspire each other more than would be expected from members who all have the same prerequisites. Hierarchical homogeneity is intended to ensure that, for example, the boss and the employee do not meet, as this can lead to blockades on both sides. The boss might be afraid of making himself look ridiculous by expressing ideas that may prove to be unproductive in the further course of the process, while employees might be afraid of falling out of favor by expressing certain ideas.
- The presence Analogy (isomorphies) are referred to as intuitive. Emotions are also addressed. Analytical methods, on the other hand, support the conscious decomposition and structuring of the problem to be solved with the help of the models explained in Chapter 2.

Intuitive Methods

Brainstorming (Clark 1958; Sell and Schimweg 2002, p. 77 f.) is probably one of the best-known intuitive creativity techniques. However, it is often understood only as spontaneous reflection or discussion about problem solving and then usually does not correspond to methodical work in the sense intended here.

The core idea is to hold an event in which a group and its moderator generate as many solution ideas as possible in a relaxed conversation, i.e. one that is only slightly guided by the moderator, on a given problem in a period of maximum 30 (Böhm 2005, p. 187) to 40 minutes (Andler 2015, p. 187). The fact that there is often confusion during the conversation and that a certain volume is unavoidable is consciously accepted, although the noise level should not exceed the trigger value of 80 decibels (dB) according to the Noise and Vibration Occupational Safety and Health Ordinance (LärmVibrationsArbSchV), if possible.

To ensure that the event delivers the desired results, both long-term and immediate preparation is required. The conclusion is a follow-up. The steps are explained in turn

[28] Groups are distinguished from teams here, with teams being understood as groups that have been put together over a longer period of time, characterized by cooperative collaboration and collective assumption of responsibility.

below, although the order may vary. For example, it is irrelevant whether the group is put together first and then a suitable room is selected or vice versa. Preparation includes defining the problem for which solutions are to be sought. Brainstorming is particularly suitable for clearly defined and not too complex problems (Ehrlenspiel and Meerkamm 2013, p. 433). For this purpose, the problem can also be formulated in the form of questions, for example, "How should product A be packaged?" (ibid., p. 434).

Depending on the problem definition, the group is to be put together, whereby the psychological principles explained in Section 4.4.4.2 and the conditions mentioned in Section 4.4.5.1, in particular with regard to technical heterogeneity and hierarchical homogeneity, are to be taken into account. In addition, organizational aspects such as temporal and spatial availability, the invitation, if necessary, accommodation and care before and after the event are to be clarified.

In particular, the location and especially the room where the event is to take place must be selected. Above all, the socio-psychological and sociological principles explained in Sections 4.4.4.2and 4.4.4.2 must be taken into account. For the room, for example, this means that it should be designed to stimulate creativity (Fig. 4.42). It should also protect against disturbances, which also includes the accessible sanitary facilities, break rooms and the route to the event location. Furthermore, food and drinks must be provided depending on the time of day and duration of the event (catering). To avoid distractions and situations that could lead to a bad mood, the participants must be able to feel comfortable.

It is often recommended to equip the room with tools that support the sketching of ideas or the moderation (Andler 2015, p. 186; Böhm 2005, p. 187; Ehrlenspiel and Meerkamm 2013, p. 433). This may include, for example, a flip chart, pin boards including cards and pens, or a blackboard. Care should be taken when using technically sophisticated aids such as so-called smart boards, as not all participants may be able to use them effortlessly and they may also cause compatibility problems when connected to other electronic devices such as smartphones or tablet PCs. In this regard, the frugal principle applies: the simpler, the better. Exceptions may be groups that are made up of particularly tech-savvy members and are therefore more generous than others at overlooking technical inadequacies or quickly solving problems independently.

With the timely invitation, the participants receive information that is as clear and sufficiently detailed as possible about the event and the problem, the expected contributions, the course of the event and, of course, all formalities such as time and place, and, if necessary, travel, accommodation and the settlement of expenses and fees. The participants should also be informed why they were selected and what role they are expected to play. Finally, it may be necessary to include an explanation of the method with the invitation. The expectation associated with explaining the problem early on is that the invitees will deal with it before the actual event and then probably develop ideas. However, this carries the risk that the participants will commit themselves to some extent and then no longer bring the openness to the event that is necessary for free, creative work (Böhm 2005, p. 187).

Finally, it must be clarified by mutual agreement whether the participants should have copyright for the ideas they contribute. The advantage of such a regulation may be that the willingness of the idea provider to contribute ideas with which they associate their own exploitation interests increases. This must then be ensured in the course of recording. A particular challenge here is to document authorship even in cases where original ideas are taken up and further developed by other participants. In this respect, the brainstorming method of brainwriting (also known as the 6-3-5 method), which is very similar to brainstorming, offers advantages that will be explained later. In addition to copyright, the right of use must also be clarified.

On the day of the event, the reception of the participants must be ensured. This includes an introduction by the moderator, including an opportunity to ask questions. After the introduction, it can be particularly useful in the event that the participants do not know each other, to enable a detailed mutual acquaintance. A simple, short round of introductions is usually not enough, because for a relaxed atmosphere, the participants have to reduce possible inhibitions and build trust.

From a systems theory perspective, this can be interpreted to mean that a *social system* should arise. This requires the development of a sense of community within the group, which should see itself as a unit. Since the elements of social systems consist of acts of *communication* (Sect. 2.6.2.2), the system is formed through communication. The problem to be solved by the contributions provides a suitable crystallization nucleus. Furthermore, the formation of a *system boundary*, i.e. the demarcation from the environment, is of particular importance. A common enemy (Bion 2018), whose existence holds a group together, can be, for example, a competitor that one intends to outbid with the desired solution. Finally, the members must be given the opportunity to find their respective roles within the group and to get to know the function of the moderator. All this must be made possible before the actual brainstorming session.

Simply providing space and time and offering opportunities is often not enough. Rather, it is usually beneficial to actively facilitate the system formation process. However, to do this, the moderator should have the appropriate qualifications and experience, especially in the areas mentioned in Section 2.6 (Böhm 2005, p. 188). Getting to know each other begins with a so-called *icebreaker*, which serves to overcome initial inhibitions. This includes, for example, the question of how previously unknown participants address each other, whether with or without academic titles, whether with first or last names, and so on. Caution is advised when it comes to a possibly hasty and sweeping suggestion of addressing each other by first name, even though this can help to create a sense of closeness in many situations. Particularly in view of the increasing bad habit of changing forms of address on a large scale to the informal "du" without the consent of those affected and often even against their explicitly stated will, a countermovement can be observed that perceives this as disrespectful and insists more than ever on the "Sie" form of address customary among strangers in German-speaking countries. Name tags are also often perceived as helpful, although they are only useful if they can still be read from a distance.

After the initial meeting, it may be useful to encourage and consolidate the formation of groups through further measures. *iCooperation games* can be used for this, which at the same time contribute to the desired relaxed atmosphere. However, many of the well-known games involve dividing the group into subgroups. If these remain in place for the rest of the event, it can be a hindrance for brainstorming, since everyone should work together in such a session. An example of a well-known game that appeals to the group as a whole is the magic carpet. This requires a sturdy tarp, which is available in hardware stores, for example. The more acceptance of physical proximity can be assumed among the participants, the smaller the tarp can be. In any case, however, there must be enough space for everyone. The level of difficulty can be increased if additional objects have to be placed on the tarp. These can be, for example, cardboard boxes. For clear identification of the individual objects, they should be marked, for example with colors, numbers or labels. Another way to increase the difficulty level is to prohibit verbal communication so that the participants can only communicate with each other non-verbally. The goal of the game is to turn the tarpaulins, whereby no participant may step next to the tarpaulin at any time. If necessary, the additional items may not be placed next to the tarpaulin. This promotes goal orientation, communication and collaboration.

As soon as the group has grown together into a social system, the preparation of the actual brainstorming session begins. To do this, the moderator first explains the problem to be solved and then the rules of the discussion (Andler 2015, p. 187; Böhm 2005, p. 186; Ehrlenspiel and Meerkamm 2013, p. 433).

One of the rules is that ideas should*not be commented* on and certainly *not criticized*. This must not be done non-verbally either, in particular through facial expressions, gestures or, in general, through body language. Since such reactions normally occur unconsciously, it helps to be aware of this fact. Nevertheless, the risk that such reactions may occur during the brainstorming session cannot be completely avoided. In such cases, it is the moderator's job to gently, and perhaps humorously, remind every one of the rule. The situation is similar with what Charles Hutchison Clark (1958, p. 91 f.) called *killer phrases*. Such phrases are also used more or less unconsciously to ward off ideas and should be consciously addressed before the actual brainstorming session. The following examples can serve as illustrations:

- We've never done it this way...
- It won't work...
- We don't have the time...
- We don't have the manpower...
- It's not in the budget...
- We've tried this before...
- We're not ready for this...
- Theoretically fine, but can you make it work in practice?
- Too academic...
- What will the customers think?
- If it was any good, someone would have suggested it already...
- Too modern...
- Too old-fashioned...
- Let's discuss this another time...
- You don't understand our problem...
- We're too small for that...
- We're too big for it...
- We have too many projects now...
- Let's do a market research test first.
- It's been the same for twenty years, so it must be good...
- Don't move too fast...
- Why start something new now? Our sales are still going up...
- Let's see what happens...

- The union will scream...
- There they go again...
- Let's put it in writing...
- I don't see the connection...
- It won't work in our industry...
- We can't do it according to the rules...
- Crazy...
- Political dynamite...
- Sounds good, but I don't think it will work...
- It's not in the plan...
- No regulations that cover it...
- We've never used this approach before...
- It's not in the manual...
- It will mean more work...
- We're not responsible for it...
- Yes, but...
- It will increase the cost...
- It's too early...
- It's too late...
- It will offend...
- It won't work...
- Our people won't accept it...
- You don't understand the problem...
- No teenager is going to tell me how to run my business...

Furthermore, Charles Hutchison Clark (1958, p. 93 f.) also warns against what Edward J. Walsh of General Foods Corporation has called *soliloquized killer phrases*. These are used to discredit one's own ideas:

- May not be true, but...
- We've only done a few preliminary tests...
- This may not work, but...
- This approach is nonsensical, but...
- It's not clear that we need this, but...
- I don't know if the money can be approved, but...
- It might be a dead end, but...
- Would it hurt if we...
- Do you think it would be possible...
- This may sound far-fetched, but...
- It could take a long time, but...
- I don't know exactly what you want, but...
- You probably have an idea, but...
- You won't like this, but...
- This is against the rules, but...
- It may not be the right time, but...
- This idea seems useless, but...

- You're probably better at this, but...
- If I were younger and had my health...
- I assume our competitors have already tried this, but...
- I don't know that much about it, but...
- I don't know what the literature says, but...
- This isn't exactly on topic, but...
- I haven't thought about it yet, but...
- You'll probably laugh, but...
- My opinion isn't worth much, but...
- I'm no genius, but...
- Maybe we can't sell this to the old man, but...
- I'm not excited about this idea myself, but...
- It may not be important, but...
- This needs further study, but...
- If you want to take a newbie's suggestion...
- I'm not aware of the complexity of the issue, but...
- Joe doesn't agree with me, but...
- I know this doesn't solve the problem, but...
- If I'm way off base, correct me, but...
- Here is a rough idea of what I have in mind, so that you can poke holes in it...

A second rule is that the aim is not to contribute well-thought-out ideas or even mature concepts. What is desired is a *free play of ideas* (Andler 2015, p. 186; Böhm 2005, p. 186; Ehrlenspiel and Meerkamm 2013, p. 433). Unusual, imaginative ideas are also in demand. Suggestions for this can already be made in the course of preparing for the actual brainstorming session. For example, a bag of money can be symbolically positioned in a clearly visible place for everyone to see, to suggest that money is no object. A slideshow that shows a series of unrelated images, for example of plants, animals, landscapes, objects and people in different, preferably pleasant, situations, has also proven useful in encouraging thoughts to wander (bisociation). It is not necessary to relate this to the specific problem.

A third rule takes up the second one to the extent that the *quality* of suggestions is not decisive, and emphasizes once again that it is their quantity that matters instead. To come up with a further-reaching idea, you may need 100 ideas that are not implemented themselves, but serve as inspiration for the one idea that is ultimately pursued (ibid.).

The fourth rule is that *ideas from others should be taken up and developed*, i.e. used as suggestions.

During the actual brainstorming session, the rules should be applied. It is to be expected that the number of ideas expressed per unit of time will initially increase sharply after a short start-up phase and reach an initial maximum after a certain period of time. After that, the number of ideas per unit of time will decrease again. However, it would be wrong to end the session at this point, because conventional ideas are usually expressed in the first phase. Picking up on and developing ideas that have already been presented only picks up speed in the second phase and finally reaches a second maximum of ideas per time unit. Although this second maximum is usually not as large as the first, it is during this phase that the more imaginative ideas are to be expected, which are often particularly attractive in terms of the hoped-for innovation. After reaching the second maximum, the number of ideas per unit of time then decreases again and does not strive for any further maximum, so that the session can be ended. How exactly the outlined time frame develops is usually not foreseeable, so it makes sense to keep the end of the event open within the time frame given above and to make it dependent on the specific development.

To support the discussion, any *tools* provided, such as a flip chart, pinboard including cards, or blackboard, can be used to sketch out ideas to clarify them, for example. In addition, the tools can also support moderation and logging. However, *written minutes* should be prepared in any case, but not by the moderator and not by one of the participants, so that they can concentrate on their actual roles.

Since the aim is to generate as many ideas as possible, participants should be given the opportunity to submit ideas even after the brainstorming session has ended. The next step is the *follow-up*, which can be carried out with the participation of the session participants or without them, by the organizers alone. The ideas are organized and evaluated, whereby the following criteria, for example, can be used as a basis (Böhm 2005, p. 188): ideas that can be implemented immediately and without significant effort (quick wins) can be distinguished from ideas whose feasibility must first be tested or that probably still need to be modified. Finally, ideas can be identified that, on closer inspection, are not suitable for solving the underlying problem, but may be useful elsewhere (idea storage).

In any case, the results should also be made available to all participants. However, it is important to clarify whether and, if so, to what extent their *confidentiality* should be ensured.

There are a number of variants to brainstorming in the form just explained. One of them is the already mentioned *brainwriting*, to which the sub-variant known as *method 6-3-5* (Andler 2015, p. 189 f.; Böhm 2005, p. 196 ff.; Ehrlenspiel and Meerkamm 2013, p. 435; Sell and Schimweg 2002, p. 78) is explained. The latter stands for the fact that six participants write down three ideas in each round and pass on the form on which the ideas are noted a total of five times, so that in the end each participant has worked on each form once. A processing time of 5 minutes is usually allowed per round. The advantages over brainstorming are that no individual participant can dominate the discussion and that the contributions are better documented, which makes it easier to protect copyrights. Above all, however, mutual inspiration is better supported because, apart from the first round, the forms already contain ideas that can be built on. Finally, the demands on the moderator during the session are significantly lower. However, compared to brainstorming, mutual inspiration has its limits due to the lack of discussion.

Another widespread sub-variation of brainwriting is the *Metaplan technique* developed by the Quickborn team (Böhm 2005, p. 198 f.). For this, several pinboards covered with interchangeable wrapping paper measuring 1,100 × 1,400 mm^2 are set up freely in the room. A moderator's case contains cards made of sturdy paper in the format 210 × 100 mm^2, felt-tip pens, pin needles, a needle cushion, and additional cards in oval or cloud form that can be used for structuring. The participants write their ideas on a card each. After the allotted processing time of about 10 minutes, the cards are collected by the moderator and initially attached to some of the pinboards in a disorderly or only loosely organized manner, so that all participants get an overview of the sum of all ideas. In doing so, identical or very similar ideas can already be summarized. If anything is unclear, those who have noted the corresponding idea can be asked. In the next step, the ideas are structured on other, still empty pinboards. The models explained in Section 2.4 (e.g. Fig. 2.6 to 2.17, 2.26) are suitable for this, as are tabular representations or those in the form of mind maps, which, although often treated as a treated as a separate creativity technique (e.g. Andler 2015, p. 124 f.), but are actually only a modified form of the cause-and-effect diagram (Fig. 4.11). In the structuring phase, clusters of similar ideas can be given headings. Furthermore, it is also possible to submit ideas at a later stage in this phase. Finally, the ideas can be evaluated. This can be done, for example, by giving each participant a quota of points, for example in the form of self-adhesive dots, which they can stick on the ideas they prefer. The ideas that receive the most points together form the result of the solution finding process.

Another class of intuitive methods aims primarily to promote bisociation and analogy formation (Sell and Schimweg 2002, p. 79 ff.). In reference to association, which denotes a mental connection on one level, Arthur Köstler (1966) understands *bisociation* to mean a mental connection between ideas, images and concepts between different levels that are not normally related. This can evoke humor, a discovery or understanding, whereby the discovery is particularly relevant here. Creativity techniques based on this initially lead away from the problem that is actually to be solved into a different context, whereby the thought path fixed on the problem is left. The thinking, which has been freed to a certain extent as a result, can be inspired by the other context, in order to finally return to the problem level with new approaches to a solution.

To stimulate bisociation, images that are unrelated to the actual problem are often used, as recommended above for preparing brainstorming sessions. The thoughts that the participants associate with the images are collected and documented by a moderator, for example using the Metaplan technique. After the collection, the thoughts are linked back to the actual problem.

Analogies (isomorphisms) can be used in a similar way. An example of a technical use of an analogy is the Network Information Service (NIS) in a UNIX network. The company Sun Microsystems, which has since been taken over by Oracle Corporation, originally called this service Yellow Pages (YP) because certain services for finding other computers in a distributed computer network are comparable to the Yellow Pages, where you could find phone numbers and addresses of companies at the time. Although Sun was forced to change the name to NIS after British Telecom successfully asserted trademark rights to the name YP, the names of NIS commands still usually begin with YP.

Semantic intuition (Schlicksupp 2004) is one of the specific creativity techniques that rely on bisociation. In this technique, instead of first developing a product and then looking for a name for it, a word is formed from various terms, and this is used to deduce what a corresponding product might look like. As the name suggests, one of the creativity techniques that primarily uses analogies is the analogy technique. After that, one aspect of the problem to be solved is selected. Subsequently, analogies are sought for it, in order to select one from which it can be observed how the comparable problem is solved there. This solution is then transferred to the original problem.

This approach, which is basically a transformation from one area to another in which the problem can be solved more easily, and finally back to the original area, is relatively common. For example, certain integrals in the real time domain cannot be solved easily. If they are transformed into the complex spectral range (image range) by a *Laplace transformation*, the integral can be solved by means of a comparatively simple algebraic operation

(Braun 2019; Ulrich and Weber 2017, p. 98 ff.). The solution is then transformed back into the time range (Westermann 2020, p. 571 ff.). An analogous approach can be found in the *Theory of Inventive Problem Solving* (TRIZ) (Sect. 4.4.5.1). There, a concrete problem is first transformed into an abstract domain and then solved in the form of a contradiction matrix using a list of innovative problem-solving principles. The abstract solution then forms the basis for solving the actual problem in the specific area (Ehrlenspiel and Meerkamm 2013, p. 453 ff.; Kirchner 2020, p. 152 ff.). Finally, an old Chinese wisdom is widespread in management theory, and in particular in *time management*, which says: If you are in a hurry, take a detour (Seiwert 2018). This demonstration of analogies to the approach of analogy technique should also have made its working method clear, because there may be previously unknown application possibilities for transformations that can be found and developed by analogy with the transformations outlined here as examples.

Analytical Methods

Analytical creativity techniques are largely based on the models explained in Chapter 2 and break the problem down into sub-problems, initially solving these sub-problems individually and then combining them to form the overall solution (Ehrlenspiel and Meerkamm 2013, p. 455). In view of the wide range of such methods, only a few of them can be explained in detail in the following. Furthermore, the explanations cannot always be complete, so that reference must be made to specific literature, which often has a similar scope to this book.

Morphological Analysis

Morphological analysis uses the experience that there are usually several solution variants for a sub-function for the purpose mentioned above. Since an overall function usually consists of several sub-functions, the solution variants for each sub-function can be combined in different ways with regard to the overall function. The aim of the analysis is then to select the optimal overall solution from the combinations of variants (Andler 2015, pp. 211 f.; Böhm 2005, pp. 212 ff.; Ehrlenspiel and Meerkamm 2013, p. 455; Wiendahl 2014, p. 120). The morphological box shown in Fig. 4.43, using the example of a coffee machine, is used for this purpose. It shows the path of a conventional overall solution by way of example. However, extraordinary solutions can also be found.

One of the reasons why morphological analysis is so widespread is that it steers thinking in an orderly direction and avoids hasty solutions or ones based on personal preferences. Formal documentation takes the pressure off short-term memory, freeing up capacity that can be redirected to creative problem solving (Ehrlenspiel and Meerkamm 2013, p. 457).

Furthermore, it is possible to understand how solutions were arrived at and what alternatives there might have been. Alternatives that are not pursued remain in the system as a store of ideas. One disadvantage, however, can be the combinatorial diversity of possible solutions, which, as the solution space increases, can quickly reach the limits of cognitive processability (ibid., p. 456).

Partial function \ Partial solutions	1	2	3	4	5	6	7
Water reservoir	Integrated tank	External tank	Passage				
Energy source	Electrical grid	Electric battery	Electric microwave	Electric induction	Gas	Oil, gasoline	Wood, coal
Hot water transport	Riser pipe	Heavy force	Pump	Manual	none		
Extraction	Filter	Cartridge	Bag	Open mixture			
Separation	none	Setting	Filter	Centrifuge	electro-magnetic		
Coffee storage	none	Integrated tank	External tank				
Heat retention	Electric plate	Flame: Candle	Flame: Gas	Flame: Gasoline	Tank insulation	none	
Withdrawal	Manual	Drain tap	scoop	suck			

Conventional solution

Fig. 4.43 Morphological matrix

Theory of Inventive Problem Solving (TRIZ)

The theory of inventive problem solving is a free translation of the Russian acronym TRIZ, which stands for 'Teoriya Resheniya Izobreatatelskikh Zadatch' (Altschuller 1986; Ehrlenspiel and Meerkamm 2013, pp. 453 ff.; Kirchner 2020, pp. 152 ff.). The methodology is based on an analysis of a very large number of successful patent specifications, from which 40 universal principles of invention were derived. These principles serve to overcome contradictions. Furthermore, it has been recognized that the development of technical systems follows determinable laws and patterns. On this basis, methods and tools have been developed that provide for a transformation of the concrete problem into an abstract space, a solution of an abstract problem there and its retransformation into the concrete space (Fig. 4.44).

The first step is to identify the contradictions that make the problem a problem. For example, a camera tripod should be high enough to allow the cameraman to film or take pictures comfortably. On the other hand, the tripod should be easy to carry when traveling, which means that it should be as small as possible to fit in a bag, for example. Fig. 4.44 shows that to do this, the abstract contradiction between the length and volume of the moving object must be resolved. To do this, a matrix with the 40 solution principles for technical problems is used. Fig. 4.45 shows only a section of this, for reasons of space, and the solution we are looking for is not included here. The relevant sources indicate that the 7th principle is suitable in this example, which, like a matryoshka, accommodates an object inside another object and allows this to be done multiple times. This abstract approach leads to the specific solution, which is known in this example and consists of a telescopic cylinder that allows the stand to be pushed together.

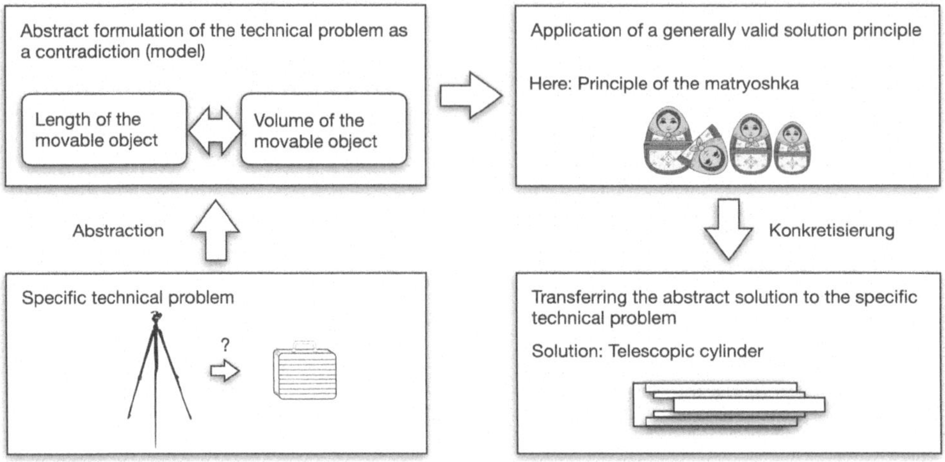

Fig. 4.44 Solving technical problems according to TRIZ by abstraction and concretization

The list of technical parameters of the contradiction matrix contains the parameters given below (Kirchner 2020, p. 425):

1. Mass of the movable object. This refers to the force caused by gravity that a moving object exerts on the support that prevents it from falling. The movable object itself can change its position on its own or as a result of external forces.
2. Mass of the immovable object. This refers to the force caused by gravity that a stationary object exerts on a support. Such an object cannot change its position either by itself or by external forces.
3. Length of the movable object. The length refers to the dimension (length, width or height) that extends in the direction of motion.
4. Length of the stationary object.
5. Area of the moving object. This refers to the plane of the object that changes position due to internal or external forces.
6. Area of the fixed object. This is the plane of the object that does not change its position.
7. Volume of the movable object that changes its position due to internal or external forces.
8. Volume of the stationary object that does not change its position.
9. Velocity.
10. Force, i.e. the ability to cause physical changes in an object. It is irrelevant whether this is done wholly or partially, continuously or temporarily.
11. Tension or pressure, i.e. the intensity of the forces acting on an object (force per unit area).
12. Form, i.e. the outer contour of an object, which can change completely or partially, temporarily or permanently.
13. Stability of the object's composition, i.e. its resistance to imposed changes in shape or instability.
14. Strength, i.e. the ability of an object to withstand loads within certain limits without being significantly altered.

		1	2	3	4	5	6		39
		Mass of the movable object	Mass of the immovable object	Length of the movable object	Length of the immovable object	Area of the movable object	Area of the immovable object		Productivity
1	Mass of the movable object			15, 8, 29, 34		29, 17 38, 34			
2	Mass of the immovable object				10, 1, 29, 35		35, 30, 13, 2		
3	Length of the movable object	8, 15, 29, 34				15, 17, 4			
4	Length of the immovable object		35, 28, 40, 29				17, 7 10, 40		
5	Area of the movable object	2, 17, 29, 4		14, 15, 18, 4					
6	Area of the immovable object		30, 2, 14, 18		26, 7, 9, 39				
39	Productivity								

Fig. 4.45 Excerpt from the TRIZ contradiction matrix (Kirchner 2020, p. 153)

15. Durability of a movable object, i.e. the period of time during which a movable object is able to maintain its function.
16. Durability of an immovable object.
17. Temperature, i.e. the increase or decrease of heat as a cause of extensions to an object during the execution of a function.
18. Brightness or visibility conditions, i.e. the light energy per illuminated area, the quality and characteristics of the light and the degree of illumination.
19. Energy consumption of a moving object.
20. Energy consumption of a stationary object.
21. Power or capacity in the sense of the relationship between work and time. This parameter is used to describe necessary but unwanted changes in the object's performance.
22. Energy losses.
23. Material losses.
24. Loss of information. Parameters 22 to 24 refer to the IPO model (Fig. 2.6).

25. Loss of time.
26. Material quantity, i.e. the number of elements or the quantity required for one element to produce the object.
27. Reliability or safety.
28. Measurement accuracy.
29. Manufacturing accuracy.
30. Damaging effects on the object from the outside.
31. Harmful effects caused by the object itself.
32. Ease of manufacture.
33. Ease of operation.
34. Ease of repair.
35. Adaptability or universality.
36. Complexity of structure.
37. Complexity of control and measurement.
38. Degree of automation.
39. Productivity or functionality.

The following 40 principles are available to resolve contradictions between the parameters just listed in the abstract (ibid., p. 426 ff.):

1. Principle of disassembly or segmentation. This allows an object (a) to be disassembled into equal, independent parts. With regard to a reduction in size of the example of a tripod selected above, a solution would also be conceivable in which the legs can be unscrewed. Furthermore, an object (b) can be designed so that it can be disassembled at any time. Finally, there is the possibility (c) of designing an object that can already be disassembled so that it can be disassembled further. In the example chosen, the tripod legs that can be unscrewed could be disassembled further by means of additional screw connections.
2. Principle of separation. This refers primarily to (a) the possibility of separating the part of the object that causes the contradiction from the rest of the object. An example of this is the power supply units of mobile devices such as smartphones, which are only connected to the device when the batteries inside need charging.
3. Principle: local quality. This can be achieved, for example, by changing from a homogeneous structure of the object or the surrounding medium to a single-homogeneous structure. An example of this is tailored blanks, which have different material properties and geometries (e.g. material thickness) depending on the local load to be expected.
4. Principle of asymmetry. At first glance, hand tools such as pliers appear to be symmetrical. However, from an ergonomic point of view, it makes sense to shape the handles asymmetrically for better handling.
5. Principle: the principle of coupling. An example of this is the single-lever mixer tap, which mixes cold and hot water in such a way that the temperature is already the desired temperature when the water comes out of the tap. However, it is not a good idea to install single-lever mixers in narrow showers, because, especially in unfamiliar surroundings such as hotels, you are likely to accidentally hit it, and then you either feel ice-cold or boiling hot.

6. The principle of universality. This principle stipulates that an object fulfills several different functions, thereby making further objects redundant. One example is multifunction printers, which also enable copying, scanning and faxing.
7. Matryoshka principle. The example of a tripod used above makes it clear that this principle involves the accommodation of one object within another object, and that this can be done multiple times.
8. The countermass principle. This can be achieved by (a) coupling an object to another object with mass, as is the case with counterweights in elevators, for example, but also (b) by using suitable media such as fluids that have a damping effect.
9. Principle of anticipated counteraction. This is the case, for example, in a mechanical watch whose spring must first be tensioned before it drives the watch.
10. Principle of anticipated action. Anticipation of an action can be total or partial. A practical example is the perforation of packaging, whereby its opening is partially anticipated.
11. The "priority cushion" principle. This refers to preventive measures such as the insertion of a fuse in an electrical circuit or a predetermined breaking point in a mechanical component.
12. Principle of equipotentiality. Suitcases with wheels are an example of this, as they need to be lifted and carried less often than suitcases without wheels. The energy required to move the object remains relatively constant.
13. Principle: Reversal of function. One way of applying this principle is (a) to counteract the desired effect rather than bringing about the effect that is actually to be achieved. A second possibility is (b) to make the object that is actually movable immovable and the object that is actually immovable movable. A third possibility (c) arises when the object is turned upside down, so to speak. Finally, the fourth possibility arises when (c) the phases of a movement are performed in a different order. An example is a wind tunnel in which the wind resistance of an object under investigation is not caused by the movement that the object performs in practical use, but by a rotor that sets the air in the object's surroundings in motion.
14. Principle of similarity to a sphere. In more general terms, this refers to (a) all curved geometries, but also (b) the use of spheres, rollers and spirals, as well as (c) the replacement of linear motion with rotational motion. A well-known example is the use of a vehicle with wheels instead of a sled that glides on runners; another is the sidecut of carving skis.
15. Principle of dynamization. This means (a) that the objects or their environment are to be designed in such a way that they automatically adapt to all operating conditions. A second interpretation is to (b) subdivide the objects in such a way that the parts can arrange themselves among themselves. Furthermore, there is the possibility of (c) making immobile objects mobile, adjustable or interchangeable. Finally, (d) the degree of an already existing mobility can be increased. One example is the computer mouse, which can be moved more freely without a cable connection than with one.
16. Principle of partial or redundant action. If an effect cannot be achieved with 100% of the effort actually required, a little less should be used, for example 90%, or a little more, for example 105%. One example is grouting the transition between a tiled floor and a

tiled wall, where the silicone is first applied generously and then smoothed, with any excess material removed.

17. Principle of transition to higher dimensions. This can include (a) the change of mutual arrangement in the plane or in space, or the transition from the second (plane) to the third (space) dimension, but also (b) the arrangement of an object in several instead of only one plane, (c) tilting or laying the object on its side, (d) using the back or the interior, and (e) using light beams that fall on the surroundings or on the back of the object. A well-known example is the car window scraper, which is also equipped with a brush to remove snow and a rubber lip that can also be used to clean the windows.
18. Principle of utilizing mechanical vibrations. This includes (a) setting an object in vibration, but also (b) changing the frequency and/or amplitude of vibrations that are already present, as well as (c) using the natural frequency. In this way, (d) mechanical vibrators can be replaced by piezoelectric vibrators and (e) ultrasonic vibrations can be useful in combination with electromagnetic fields. One example of this is electric toothbrushes.
19. Principle of periodic vibration. This can mean (a) replacing a continuous action with a periodic or pulsating one, but also (b) changing the frequency of an already existing periodic effect. Furthermore, (c) a different or additional effect can be triggered during pauses and (d) the periodicity of the effect can be aligned with the natural frequency. Depending on the application, both a match with the natural frequency and a deliberate avoidance of a match are possible. One example is an electric toothbrush that suggests a change in the tooth region to be brushed by briefly changing its vibrations.
20. Principle of continuity of effective processes. This should (a) ensure a constant load and (b) avoid idle time and discontinuities. An example is the use of an assembly line to manufacture mass-produced products.
21. Principle of rushing or skipping. According to this, dangerous or harmful stages should be passed through at high speed. Lancing devices for obtaining capillary blood, as required by diabetics to determine their blood sugar level, are an example of this. The alternative use of lancets would be more painful.
22. Principle of converting something harmful into something useful. This can be done (a) by the direct utilization of an actually negative effect of a medium or (b) by superimposing it with the effect of another harmful medium, as well as (c) by amplifying a harmful effect beyond the point where it ceases to be harmful. One example is the turbocharger in a combustion engine, where the exhaust gas flow is used to drive a turbine wheel to compress the intake air in the combustion chamber in such a way that both the power and the efficiency of the engine are increased.
23. The principle of feedback. This actually refers to the closing of control paths to control loops, for example by using thermostats to control room temperature.
24. The principle of the "mediator." A mediator is an object that temporarily surrounds another object or a medium, for example, packaging that enables storage and transportation, but also, for example, a catalyst in the course of a chemical reaction.
25. Principle: self-service or self-sufficiency. Energy harvesting (Ylli and Mansoli 2022) is one area of application for this. A technically outdated example, but one that is still highly valued among enthusiasts, is a mechanical watch with an automatic winding mechanism that draws the energy it needs from the wearer's movements.

26. The principle of copying. This can include (a) the replacement of expensive, sensitive or difficult-to-handle objects with cheap, insensitive or easy-to-handle copies, but also (b) the replacement of an object with an image (which may also be enlarged or reduced) and (c) in particular with digital copies (e.g. virtual reality). A flight simulator is a well-known example of this, replacing expensive training in an aircraft with a still expensive simulator that is, however, significantly cheaper in comparison.
27. Principle of cheap short-lived instead of expensive long-lasting. Disposable packaging is a bad example in this respect, especially for ecological reasons.
28. Principle of the use of mechanical principles. On the other hand, Karakuri is a promising example of a mechanical automation that uses mechanisms such as levers, cables or cams to transmit power and weights or springs to store energy (Ahrens 2022, p. 119 ff.).
29. Principle of application of pneumatic and hydraulic designs. This involves (a) replacing solid, heavy objects with liquid or gaseous substances. This principle also involves (b) the use of negative pressure (vacuum). For example, when building wooden boats, veneers are glued over a mold and pressed on with the help of a plastic film under which a vacuum is created.
30. Principle of the use of flexible casings and thin foils. The example given for the 29th principle can also be applied to this one.
31. Principle of using porous materials. This can be seen in house construction, where, for example, bricks are used for thermal insulation whose cavities are filled with perlite, a mineral insulating material made of volcanic glass.
32. Principle of color change. One example of this is dark-colored wine bottles, which protect the wine from light.
33. Principle of similarity or homogeneity. This principle is applied, for example, in riveted joints, where rivets made of the same material are chosen as the material of the components to be joined, since this prevents loosening due to the same thermal expansion and also prevents electrochemical corrosion.
34. Principle of removal and regeneration of parts. This is how self-dissolving sutures are used to close wounds, avoiding the need to remove stitches.
35. Phase change. This principle is used, for example, in the demagnetization of ferromagnetic materials by heating above the Curie temperature.
36. Principle of the application of phase transitions. This principle is applied, for example, when demagnetizing ferromagnetic materials by heating them above the Curie temperature.
37. Principle of application of thermal expansion. Liquid thermometers are a well-known example of this.
38. Principle of using strong oxidants. This principle is used, for example, in laser cutting, where the cutting performance is increased by adding oxygen. Another example is in biogas plants, where the bacteria increase their activity when supplied with oxygen.
39. Principle of using an inert medium. Damage can be limited by using argon extinguishing systems to fight fires, compared to using water or powder.
40. Principle of the application of composite materials. Carbon fiber reinforced plastics (CFRP) are an example of this (Haka 2022).

TRIZ encompasses much more than the aspects explained here, so a presentation that is even close to complete would be too much to go into here. However, the selected aspects can give an impression of the way of working associated with TRIZ.

Bionics

Bionics is a portmanteau word that is made up of the partial terms biology and technology and is an example of how *analogies* are a good basis for interdisciplinary, but above all creative, work. In this context, natural phenomena serve as inspiration for technical solutions. One of the first applications usually cited, and this is also the case here, is Leonardo da Vinci's idea of building flying machines modeled on the flight of birds. Current examples include the just as frequently mentioned Velcro (Wawers 2022, p. 1), which George de Mestral developed based on the model of burdock (Arctium) and applied for a patent in Switzerland in 1955, or so-called wing lets, usually upward (rarely downward) extended outer wings at the end of an aircraft's wings, which, modeled on the wings of birds, reduce induced drag, among other things, and have since been further developed into spiroids (split-hinged loops).

The underlying method of bionics is the observation that nature has produced a wide variety of solutions and optimized them in the course of evolution, and that it is possible to apply such solutions successfully in technical systems (VDI 6220 Sheet 1:2021-09). However, since the solutions that are fundamentally suitable for this purpose can be found in all areas of nature, a correspondingly broad and in-depth biological knowledge is required, which is usually only available from experts in this field and is also spread across various specialized areas, such as botany, zoology, microbiology, biophysics and genetics. Conversely, technology is also a very broad field of knowledge for biologists. This makes intensive cross-disciplinary collaboration necessary. Systems theory can be helpful here, since it is applied in both technology (e.g. Kaffenberger et al. 2012) and in parts of biology (e.g. Maturana and Varela 2018) and thus offers a unified basic understanding.

This can be seen, among other things, in the intermediate step of *abstraction*, which is comparable to the approach of TRIZ (Fig. 4.44) and the other approaches mentioned in Section 4.4.5.1. In bionics, this can be done in two directions: (1) *Top-down* refers to the direction in which a technical problem is first identified, for which analogies in nature are then sought, in order to transfer any solutions found there to the technical problem. To do this, however, the specific technical problem must first be abstracted with regard to the functionality sought. The search for an analogy then takes place at this level. To solve the actual problem, a further concretization is then required. The winglets mentioned above are an example of this. (2) In contrast, bottom-up refers to a process that starts with biology. Technical application possibilities are sought for findings that have been obtained in this discipline. An example of this is the use of the lotus effect discovered by Wilhelm Barthlott (Cerman et al. 2005), which minimizes the adhesion of dirt particles due to a complex microscopic and nanoscopic surface structure.

Klaus Ehrlenspiel and Harald Meerkamm (2013) offer worksheets and checklists for methodical construction in addition to their book on the Internet, which contain an *association list*, i.e. a list of search terms for finding natural phenomena that lend themselves to technical use. To protect copyrights, only a few examples will be cited below to give an impression of whether and, if so, to what extent the list can be helpful to the reader of this book in a specific case. In general, studying the original source seems just as recommendable as reading the book on which it is based.

The starting point for the list of associations are the following technical functions:

- modifying,
- depositing,
- absorbing,
- storing,
- measuring and detecting,
- generating,
- vaporizing,
- condensing,
- melting,
- drying,
- vibrating,
- rotating,
- lifting,
- moving,
- smoothing,
- forming,
- cleaning,
- dismantling,
- destroying,
- separating,
- removing,
- extracting,
- protecting,
- stabilizing,
- aligning,
- merging and
- distribute.

These functions are differentiated into subcategories, for which analogies to corresponding biological solutions are then given. The following examples give an impression of what can be expected.

- A simple example is the change of dimensions that occurs in biology as a result of growth through cell division.
- With regard to the change in the electric current, the electroplax of the electric eel (Electrophorus) is offered as a possible analogy. The organ in question consists of non-contracting muscle fibers. On one side, they have papillae, i.e. small protrusions that arise from the bulging out of cells or cell groups, and on the opposite side, they have synapses. The release of acetylcholine excites the electroplax cells, which can briefly generate electrical voltages of up to 860 V.

- An example that requires a comparatively large amount of biological knowledge can be found in the case of a change in the sense of a deformation. The associations offered with biological phenomena are the turgor movement, the nutation movement, myofilaments, growth, turgor, abducin (bivalves, cross-linked polymer), resilin (insects, cross-linked polymer), elastin (aorta, more global protein complexes), collagen (skin, contraction by ions). In such cases, close cooperation with experts in biology seems inevitable, but at the same time it is also a promising approach.

Experiment

Experiments offer the possibility of empirical investigations, not in the field (Sect. 4.2.4), but on the basis of a model (Sect. 2.2). In the *deductive form* (Sect. 4.1.4.2), a *hypothesis* is first formulated about a fact to be investigated. Those parameters that are assumed, based on the hypothesis, to have a significant influence on the phenomenon under investigation are included in the model. Other influences are excluded as far as possible. For this, it is particularly important to select the *samples randomly*[29] (Diekmann 2020, p. 337; Eifler and Leitgöb 2019, p. 209). By deliberately changing variables (Sect. 2.4.6), their influence (cause) on the investigated fact (effect) is observed. This verifies or falsifies the hypothetically assumed regularity and, if necessary, specifies it. In the course of such a model-based cycle of knowledge, the insights gained in the experiment must finally be transferred to reality (Fig. 2.5).

In terms of discoveries and inventions, the experiment itself is therefore not a creativity technique in the strict sense, since the actual idea is already expressed in the preceding formulation of the hypothesis. The experiment then serves only to test this idea. Nevertheless, experiments can contribute to a creative process to the extent that they support a targeted examination of the modeled facts. In this sense, they are often understood as *questions to nature*, or, as Carl Friedrich von Weizsäcker formulated it in reference to Immanuel Kant, as an *interrogation of nature*. A suspect is also subjected to an interrogation in a specially designed interrogation room, which is designed to exclude falsifying influences and promote the recognition of the essential, because there is a suspicion (hypothesis) that he has committed the offense in question.[30] In this case, too, the interrogation serves to verify or falsify the suspicion, or to specify it if necessary. Therefore, the analogy seems quite suitable for conveying a general understanding of experiments.

In addition to the deductive form, there is also the inductive form of the experiment (Sect. 4.1.4.2), which is not based on a hypothesis. Instead, one lets oneself be surprised by its course and outcome[31]. An example is the discovery of graphene, a single-layer carbon material (Hornbogen et al. 2019, p. 141 ff.), which Andre Konstantin Geim and Konstantin Sergejewitsch Novoselov discovered in 2004 during after-work experiments and for which they received the Nobel Prize in Physics in 2010. In this way, experiments themselves can be a source of invention and innovation. The heuristic of trial and error is closely related to this.

[29] This should not be confused with random sampling.

[30] Surveillance without cause, as provided for by the ePrivacy Regulation (ePVO) of the European Union (EU), for example, is widely disputed.

[31] However, this is riskier than conducting deductive experiments, since the surprise can also be unpleasant.

The main advantages of experiments compared to field research are that the laws (cause-and-effect relationships, causalities) can be better isolated from other influences. Furthermore, they are *reproducible* and can therefore be *independently verified* (Eifler and Leitgöb 2019, p. 204). However, deriving objectivity from this is subject to the fundamental limitations of constructivism (Sect. 2.1). Nevertheless, experience shows that viable insights can be gained (von Glaserfeld 1996, p. 43).

In science and technology, causalities are often relatively unambiguous: if, for example, a ball is pushed and then starts to move, it can be assumed with some certainty, especially in a corresponding experiment in which all other possible causes can be largely excluded, that the push was the cause. In the social sciences, however, statements often cannot be made with this degree of certainty. Therefore, a cause is often referred to as an *INUS condition*, where the acronym stands for "insufficient but non-redundant part of an unnecessary but sufficient condition". This takes into account that causes other than those assumed can also bring about an effect and that identified regularities are not deterministic. Rather, at best probabilities can be given (Eifler and Leitgöb 2019, p. 205).

Three necessary conditions apply to the demonstration of causality in all areas of application (ibid., p. 206):

- For a causal relationship between a hypothetical cause and a hypothetical effect to be assumed, the cause must precede the effect in time and
- *co-vary* with it.
- Furthermore, it must be possible to *exclude alternative explanations*.

The effect to be explained is referred to as the *dependent variable* (outcome), and the cause whose influence on this variable is to be examined is called the *independent variable* (treatment). Other possible influences are treated as *confounding variables* and must be controllable in the experiment. Despite all measures for the most reliable modeling possible, every experimental proof of an effect faces the *fundamental problem of causal interference*, i.e., the problem that it always remains unknown whether and, if so, how the dependent variable would change if the independent variables were not changed or not changed in the manner undertaken. In the social sciences, parallel studies on a group of subjects and on a control group (comparable to the use of placebos in medicine) are often carried out to limit this problem (Diekmann 2020, p. 337 ff.; Eifler and Leitgöb 2019, p. 209).

The extent to which an effect can be assumed to be clearly attributable to a specific cause is referred to as internal validity (ceteris paribus validity). Conversely, *internal validity* refers to the possibility of excluding confounding variables. The most important *sources of error* in this regard (Diekmann 2020, p. 344 ff.; Eifler and Leitgöb 2019, p. 207) are abbreviated in English with the acronym THE MESS:

- Testing (reactivity): This can occur, for example, if the measurement is perceived as unpleasant, so that a reaction also occurs to it.
- History (temporal influences): Such influences can occur, for example, if an environmental scandal has become public knowledge immediately before a survey on environmental protection.
- Instrument Change (change of measuring instrument): Since measuring instruments themselves show a degree of dispersion, measuring with two different measuring instru-

ments can lead to different measurement results, which can be attributed to this and not to the measured effect.

- Statistical Regression toward the mean: An example of this can be found in medicine when patients are examined who all show extreme values of a characteristic, for example high blood pressure, in the course of screening. If the same group is examined again at a later point in time, most patients will then show a lower blood pressure, regardless of whether their blood pressure was treated or not (Kleist 2006, pp. 1023 ff.). In this respect, it is a statistical artefact.
- Maturation: This can occur when respondents engage more deeply with the subject of the survey between two survey dates.
- Experimental mortality (systematic loss): If subjects are no longer available for a follow-up examination, this may be due to systematic influences, such as a reaction to the first examination, and may influence the result.
- Selection: This refers, for example, to the fact that only subjects are observed who have similar relationships with regard to the facts being examined, for example football fans at a football game.
- Selection interaction: This refers to the interaction between the selection and another factor. Other factors may be added to these main influencing factors:
  - Exchange of information: If and to the extent that communication about it takes place between subjects during a study, this can lead to behaviors or actions in follow-up studies that influence the study result. For example, the group of subjects may strive for conformity (e.g. formation of a social system), be particularly motivated or be demotivated or try to compensate for certain conditions.
  - Rosenthal effect: In the sense of autosuggestion, the experimenter may behave unconsciously or subconsciously towards the subjects in such a way that he influences them in the sense of the expected results, despite consciously trying to be neutral. In the sense of suggestion, he can transmit information to the subjects through unconscious or subconscious behavior that informs them of the expected result. The subjects may then feel encouraged to act in the experimenter's favor.
  - An ex post facto design is also frequently chosen, in which an existing group of subjects is examined for a dependent variable. The group assignment is based on events in the past. The independent variable is only defined after the data collection. Such an approach is not an experiment in the true sense of the word, but it can be useful when conducting a real experiment is not practical or is not justifiable for moral reasons.

In addition to the internal validity of causal conclusions, which is considered to be of the utmost importance (Eifler and Leitgöb 2019, p. 213), external validity indicates the extent to which the experimentally determined laws can be transferred to other situations. This possibility can be affected by the following relationships:

- Mutual relationships between the examined causality and the examined units. This may be the case, for example, if certain effects can only be demonstrated for people with certain characteristics, such as adolescents.
- Mutual relationships between the examined causality and the variation of the experimental impact (treatment variation). This may be the case, for example, if the influence

of media on a group can only be proven for internet offerings, but not for traditional media such as radio and television or for newspapers and magazines.

- Mutual relationships between the examined causality and the type of dependent variables (outcomes).
- Mutual relationships between the causality under investigation and the setting. This may be the case if effects can only be demonstrated in certain settings, for example only in commercial enterprises and not in public administrations.
- Mutual relationships between the causality under investigation and context-dependent effects (context dependent mediation).

An experiment usually involves several *attempts*. From attempt to attempt, the independent variables (IV), which are also referred to below as *factors* (Fig. 2.7, bottom right), are varied so that their effects on the dependent variable (DV) can be recognized and described in the form of laws. For example, the aggregate state of water (DV) is solid (ice) at temperatures below 0 °C (IV), liquid at temperatures between 0 °C and 100 °C, and gaseous (vapor) at temperatures above 0 °C (UV), under otherwise identical conditions (including atmospheric pressure p = 1,013.25 hPa). If an experimental test starts from a liquid state at a usual room temperature of, for example, 20 °C, the temperature factor must be set in at least two stages: t = 0 °C (freezing) and t = 100 °C (boiling).

In many cases, however, the behavior of systems to be investigated is complex and requires a test plan that is tailored to it. There are basically three options for doing this (Schneider 2020, p. 255 ff.; Siebertz et al. 2010, p. 6 ff.):

1. *Full factorial experimental designs*. In this type of design, all combinations of the factors are examined. The number of experiments required for this, n_r, is calculated from the number of factors, n_f, and the number of levels, n_l

$$n_r = n_l^{n_f} \tag{4.22}$$

   Fig. 4.46 shows experimental designs for two (22) and three (23) factors, each with two levels, indicated by + and –. Alternatively, other codes such as numbers are often used. In the first case, $n_r = 4$, and in the second case, $n_r = 8$ experimental runs yn are required. However, in practice the number of experiments is very large in many cases and thus quickly becomes uneconomical. Therefore, a key objective of statistical experimental design is to minimize the number of required experiments while maintaining defined accuracies.
2. *One-factor-at-a-time*: Only one factor is changed at a time, while all others are held constant (ceteris paribus). If and to the extent that several factors (IV) affect the dependent variable (DV), this approach assumes that the effects of the IV are independent of each other. If this is not the case, such experimental designs can lead to false conclusions (Siebertz et al. 2010, p. 7). For example, if an effect depends on two factors, varying one factor while keeping the other constant can result in a certain optimum. If the initially varied factor is then fixed at its optimum in order to vary the second factor from there, an optimum may also result for the second factor. The superimposition of both optima then leads to an apparent overall optimum. However, the actual overall optimum can result from a completely different factor combination.

2^3-Experimental plan			
2^2-Experimental plan			
Attempt	x_1	x_2	x_3
y_1	−	−	−
y_2	+	−	−
y_3	−	+	−
y_4	+	+	−
y_5	−	−	+
y_6	+	−	+
y_7	−	+	+
y_8	+	+	+

Fig. 4.46 Experimental design for two (22) and for three (23) factors, each with two levels, which are indicated with + and –

3. The statistical experimental design can identify the actual overall optimum by varying several factors simultaneously. To do this, measurement points are laid over the entire factor space according to certain statistical specifications. This space covers the setting range of all factors and its dimension corresponds to the number of factors. Fig. 4.47 shows the corresponding factor space for a full factorial experimental design for three factors with two levels each (2^3 experimental design). A *fractional factorial experimental design* reduces the number of experiments to be carried out according to a full factorial plan by a certain fraction. For example, of a 2^4-run plan, only 2^{-2} runs need be carried out, namely $2^{4-2} = 16/4 = 4$ runs, if factors can be combined in such a way that no significant effects are mixed together.

Both commercial and free software tools are available for designing specific statistical experiments. DoE packages are also available for the open-source programming language R[32] for statistical calculations and graphics.

Since the previous remarks were primarily based on experiments that can serve to develop technically oriented systems, the following is intended to outline how changes can be brought about in a psychological and social context that also proceeds experimentally in a broader sense. Systemic *structural constellation work* (SySt) is chosen as an example here (Sparrer 2016), as it is rooted in systems theory. The main focus here is on its use in social systems and in particular in organizations. The aim is to solve problems in groups such as teams or departments. The hurdles that need to be overcome in this process have already been explained in Section 2.4.4, among other places. SySt takes into account the insight that social systems can only change themselves, while attempts to influence them from the outside are often counterproductive and, at best, largely ineffective. This is evident, among

[32] If at all, then here and in the following, in order to maintain neutrality with regard to commercial software offers, only freely usable software is referred to. Mentions of this kind are not to be understood as recommendations.

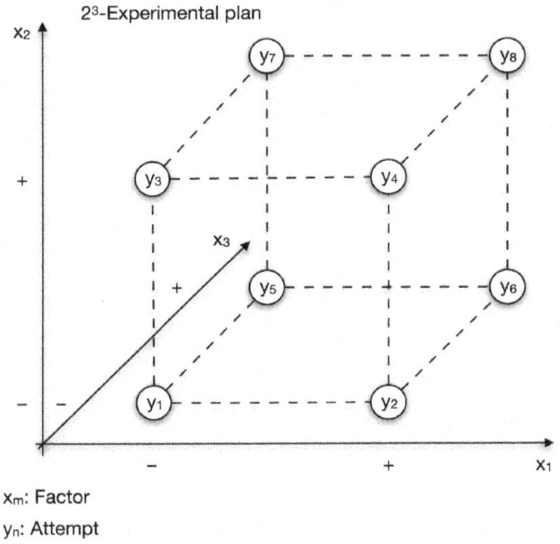

Fig. 4.47 Graphical representation of the factor space for a full factorial experimental design with three factors, each with two levels (23 experimental design)

other things, from the fact that the person facilitating the constellation sees themselves as a host and does not identify and interpret problems or work towards specific solutions. Rather, their role is to support the social system in doing so itself. To this end, they contribute methodological expertise.

Furthermore, the SySt assumes that initially only the structures of a system (Sect. 2.4.2) can be made transparent, i.e. the elements and relations as well as the patterns (topologies) that form them. These then reveal which processes take place within them. The basic principle is to externalize internal structures that are difficult for consciousness to access. To do this, the structures are represented by spatial arrangements of elements. This can be done, for example, by people moving into positions to be defined in a room, or on a smaller scale on boards such as the so-called system board (Polt and Rimser 2020), on which, among other things, figures are arranged.

As a rule, the people setting up the constellation and those represented in it are not the same. Therefore, if a constellation is to be set up by people, independent third parties are needed who are available for this purpose. These third parties should be sufficiently familiar with the method and bound to confidentiality. The people represented then represent individuals and their roles in the social system under consideration. In addition, other relevant aspects such as goals, obstacles to achieving goals, resources or missing elements can be objectified, whether through additional people or through objects that are also arranged in the structure. The spatial arrangement of the elements in relation to each other and the relations between them (including missing ones) should make aspects of the problem to be solved tangible.

Initial insights can be gained from the process of setting up the constellation, i.e. selecting the relevant elements, naming their characteristics (role or function in the system, behavior, etc.) and relating the elements to one another. In the further course, for example, the as-is situation perceived as problematic and, subsequently, changed situations in the sense of so-called *positional work* can be set up. Corresponding experiments are usually carried out according to the principle of trial and error. Sensory perception promotes the progress of knowledge. Further contributions are made by the discussion among those setting up. The people being represented can also contribute to this in the form of their so-called *represented perception* (Neumann-Wolff 2022, p. 101 ff.), by trying to describe how the element they represent would interpret the established structure from their perspective and what could result from it. The aim is always to raise the subconscious and unconscious to the level of consciousness and thus make it transparent, in order to encourage the social system to develop in a desired direction.

4.4.5.2 Optimization Methods

Optimizations do not always take place, but often in two stages: In the first stage, an *opening procedure* is used to find the best possible starting position. This position is particularly good if the *improvement measures* to be carried out in the second stage do not involve too much effort. Many optimizations do not achieve an absolute optimum in this way, and even if they do achieve one, it is not always possible to prove that it is an absolute optimum. One reason for this is that many optimization problems are combinatorial problems, which are often NP-hard or even NP-complete. The latter, for example, has the consequence that all known deterministic algorithms for solving a corresponding problem require an exponential amount of computing time, so that they can be solved in principle for a small number of problem elements and in principle for a large number of such elements, but that the latter cannot be achieved within the lifetime of the problem solver.

A model example of this is the towers of Hanoi shown on the left in Fig. 4.48 (Funke 2003, p. 109 f.). The problem is to restack differently sized disks, which are stacked in order of size on an initial rod, by using an auxiliary rod arranged in the middle, onto the target rod arranged at the other end, in such a way that only one disk is moved at a time and that a larger disk must never be placed on a smaller one. If only three discs are used, this problem is so easy to solve that it can be used as a children's toy. The shortest path requires $2^3 - 1 = 7$ steps. Legend says that after moving 100 discs, the end of the world will be reached. If this is optimally achieved with each step, $2^{100} - 1 \approx 10^{30}$ steps are required. Legend has it that monks in a monastery in Hanoi have been doing this for generations. Assuming, for the sake of argument, that each step takes one second, then about 10^{18} disks, or just over half, have been flipped since the Big Bang. From an academic point of view, it is interesting to note that the solution space of this problem is spanned by the Sierpinski triangle (Fig. 4.48, right), which has a fractal structure (Schuster 1994, p. 55). The optimal solution forms a trajectory along the outer edge of this triangle.

In order to apply analytical methods despite such difficulties, *heuristics* are used. These are methods that use less than the theoretically available information to arrive at a viable result with an acceptable amount of effort. Although their rule-based nature does not always allow for *reproducibility* and *intersubjective verifiability* of the results, in many cases it does allow for it, but in most cases it does not deliver an absolute optimum.

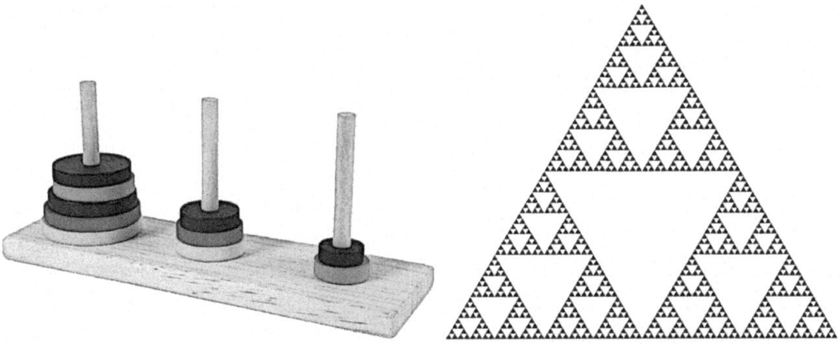

Fig. 4.48 Sierpinski triangle as a visualization of the solution space for the optimal (shortest possible) solution to the problem of the towers of Hanoi

Dietrich Dörner (1979, p. 38) classifies the procedure according to *trial and error* as the simplest form of heuristics. For this purpose, a hypothesis is formulated and then subjected to a falsification attempt in the sense of critical rationalism. If this is successful, a new hypothesis is formulated. The experiences with the previous hypothesis can be incorporated into this. The new hypothesis is then subjected to a falsification attempt and so on. This process is terminated as soon as a hypothesis has been verified or when the available time has elapsed. In the latter case, one can try to use the insights gained so far to solve the problem.

In computer science, this approach is known as the *brute force method.* It is used, for example, to decode passwords. The exponential increase in the number of arithmetic operations with an increasing number of characters used is one reason why long passwords are preferable. Since the computing power of modern computers is constantly increasing, methods such as key stretching are now required to stretch short passwords. Other methods are also used, such as limiting the number of attempts to enter passwords. The necessity of such measures can be seen as proof that a process of trial and error can be very successful for certain problems.

One can also come to such a conclusion if one takes into account that the *evolution* of life is also based on the principle of trial and error. In the course of reproduction, genes and traits are recombined and must subsequently prove themselves in the organism's habitat (*fitness*). If this does not succeed before the organism can reproduce, the corresponding combination does not continue. On the other hand, the better the organism is adapted to its habitat, the more likely it is that its reproduction and that of its offspring will be successful (*selection*). However, since this form of evolution can lead to a dead end from the point of view of optimization, recombination is supplemented by *mutation*, which now and then, spontaneously or triggered by external influences mutagens such as ionizing radiation or mutagenic chemicals), leads to more far-reaching changes in genes and traits. Although this increases the risk of non-adaptation to the conditions of the habitat, it also gives a species the chance to escape from an evolutionary dead end. At the same time, new species arise from this polymorphism. Finally, the Baldwin effect must be taken into account, according to which not only the changes taking place in the genotype of a population influence the progress of evolution. Rather, the *learning* of individuals (phenotypes) also has an indirect influence on it.

These principles of biological evolution form the basis for corresponding algorithms for mathematical optimization (Weicker 2015, p. 19 ff.; VDI/VDE 3550 Blatt 3:2003-02). Computer programs use them to simulate evolutionary processes. To do this, a possible solution to a problem is treated like an organism. The problem provides the fitness criterion against which the solution must prove itself. To do this, it is subject to the changes outlined above, i.e. recombination and mutation. Through selection, only those solutions that prevail against other solutions have a chance of reproducing. For practical applications, there is currently no universal software that can be used for problems of all kinds. Rather, *evolutionary algorithms* are occasionally part of application-specific software solutions for certain problem categories or industries, for example in logistics to optimize the loading of means of transport (Weicker 2015, p. 254). Furthermore, libraries are available for programming languages such as Python or R (both open source) and for software packages for applications in numerical mathematics such as Scilab (also open source).

As an example to illustrate quantitative optimization methods, Varignon's apparatus will be presented below in order to then mathematically solve the *location problem* that can be solved mechanically with the help of the Steiner-Weber model (also: Fermat problem). The mechanical functioning of Pierre de Varignon's (b. 1654 in Caen, d. 1722 in Paris) table (Fig. 4.49) is first illustrated. Given a number of locations, the aim is to find a new location that minimizes the transportation effort to all given locations. To do this, holes are drilled to scale in a table at the given locations, through which threads are passed that are knotted together on the upper side. Weights are hung from their lower ends, corresponding to the respective transportation effort. What limits the transport effort is irrelevant in this respect, as this can be, for example, the weight, volume or value of the goods to be transported. However, only one criterion of this kind can be used. If there is sufficient low friction, the knot is pulled to the optimal location.

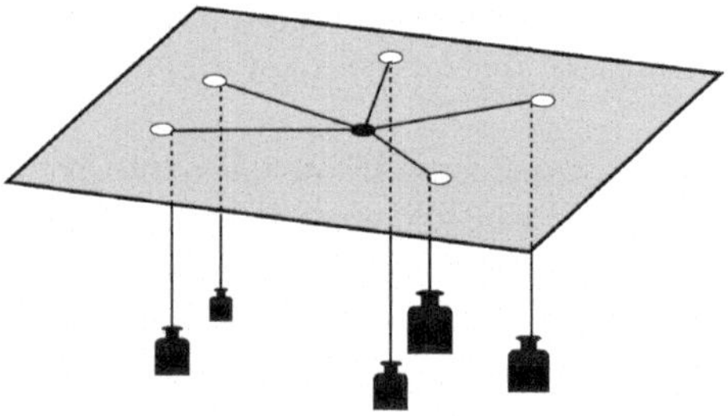

Fig. 4.49 Varignon's apparatus for optimizing locations

This mechanism can also be formulated mathematically as a so-called Steiner-Weber problem (also: Fermat problem) (Domschke et al. 2018, p. 18 ff.). For this purpose, the coordinates (x_i, y_i) of the $i = 1, \ldots, n$ locations are marked in a Cartesian coordinate system spanned by the axes x and y (Fig. 4.50). The distances can be measured both at right angles and directly as the so-called Euclidean distance, which corresponds to the

Pythagorean theorem (hypotenuse theorem) for the special case of $n = 2$ that is relevant here. This results in the optimization problem

j	u_j	v_j	b_j
1	2	4	3
2	8	10	2
3	10	2	2
4	16	8	1
5	22	10	1

u_j, v_j: Coordinates

b_j: Transportation volume

Abort barrier α = 0,001

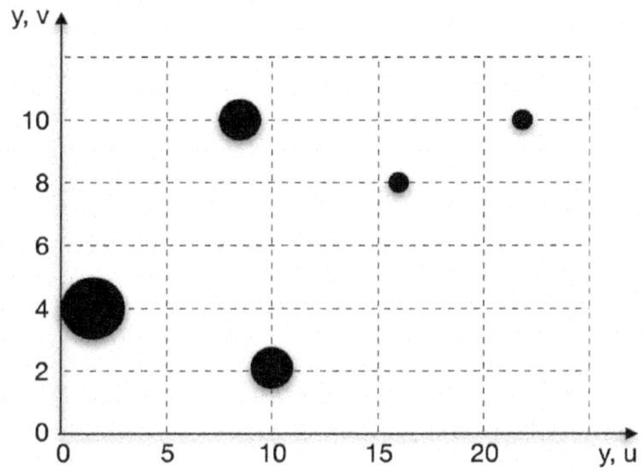

Fig. 4.50 Mathematical modeling of Varignon's apparatus using the Steiner-Weber model (Domschke et al. 2018, p. 18 ff.)

$$F(x^*, y^*) = \sum_j b_j \sqrt{(x^* - u_j)^2 + (y^* - v_j)^2} \tag{4.23}$$

An analytical solution requires partial differentiation and setting equal to zero:

$$\frac{\partial F}{\partial x} = \sum_j \frac{b_j(x^* - u_j)}{\sqrt{(x^* - u_j)^2 + (y^* - v_j)^2}} = 0 \tag{4.24}$$

$$\frac{\partial F}{\partial y} = \sum_j \frac{b_j(x^* - v_j)}{\sqrt{(x^* - u_j)^2 + (y^* - v_j)^2}} = 0 \tag{4.25}$$

The respective solutions are

$$x^* = \frac{\sum_j \left[\frac{b_j u_j}{\sqrt{(x^* - u_j)^2 + (y^* - v_j)^2}}\right]}{\sum_j \left[\frac{b_j}{\sqrt{(x^* - u_j)^2 + (y^* - v_j)^2}}\right]} \tag{4.26}$$

$$y^* = \frac{\sum_j \left[\frac{b_j v_j}{\sqrt{(x^*-u_j)^2+(y^*-v_j)^2}} \right]}{\sum_j \left[\frac{b_j}{\sqrt{(x^*-u_j)^2+(y^*-v_j)^2}} \right]}$$

These solutions show that a complete isolation of the coordinates x^* and y^* is not possible. Therefore, a solution by analytical means is not possible. As an alternative, iterative approximation according to Miehle is offered (Fig. 4.46). In a first step, the centroid coordinates (x_0, y_0) are calculated as follows:

$$\left|x^\mu - x^{\mu-1}\right| \leq \alpha \qquad \left|y^\mu - y^{\mu-1}\right| \leq \alpha \tag{4.27}$$

In the next step, iterations μ are performed for (x^μ, y^μ). The added ε prevents a division by zero if $x^\mu = u_i$ and $y^\mu = v_i$.

$$x^\mu = \frac{\sum_j \left[\frac{b_j u_j}{\sqrt{(x^*-u_j)^2+(y^*-v_j)^2+\varepsilon}} \right]}{\sum_j \left[\frac{b_j}{\sqrt{(x^*-u_j)^2+(y^*-v_j)^2+\varepsilon}} \right]} \tag{4.28}$$

$$y^\mu = \frac{\sum_j \left[\frac{b_j v_j}{\sqrt{(x^*-u_j)^2+(y^*-v_j)^2+\varepsilon}} \right]}{\sum_j \left[\frac{b_j}{\sqrt{(x^*-u_j)^2+(y^*-v_j)^2+\varepsilon}} \right]}$$

The iterations are carried out until a termination criterion is met, which is set with the parameter α (Fig. 4.51):

$$\left|x^\mu - x^{\mu-1}\right| \leq \alpha \qquad \left|y^\mu - y^{\mu-1}\right| \leq \alpha \tag{4.29}$$

For the example outlined in Fig. 4.50, the coordinates of the optimal location for $\varepsilon = 0.0001$ (taken into account by F') are calculated according to Fig. 4.51.

In many cases, it is not possible or useful to optimize quantitatively. One reason for this may be the small number of influencing variables that can be taken into account. And the high degree of abstraction of the results may also be a reason against it. For example, only the transport costs and the Euclidean distances were included in the calculation of an optimal location just outlined. In terms of a practical realization of the solution, however, a large number of further questions need to be answered. These include whether there are actually passable roads near the direct connecting lines between the locations, whether a buildable property can be found at the optimally identified location at an acceptable price, whether the necessary labor force is available there, and so on.

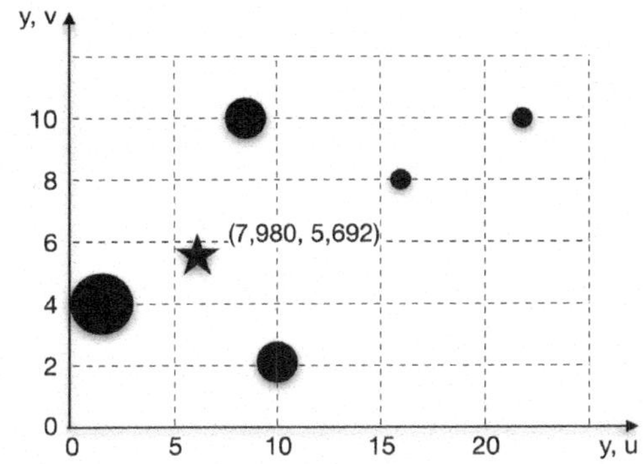

μ	x^{μ}	y^{μ}	$F(x^{\mu}, y^{\mu})$	$F'(x^{\mu}, y^{\mu})$
0	8,889	6,000	59,1128	59,1129
1	8,250	5,844	58,7614	58,7615
2	8,138	5,770	58,7035	58,7036
3	8,049	5,732	58,6926	58,6927
4	8,010	5,713	58,6899	58,6904
5	7,993	5,702	56,6898	56,6900
6	7,985	5,697	56,6868	56,6899
7	7,982	5,694	56,6868	56,6898
8	7,981	5,692	56,6868	56,6898
9	7,980	5,692	56,6868	56,6898

Fig. 4.51 Solution of the Steiner-Weber problem by iterative approximation according to Miehle (Domschke et al. 2018, p. 18 ff.)

In this context, quantitative optimization methods are often used in the early design phases, when the design object is only roughly outlined anyway in the course of a process from rough to detailed (Sect. 3.3.1). As the level of detail increases, further criteria are added, so that a *transition to qualitative optimization methods* may then become necessary.

Unlike analytical methods and many heuristics (quantitative methods), qualitative methods do not provide calculable and reproducible results: other people would most likely come up with different solutions for the same problem at a different time and in a different place, even if the same methods were used. Nevertheless, documenting the solution process and the result allows for intersubjective verification, even with qualitative methods. While it is not possible to achieve an absolute optimum, it is possible to achieve a *relative optimum*, i.e. one that proves to be the best solution in comparison (in relation) to the given situation and to alternative solutions.

This is where *variant planning* comes in (Haberfellner et al. 2018, pp. 57 ff.; Naefe 2018, pp. 117 ff.). For a more detailed explanation, this approach should immediately be combined with the procedure *from rough to detailed* (Sect. 3.3.1) (Fig. 4.52). In the course of designing a solution, for example, creativity techniques (Sect. 4.4.5.1) are used to develop various possible solutions. In line with the *principle of deferred judgment* (Haberfellner et al. 2018, pp. 75, 258; Sell and Schimweg 2002, p. 77; Traut-Mattausch and Kerschreiter 2018, p. 274), these solutions are only evaluated comparatively afterwards. If necessary, the initial situation can be included (omission variant) to ensure that a new solution is significantly better than this in any case.

In Fig. 4.52, various ideas for solutions are developed on the basis of the problem. A comparative evaluation then reveals one idea as the relatively best idea. In the course of designing a work system (Bokranz and Landau 2012; Schlick et al. 2018), it can be discussed at this level of detail, for example, whether an assembly should be carried out on one work system or distributed across two work systems in such a way that pre-assembly is carried out at the first station and final assembly at the second. Assuming that a single work system is chosen, the next level of detail would answer the question of the direction in which assembly should take place. At first glance, working from left to right often seems obvious.

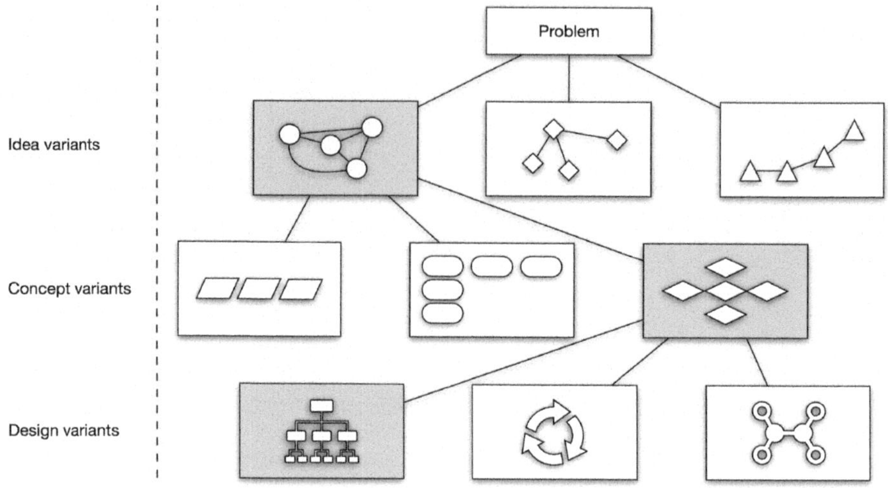

Fig. 4.52 Relative optimization through variant planning from rough to detailed (Haberfellner et al. 2018, p. 58)

In the horizontal, however, if you take a circle with 360 degrees as a basis, in principle 359 other directions are also possible, including, for example, from right to left, from back to front or from front to back. Furthermore, an addition in the third dimension is also conceivable, which then also allows, among other things, working from above or below, or from bottom to top. On closer inspection, working in a circle also appears to be particularly interesting for reasons of time management, since the pick-up and drop-off points can then coincide, which eliminates the need to cover distances with the hand-arm system without value-adding activity (Bokranz and Landau 2012, Vol. 2, p. 221, Fig. III-198). With such considerations, it becomes clear how numerous variants can be. In addition, a systematic search for variants may also lead to those that would not be noticed by anyone at first glance.

The concepts developed at this level can now in turn be subjected to a comparative evaluation. At the next level of detail, several designs are then produced for the relatively best concept, which are then compared with each other again, and so on.

Since the selection of a variant at a certain level of detail is made without knowledge of the possibilities and limitations at lower levels of detail, it is not certain that the best variant will be found in this way. If at a certain stage the impression arises of having reached an impasse, it is possible, albeit laborious, to jump back to a higher level in order to continue the search from there in a different direction (Haberfellner et al. 2018, p. 256). The limits of the possibility of finding the true optimum in this way are vividly illustrated by the example of the awarding of contracts. For this purpose, the usual procurement regulations, such as those for the public sector (Sect. 12 UVgO), stipulate that three[33] offers be obtained. However, it is obvious that, in view of the usually significantly larger

[33] Although at least three offers are usually requested, in practice this leads to no more than the minimum being implemented, given the usually limited resources.

number of potential providers, the best possible providers are not necessarily chosen. The hope, which is certainly not entirely unjustified, rests solely on the fact that the competition to which the providers are exposed in this way will lead to offers that are in line with the market. If this is not confirmed in the course of the increasingly detailed selection process, further providers can be consulted.

Reinhard Haberfellner et al. (2018, p. 243 ff.) describe the alternation between creative-design and destructive-critical phases as an *alternation between synthesis and analysis*. However, methods of analysis are only explained in detail in Sections 4.5.6 and 4.6.4. Therefore, at this point, attention should be drawn to the occasionally observed difficulty of distinguishing between the evaluation phases. To do this, the level of detail can be used initially and, in the course of the more detailed elaboration, the respective object of consideration, whereby the latter is ultimately also an expression of the level of detail from a systematic perspective. However, for the sake of better comprehensibility, a distinction should be made between the two.

In the above-mentioned example of work system design, a single- and a multi-user work system were provided at the highest level of detail. In this case, it is the entire work system, but in a very rough (abstract) form. At the next level of detail, the direction of work was discussed. On the one hand, this corresponds to a more detailed consideration, but at the same time it only deals with a partial aspect of the entire work system. So while the level of detail of the object of observation increases from level to level, the respective object of observation represents an ever smaller section of the overall system. In the selected application example, it is even possible to compare variants of individual hand movements (Bokranz and Landau 2012, Vol. 2, p. 216) at the lowest level, which can be compared with an accuracy of 0.036 s (1 TMU) in terms of their cost-effectiveness with the help of MTM. As the design progresses, the view zooms in deeper and deeper until it reaches the level of the elements, i.e. the units that cannot be broken down any further in a meaningful way.

One advantage of the V-Model (Fig. 4.28) is that it visualizes this approach from high-level to detailed requirements particularly clearly on the left, descending branch. At the same time, it can be seen in anticipation of the following Section 4.4.5.3 that the level of detail in the following phases of integration decreases again, especially in connection with the tests required at each level of detail, while the objects of consideration become more extensive from level to level, until finally the design object is considered as a whole again. In these phases, the analyses serve the purpose of *verification*, i.e., answering the question whether the design objects fulfill their respective functions without defects (Sect. 4.5.6). The final assessment in the sixth phase of the problem solving cycle (Sect. 4.6) serves the purpose of *validation*, i.e., answering the question whether the right object has been designed. This is also expressed in the V-model (Fig. 4.28).

4.4.5.3 Test Methods

In the spirit of Test-driven Development (TDD), tests are developed and implemented in parallel with the development and realization of the design objects. This is done independently of the development process to ensure *objectivity*. Tests are used for *verification* and *validation* (Informationstechnikzentrum Bund 2019, p. 34; Wenzel et al. 2008, p. 33 ff.).

- So-called black-box tests (Fig. 3.19) are required for verification, which serves to check whether a design object meets the requirements (goals, specifications). In software development, these are tests that are carried out without knowledge of the inner workings of the design object. Rather, only the external effects may be known. The independence of the test development in connection with the black-box view is intended to prevent tests from being designed around possibly known or feared errors.
- In addition to black-box tests, white-box tests are also required. Although these should also be defined and carried out independently of the development and realization of the design objects, they should be carried out with knowledge of the internal functioning in order to be able to check them for errors.
- Finally, tests must be defined that serve to validate whether the design object is suitable for the intended use case. This also corresponds to the review of the specification derived in the course of the situation analysis P 2 and the formulation of objectives P 3.

The characteristics and behaviors of design objects (products, services) to be tested vary greatly from case to case and are very diverse overall. In the automotive industry, which is used as an example below, these include static tests (e.g. strength, tightness, blocking, pull-off forces), dynamic mechanical tests such as shaking, swinging, rotating or falling, climatic tests (e.g. temperature, humidity, corrosion, fuel, ozone), electrical tests and tests with regard to electromagnetic amplification, noise tests and many more (VDA volume 3:2008, p. 206). In other industries, other criteria may be relevant. Therefore, it is not possible here to provide a general and comprehensive overview and insight into the definition of specific tests. However, in order to at least provide an impression of what is specifically involved in this context and what work is associated with it, the *planning of product reliability* will be outlined below as an example (e.g. Bracke 2022). The source cited above deals with this topic in much greater detail and in a very comprehensible way, and we expressly recommend reading it as a supplement at this point.

In order to be able to sketch out which tasks are usually involved in the course of defining tests, using the example of planning the reliability of products, this is first defined as the ability of a product to fulfill the requirements determined by the intended use without error during the entire period of use (ibid., p. 33). Since reliability only becomes relevant in the L2 usage phase, the example of a definition of tests based on this does not include the L1 development phase. Therefore, it should be noted that numerous tests are also required to ensure the success of this phase, for example, with regard to the error-free installation of parts or assemblies in the end product.

The basis of every test is knowledge of the expected results. In this regard, tests have already been distinguished from experiments, the outcome of which can at best be assumed hypothetically, but is often also more or less unknown. Knowledge of the expected results is based on knowledge of the corresponding laws. For this purpose, reliability is to be modeled on the basis of the failure rate $\lambda(t)$. This rate indicates the number of failures at a specific point in time t in relation to the total number of units that have remained intact up to that point (ibid., p. 33). Let us assume that 12 components were subjected to load changes in 12 experiments. The number of load cycles until the respective failure is shown in the upper diagram of Fig. 4.53. In the lower diagram, the bar heights represent the relative frequencies h_{rel} as a quotient of the number of failures in the respective class n_A and the total number n of failures.

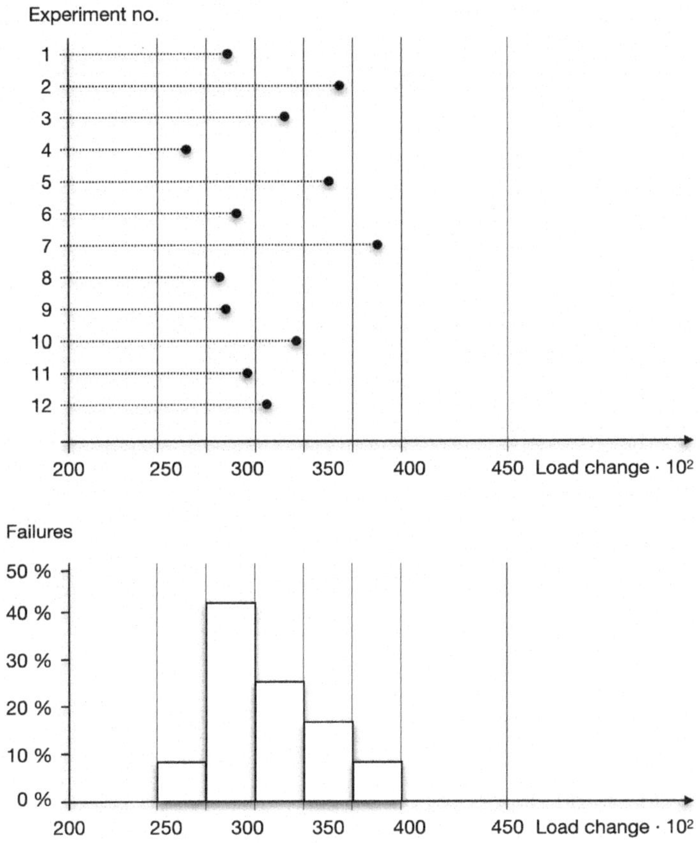

Fig. 4.53 Downtime and downtime histogram (VDA Vol. 3:2008, p.)

$$h_{rel} = \frac{n_A}{n} \tag{4.30}$$

For this purpose, the number of classes n_k was estimated according to the following equation 4.31:

$$n_k \approx 1 + 3.321\,\lg(n) \tag{4.31}$$

The class width Δb then results from the range R, which in turn is calculated from the difference between the largest and smallest measured values, as follows:

$$\Delta b \approx \frac{R}{1 + 3.321\,\lg(n)} \tag{4.32}$$

As shown in Fig. 4.54, the empirical distribution function $F^*(t)$ results from the cumulation of the frequency distribution (cumulative frequency). For an infinitely large number of classes $n \rightarrow \infty$, the distribution function $F(t)$ (probability of default) is obtained as an integral over the density function:

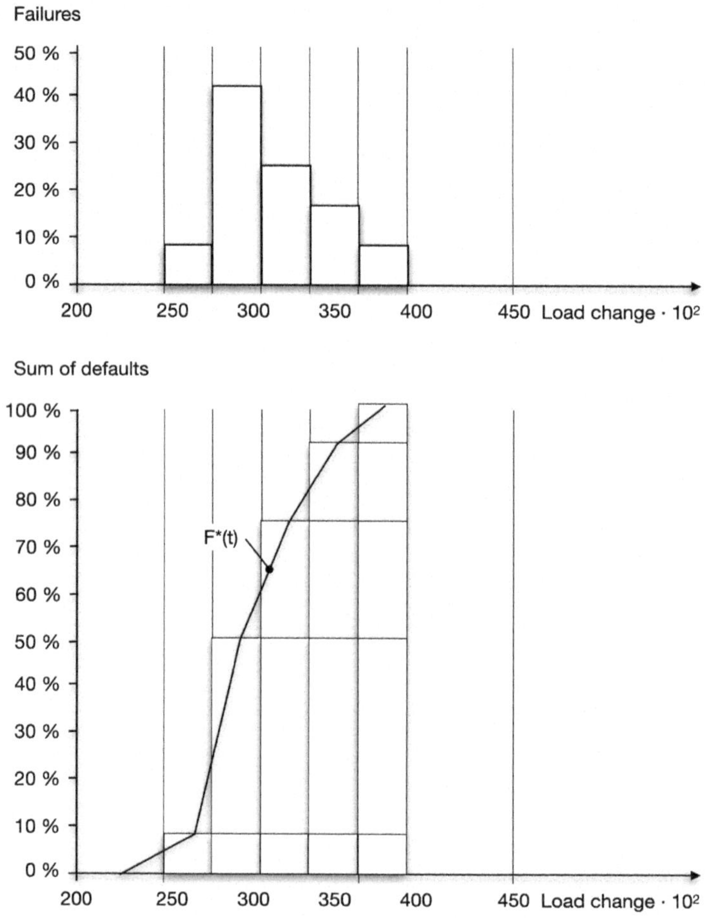

Fig. 4.54 Cumulative frequency and distribution function (VDA Vol. 3:2008, p. 30)

$$F(t) = \int_0^t f(\tau)d\tau \tag{4.33}$$

The density function $f(t)$, whose significance has already been explained in Sect. 2.4.5 using examples, is derived from the distribution function $F(t)$ by differentiation. This is only mentioned again briefly because the density function is needed in the further course.

$$f(t) = \frac{dF(t)}{dt} \tag{4.34}$$

In practice, however, one is less interested in units that have already failed and more in those that are still intact. These are represented by the survival probability (reliability) $R^\star(t) = 1 - F^\star(t)$, which is the complement of $F^\star(t)$ (Fig. 4.55).

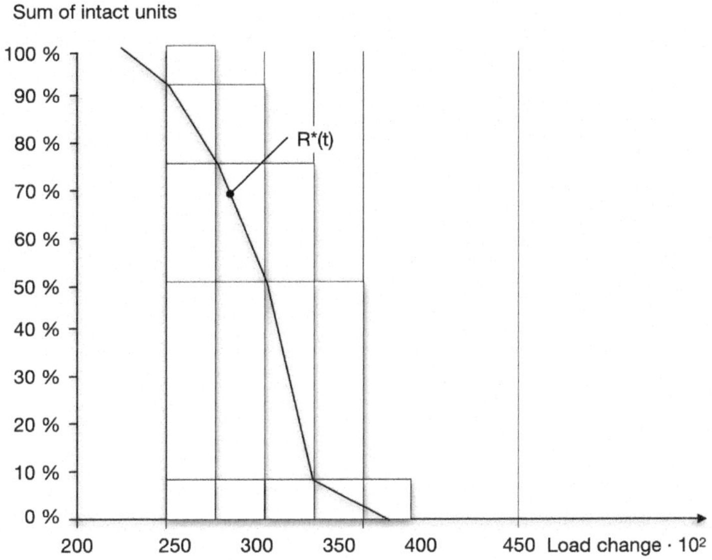

Fig. 4.55 Representation of the failure behavior with the histogram of the survival frequency or with the empirical survival probability R*(t) (VDA Vol. 3:2008, p. 32)

On this basis, the failure rate $\lambda(t)$, which is used to model the reliability of products, can be calculated as the quotient of the density function $f(t)$ and the survival probability $R(t)$.

$$\lambda(t) = \frac{f(t)}{R(t)} \tag{4.35}$$

It indicates the probability with which a unit that has not yet failed by the observation time t_x will fail in the remaining time interval $t_x + dt$ until the end of the service life t.

The failure rate $\lambda(t)$ can be used to model the failure behavior of products throughout their entire period of use. The corresponding function is shown in Fig. 4.56 and is also referred to as a *bathtub curve* due to its shape. The risk of failure is initially high during the start of regular operation, since in most cases errors occur that were not detected during the design phase. The failure rate then decreases as improvements are made, entering a phase in which only random failures are to be expected. These can result, for example, from operating errors (human error) or from inadequate maintenance. At the end of the service life (palliation L 3), the failure rate then increases again due to wear and tear and fatigue.

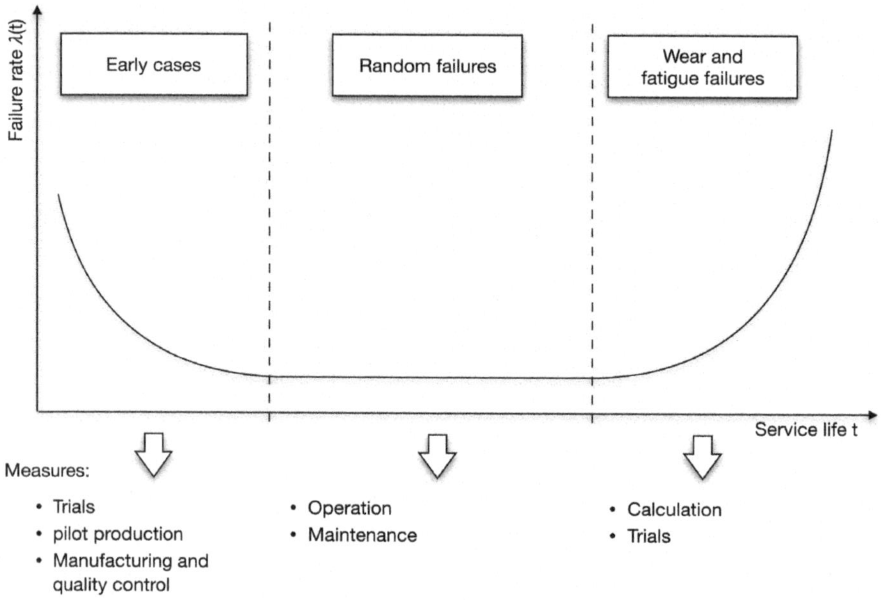

Fig. 4.56 Principle sketch of the failure rate of technical products during the usage phase (bathtub curve) (Bracke 2022, p. 34; VDA Volume 3:2008, p. 35)

In Mechanical Engineering, in particular, the *Weibull distribution* is used to model this failure behavior (ibid., p. 46 ff.). Its advantage is that it can take on the forms of the three phases just explained by varying its shape parameter b (Sect. 2.4.5.1): for $b < 1$, the density function of the Weibull distribution resembles the density function of a negative exponential distribution (Fig. 2.22); for $b = 1$, it corresponds exactly to the exponential distribution. Its essential characteristic of a constant failure rate λ can be interpreted to mean that, relative to the units that are still intact at a particular point in time, a constant percentage of units will always fail (ibid., p. 44). For $b > 1$, the density function of the Weibull distribution always begins at $f(t) = 0$, reaches a maximum with increasing lifespan and then falls away at a flat rate. To approximate reliability in the last phase of the system's life cycle, the shape parameter $b = 3.5$ is used (ibid., p. 46).

The Weibull distribution is used in both a two-parameter and a three-parameter form. In addition to the above-mentioned shape parameter b as a location parameter, the characteristic lifetime T is added for both distribution forms. The distribution becomes three-parameter when a failure-free time t_0 must be taken into account. This is the case, for example, when a specified wear margin is used up first, as is the case with the brake disks and brake pads of a disk brake. In the following, only the two-parameter case is considered for the sake of simplicity.

When using the Weibull distribution, the survival probability (reliability) of a unit $R(t)$ is calculated as follows:

$$R(t) = e^{-\left(\frac{t}{T}\right)^b} \tag{4.36}$$

$$F(t) = 1 -^{-\left(\frac{t}{T}\right)^b} \tag{4.37}$$

The density function $f(t)$ is obtained by differentiation to

$$f(t) = \frac{dF(t)}{dt} = \frac{b}{T} \cdot \left(\frac{t}{T}\right)^{b-1} \cdot e^{-\left(\frac{t}{T}\right)^b} \tag{4.38}$$

and the failure rate $\lambda(t)$ according to the generally valid equation given above to

$$\lambda(t) = \frac{f(t)}{R(t)} = \frac{b}{T} \cdot \left(\frac{t}{T}\right)^{b-1} \tag{4.39}$$

In these equations, the static variable t is related to the characteristic lifetime $T(t/T)$. It follows that the quotient takes the value 1 at $t = T$. Thus, the failure probability is calculated as $F(t) = 1 - e^{-1} = 0.632$. The characteristic lifetime T thus corresponds to a failure probability of $F(T) = 63.2\%$ or a survival probability of $R(T) = 36.8\%$. Since the formation of an arithmetic mean is not meaningful in this case, these values can be understood as specific forms of mean values (Bracke 2022, p. 50). Furthermore, a q-percent lifetime B_q is often used to characterize the failure behavior, which expresses the time until q percent of the units from a total quantity have failed. The B_{10} lifetime is common.

$$B_{10} = F-1(q) = T \cdot \sqrt[b]{-\ln(1-q)} = T \cdot \sqrt[b]{-\ln(1-0,1)} = T \cdot 0.1054^{\frac{1}{b}} \tag{4.40}$$

The non-linear progression of the failure probabilities $F(t)$, which Fig. 4.54 shows, is not well suited for practical application. Instead, the two-parameter distribution is transformed into a straight line by using a special Weibull probability paper (Weibull mesh). For this, the abscissa is divided logarithmically and the ordinate is divided in double logarithmic scale (ibid., p. 75 ff.):

$$x = \ln t \tag{4.41}$$

$$y = \ln\{-\ln(1 - F(t))\} \tag{4.42}$$

Fig. 4.57 shows the Weibull probability paper with different failure probabilities F(t). The slope of the straight line is influenced by the shape parameter b, which can be read directly on the right-hand ordinate if the straight line is shifted in parallel through the pole (ibid., p. 52 f.). It should be noted that the three areas of the bathtub curve (Fig. 4.56) are characterized by different shape parameters b. This results in a separate line for each area in the probability paper (Bracke 2022, p. 57). Standard software such as spreadsheet programs do not support this. However, specific software is available on the market for this purpose.

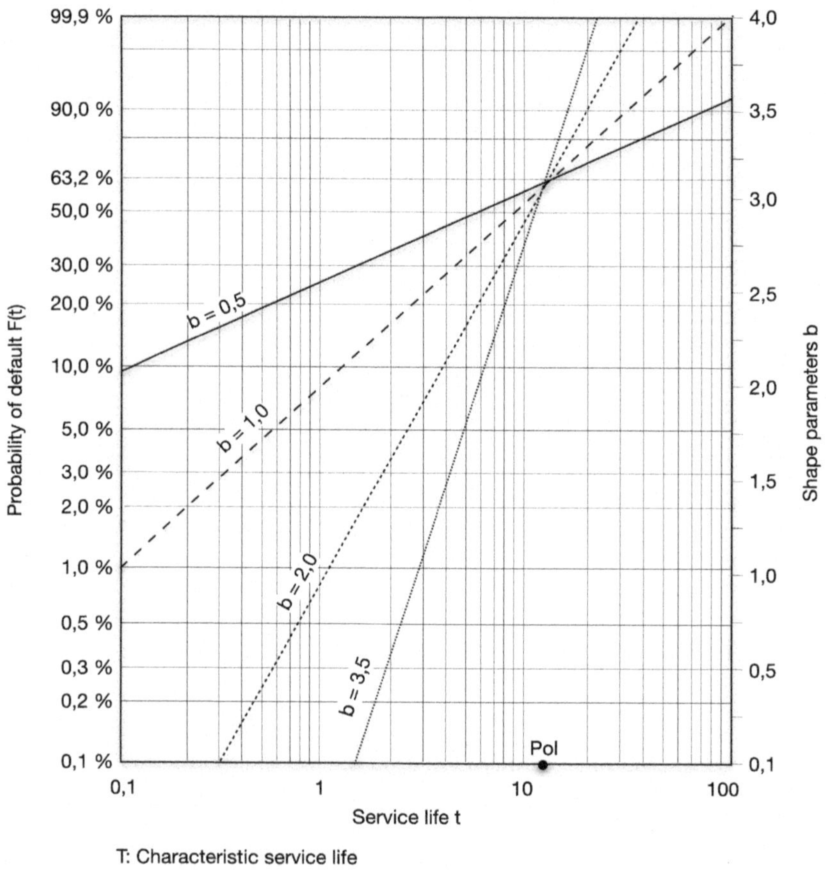

Fig. 4.57 Weibull probability paper with different failure probabilities (VDA Vol. 3:2008, p. 52)

Many tests are not carried out completely, but only on the basis of samples. Although this reduces the effort involved in conducting tests, it does, however, make it necessary to extrapolate from the sample to the population. This is not possible with absolute certainty. Nevertheless, a confidence interval can be specified with a relatively high degree of certainty, which indicates the probability with which the findings obtained on the basis of samples also apply to untested products. Since the determination of the *confidence interval* is one of the essential tasks of defining tests, it will be explained below, at least in outline, using an example taken from the supplementary recommended literature (ibid., p. 86 ff.) for a service life test of gear wheels. While this example is initially only about planning tests, it also includes their execution and evaluation of the results. However, this helps to understand what needs to be considered during the planning stage, because, as has been explained several times, tests differ from experiments in that their result is basically already known in the form of relevant requirements and only serves to verify and validate them, so it often helps to think of them in terms of their result.

In the example, the sample size is set to n = 10 gears. The failure behavior caused by pitting is to be analyzed. For this purpose, the gears are subjected to a load of $\sigma_H = 1,528N/mm^2$. In the first test, the corresponding sample fails after 15,100,000 load cycles (LC), in the second after 12,200,000 LC, in the third after 17,300,000 LC, in the fourth after 14,300,000 LC, in the fifth after 7,900,0 00 LC, in the sixth after 18,200,000 LW, in the seventh after 24,600,000 LW, in the eighth after 13,500,000 LC, in the ninth after 10,000,000 and in the tenth attempt after 30,500,000 LC.

In the example, it should remain a single sample, but to understand the further explanations, it is temporarily helpful to consider that tests often include several samples. These do not provide exactly the same results, but are subject to a distribution. To take this into account, the test results available in the order of the tests are resorted in ascending order according to the number of load cycles, i.e. they are *ranked* t_i so that the smallest value takes the first rank t_1 = 7,900,000 LC, and the largest value takes the highest rank, in this case t_{10} = 30,500,000 LC. With $m > 1$ samples, all the values (rank sizes) of the first rank would generally be around the order of 7,900,000 LC. The dispersion of the test results of a rank in more than one sample, known as the rank size distribution, corresponds to a trinomial distribution, whose density function is given by

$$\varphi_n(t_i) = \frac{n!}{(i-1)! \cdot (n-i)!} \cdot F(t_i)^{i-1} \cdot f(t_i) \cdot [1 - F(t_i)]^{n-i} \tag{4.43}$$

This results in a separate form of distribution for each rank. The respective median value $F(t_i)_{median}$ is first determined from each of these distributions. Since there is no closed solution for this, the values are taken from a corresponding table (ibid., p. 392 ff.). The following median values can be read from this for the 10 ranked values of the exemplary test series: $F(t_1) = 6.7\%$, $F(t_2) = 16.3\%$, $F(t_3) = 25.9\%$, $F(t_4) = 35.6\%$, $F(t_5) = 45.2\%$, $F(t_6) = 54.8\%$, $F(t_7) = 64.4\%$, $F(t_8) = 74.1\%$, $F(t_9) = 83.7\%$, $F(t_10) = 93.3\%$. The value pairs $(t_i, F(t_i))$ are transferred to the Weibull probability paper (Fig. 4.55) and ideally yield the expected straight line. Due to the scatter that is usual in practice, this is normally to be drawn as an equalization straight line through all values.

This straight line now provides the first results of the test series, because the point of intersection of the straight line with the 63.2% line indicates the characteristic lifespan T of the gears on the abscissa, in this case specifically with T = 20,000,000 LC, and by a parallel translation of the best-fit straight line through the pole, its slope can be read as $b = 2.6$. This results in the specific Weibull distribution of

$$F(t) = 1 - e^{-\left(\frac{t}{20 \cdot 10^6 LC}\right)^{2.6}} \tag{4.44}$$

Now the confidence interval still needs to be added. If this is set to 90%, as is usual, then a 95% limit above the regression line and a 5% limit below it must be drawn. These values must also be taken from the corresponding tables (ibid., p. 388 ff.) for each individual rank size of the sample. For example, for the smallest rank size value $t_1 = 7{,}900{,}000$ LC, the values $F(t_1)_{5\%} = 0.51\%$ and $F(t_1)_{95\%} = 25.9\%$ are given. These values are also entered into the Weibull probability paper in order to draw the respective best-fit curves through the points obtained (Fig. 4.58). This now allows us to deduce from the sample that the straight line of the median values of the population lies between the confidence limits with a probability of 90%. If we were to apply the straight lines of best fit to the confidence intervals, which we will not do here, you could also read the minima and maxima of the key figures for the probability of failure (ibid., p. 93).

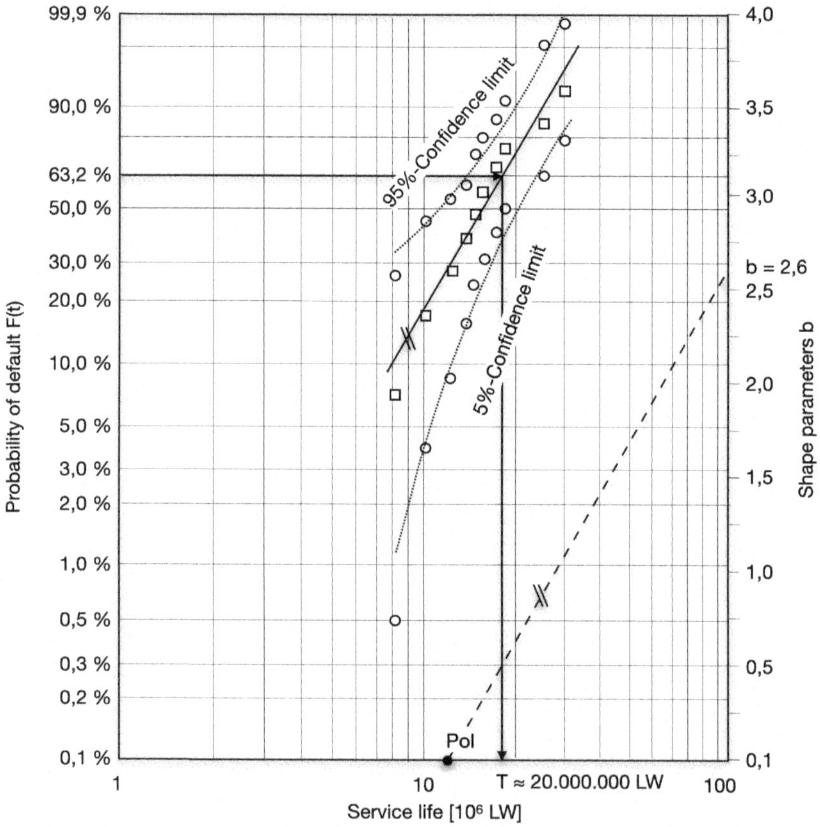

Fig. 4.58 Weibull line and 90% confidence interval (VDA Vol. 3:2008, p. 92)

In view of this expected result of the reliability test, its planning begins by defining the parameters of the tests on the basis of the outlined statistical laws. This includes, in particular, the *size of the sample*, which, on the one hand, has a significant influence on the accuracy of the result, but on the other hand, also affects the effort required for the tests. Furthermore, it must be ensured that the selection of samples is carried out and how, because it is important to ensure that the *random principle* is adhered to.

The next step is to define the *test strategy*. Only in the case of complete tests is it not necessary to define sample sizes, since in such cases all components are tested. However, this is rarely the case in practice. As a rule, incomplete (censored) tests are carried out to limit the effort involved. While these are only terminated after the failure of the samples, there are also test strategies that only perform the tests up to a predefined service life or until a previously defined number of components fail. Particularly extensive reductions in testing are possible by applying the so-called sudden death strategy or by increasing the stresses. Finally, the expected *confidence level* is to be determined with regard to the statistics of the tests.

However, test planning does not end with the definition of the statistical parameters. The *test setup* must also be planned, which in turn requires components for applying the load as well as measuring and testing technology. Test benches such as these are available on the market for many recurring tests, such as the test used here as an example to determine the failure behavior of gears. In such cases, the question arises, among other things, as to whether these should be operated in-house or whether external services should be used. For less frequently occurring applications, test stands often have to be developed, designed, built and put into operation first, which then entails an independent design project that should in turn be carried out with the help of the methods, procedures and techniques of Systems Engineering.

The remarks on the definition of tests have shown, using a very small but very relevant example from practice, what is required in this phase of the problem-solving cycle to do justice to *Test Driven Development* (TDD), as is also rightly required by agile approaches and particularly emphasized in the V-Modell (Fig. 4.28). At the same time, it should be clear that this aspect is also an integral part of Systems Engineering, and in which place.

4.4.6 P 4.6: Evaluation

The situation analysis, the formulation of objectives and the now completed definition of methods together form the essential confidence-building measures for those stakeholders who are not directly involved in the subsequent development of the solution. Therefore, it is also important in the definition of the method that all participants and stakeholders are convinced that the underlying theories (hypotheses, assumptions) and models, as well as the defined methods, procedures and tools, are suitable for solving the problem efficiently and effectively (economically). Therefore, the specifications, which are written by the contractor and essentially document how and with what the problem is to be solved, are usually also signed by the client.

What follows is the hard work. Avoiding this work is one of the main reasons why the client hands the solution over to the contractor, whether it is part of their own organization or a third party. Also, the client does not always have all the necessary skills and resources to solve the problem themselves. This is another reason why the client hands over the solution. However, what a client always wants to understand, at least to some extent, is how and with what the contractor believes they can successfully solve the problem. In addition to a shared understanding of the situation and a common objective, this is an essential basis for the trust that the client must place in the contractor. Therefore, a joint evaluation of the defined methods, which leads to a mutually agreed result, is the last hurdle that must be overcome before the actual work can begin.

4.5 P 5: Development of a Solution

The search for solutions usually requires a cycle of synthesis and analysis to be repeated several times (e.g. Bender and Gericke 2021, p. 174; Haberfellner et al. 2018, p. 243 ff.). The strict separation between creative and critically reflective phases corresponds to the *principle of deferred judgment* (Haberfellner et al. 2018, pp. 75, 258; Sell and Schimweg 2002, p. 77). As already mentioned in Sect. 4.4.4, this is based on the insight that with almost every idea of how something could work, the idea is also immediately born of why it probably won't work after all. As a result, many ideas are lost that could prove valuable on closer inspection. Since such an analysis is usually done intuitively, it can neither be avoided nor guided intuitively. This requires a conscious and methodically supported approach. In addition to the use of creativity techniques (Sect. 4.4.5.1), the strict separation between synthesis and analysis also contributes significantly to this.

Due to the self-similarity of the problem-solution cycle (Fig. 3.32), it is also run through within the solution development phase. To make this clear, let us assume, as in the explanation of previous phases, that the development is an independent project that may even be outsourced to another company. Then it becomes clear that this company also carries out the order in the form of an independent project and must go through all the phases to do so.

If the explanations of the contents of the individual phases are comparatively short, there are two reasons for this: Firstly, much has already been said in connection with the corresponding phases of the overarching problem-solving cycle, which is just as valid here, but of course to an appropriate extent depending on the scope of the project. On the other hand, and this appears to be the more important reason, the following section is now primarily concerned with actually doing everything that has been planned and prepared so far. However, no matter how many words are used, the reader cannot be relieved of the task of actually doing it. Nevertheless, the explicit identification of the individual phases by means of corresponding sections should not be dispensed with for reasons of systematics.

4.5.1 P 5.1: Occasion

In principle, everything that has already been explained in detail in Section 4.1 applies to the initiation of the specific solution to the problem. One of the strengths of recursive thinking lies in the transferability of insights once they have been gained and in the resulting possibility of being able to avoid repetition. Therefore, as just announced, this section can now be kept short here. In practice, however, this should not tempt you to pay too little attention to the solution development. Even if it seems to fit seamlessly into the overall project, the beginning of this new phase should be consciously initiated.

4.5.2 P 5.2: Situation Analysis

For the analysis of the initial situation of the actual problem solution, the same applies as for the occasion: everything essential has already been explained in Section 4.2 and also applies here. However, in this phase of the problem solution, it is also important to clarify the actual state, which is to be transformed into a target state in the course of the solution development, so that the barrier that has to be overcome is clearly apparent. In particular, it is important to check whether the results of the previous situation analysis (P 2) and the formulation of objectives (P 3) are complete and clearly understandable. If this is not the case from the point of view of those who will be entrusted with the subsequent problem solving, this may be a reason to jump back to the corresponding phases and to deliver or improve the missing or unclear results. This, too, can perhaps be best imagined if one starts from the point of awarding a contract. In such a case, the client and the contractor would carry out a corresponding clarification of the contract.

4.5.3 P 5.3: Target Definition

Now that the introductory remarks and the two preceding Sections have been referred to in each of the higher-level Sections, it should be clear at this point that this also applies to phase P 5.3 of the formulation of objectives in the course of solution development (P 5). So what has already been explained in Section 4.3 on the formulation of objectives also applies here in principle and should therefore not be repeated in full here either.

4.5.4 P 5.4: Method Planning

The methods to be used in the further course of the problem solving, i.e. in synthesis P 5.5 and in analysis P 5.6 as well as for the final evaluation P 6, have basically already been planned in phase P 4 and will not be repeated here. Nevertheless, there is usually still a lot to be done in the phase explained here in the concrete preparation of the application of the planned methods in the sense of the approach from the rough to the detailed (Sect. 3.3.1).

For example, many methods require *configuration*[34] and *organizational preparation*. For example, the specific determination of the members of groups, the appointment of moderators, and the planning of space and time should be considered (Sect. 4.4.5.1). And statistical experimental design (with regard to the execution of experiments is often only undertaken shortly before the start of the experiments (Sect. 4.4.5.1). Phase 5.4 is intended for this.

4.5.5 P 5.5: Synthesis

Synthesis, alternating with analysis, is now at the heart of solution development and should therefore be explained in detail. Reinhard Haberfellner et al. (2018, p. 251 ff.) distinguish between the *from-outside-to-inside strategy* and the *from-inside-to-outside strategy* as conceivable approaches. The first strategy corresponds to the approach from the general to the specific (Sect. 3.3.1) and is suitable when complex problems are to be solved and there is a great deal of leeway for design, as is ideally the case with new developments or extensive further developments. The second strategy, on the other hand, starts from the core of the problem and seeks improvements. It is suitable when the problems are manageable and the scope for design is small, which is particularly the case with existing systems. In Production Management, this approach is referred to as continuous improvement (Kaizen) (e.g. Belz et al. 2015, p. 50 ff.) and in Project Management it corresponds to the basic idea of agile methods (e.g. Goll and Hommel 2015, p. 79). However, Reinhard Haberfellner et al. (2018, p. 252) point out that Systems Engineering primarily deals with complex problems and therefore prefers the from-outside-to-inside strategy.

In the following, *design principles* will be mentioned and explained – if this has not already been done elsewhere – that Reinhard Haberfellner et al. (2018, pp. 135 ff.; 246 f.) recommend with reference to A. Terry Bahill and Rick Botta (2008, pp. 9 ff.). The latter authors explicitly point out that these principles apply not only to the development of technical products, but to all types of problem solving, including, therefore, the organizational problems that are always meant here.

(1) The principles already explained in detail include the *use of models*, i.e., *abstraction* (Sect. 2.2). In particular, the Systems Modeling Language (SysML) provides tools (Weilkiens 2014) that promote technically oriented system thinking. With regard to the solution that comes into focus in the course of the concrete development, reference is also made to the difference used by Reinhard Haberfellner et al. (2018, p. 245 f.), with reference to Alfred Büchel (1969), between the *image* as a model of the initial state and the *role model* as a model of the desired state. While the image always fades, the model should be further concretized until the solution, which is always based on it, is finally released into reality, where it must prove itself.

[34] This should not be confused with so-called Configuration Management, which serves to maintain the consistency (integrity) of a product or product group throughout its entire life cycle. There are several standards for this, such as DIN ISO 10007:2020-10.

(2) The procedure *from the rough to the detailed* (top-down) was also explained in detail in Section 3.3.1 and just again. A. Terry Bahill and Rick Botta (2008, p. 9) point out in this context the necessity of structuring the system to be developed hierarchically and, if possible, to remain at one hierarchy level until the development of all sub-systems arranged at the respective level has been completed. If development on one branch runs ahead, the result may be that more detailed and therefore usually more expensive development results have to be discarded if it turns out on another branch that a development result cannot be achieved. This is especially a risk in Simultaneous Engineering (Fig. 3.21), for example, if a production process for a product on which significant changes will be made later is already largely completed.

(3) A design principle that has not yet been sufficiently expressed calls for starting with the development of those sub- or part-systems or elements that pose the *greatest risks* (ibid.; Haberfellner et al. 2018, p. 247), because these are more likely to lead to the failure of solution attempts. If this happens, it does not also affect those solution approaches that appear promising in themselves but depend on the failed approaches and consequently cannot be pursued further.

(4) A. Terry Bahill and Rick Botta (2008, p. 10) also attach great importance to the ongoing setting of *priorities*. How this is to be done in practice is left open by the authors mentioned so far. Therefore, at this point, Ulrich Thonemann's (2015, p. 374 ff.) comments on this topic will be explained. In addition to the prioritization of the development of risky elements already recommended above for the third principle, there are a number of other priority rules that, while they should be generally known, experience shows that their respective effects cannot always be explicitly stated.

- The quasi-natural priority rule is to process work in the order in which it arrives (FIFO: First In, First Out or, synonymously, FCFS: First Come, First Serve). This rule provides the optimal sequence when, for example, aging processes, as they occur with perishable goods such as food, are to be taken into account, or when fairness is to be ensured, for example when serving customers who can observe the service process. This priority rule is used in particular in legal matters, in that if there are several equally ranked processes, the one that occurred earlier always takes precedence (prior tempore potior iure). This is based on the legal concept handed down from ancient times and still valid today, that you cannot transfer more rights than you yourself have (nemo plus iuris ad adium transferre potent quam ipse haberet).
- The expectation that compliance with the FIFO rule will also minimize waiting times for customers is only fulfilled for the moment, however, while the average waiting time for repeated service processes, such as the time spent at the supermarket checkout, is not minimized in this way. To do this, you always prioritize the work that causes the shortest processing time (SPT). While this processing rule would probably not be enforceable at supermarket checkouts for obvious reasons, although it would actually be in the interest of customers, it is very effective where a handling process does not have to take into account either aging processes or fairness requirements and is intended to minimize the average completion time.
- Another concern may be to prioritize work in such a way that the average maximum delay is minimized. The EDD rule (Earliest Due Date), according to which the orders with the earliest due date are processed first, is suitable for this.

- Finally, the average number of late jobs can also be minimized. However, the Moore's algorithm to be used for this is somewhat more complex than the priority rules mentioned above. In a first step, the orders are to be sorted according to the EDD rule. Ulrich Thonemann (ibid., p. 378 ff.) refers to the set of orders sorted in this way as R. He also defines a set $J = \{\emptyset\}$, which is used to collect the delayed orders in the sequence. In the second step, the first delayed order i is selected from the set R (if no order is delayed, step four is carried out). In the third step, of all orders in the set R that do not start after the first delayed order, the one with the longest processing time is transferred to the set J. After that, a jump back to the second step occurs. In the fourth and final step, the set J is appended to the set R, whereby it can be shown that the order of the appended orders is not important.

Priority rules like this one can be applied universally, for example, to production planning and control (PPC) as well as to data processing when assigning computing operations to processors.

Another prioritization option is known as the *Eisenhower Matrix*, which emphasizes the difference between *importance* and *urgency*. The name refers to a speech given by US President David Dwight Eisenhower on August 19, 1954 at the second plenary assembly of the World Council of Churches in Evanston, Illinois, in which he quoted an unnamed university president as saying: "I have two kinds of problems, the urgent and the important. The urgent are not important, and the important are never urgent". The matrix shown in Fig. 4.59 is derived from the combination of the two-level criteria of importance and urgency.

		Urgency	
		urgent	not urgent
Importance	important	do it yourself immediately	do it yourself later
	not important	Delegate to employees	Do not complete

Fig. 4.59 Eisenhower matrix for prioritization

(5) Elements or sub- and partial systems should, if possible, only interact on the same hierarchical level as well as with the immediate superlevel and with the immediate sublevel. Interactions across multiple hierarchy levels should be avoided to minimize complexity. The *automation pyramid* (Heinrich et al. 2020, p. 4; Seitz 2015, p. 24) outlined in Fig. 4.60 with its strictly hierarchical structure has been a successful example of this for a long time. While the physical levels are named on the left side of the pyramid, the respective integration into data processing and communication technology can be found on the right side, based on the Open Systems Interconnection Model (*ISO/OSI model*) of the International Organization for Standardization (ISO), which is also hierarchically structured and outlined in Fig. 4.61. This is an equally good example of a transparent hierarchical structure.

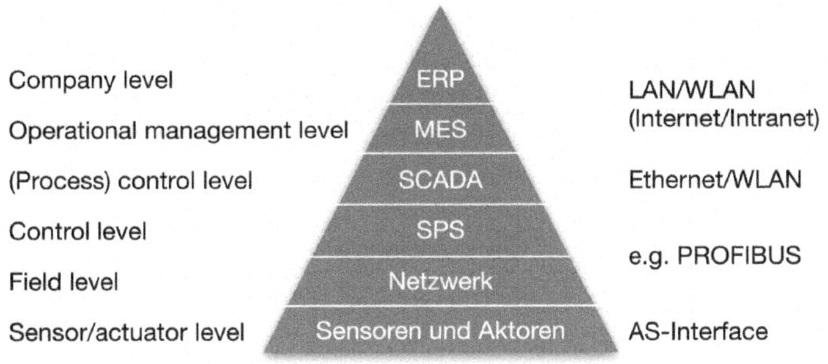

Fig. 4.60 Automation pyramid

- At the lowest level of the automation pyramid (Fig. 4.60) are the devices equipped with sensors, actuators and displays, which are connected to each other via the Actuator Sensor Interface (AS-i) at the field level above (DIN EN 62026-2:2020-12) in order to exchange binary data and to obtain an electrical DC voltage of 30 V, which is sufficient to operate sensors such as light barriers (yellow cables). Actuators receive a DC voltage of 24 V, which is necessary for their operation, via an additional line (black). The devices are connected as so-called slave devices to a master device that controls communication. Alternatively or in addition, IO-Link according to DIN EN 61.131-9:2015-02 can be used. This standard uses bidirectional data exchange as an alternative to a field bus and is able to change slave device parameters during operation. This is occasionally cited as a reason to describe such devices as intelligent.
- AS-i and IO-Link masters are connected to the higher-level bus and used to connect to a programmable logic controller (PLC) and to human-machine interfaces (HMI). In industrial applications, this higher-level bus is often designed as a process field bus (PROFIBUS) or as a process field network (Profinet). In addition to a master-slave process, PROFIBUS uses a mechanism known as token passing (Sect. 2.6 2.4.1), which allows the use of several PLCs on a common bus by giving access to the field bus to whichever PLC has the token. Data that is to be forwarded to another connected device is attached to the token and provided with the recipient's address. In a ring topology (Fig. 2.11 c) according to IEEE 802.5, the token is passed on to the physically next station; in a bus topology (Fig. 2.11 b) according to IEEE 802.4, it is passed on to the addressed network card. The receiving device reads the data and marks the token as free so that it can be rewritten. Profinet is an open Industrial Ethernet standard according to DIN EN IEC 61158-1:2020-04 and DIN EN IEC 61784-1:2020:08 and is a further development that enables significantly higher transmission speeds. In the ISO/OSI model (Fig. 4.61), PROFIBUS can be assigned to the first and second layers, while PROFINET can be assigned to the seventh layer as an application, which is based on the standard Ethernet of the first and second layers. PROFINET therefore accesses all layers of the ISO/OSI model.
- The control or regulation of the devices is carried out with the help of programmable logic controllers (PLC). In addition, further tasks such as alerting or recording operating messages (data logging) can be taken over. Programming is carried out on a separate

programming device, from which the finished program is then transferred to the PLC. In addition to the PLC, user interfaces (HMI) are also integrated at this level of the automation pyramid (Fig. 4.59), for example control panels with signal lamps, display panels with buttons or visualization software on a terminal, which allow monitoring and targeted intervention by the operating personnel.

- Further monitoring and control takes place at the fourth level of the automation pyramid, the process control level (Supervisory Control and Data Acquisition: SCADA). At this level, the production data acquisition (PDA) or the machine data acquisition (MDA), among other things, can be connected. Communication takes place in the ISO/OSI model at the fourth level.
- The fifth level of the automation pyramid is the Operations Management level. This is where manufacturing execution systems (MES) are used in accordance with Sheet 1 of VDI Guideline 5600, which serve to control production, including the release and monitoring of production orders. In the ISO/OSI model, data exchange with the higher-level system takes place in particular via TCP/IP.
- At the top level of the automation pyramid are enterprise resource planning systems (ERP), which support the company-wide control of all value-added processes. Since the corresponding software also includes cross-company supply networks, for example in the context of Supply Chain Management, and is also offered on a cloud basis, it is connected not only to the local area network (LAN) but also to the Internet.

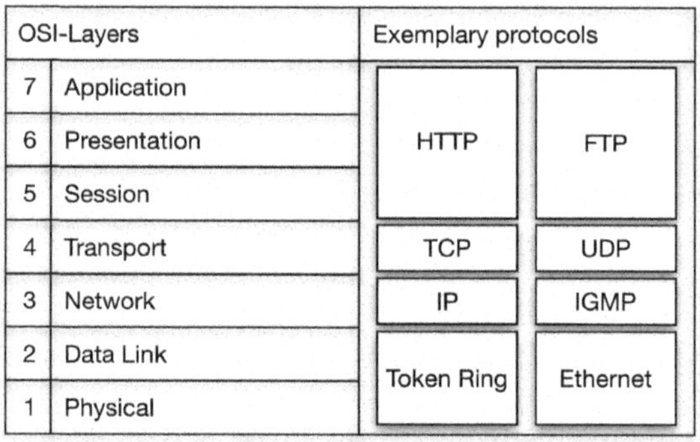

Fig. 4.61 Open Systems Interconnection Model (OSI) of the ISO

(6) A. Terry Bahill and Rick Botta (2008, p. 10) mention the necessity of paying particular attention to the *interfaces* as the sixth design principle. For example, a datum that needs to be exchanged between elements or sub- or part-systems multiple times may indicate unnecessary fragmentation. The number of required interfaces should be minimized. The remaining interfaces should be standardized. Finally, if possible, it should not be necessary to change an interface when the element behind it changes.

(7) The goal of a design does not always have to be an absolute optimum. Following the pragmatic *Pareto principle*, it is usually sufficient to develop satisfactory (viable) solutions. The optimization methods explained in Section 4.4.5.2 take this into account, among other things, by usually proceeding *heuristically*.

(8) In the sense of Model-based Systems Engineering (MBSE), a development should be based on a holistic and always up-to-date *model* that supports communication between all parties involved and affected and, to this end, not only reflects technical but also economic, social, legal, psychological and ecological aspects (Sect. 2.4.6). Before any implementation, the effects of planning and changes to the model should be tested (simulated).

(9) Developments should establish stable *intermediate states* (baselines). For example, a new road can be built in sections. If it turns out over time that it cannot or should not be completed because, in the meantime, it has become more widely accepted that the alternative expansion of the rail network is preferable for ecological reasons, the sections completed so far can still be used.

(10) Developments should be *evolutionary*. A. Terry Bahill and Rick Botta (ibid.) explain what this means using the example of the B-52 Stratofortress developed by Boing for the US Air Force, of which only a basic version was delivered in a first step, in order to then add further capabilities. In this respect, this design recommendation is compatible with the one mentioned above.

(11) Developments should be appropriate for the company that is responsible for the development. This is taken into account, for example, by the SWOT analysis (Sect. 4.2.4.8), which not only looks at opportunities and risks externally, but also at the company's own strengths and weaknesses internally.

(12) Problems should be formulated in a *solution-neutral* way (Sect. 4.3.4.5). Instead of demanding the introduction of a software (means), for example, the goal that is expected from the introduction should be formulated. In some circumstances, going through the problem-solving cycle will lead to a solution that does not require software, thus saving the associated costs.

(13) Requirements should be formulated in the form of *use cases*. SysML provides the use case diagram for this purpose (Alt 2012, p. 49; Weilkiens 2014, p. 234 ff.). This describes a set of actions of a system that lead to an observable result, which offers added value for the system itself, for other systems or for subjects. The difference between a user story used in agile Project Management (Pichler 2008, pp. 34 ff.; Sutherland 2015, pp. 124 ff.; Wirdemann 2017, pp. 50 ff.) and a use case is that user stories (Sect. 2.1) are formulated more concisely and thus tend to be a brief description of use cases.

(14) Each function should only be assigned to one component. This corresponds to *component-based development* in the context of object-oriented programming. In software development, different standards such as Jakarta Enterprise Beans are offered for this purpose. Comparable standards should also be developed for the development of hardware of all kinds.

(15) All functions should be transparent and documented. So-called Easter eggs, such as the flight simulator hidden in the spreadsheet program Excel 97, allow insiders to distinguish themselves from the common people, but promise no further benefit.

(16) The call by A. Terry Bahill and Rick Botta (2008, p. 11) to build prototypes (*rapid prototyping*) in the early development phases shows that plan-based Systems Engineering is not as much at odds with agile approaches as is sometimes claimed (e.g. Goll and Hommel 2015, p. 56 ff.). *Cardboard Engineering* (Gorecki and Pautsch 2012, pp. 29 ff.; Hagen 2009, pp. 10 ff.), explained in Sect. 3.5.5, is an example of this. And *additive manufacturing*

processes (synonym: 3D printing) are established to quickly produce physical models from CAD models with reasonable effort (e.g., Grund 2015).

(17) *Open standards* are to be preferred to proprietary standards. Open standards are publicly available specifications for the fulfillment of a specific task. They can be read and implemented by everyone. Examples are ANSI/EIA-632, IEEE 1220-2005-09, ISO 15288:2023, UML, SysML, OMG, CORBA or USB. In contrast, proprietary standards are controlled by a single company. Java Server Faces can be the interface between HTML code and logic code. Although Java Server Faces is a scalable mass product, it is still controlled by a single commercial company, as is the Source Code Control Interface Specification (SCCI) from Microsoft.

(18) Data and their processing should be *encapsulated* together. Software achieves this, for example, with header files and with class diagrams that contain attributes (data) and operations (behavior). Hardware achieves this through distributed control, in which each card has a processor and memory. Data should be hidden from objects that do not need it. If the data structure is changed, the other objects do not need to be informed of this change. This principle is also referred to as information hiding and function hiding. In the Unified Modeling Language (UML), model elements such as attributes and operations can be marked with the visibility indicators "public", "protected", "private" or "package". Public documents describe only the interfaces and the public functions and data. This principle has nothing to do with security, but serves only to improve the efficiency of changes.

(19) A. Terry Bahill and Rick Botta (ibid., p. 12) also recommend creating a specific *glossary* to accompany a development, in which relevant technical terms are collected and defined. This also includes designations for functions, classes, methods and overarching organizational principles.

(20) The reusability or recyclability of new products should be taken into account during their development. This also applies retroactively to products from older generations that can be incorporated into a new generation. This is associated not only with ecological advantages, but usually also with economic ones.

These recommendations by A. Terry Bahill and Rick Botta (ibid.) should be supplemented by the observation that the results of problem solving often serve to accomplish tasks (Sect. 3.2). Whether the individual case involves technical systems or organizational aspects or technical support for organizational aspects, i.e., a combination of both, is of secondary importance. What is important is the insight that the completion of tasks should be automated as much as possible, whether through the use of technology or the establishment of routines for human action. Therefore, the result of a problem solution is usually a *control loop* in one way or another. This is why cybernetic knowledge and skills are also part of the scope of expertise in Systems Engineering.

4.5.6 P 5.6: Analysis

In the course of solution development P5, the analysis of solutions follows the *principle of deferred judgment* (Haberfellner et al. 2018, pp. 75, 258; Sell and Schimweg 2002, p. 77; Traut-Mattausch and Kerschreiter 2018, p. 274). It serves to ensure, in the course of verification, that each solution variant is fundamentally suitable in itself, in order to derive starting points for improvements in the subsequent synthesis phase from the critical examination required for this, or to eliminate variants that have been definitively found to be unsuitable (Haberfellner et al. 2018, p. 257). On the other hand (2), comparative evaluations are necessary in the course of the process from the rough to the detailed in connection with variant planning (Fig. 4.52) in order to subsequently reduce the solution space, which is enlarged in the course of the synthesis, back to a manageable size and to further pursue the relatively best solution variants in each case. This corresponds to a *validation* in which the question is to be answered as to which solution variant best fulfills the respective purpose. However, the analysis is limited to the verification, while the validation is the subject of the evaluation (P 6) (Haberfellner et al. 2018, p. 74). This interplay between verification and validation is particularly evident in the V-model (Fig. 4.28), which, as already explained in Sect. 4.3.3.3, is comparable to the problem solution cycle (Fig. 3.23).

Section 4.4.5.3 already explains how concrete tests can be carried out. Furthermore, test methods, procedures and tools are so application-specific and diverse that it is not possible to provide a comprehensive description here. The only thing that can be done is to refer to the specific literature. The following comments therefore concentrate on general aspects of analysis, especially in their interplay with synthesis. Following Reinhard Haberfellner et al. (2018, p. 258 ff.), analysis includes, in particular,

- a formal analysis,
- an analysis of integrability in the sense of looking outward,
- an analysis of the structures, processes and functions in the sense of an internal view,
- an analysis of operational efficiency,
- an analysis of the conditions and
- an analysis of the consequences.

Formal analysis includes the question of assessability and the fulfillment of formal requirements. A solution can be assessed if well-founded statements can be made about all relevant aspects. In no way should one, as they say, leave it to chance, i.e. take uncalculated risks or not know what one is doing or what the consequences of one's actions will be. The breadth and depth of assessability depends on the respective level of detail of the object of the assessment. An example of a formal requirement may be the existence of meaningful documentation, approval by an internal body or official authorization.

The *analysis of integratability* refers to the behavior of the design object in the context of a supersystem (Fig. 2.17), in relation to other systems and elements and in its environment (Fig. 2.10). From the perspective of the focal system, the focus is thus on the outside world. Particular attention should be paid to the interfaces between the focal system and other systems, as well as to its environment, since disruptions regularly occur there due to the operational closedness of psycho- and sociotechnical systems (Sect. 2.6). Against this background, it is unrealistic and misleading to replace the term interface with that of a seam (Haberfellner et al. 2018, p. 259) and thus to suggest that the principle-related bridge can be avoided. In fact, they can at best be dealt with in an appropriate way, but losses will always have to be taken into account. The IPO model (Sect. 2.4.1) is therefore quite suitable for balancing material (mass) and energy flows; however, the use of the extended function models (Sect. 2.6.2) is recommended for balancing information flows. Process models (Sect. 2.4.5) are used to analyze the integrability into higher-level *processes*.

When *analyzing structures, processes and functions*, Reinhard Haberfellner et al. (ibid.) mean the functions and processes within the focal system (from the perspective of the focal system, looking inward). The analysis of the structures should, in particular, answer the question of whether the topologies (Fig. 2.11) fulfill their purpose. This can be seen, among other things, from whether the processes that take place on the structures are efficient and effective. This interaction of structures and processes causes the function of the focal system. Fig. 2.32 provides a vivid example of this. What is modeled in it reveals the structures and processes as well as the resulting function, in this use case the logistical function (Eq. 2.40).

The analysis of *operational efficiency* focuses on *user-friendliness*[35] (usability) as well as on safety and reliability. This is examined, among other things, in the context of *ergonomics* as part of occupational science (e.g. Bokranz and Landau 2012; DIN EN ISO 26800:2011-11; DIN EN ISO 9241-1:2002-02 ff.; Schlick et al. 2018). Anthropometric work design is concerned with the ergonomic design of work systems, energetic work design ensures that tiring activities are balanced with phases of recovery to avoid overexertion of the workers, and informational work design ensures that psychological and sociological aspects are taken into account. The three sub-areas of occupational science mentioned above thus cover the three flow variables of material (mass), energy and information, which are usually distinguished by Systems Engineering (Fig. 2.6). Closely related to ergonomics is *occupational safety* (e.g. DIN 45001:2018-11; Kern and Schmauder 2005; Kohlert and Cooper 2018), which is intended to ensure an injury- and accident-free workplace. Both are primarily aimed at employees within the framework of Stakeholder Value Management (SVM), who can and should be involved in analyses themselves, but who are also represented by several instances whose involvement is often mandatory. These include the employers' liability insurance association, the works council or staff council, or, in the case of church institutions, the employee representation in accordance with § 16 ASiG, the safety officer in accordance with § 22 ASiG, the company doctor and occupational safety specialists. In addition, depending on requirements, there may be a representative for severely disabled employees in accordance with § 178 (4) SGB IX and experts from the respective fields of work. In accordance with § 11 ASiG, they all form the Occupational Safety Committee (ASA), together with employer representatives, among others. A very large number of regulations (laws, ordinances, implementing provisions, standards, guidelines, etc.) must be observed for this purpose. The same applies to product safety. This makes it clear that the analysis of user-friendliness, safety and reliability is usually very extensive and also risky in that it can easily happen that binding regulations are overlooked or misjudged. Typical measures to ensure safety and reliability are protective devices, redundancies and the avoidance of central units. These can be both technical and organizational measures.

Another aspect of the operational safety analysis is ease of *maintenance* (DIN EN 13306:2018-02; DIN 31051:2019-06). This includes *servicing*, which comprises measures

[35] In this context, user-friendliness is often mentioned (e.g. Haberfellner et al. 2018, p. 260); however, this says something about the friendliness of the user, which is usually not meant. The situation is similar with the user interface, which usually refers to the user interface.

to delay the reduction of the existing wear margin and serves to maintain it, *inspection*, which serves to record and assess the condition of the object to be maintained and to predict, the *repair* to restore a defective object, and the *improvement* to increase safety and reliability and to eliminate weak points, while maintaining the function of the original object. To this end, preventive and corrective measures are usually taken. The object of the analysis is to what extent the measures can be carried out efficiently and effectively. In particular, risk calculations (Sect. 4.2.4.8) provide an important basis for this.

The analysis of the *conditions* under which the design object can be operated covers aspects that are not themselves the subject of the design. This may be, for example, a restricted ambient temperature, but also the existence and a specific characteristic of other systems. For example, a fully occupied Airbus A380 with 168 tons of kerosene on board requires a runway length of 3,560 m, while Neopanamax ships require a fairway with a depth of more than 16 m. What can happen if aspects such as these are not considered at an early stage is shown by the historical example of the S-Class Mercedes W140, which was available from 1991, and which, at 2.16 m wide, did not fit onto the car train to Sylt and could only be brought there on truck wagons.

The analysis of the *consequences* relates in particular to effects that lie far in the future and far away, but also to the consequential costs that are closer in time and space. For large and/or very risky projects, such as the use of nuclear energy and the associated disposal problems or genetic engineering projects such as the release of genetically modified organisms (GMOs), this results in a *Technology Assessment* (TA). A particular problem arises when consequential costs or risks are externalized, i.e. are not borne by those who benefit from the use. This is the case, for example, with the disposal of radioactive waste, the costs and risks of which will be borne to a considerable extent by future generations, without them deriving any benefit from it.

4.6 P 6: Evaluation

In Sect. 4.3.3.3, in connection with the V-model (Fig. 4.28), it has already been indicated that the transition between the analyses, which take place in alternation with syntheses in the course of solution development P 5, and the final evaluation P 6 is not always clear. Rather, it is to be determined on a project-specific basis. The following should provide some guidance in this regard.

(1) A first clue can be found in the distinction between *verification* and *validation*. As already explained in Section 4.4.5.2, analyses are primarily carried out in the form of tests and these are mainly used for verification, i.e. to answer the question of whether the respective objects function without errors. Assessments, on the other hand, assume the existence of fundamentally functional solution variants. These variants are then compared with each other in order to identify, in the sense of a validation, the variant that best fulfills the intended purpose.

(2) A further consideration, closely related to the first, is whether the need is to primarily involve experts or also laypersons, whereby the latter is an essential prerequisite for the possibility of participation in the course of Change Management, as explained in Section 3.5.3.

Experts are particularly needed in the context of verification, when rational criteria such as technical functionality or costs are applied. In addition, validation is also based on criteria that cannot be rationalized or only to a limited extent, such as cultural norms, moral values and even personal taste. Not only experts, but also involved and affected individuals and groups can and should express their views on these, even if they are classified as laypersons with regard to the subject of the design.

(3) A further point of reference for distinguishing between analyses P 5.6 alternating with syntheses P 5.5 and assessments P 6 is provided by the necessity of political decisions, which themselves are not the subject of the design project (Sect. 2.6.2.3), but follow other, in particular political, principles and which need to be prepared with a technical assessment P 6. Even if decisions do not necessarily and not in every respect have to follow a logical sequence, they should be as factually sound as possible (Haberfellner et al. 2018, p. 272). For example, the concept of Lean Production is based on the twelfth principle, which requires (Liker 2016, p. 315) that managers should visit the site before making a decision (genchi genbutsu) in order to understand the problem themselves and not just rely on experts. And this is also how the requirement of the Quality Management standard DIN EN ISO 9000:2015-11 (p. 19 f.) for *fact-based decision-making* is to be interpreted. Conversely, the present remarks, which are primarily addressed to experts, also serve, among other things, to convey an understanding of the principles in the political sphere of action (Fig. 2.4) and of the interactions between these spheres (Fig. 2.57). Ideally, this results in what was described and explained in Paragraph 2.6.2.3 as *interpenetration*.

Apart from tests (Sect. 4.4.5.3), Systems Engineering offers comparatively little methodological support for assessments in preparation for political decisions (e.g. Haberfellner et al. 2018, pp. 269 ff.). In contrast, *decision theory* (Göbel 2018; Laux et al. 2018) plays to its strengths in this regard. Therefore, the further explanations are essentially based on this, but not without the repeated reference that this theory interprets the concept of decision differently (Sect. 2.6.2.3) and also, in contrast to the insights on which this paper is based, possibility that, in contrast to risk, there can be certainty on the one hand and uncertainty on the other (Göbel 2018, p. 56; Laux et al. 2018, p. 35). If, apart from these limitations, decision theory is used as a basis in the following, this is limited by the fact that it is very extensive, so that with regard to a possibly more extensive interest, reference must be made to literature such as the examples given here.

In addition to the classification of the evaluation phase in the context of Systems Engineering and decision theory, the connection to agile concepts such as Scrum should be pointed out in advance. An empirical study has led to the finding that experts see an advantage in the *incremental learning* of agile concepts such as *Scrum*, compared to conventional Project Management. In fact, the DIN 69901-2:2009-01 (p. 50) standard only provides for project experiences to be secured at the end of a project (lessons learned). Scrum, on the other hand, recommends both a *review* and a *retrospective* at the end of each sprint (Pichler 2008, p. 107 ff.); while the review serves to evaluate the work results achieved in terms of content and form, the retrospective is used to assess the team's collaboration in order to derive improvement measures.

The problem-solving cycle incorporates this incremental learning thanks to its self-similarity, in that each individual phase is also concluded with an evaluation of what has been achieved so far. In addition, SE also includes the recommendation to evaluate intermediate results in even shorter cycles if necessary, for example in regular team meetings (Kaffenberger et al. 2010, p. 190). However, the SE focuses more on what Scrum refers to as a review (Haberfellner et al. 2018, pp. 269 ff.; Kaffenberger et al. 2010, pp. 186 ff.), while the retrospective has actually been neglected so far. This is due to the fact that SE does not see itself as Project Management, but rather as organizing Project Management in parallel with content-related problem solving (Haberfellner et al. 2018, p. 27). Nevertheless, a retrospective can easily be integrated into the SE. This includes self-critical reflection by each team member (psychological perspective) as well as open discussions between members (socio-psychological and sociological perspective), which promote social cohesion but are also intended to bring conflicts to light and, if necessary, resolve them as far as possible (e.g. Nerdinger et al. 2019). The retrospective can be supported, for example, by *supervision* or coaching (Belardi 2018).

4.6.1 P 6.1: Occasion

The evaluation phase P 6 is primarily linked to the goal formulation phase P 3, since the goals developed in this phase are now used as criteria for the evaluation. Using a mathematical metaphor, one could say that the target formulation opens a bracket and that this is closed again with the evaluation, because setting targets and not checking whether and to what extent they have been achieved after the measures have been completed would not make sense. Therefore, when arranging an assessment, it is helpful not only to start with the solution development phase that immediately preceded it, but also to refer back to the formulation of objectives.

This applies in particular to the involvement of the stakeholders. The extent to which this is necessary has already been explained in Section 3.5.3. Depending on the phase in the system life cycle (Fig. 3.1) in which the assessment takes place, the stakeholders or their representatives were involved to varying degrees in the solution development phase (Fig. 3.31). The more extensive their involvement as the level of detail progressed, the more time-consuming it can be to get the stakeholders back around the table and familiarize them with the subject of the evaluation. Furthermore, the longer the previous development phase lasted (temporal distance) and the greater the spatial distances, the greater the challenges associated with this. If and to the extent that there are concerns that the involvement of stakeholders could lead to frictional losses, it may be helpful to carry out continuous Stakeholder Value Management in parallel with development, with a view to the final joint evaluation, and to keep stakeholders informed of progress at appropriate intervals.

4.6.2 P 6.2: Situation Analysis

The content and scope of the situation analysis depend on various factors. If the evaluation follows on directly from the solution development and is carried out by the same people, it will tend to be brief. If, on the other hand, it is a more or less independent undertaking to be carried out by people who were not involved in the previous phases, the status of the overall project will first have to be determined and communicated before the actual evaluation can take place.

A key aspect of the situation analysis required in this phase will have to address the question of which stakeholders are to be included in the evaluation and which target groups it is aimed at. From a systems theory perspective, it is important to enable *interpenetration* between the spheres of action of the social system in which the project is being carried out, or even between different social systems, if, for example, customers need to be involved (Parag. 2.6.2.3). The more open and trusting the communication required for this, the better it will succeed. However, there are limits to open communication, especially when it comes to making political decisions (Machiavelli 2017).

4.6.3 P 6.3: Target Definition

The formulation of the objectives of the evaluation is similar to the analysis of the situation: if it is directly linked to the preceding phase of solution development P 5 and is carried out by the same people, it will be clear to everyone what it is about. If, on the other hand, it is a more or less independent project to be carried out by people who were not involved in the previous phases, some effort will be needed to achieve a common understanding of the objectives of the evaluation of the available solution variants.

In particular, it is important to establish agreement between all participants in the course of formulating the objectives and to create the appropriate awareness that at the end of a development, it can no longer be a matter of introducing completely new evaluation criteria. The longer the development has taken, i.e. the longer it has been since the objectives were formulated, the more extensively circumstances may have developed and changed requirements may result. This can and should be countered as far as possible with discipline. This can be achieved, for example, by taking changed or new requirements as an opportunity for a follow-up project in order to be able to complete the current project first. If no agreement can be reached on this, it must be clear to all parties involved that this requires a return to the phase of formulating objectives (P 2), with the consequence that all subsequent phases have to be gone through again more or less completely. This in turn carries the risk that, in the course of the evaluation, changed or new requirements will be imposed and so on.

The necessity of falling back on the objectives defined in the phase of formulating objectives P2 as far as possible does not necessarily mean that objectives have to be defined with regard to the use as evaluation criteria. In fact, it may well be useful and necessary to modify or specify criteria in the light of new insights, which may also be a result of the problem-solving (Haberfellner et al., 2018, pp. 269-270).

Finally, one of the aims of the evaluation is to reflect on the course of the design project in terms of *lessons learned*. This can concern aspects that have not proven themselves, but also those that can be recommended as *best practices* for future design projects. Both relate primarily to approaches, methods, tools and models. The *retrospective* can be distinguished from this with reference to agile approaches to Project Management, which is concerned with evaluating the collaboration between team members and with other participants and stakeholders.

4.6.4 P 6.4: Method Planning

Unlike in decision theory, this is exclusively about evaluations by groups (Göbel 2018, p. 159 ff.; Laux et al. 2018, p. 577 ff.). This results from the consistent application of Stakeholder Value Management and the resulting insight that projects that require the application of Systems Engineering always have an impact on more than just one stakeholder group and that these must always and without exception participate (Sect. 3.5). This is not intended to contradict the fundamental possibility and necessity of model simplifications (Sect. 2.2) and, to that extent, not to contradict approaches that place decisions in the hands of individuals either. However, since only a selection can be included here anyway, given the large scope of decision theory, the criterion of alignment with groups should be decisive.

Since stakeholders usually have different interests, some of which may conflict with each other, it makes sense from a decision-theoretic point of view to establish a connection to game theory (Sect. 4.2.4.8) (Göbel 2018, p. 159 f.), because this is actually concerned with such situations. Nevertheless, it is not applied here, although the reason for this is not entirely straightforward, because *game theory* primarily deals with actors who are in *conflict* with each other, whereas here the aim is for actors to reach a *consensus*. A justification based on this difference is difficult insofar as, on the one hand, game theory also includes *cooperation* as a strategy and, on the other hand, attempts to assert one's own interests against others in the evaluation of variants for solving problems are more likely to be the rule than the exception in practice.

James Gardner March and Herbert Alexander Simon (1993, p. 149 ff.) distinguish four conflict situations (Göbel 2018, p. 160 f.):

- A group interested in *solving a problem* shares common goals and seeks the best solution[36] by mutual agreement. The basis of this conflict resolution strategy is consensus. However, if this goes too far, there is a risk that group cohesion will have a stronger effect than the pursuit of an optimal solution to the problem (Janis 1972).
- There are groups in which there is a common *conviction* regarding certain fundamental goals, but there is disagreement about the right way to achieve them, i.e. about the instrumental goals. However, these differences can be overcome through argument. The basis for resolving such conflict situations is therefore the power of *argumentative persuasion*.
- Often, conflicts of interest cannot be resolved by persuasion. This quickly leads to threats, false allegations or horse-trading. In such a situation, the only way forward is through *negotiations* aimed at *compromise*. These are characterized by the fact that all sides make concessions on their own ideas, without, however, being convinced that

[36] In this context, one is tempted to speak of an objectively best solution. The only thing that speaks against this is that, in the light of constructivism (Sect. 2.1), humans in principle have no access to real reality.

these contribute to an overall optimal solution. The overall solution is ultimately not considered to be the best solution in objective terms, but rather the best solution that can be enforced. Negotiations follow defined rules that are intended to avoid possible, but undesirable, negotiation outcomes under the applicable legal and moral order, such as overreaching (e.g. § 138 BGB on immoral legal transactions). In this context, so-called *win-win solutions* are desired. This does not mean that there are no situations in which one side loses what the other wins (see below). However, this is subject to different rules. The basis of negotiations is *influence*: the negotiating partners can influence the outcome of the negotiations, but usually cannot determine it unilaterally (dominantly). A practical example at the state level is negotiations between coalition partners.

- *Political* solutions are also based on transparent conflicts of interest, but they do not aim for compromise, but for one party (government) to prevail against another (opposition); one side wins, the other loses. In order to protect the losers from extreme disadvantages, however, such constellations usually provide for the protection of minorities. In particular, no inalienable rights (ius cogens) may be at stake, only rights that can be waived (ius dispositivum). As a rule, a higher instance watches over this, for example the constitutional court at the state level. The basis of political decisions is power. Therefore, it can be assumed that those who have power do not need a factually sound assessment, since they can ignore it because of their power. However, power is also based on success, at least in the long term, and for that, problem solutions that work are a necessary, if not sufficient, condition.

For further considerations, only the first two conflict resolution strategies are relevant, since they are essentially technically sound and can therefore be assigned to Systems Engineering. The last two strategies, which can be assigned to game theory, build on the assessment provided by SE, but then go beyond it by adding extraneous aspects such as influence and power in order to absorb the contingency that cannot be absorbed on an issue-based basis , if only because every solution is made for the future, which is principally unknown (temporal dimension), and because every solution can have long-range effects and be dependent on long-range influences (spatial dimension) that cannot be fully included in solutions. Nevertheless, it is also important in SE to recognize and take into account the principles that apply in connection with the two conflict resolution strategies mentioned last, because since the evaluation of solution variants that have emerged from a technically sound, systematic and methodically supported problem solution takes place at the interface between the economic technical and political spheres of action (Fig. 2.57) and should enable *interpenetration* between these spheres of action (Parag. 2.6.2.3), the political strategies always have at least some effect in the economic-technical sphere of action and vice versa.

Despite the fundamental willingness to cooperate, which is assumed in the following, conflicting goals remain between different stakeholders and thus the respective interest in finding one's own goals as far as possible in a solution. Decision theory knows two basic approaches to this (Göbel 2018, p. 161 ff.).

(1) All groups or their representatives evaluate the available solution variants *together*. To do this, they agree on a scheme, i.e. on rules according to which the evaluation is to take place, and on the criteria and their weightings, which, however, are essentially already derived from the objectives developed in phase P 3. They make joint assumptions about the probability and extent of events that have an influence on the evaluation, and

discuss differences until a consensus or a compromise is found or a priority is set. To a certain extent, even in a fundamentally cooperative constellation, there can be negotiations that were actually attributed to the conflict-oriented strategies above. This shows the fluid transition that arises from the interpenetration of different spheres of action and is therefore entirely intentional.

(2) Alternatively, the groups or their representatives can, in a first step, develop a position *among themselves* and then, in a second step, incorporate it into the collective evaluation.

Both approaches can also be combined, for example by first agreeing on a procedure (scheme or rules, criteria, etc.) and then working out one's own proposals for weighting the criteria, which are then introduced into a discussion or negotiation in plenary.

In view of the expected conditions under which decisions will have to be made following the assessment, the technical basis for decision-making can be designed in a targeted manner. For example, if negotiations are expected in which there is a chance that those who present the better factual arguments will prevail, a *balance of arguments* can be developed in the course of the assessment (Haberfellner et al. 2018, p. 274 ff.). At the same time, creating such an assessment is relatively simple and is therefore also suitable in early design phases, even if the solution variants developed so far are still available in a relatively rough and therefore simple form. The assessment lists the advantages and disadvantages of the compared variants in verbal form – usually in the form of a bullet point – and can also contain a recommendation at the end.

However, quantifying methods provide more reliable arguments. Their explanation begins with the possibility of reducing the number of fundamentally possible solution variants and thus the effort required for a detailed comparison in the event that it is very large, by means of a rough pre-selection (Göbel 2018, p. 75 ff.). This is explained below using the example of an early *sorting out of inefficient variants*. For this purpose, the numerical example chosen by Elisabeth Göbel (ibid.) is adopted, while her scheme in Fig. 4.62 has been transferred to one that is more common in Systems Engineering. For the sake of simplicity, it is assumed that the expected costs and services associated with the purchase of a machine can be expressed in monetary terms. The net present value is to be maximized, while the maintenance costs and the costs for environmental protection are to be minimized. In contrast to the Von Neumann-Morgenstern expected utility function (Eq. 2.41), the environmental conditions, such as the overall economic development and the competitive conditions, are assumed to be constant (certain). To calculate the maximums of the utility Nij, costs are entered as negative values in the scheme. In this, variant V_4 proves to be inefficient even without summing up the individual utility values N_{4j}, because it shows a lower benefit than all other variants with regard to the net present value Z_1 and the maintenance costs Z_2, and with regard to the costs for environmental protection Z_3, it does not promise a greater benefit than the worst of the other variants V_3 in this respect.

Another possibility for roughly reducing the number of variants that are initially too large is to *prioritize* one target criterion (target dominance, knock-out criterion). In the example chosen above, this could be the capital value Z_1. This would mean that only variants V_1 and V_2 would be considered (ibid.).

If this is not enough to reduce the number of variants to a manageable level, *lexicographic order* can help (ibid.). To do this, the next most important objectives in terms of their significance are used as the only criterion (aspect-wise elimination). In the example chosen

Targets (Z)	Variants (V)			
	V_1	V_2	V_3	V_4
Z_1 Net present value	N_{11} = 1.000	N_{21} = 1.000	N_{31} = 600	N_{41} = 200
Z_2 Maintenance costs	N_{12} = -100	N_{22} = -110	N_{32} = -120	N_{42} = -130
Z_3 Environmental costs	N_{13} = -10	N_{23} = -5	N_{33} = -4	N_{43} = -4
Total target fulfillment (ZE)	↓	↓	↓	↓
ZE_1	Σ_1 = 890			
ZE_2		Σ_2 =885		
ZE_3			Σ_3 = 476	
ZE_4				inefficient

Fig. 4.62 Preselection by sorting out inefficient variants (Göbel 2018, p. 75 ff.)

above, the costs for environmental protection could be used as the second most important criterion after the net present value as the most important criterion. This would leave only variant V_2.

As a rule, it does not make sense to eliminate all variants in this way, especially when there are a large number of target criteria. Rather, the aim is merely to reduce what may be an excessive number of variants to a manageable number. The remaining variants are then compared with each other in the manner shown in Fig. 4.62 for the three variants V_1, V_2 and V_3, by adding up the individual utilities N_{ij} column by column, i.e. for each variant V_i. The variant with the greatest total target fulfillment ZE_i is then the best variant (relative optimum).

In the present case, an *ordinal utility function* is used as the variants are only ranked. This is simpler than taking a *cardinal utility function* as a basis, in which the differences between the utility values can be interpreted. However, it carries the risk that, if the differences N are slight with regard to the respective dominant criteria Z, variants are eliminated that may prove to be more advantageous on closer inspection, i.e. when further criteria are added. In the example chosen, if the net present value were increased slightly to N_{11} = 1,001, the choice would not have fallen on variant V_2, but on V_1. In the case of very small differences between variants, whose benefits are measured on an ordinal scale, *sensitivity analyses* can help to identify effects such as those just demonstrated. To do this, the utility values are varied within the assumed ranges of fluctuation.

The *utility analysis* (UA) extends the scheme that is currently only recommended for a pre-selection by expressing the significance (importance, relevance) of a criterion Z_p by a weighting factor g_p (Göbel 2018, p. 78 ff.; Haberfellner et al. 2018, p. 276 ff.; Kühnapfel 2021). For this purpose, in addition to the weighting factors, the utility values N_{ij} assigned to the objectives Z_j in each case are to be determined on the basis of a value or utility function. The products of the weighting factors G_j and the utility values N_{ij} are added up to the preference value ZE_i (total utility value) of each solution variant (n: index of the target values).

$$ZE_i = \sum_{j=1}^{n} G_j N_{ij} \tag{4.45}$$

Before explaining how the individual factors are determined, it should be noted that although their quantification suggests the appearance of objectivity, the determinations actually do not lose their subjectivity as a result. This carries the risk of an apparent objectivity. The actual advantage of the method does not lie in its objectivity, but in making subjective evaluations transparent, thus making them accessible for discussion and supporting the finding of compromises. Furthermore, the result of the evaluation is comprehensible.

In a first step, the weighting factors G_j of the objectives Z_j are to be determined. To do this, it makes sense to express these as percentages. If there is a manageable number j of objective criteria Z_j, a single-stage procedure is sufficient. However, if there is a large and therefore confusing number j of criteria Z_j, a multi-stage approach is helpful (Kühnapfel 2021, p. 40 f.). This avoids the so-called leveling distortion, which can lead to a creeping shift in importance towards unimportant criteria as the number of weightings increases. To minimize this risk, the criteria are first categorized. Then the categories are given weightings that add up to 100%. Subsequently, the target criteria within each category are given weights that again add up to 100%. Finally, the weight of each target criterion is multiplied by the weight of the category, thus yielding the actual criterion weight. The sum of these criterion weights again yields 100%. This procedure can be further scaled depending on the number j of target criteria Z_j. If this appears necessary, however, it should first be checked whether the number j of criteria Z_j cannot and should not be reduced instead.

The second step is to determine the utility values N_{ij}. This is not done per variant V_i, in the scheme outlined in Fig. 4.63 (Haberfellner et al. 2018, p. 277), i.e. not from top to bottom (vertically), but per target criterion Z_j, i.e. horizontally. The endeavor to achieve the highest possible degree of rationality has led to the development of a very large number of methods for doing so (e.g., Göbel 2018; Kühnapfel 2021; Laux et al. 2018). Only a few of them can be presented below.

- One possibility is to assign fixed utilities to fixed targets, for example a 1 if a criterion is met and a 0 if it is not met. This corresponds to a *nominal scale* (Göbel 2018, p. 80).
- A second option is to rank variants with respect to a criterion and to give the ranks corresponding points (ibid.). On the basis of the preference for height maximization (Sect. 4.3.3.2), for example, a crane with a higher load capacity of, say, 900 kg can be rated better than one with a load capacity of 300 kg. The evaluation of a crane with a load capacity of 500 kg would be somewhere between the other two. So you can give the first crane 3 points, the second 1 point and the last 2 points (alternatively, for example, 0, 2 and 1 point). This is an *ordinal scale*. The advantage of this method is that it is very easy to determine the utility values and that it allows a ranking between different solution variants. The disadvantage, however, is that the degree of target fulfillment is not expressed. For example, a crane with 350 kg lifting capacity would also receive 2

points, although it is much closer to the crane with the minimum lifting capacity than to the one with the maximum lifting capacity.

- For measurable characteristics such as the lifting capacity of cranes, the degree of differences can be expressed more precisely, but this requires a somewhat greater effort. Thus, the maximum (900 kg) and the minimum (300 kg) define two points. Assuming a linear utility function, these define a straight line. This means that a crane with a load capacity of 500 kg would receive $N = 1/(900kg - 600kg) \cdot 500kg = 1.67$ points instead of 2 points, and one with 350 kg would only receive 1.17 points. This way, you get a *cardinal scale*.

Usually, utility values are normalized, i.e. expressed as a percentage. So if a load capacity of 300 kg corresponds to a utility of 0% and one of 900 kg to a utility of 100%, the linear equation is $N = 1/(900kg - 300kg)x - 0.5$, where x represents the load capacity of the respective crane variant to be evaluated. A crane with a load capacity of 500 kg therefore has a utility value of $N = 33\%$.

Once all the objectives Z_j and their weightings G_j have been determined, the total utility value ZE_i of the variants V_i can be calculated. Fig. 4.63 (Haberfellner et al. 2018, p. 277) shows a possible scheme for this; however, other schemes can also be found in the literature on decision theory. Which of these is preferred is a matter of taste.

Targets (Z)	Weight (G) $\Sigma = 100\ \%$	Variants (V)					
		V_1		V_2		V_3	
		N	G · N	N	G · N	N	G · N
Z_1	G_1	N_{11}	$G_1 \cdot N_{11}$	N_{21}	$G_1 \cdot N_{21}$	N_{13}	$G_1 \cdot N_{31}$
Z_2	G_2	N_{21}	$G_1 \cdot N_{21}$	N_{22}	$G_1 \cdot N_{22}$	N_{13}	$G_1 \cdot N_{32}$
Z_3	G_3	N_{31}	$G_1 \cdot N_{31}$	N_{23}	$G_1 \cdot N_{23}$	N_{13}	$G_1 \cdot N_{33}$
Total target fulfillment (ZE)			↓		↓		↓
ZE_1			Σ_1				
	ZE_2				Σ_2		
	ZE_3						Σ_3

Fig. 4.63 Scheme of a utility analysis (Göbel 2018, p. 78 ff.; Haberfellner et al. 2018, p. 276 ff.; Kühnapfel 2021)

In addition to the utility analysis explained here, there are a number of other methods that are not presented in detail due to their similarity, but should at least be mentioned. These include the *Analytic Hierarchy Process* (AHP), which is particularly common in the Anglo-American world (Kühnapfel 2021, p. 87 ff.). The core of this method is a comparison of the evaluations in pairs, but this is so time-consuming that it requires the use of specific software. Furthermore, the *PROMETHEE method* should be mentioned as a variant of NWA (Kühnapfel 2021, p. 96 ff.). In contrast to the NWA, it does not calculate an overall utility ZE_i per variant V_i, but initially leaves it at the evaluation of the respective advantage of a variant over others for each criterion Z_j. The sums of the advantages are then formed. This can lead to a clear preference for a solution variant if both an NWA and a subsequent sensitivity analysis do not lead to a clear preference. However, since this is less comprehensible than the result of a NWA, this method is also rarely used in practice (Kühnapfel 2021, p. 98). Nevertheless, these two examples show that the NWA is rightly the preferred method for the comparative evaluation of solution variants.

A utility analysis is only suitable[37] for criteria that cannot be quantified in monetary terms. If and to the extent that a *monetary evaluation* of variants is reasonably possible and appears necessary, which is increasingly the case with an increasing level of detail, profitability calculations should also be carried out in addition to the qualitative evaluation (Burggräf et al. 2021, pp. 336 ff.; Haberfellner et al. 2018, pp. 279 ff.). In business administration, *investment accounting* is particularly widespread for this purpose (Thommen and Achleitner 2006, p. 613 ff.). A distinction is made between static and dynamic methods, as well as those from the field of Operations Research (OR). In static methods, the timing of the calculation variables is not taken into account, so that, in particular, no discounting or compounding occurs. This means that the calculations can be limited to one period, but are then only based on average values. A major advantage of such methods is that they are easy to carry out. Dynamic methods are more complex because they try to take into account the temporal occurrence of operands during the entire useful life. Comparability of payment flows that accrue at different times is achieved by discounting to a specific point in time. Methods of the German Commercial Code (OR) ultimately attempt to model interactions between various factors such as procurement, production, sales, investment and financing.

textitStatic investment calculations include textitcost comparison accounting (Thommen and Achleitner 2006, p. 615 ff.), which compares the costs of the variants to be compared. The costs per accounting period (e.g. financial year) or the costs per unit of output can be used as a basis. Revenues are not taken into account because it can be assumed that these will be the same for all variants, that they cannot be attributed to a single investment I, or that they cannot be determined at all. Of the costs K, only those that are caused by the solution variants and are not approximately the same for all variants are taken into account. These include the operating costs K_b and the output-independent capital costs (fixed costs), which are calculated in particular from the depreciation per time period K_a and from the interest costs K_{zi} of the average tied-up capital. Fig. 4.64 shows an example of this (ibid., p. 616 f.). In this example, the total costs K are calculated as

$$K = K_b + K_a + K_{zi} \tag{4.46}$$

With the investment amount I, the duration n of the investment project and the liquidation proceeds L at the end of the useful life, the depreciation costs K_a are calculated as follows:

$$K_a = \frac{I - L}{n} \tag{4.47}$$

The annual interest rate p is used to calculate the interest costs K_{zi} as follows

[37] In this respect, the example adopted by Elisabeth Göbel (2018, p. 75 ff.), shown in Fig. 4.59, is not entirely consistent.

A. Costs per year	Variant 1		Variant 2	
- Output data				
- Acquisition cost	260.000		190.000	
- Useful life	5		6	
- Liquidation proceeds	10.000		10.000	
- Capacity/period	12.000		10.000	
- Capacity utilization/period	10.000		10.000	
- Capital costs/year				
- Depreciation and amortization	50.000		30.000	
- Interest (10 %)	13.500	63.500	10.000	40.000
- Operating costs/year				
- Labor costs	30.000		40.000	
- Material costs	25.000		26.000	
- Maintenance costs	10.000		12.000	
- Energy costs	4.000		6.000	
- Other operating costs	15.000	84.000	18.000	102.000
- Total costs/year		147.500		142.000
B. Costs per unit of service	**Variant 1**		**Variant 2**	
- Output data like A, but utilization/period	10.000	12.000		10.000
- Capital costs/unit of output	6,35	5,29		4,00
- Operating costs/performance unit	8,40	8,40		10,20
- Cost/performance unit	14,75	13,69		14,20

Fig. 4.64 Example of a cost comparison calculation (Thommen and Achleitner 2006, p. 617)

$$K_{zi} = \left(L + \frac{I - L}{2}\right) \cdot \frac{p}{100} = \frac{I + L}{2} \cdot \frac{p}{100} \tag{4.48}$$

The total period costs K are therefore

$$K = K_b + \frac{I - L}{n} + \frac{I + L}{2} \cdot \frac{p}{100} \tag{4.49}$$

And the cost per unit of output, k, for a quantity x to be manufactured

$$k = \frac{K}{x} \tag{4.50}$$

Furthermore, the application rate at which the two variants have the same cost is also relevant. The critical rate x_{crit} is determined using the *break-even analysis*:

$$K_1 = K_2 \tag{4.51}$$

$$K_{zi1} + K_{a1} + K_{b1} \cdot x_{crit} = K_{zi2} + K_{a1} + k_{b2} \cdot x_{crit} \tag{4.52}$$

$$x_{crit} = \frac{K_{zi2} + K_{a2} - K_{zi1} + K_{a1}}{k_{b1} + k_{b2}} \tag{4.53}$$

As long as the produced quantity is smaller than x_{crit}, variant 1 offers advantages. If the output quantity exceeds the critical quantity, variant 2 is advantageous.

As the example calculation shows, a *cost comparison calculation* is relatively simple. However, the disadvantage is that revenues are not taken into account. So whether a profit can be made at all cannot be determined on this basis. Furthermore, no relationship is established with the amount of capital employed. Finally, no possible changes in costs, such as those for wages and materials, are taken into account. Including revenues is part of comparative profit calculation (Thommen and Achleitner 2006, p. 618 ff.). However, a meaningful determination of the expected profit is often difficult because more resources are usually involved in the production of goods or the provision of services than in the capital good under consideration, which makes it difficult to allocate the profit between the capital good and the other resources. In addition, the capital good is often not used to produce a single product. Finally, profit does not result from costs, but from competition in the market, so it depends on external factors that are difficult to calculate.

If different capital investments are required for different variants, a *profitability calculation* is appropriate (ibid., p. 621 f.). It is based on the cost and profit comparison calculation and sets the average annual profit in relation to the average capital employed. However, it does not take into account how long the capital remains tied up, whether the capital differences can be used elsewhere, or under what conditions this could be done. The amortization calculation is again useful for this (ibid., p. 622 ff.).

Dynamic investment calculation methods go beyond static methods in that, on the one hand, they do not work with average values, but rather with time-dependent cash flows until the planned useful life is reached, and in that they take into account incoming and outgoing payments that occur at different times. The latter is done by discounting the payments to the point in time at which the first payment occurs. The discount factor v is calculated as

$$v = \frac{1}{(1+i)^2} \tag{4.54}$$

whereby

$$i = \frac{p}{(100} \tag{4.55}$$

is the discount rate and t is the year in which the payment is made.

In practice, the discount rate is relatively difficult to determine because the risk associated with the investment must be included in the calculation. Often, the return on a low-risk capital market investment with a long term (e.g. a 10-year German government bond) is used as a basis and increased by a risk premium. However, the latter must be determined subjectively.

The present value Z_0 of a future payment Z_t in t years at the current point in time t_0 is calculated as

$$Z_0 = Z_t \frac{1}{(1+i)^t} \tag{4.56}$$

In practice, these discount factors are taken from tables (ibid., p. 628), in which they are given for certain periods and for different interest rates.

These principles of dynamic processes are applied, among other things, in the *Net Present Value* (NPV) method. According to this method, all incoming and outgoing payments caused by an investment are discounted to a defined point in time. Specifically, the net present value K_0 is calculated from the difference between all discounted incoming payments E0 and outgoing payments A_0:

$$K_0 = E_0 - A_0 \tag{4.57}$$

The deposits E_0 are calculated as follows:

$$E_0 = \sum_{t=0}^{n} \frac{e_t}{(1+i)^t} + \frac{L_n}{(1+i)^n} \tag{4.58}$$

This includes the individual cash inflows during the useful life n [years], which are due at the end of each time period t ($t \in \{0, 1, 2, \ldots, n\}$: time index). In most cases, these are the revenues from the sale of the services rendered. L_n is the liquidity revenue at the end of the useful life.

The payments A_0 are calculated as follows:

$$A_0 = \sum_{t=0}^{n} \frac{a_t}{(1+i)^t} + I_0 \tag{4.59}$$

This includes at the individual deposits during the useful life n [years], which are due at the end of each time period t. Specifically, these are consumption factors such as auxiliary and operating materials. I_0 are payments in connection with the investment, for example, the development and material costs.

The net present value K_0 is now calculated as follows

$$K_0 = \sum_{t=0}^{n} \frac{e_t - a_t}{(1+i)^t} + \frac{L_n}{(1+i)^n} - I_0 \tag{4.60}$$

If the net cash inflows $g_t = e_t - a_t$ accrue evenly over the entire useful life, this equation can be represented as follows using the *net present value calculation*:

$$K_0 = ASF \cdot g + \frac{L_n}{(1+i)^n} - I_0 \tag{4.61}$$

In this, ASF is the discount sum factor (synonym: annuity present value factor, capitalization present value factor), which is calculated as follows:

$$ASF = \sum_{t=0}^{n} v_t = \frac{(1+i)^n - 1}{i \cdot (1+i)^n} \tag{4.62}$$

In practice, this factor is also usually taken from corresponding tables (ibid.).

For an exemplary calculation, let us assume that the investment I_0 = \$60,000. The liquidity proceeds at the end of the useful life are set at L_0 = \$10,000. The interest rate per year is p = 10%, the time index t = 3. The cash inflow surplus is expected in the first year at g_1 = \$20,000, in the second year at g_2 = \$30,000 and in the third year at g_3 = \$25,000. The net present value is thus calculated as K_0 = 0.909 · \$20,000 + 0.826 · \$30,000 + 0.751 · (\$25,000 + \$10 000) - \$60,000 = \$9,245. The discount rate can be calculated by taking the financing costs as a basis and expecting the investment to generate a return at least equal to the cost of the capital employed. A second option is to use the return on an alternative investment opportunity. Finally, a target return can also be specified.

The net present value method is used to derive the *Internal Rate of Return* (IRR) method as a further possible method of dynamic investment calculation (ibid., pp. 631 ff.). This is the interest rate at which a net present value of K_0 = \$0.00 is obtained. This interest rate thus indicates the internal or effective interest rate of the investment. And the *annuity method* is also a modification of the net present value method (ibid., p. 633 ff.). It does not show the incoming and outgoing payments over the entire useful life, but converts this net present value into equal annual incoming surpluses (annuities). This results in a periodization of the net present value over the entire useful life, taking into account compound interest.

Fig. 4.65 shows how the evaluation matrix (utility analysis) and the investment calculation can be brought into a clear context. This is demonstrated using the example of a *cost-effectiveness graphic* (Haberfellner et al. 2018, p. 280).

What has been developed so far is based on what can be described as facts, as long as the limitations resulting from the constructivist basic understanding of being able to grasp reality as such at all are taken into account (Sect. 2.1). This creates a basis on which *decisions* can ultimately be made (Sect. 2.6.2.3). However, it should be noted that although decisions should be based on facts, they go beyond them, otherwise they would not be decisions in the true sense of the word, but also calculations. In fact, however, decisions serve to absorb the contingency that usually remains even after the most extensive, in-depth and careful calculations. Therefore, decisions can only be justified professionally to the

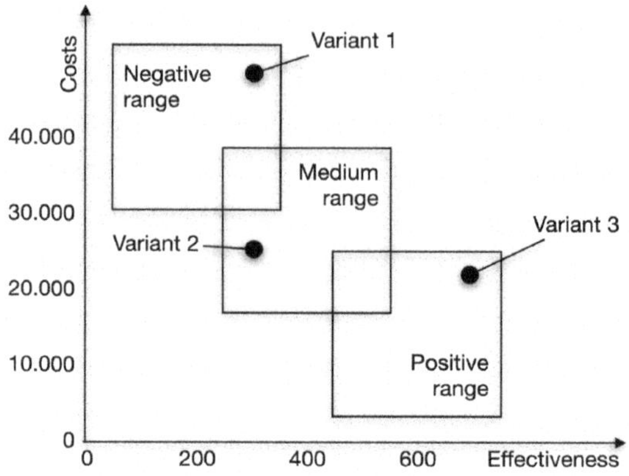

Fig. 4.65 Visualization of cost-effectiveness (Haberfellner et al. 2018, p. 280)

extent that those who have to make them should be expected to have a certain understanding of the facts and the calculations presented by experts. Above all, however, decisions are to be justified politically. If and to the extent that significant uncertainties remain despite the methodically supported calculations, these are to be transferred into certainty by decisions for those who will subsequently continue to work on the technical side and need a secure basis for doing so.

4.6.5 P 6.5: Conducting the Evaluation

Analyses are carried out in conjunction with their respective preceding syntheses, initially on the left, descending branch of the V-model (Fig. 4.28) from the rough to the detailed and, after the solution elements have been completed, on the right, ascending branch from the detailed to the rough in the sense of an increasing integration of the elements and sub- or partial systems into higher-level systems, i.e. cyclically. This leads to the fact that, especially in projects that have the design of technical systems as their subject, only one solution is available at the end. For example, it would not be economical to develop three variants of an aircraft up to their operational maturity in order to then present them to decision-making bodies, which would then decide which of the variants should go into series production. In such cases, the phase transition from analysis P 5.6 to evaluation P 6 is fluid. The latter then represents a very comprehensive analysis, as it is accompanied by the progressive concretization of a project.

However, there are also many design projects that result in several solution variants, each of which is suitable in principle, from which one variant is then selected in the context of a political decision-making process. This is particularly necessary when significant risks (contingencies) remain despite the competent and conscientious work of the experts. This is particularly the case in the early planning and design phases, for example when developing concepts or rough drafts. This in turn is associated with the fact that the effort involved in developing variants, which are discarded after a political decision in favor of only one of these variants, is limited and is usually offset by the greater benefit that arises from the choice between different variants. The further advanced a design project is in terms of its realization, the less sense it usually makes to ultimately terminate it for political reasons. As a project advances, political influence should therefore decrease. Shortly before commissioning, only the experts are ultimately required. The role of politicians is then limited to ceremonially giving the go-ahead, thus officially assuming political responsibility.

As a rule, committees are set up to make decisions within the framework of Project Management (Haberfellner et al. 2018, p. 173 f.). In addition to the client itself, this may include a project committee with representatives of the stakeholders. The project manager serves as the interface between such a committee and the project team that carries out the work. Above all, it is up to him to mediate between the factual logic and the political logic. An important tool for this is the *processing and condensing of information*, as well as its appropriate *presentation* (e.g. ibid., p. 216 f.). Just as the bait must always taste good to the fish and not to the angler, the information provided by political bodies must above all be presented from a political perspective. In this context, less is often more:

- *One-pagers*, for example, are welcome. These are documents in which all the information relevant to the decision is summarized on one page. An alternative to this, which should also be kept short but does not necessarily have to fit on one page, is the *management summary*. Such summaries can be structured in the same way as the solution itself: (1) What is the reason for the decision? (2) What characterizes the current situation? (3) What should be achieved? (4) How should the goal be achieved? What will it cost? (5) What are the decision alternatives? What can be saved or improved in each case? (6) What do the experts recommend? What are the reasons for this?
- This can also include a traffic light, for example, which uses colors to visualize how the status of a project is assessed from an expert point of view.
- *Key Performance Indicators* (KPIs) are also well suited for the condensed presentation of decision-relevant information. However, caution is advised when using monetary values. Since key figures are nothing more than highly aggregated models, they cannot be readily related to specific business figures such as profit and loss statements (Bokranz and Landau 2012, vol. 1, p. 12 ff.).
- Finally, *visualizations* are well suited to persuade political decision-making bodies to follow the recommendations of experts. The system representations shown in Chapter 2 can provide suggestions for this. Computer programs for presenting facts offer forms of visualization that can aesthetically enhance such representations.

However, the compaction of decision-relevant information and the associated desire from the specialist side that the decision is made in the sense of their recommendation for action must not be allowed to tempt anyone to manipulate decisions. It is important to clearly highlight *decision alternatives* and to present their respective consequences from a specialist point of view as clearly and objectively as possible. The political *influence* of experts should not be underestimated. Even the answer to the question of whether or not a matter will be submitted for decision has a significant influence on what will ultimately be decided. Furthermore, this applies to the selection of information that is associated with a decision proposal, and how it is presented is also important. Experts who enter the political arena with the aim of exerting targeted influence should be aware of the risk that they can quickly slip up if they do not have a sufficient command of the political rules of the game.

It is equally unhelpful for political leaders to make *decisions in isolation*, without first consulting experts (Weber 1972, p. 125 ff.). Contrary to popular belief, political leaders do not qualify for this role by proving themselves to be the best professionals. If such an assumption is made, it usually leads to what is known as the *Peter Principle* (Peter and Hull 2001): every employee rises through the chain of command until they reach the level at which they no longer contribute anything of value to the organization. On this basis, organizations can determine the so-called Peter level, which indicates the ratio of management positions filled by suitable people to those filled by people who have reached the point where they can no longer prove themselves, but are not demoted back to the position in which they last proved themselves and thus seemingly qualified for the current management position.

The only exception is the company founder who implements their own idea and usually also has the necessary expertise for it[38]. However, they usually only manage very small companies at first, where political decisions have only a limited impact. As soon as a company reaches a size where *functional differentiation* is required to cope with the associated complexity (Sect. 6.1), political leaders primarily have political skills and only limited technical skills. This also applies even if they have acquired a professional qualification. Every (political) decision should be based on a professional solution to the problem and a recommendation for action resulting from it. In the sense of the variant planning explained above (Sect. 4.4.5.2), this should show several possible solutions (alternatives). Otherwise, a decision is reduced to "do" or "stick with the current approach".

The consequence of a decision at a phase transition can lead to the *continuation* of the project with the subsequent phase. However, it is also possible to *jump back* to one of the previous phases or even to *abort* the project (Haberfellner et al. 2018, p. 154 ff.). This is why we speak not of a problem-solving process, but of a problem-solving cycle. Since problems are characterized by a barrier that needs to be overcome, a consistent, straightforward solution can only be expected in exceptional cases and with good luck. Therefore, setbacks and failures are common and not an expression of incompetence or lack of motivation (Haberfellner et al., 2018). However, since the latter corresponds to a widespread assumption, those in charge are tempted to complete problem solutions at almost any cost. In fact, however, it is a sign of competence and sovereignty when those in charge end projects that prove to be less promising as early as possible or to restart them in whole or in part. To put it in a popular saying, it is better to end with terror than to live in terror.

4.6.5.1 P 6.5.1: Lessons Learned

In addition to the content of the design project, the project itself must be evaluated. This concerns the following in particular:

---

[38] One example is William Henry (Bill) Gates III, an exceptional Steven Paul Jobs.

- The approach, which is well structured by the underlying *process models*, such as the system life cycle at the macro level and the problem-solving cycle at the micro level within the relevant detailing phases, but should still be regularly reflected upon. A first question could be whether the models have been consistently applied. If this is not the case, which may well make sense in the course of the so-called tailoring, the reasons for this and the success should be scrutinized. A further question may be whether the approaches are understood and accepted by all those involved and affected. The resulting measures could, for example, consist of tailoring or training in Systems Engineering. Ultimately, the current state of science and technology cannot guarantee that Systems Engineering itself is complete, sufficiently clear and free of errors, so that it should also be the subject of critical reflection.
- The *methods* used. In this regard, questions arise such as whether the right methods have been applied and whether they have been applied correctly. The question may also arise as to whether the dosage of the application of methods appears appropriate in the light of the specific design project. Too much can be just as suboptimal as too little.
- The underlying *models and theories*. In addition to using the right models and theories and applying them correctly, it is particularly important to choose a level of abstraction that is appropriate to the problem. The *tools* used, which includes both hardware and software. This can include experimental setups for conducting experiments, manufacturing processes for producing prototypes or samples, computer programs for evaluating data, and much more.
- *Collaboration* within the team and with external parties and stakeholders. In this context, the question may arise, for example, as to whether and, if so, what difficulties there were with the authorities in approving measures, whether suppliers complied with delivery conditions (deadlines, quantities, quality, etc.), whether customers provided the necessary information in good time, in full and in a comprehensible form, and whether the organization of project-related processes was effective.
- The *cost-benefit ratio*. This includes, among other things, a commercial post-calculation with regard to the question of whether the financial budget was adhered to. In detail, however, it concerns the use of resources as a whole, however this can and should be broken down in a specific case. This can concern time and space in general, but also, for example, human resources or the consumption of energy and materials, right up to the impact of the design project on the natural environment (ecological balance).

Normally, questions like these do not arise only at the end of a design project. Rather, they are asked on an ongoing basis, particularly in the course of an accompanying Project Management, and answered in intermediate stages. Therefore, the final evaluation discussed here is in such cases merely a summary of what has already been collected and evaluated during the implementation. It should now be self-evident that the interests of all relevant stakeholders must be taken into account.

4.6.5.2 P 6.5.2: Retrospective

Based on corresponding approaches from agile Project Management, the retrospective serves to reflect on *personal collaboration* within the team and with external parties involved and affected (Jakoby 2021, p. 377; Kusay-Merkle 2021, p. 307 f.). Michael A. West (2012, p. 92 ff.) recommends holding such meetings away from the regular workplace. The physical distance and unfamiliar surroundings can positively influence the team's view of itself and promote the effectiveness of the collaboration. Furthermore, such meetings should not be completely structured to provide opportunities for spontaneous conversations. As topics, he recommends, among other things,

- a reflection on the team's goals and their appropriateness,
- an alignment regarding the roles of the individual team members, both from their own perspective and from the perspective of everyone else,
- an assessment of the quality of communication (written, oral, formal, informal, etc.),
- a reflection on the frequency and duration of meetings, as well as on the discipline during the meetings (agenda, punctuality, adherence to explicit and implicit rules, efficiency and effectiveness, etc.),
- an assessment of decision-making processes,
- an investigation of possible needs for further training or personal development,
- a reflection on conflicts,
- a reflection on one's own strengths and weaknesses, and
- the appreciation of successes.

It can also be important to address the issue of cooperation with external parties involved and affected, for example, how to deal with customers perceived as difficult or how to behave towards unreliable suppliers or service providers. In this respect, for example, agreement in the team can give individual team members support when needed.

The retrospective can be supported, for example, by *supervision* or *coaching* (Belardi 2018).

4.6.6 P 6.6: Evaluation of the Evaluation

In the course of evaluating the evaluation, it is important to examine whether the bases for decision-making appear suitable for providing decision-makers with the right information in a suitable way as a basis for decision-making. Key questions such as the following can be helpful for this:

- Does the documentation to be provided to decision-makers include all relevant information?
- Is the scope of the documents to be handed out appropriate? If a scope that is actually considered too large cannot be avoided, an abstract can be added at the beginning. Supplementary information can also be added as an appendix.
- Does the order serve a purpose? In the so-called newspaper style, the importance of the content decreases towards the end, while the novel style tries to maintain a sense of suspense until the very end.
- Are facts that need to be emphasized visualized?

- Are the explanations written in a language and form of presentation that is understandable for those who have to make decisions and who are often to be classified as laypersons from a technical point of view? For example, an exploded view can be clearer than a two-dimensional technical drawing.
- Are suitable discussion partners available for technical discussions who are able to present the results of technical work in a political context? In addition to expertise, this may also include, for example, a confident manner and the ability to express complicated facts in a simple way and to get to the heart of the matter.
- Can additional information be provided at short notice if necessary?
- Can additional discussion partners be called in at short notice if required?

Chapter 5
L 3: Phase Transitions

Phase transitions occur both in the course of the system life cycle (Fig. 3.1) and in the problem-solving cycle (Fig. 3.23) and are referred to differently depending on the context: In Project Management (DIN 69901-2:2009-01, p. 11), for example, they are referred to as *milestones*, in Systems Engineering also as *decision thresholds* (Walden et al. 2023, pp. 29 ff.), Robert G. Cooper (2002, pp. 128 f.) calls them *gates* and Nadin Ebel (2011) *control points*. Clear definitions and clear demarcations from one another or references to one another are only partially found. In essence, however, all designations mean the same thing.

The following characteristics are common to all phase transitions:

1. Phase transitions are *crises*. Proven procedures and methods are available to overcome them.
2. After a phase transition, the prevailing laws are fundamentally different from those that applied before.
3. The changes that come with them are *irreversible.*
4. Due to the changes, a phase transition is always *unstable*.
5. The phase transition itself follows its *own laws*, which are not comparable to those of the preceding or subsequent phase.
6. Although a phase transition is based on facts that have been established by experts, it is essentially characterized by *decisions*.

(1) Etymologically, the term crisis is derived from the Greek verb krínein, which means to separate or to divide. In this sense, a crisis separates two phases from each other, and thus essentially indicates a difference in time between a before and an after. A crisis can therefore be understood as an intermediate situation that is brief in comparison to the preceding and subsequent conditions. In this respect, it also represents a culmination, a kind of peak in which the change takes place. One can therefore also speak of a kind of tipping point, tipping because an instability is associated with it in addition to a fundamental change, and of a moment because of the relative shortness of time.

V. Ahrens, *Systems Engineering & Management*, https://doi.org/10.1007/978-3-658-46875-0_5

Crises occur regularly and repeatedly in all life courses, be it in living beings, in inanimate nature, in artifacts, in business ideas, and so on. One example is the adolescent crisis in the life of every human being. Since crises are both foreseeable and unavoidable, humans and nature have taken precautions to cope with them. In the life of a human being, in the Christian context, these are, for example, initiation rituals such as baptism at birth or Protestant confirmation or Catholic communion at the transition to Christian adulthood. In Judaism, these are Bar Mitzvah for boys and Bat Mitzvah for girls, and those who do not belong to any religion sometimes celebrate Jugendweihe. In the case of man-made objects, for example, the transition from development to operation, known as the Start of Production (SOP), is where such initiation rituals take place. These transitions are also monitored, for example, in the Quality Management of the automotive industry in Germany by the production process and product approval (PPA) in accordance with VDA 2020 in Germany and by the Advanced Product Quality Planning (APQP) of the Automotive Industry Action Group (AIAG) in the USA.

A crisis is significantly different from an exceptional situation, which can also be referred to synonymously as an emergency, a disaster or a worst-case scenario (Sect. 3.7). The term "exceptional situation" makes it clear that, unlike crises, such situations do not occur regularly, but only exceptionally. This does not mean that exceptional situations do not have to be anticipated, and that suitable precautions are not also needed to deal with them. However, these differ considerably from those of crisis management. For example, in exceptional situations it is important to return to the previous state if possible, so that they can also be described as turning points. While the aim in a crisis is to overcome it overcome by moving forward, in exceptional situations it is necessary to change direction, to turn back: after an earthquake, collapsed houses should be rebuilt. In contrast, however, a person who has become an adult should not become a child again, however much one may regret the lack of care often associated with childhood with increasing age.

Best practices include, for example, education and training in the lives of people, and in the development of products, for example, Advanced Product Quality Planning (APQP) or the introduction of new business processes, Change Management (Sect. 3.5). The latter is contrasted with Business Continuity Management (BCM) to cope with exceptional situations (Sect. 3.7), which once again highlights the difference between a crisis and an exceptional situation.

(2) After a crisis, many things are fundamentally different: children have become adolescents, adolescents have finally become adults, ideas, concepts, prototypes and patterns have finally become saleable products. While you first had to invest, you can now consume: a good education and training is a good basis for securing one's own existence; a good investment now brings a profit. However, this is also accompanied by fundamentally different principles: as an adult, you now have to earn your own income and products are no longer developed, but routinely used. The corresponding approach no longer follows the problem-solving cycle, but the control loop; a problem has become a task (Sect. 3.2). The aim is no longer change (genesis), but to maintain a dynamic equilibrium (homeostasis) for as long as possible. In this phase, genesis primarily means decay, and this is something to be resisted as much and as long as possible.

(3) While the opportunities that arise in the wake of a crisis, such as coming of age in adulthood or making a profit in business, are naturally welcomed and usually make

the efforts associated with overcoming the crisis seem bearable, the irreversibility of the changes brings the associated losses into focus. These can be manifold. For a project team, the completion of a project can result in its dissolution, which can be regretted if the team has worked well together. It may even lead to the loss of the previous job. Furthermore, decisions in favor of one path are always also decisions against other possible paths: Those who pursue a career as a systems engineer may have to give up hope of a career as a saxophonist for good. This is accompanied by *ambivalence*, i.e. inner tension and conflict. If the paths for which only one can be chosen are perceived positively, this is referred to as an *appetence-appetence conflict*; if not, it is referred to as an aversion-aversion conflict.

(4) Due to its instability, a phase transition in systems theory is also referred to as a singularity. In the context of natural science, James Clerk Maxwell was the first to use it to describe the behavior of non-linear dynamic systems, and Jules Henri Poincaré also made significant contributions in this regard. In an at least analogous sense, the phenomena observed in this context can also be transferred to social systems (Argyris et al. 2017, pp. 456 ff.). Among other things, the unreliability of the causality principle is characteristic. Thus, even small causes (disturbances) can have a large effect and vice versa. This is also known as the *butterfly effect*. In a somewhat mitigated variant, one can also speak, especially in a scientific and engineering context, of a *transient phase* that denotes a *non-stationary process*. For example, it is known that switching operations in power supply networks can trigger transients (bursts) or pulse packages and thus have a considerable damage potential.

Phase transitions are examined, among other things, by *synergetics* (Argyris et al. 2017, p. 448 ff.). According to this, a condition far from equilibrium is a prerequisite for a phase transition to occur; Ilya Prigogine has demonstrated this using thermodynamic systems. Furthermore, *systems organize themselves* in the course of phase transitions. Although they are permeated by flow variables such as energy, mass and information, as explained in Section 2.6.2 on the basis of extended functional models for psycho- and socio-technical systems (Fig. 2.6), their reaction to this is *autonomous*. For observers, the emergence of new patterns of order appears to be emergent.

Emergence is characterized by the fact that a developing order cannot readily be traced back to the action of identifiable elements (Sect. 2.1). This does not mean that the formation of orders cannot be described, designed or controlled; that is precisely the subject of synergetics. However, emergence is one reason why complex systems so successfully resist simple attempts at interpretation, such as those often found in the context of Change Management. Even proponents of this approach attest to a failure rate of around 75% (Lauer 2019, p. 49). The main reason given for this is resistance from employees and parts of middle management. However, this is often only supported by anecdotal evidence. Thomas Lauer (ibid., p. 51 f.), for example, simply cites two exemplary incidents in companies. Beyond that, however, he also refers to a survey.

The idea that the causes of the overwhelming failure of change projects could also lie in the basic assumptions and the Change Management measures derived from them does not occur to representatives of this field. And so it seems only logical that the Project Management standard DIN 69901- 5:2009-01 in paragraph 3.102 and textbook authors such as Michael Nagel et al. (2020, p. 302 ff.) divide those involved and affected into friends and foes as early as the stakeholder analysis. From this, it is then deduced that enemies must be turned into friends or fought. However, it is clear who is friend and who is foe: a friend is someone who realizes that those who want to change something and determine the path of change are right about everything, while every objection is factually unfounded. The causes for this are then stupidity (misunderstandings), fear, unjustified urge for freedom (reactance), inertia, resistance or adverse circumstances. To counter this, more of the same is usually recommended.

Now, we do not want to repeat the mistakes that others are reproached for. In this respect, it is not about refuting everything that research on Change Management has revealed over decades, including on a scientifically sound basis. The fact that phase transitions, for example, are accompanied by uncertainty and instability and that this does not leave those involved and affected untouched has already been admitted above and is not to be expected otherwise. The approach that Change Management recommends not only falls short, but sometimes borders on overconfidence and a lack of respect for others, and thus on arrogance. If you seriously want to improve the success of changes significantly and sustainably, you will have to engage more deeply with the characteristics of complex systems, and then there will be no way around phenomena such as emergence. In this respect, Change Management will have to look for the causes of its failures, especially in itself.

As is now known, emergence results from the interaction of the elements of systems. It should be recalled that in social systems these consist of acts of communication (Parag. 2.6.2.2). Furthermore, it is important to emphasize once again that it is the elements of the focal system that are important. Thus, anything that affects the system from the outside is merely perceived as irritation (perturbation, disturbance). Whether and, if so, how the system reacts is *unpredictable*. It is precisely this unpredictability that is an essential characteristic of phase transitions. Those who are actively involved and want to apply the result of their work are perceived as outsiders by those who are intended to apply it. Like a control loop, they absorb the concerns of the troublemakers. From such a microscopic perspective, this system behavior is not to be expected otherwise. Whether and to what extent individual communication acts that can be classified as incapable, sluggish or unruly contribute is largely irrelevant, because these elements alone do not sufficiently explain the collective behavior as a system.

This can be substantiated, for example, by the consideration that communication acts that are directed against a change have basically the same probability of being accepted as communication acts that provide good technical arguments. At least in a cooperative, emotionally balanced atmosphere, the better should prevail against the good. Since Systems Engineering, with its systematic approach and consistent support through appropriate methods and tools, provides a good basis for convincingly representing solutions developed on this solid foundation, there is no reason to believe that those who object will prevail. Unfortunately, however, in the face of emergent phenomena, this is only of limited use: solidly developed solutions are necessary, but not sufficient for their successful introduction.

(5) Self-organization, which is most effective during phase transitions, requires the focal system to change itself, essentially from the inside out. How this can be done in the context of Systems Engineering has already been explained in detail in Section 3.5 and will not be repeated here. However, it should be emphasized once again that solutions to problems in the development of which those affected have participated are not better accepted for that reason. In fact, one must take a step back from this often-heard argument: solutions to problems in which those affected have played a significant role are usually significantly

better than those developed exclusively or predominantly by experts, since, in addition to their explicit knowledge, the implicit knowledge of the users is also crucial for success. And because the solutions are better as a result, they are also better accepted. It is therefore not about giving those affected a feeling of being taken seriously, but about actually taking them seriously and consistently drawing on and incorporating their know-how, which is largely based on practical experience. This is no guarantee that you won't occasionally encounter a lack of interest or even outright rejection, but these will be the exception rather than the rule. On the contrary: recognition and participation almost always mobilize engagement and commitment.

Since specific recommendations for participation can already be found in Section 3.5, the aim here is to deepen the understanding of fundamental principles in order to provide those who have had as little success with the previous methods of Change Management , as well as the 75% of those reported by Thomas Lauer (2019, p. 49), approaches that appear to be more effective in the light of system-theoretical insights. After all, these have already proven themselves in many other areas of application and can thus claim a certain general validity.

One of these areas of application is the laser, which is chosen here not only because it is likely to be quite close to systems engineers in terms of technology, but also because Hermann Haken used it to mark the historical beginning of synergetics and created important foundations (Argyris et al. 2017, pp. 448 ff.). First of all, he found that the usual approach to explaining and controlling physical processes, i.e. to model individual states of a system in the form of differential equations, and thus to describe them as the ratio of the respective state equation and its temporal derivative, does not lead to success in complex systems because the equations are coupled to each other, i.e. they interact. In the case of the laser, the initial situation is characterized by a multitude of overlapping wavelengths. Only above a certain pump power (laser threshold, pump threshold) do the gain from the energy supply and the loss from absorption in the laser medium and from the decoupling of the generated emission equalize. From this intensity threshold on, a laser begins to work as such. The wavelengths of all atoms are now approximately the same.

This common mode of vibration considerably simplifies the differential equations, because now only three variables are relevant: the time-dependent amplitudes of the electromagnetic vibration states (modes), the difference in the population numbers of the energy levels (atomic inversion), and the dipole moments of the individual atoms. These variables change at different speeds: the modes change most slowly, the dipole moments most quickly. Hermann Haken established that the slowest variables, known as order parameters, determine the emerging order. In this case, the modes initially compete with each other. The winning mode then enslaves all the others. If you don't want to use this martial term, especially with regard to an application in social contexts, you can also say that the defeated fashions align themselves with the winner. This makes it possible to calculate the behavior of the overall system based on the fashion that sets the pace.

Of course, such scientific findings can only be transferred to social or psychological systems by *analogy*. What exactly one has to look for if one wants to find variables in such systems that change slowly and are therefore possibly also those from whose ranks the element emerges that everyone else will orient themselves towards and by which it can be determined which order will ultimately emerge, cannot be deduced unambiguously either analytically or empirically. Nevertheless, practical experience at least suggests that such opinion leaders also exist in social systems. Experience also suggests that it is not always clear to outsiders who has the informal say within a group, and informal communication can be just as important as the formal structure. This can be particularly true in the case of change projects when formal superiors are suspected of making common cause with external and, in particular, higher-ranking authorities.

Now, one can undoubtedly come up with the idea of identifying the informal opinion leaders in order to draw them too to the side of those seeking to change the system; that can actually be a starting point. However, there is no getting around measures like those explained in Section 3.5. Falling back on the sometimes primitive repertoire of Change Management, which can be roughly summarized as persuasion, outwitting, taking by surprise, bribery, and the like, will most certainly not lead to success in this way either.

(6) Finally, phase transitions always require decisions in the sense of what was explained in Paragraph 2.6.2.3. Although the solutions developed by experts in close cooperation with users (participation) form the basis for the following phase, every solution, however competently and conscientiously developed, leaves questions open, at least to the extent that

- every solution unfolds its effect only in the future, which in principle cannot be predicted with sufficient certainty, and
- every solution is subject to the constructivist limits of access to real reality (reality), and is therefore based on abstractions (models) (Sect. 2.2) and thus cannot, in principle, be prepared for everything that may arise in practical use. In particular, this also applies to influences and effects that are geographically distant.

The uncertainty (contingency) resulting from these and further limits of logical work can only be transferred into certainty by decisions, in that those who make the decisions take the political responsibility for it. In this process, the technical results are naturally a decisive basis. Beyond that, however, decisions are always made according to political rules. From a technical point of view, this also results in a considerable degree of uncertainty, which is characteristic of phase transitions anyway; in this context, this becomes clear once again. At the same time, the coincidence of factual and political logic reveals the interpenetration between the spheres of activity of social systems explained in Section 2.6.2.3 using Talcott Parsons' AGIL scheme.

While the approach and methodology explained in Section 3.5 on Change Management is very concrete, the additional explanations in this Section necessarily remain quite abstract. This is due to the fact that research has not yet progressed further and in particular is still struggling with emergence. Despite the success of the practical application, Hermann Haken has not succeeded in scientifically substantiating his theory either, according to Achim Stephan (1999, p. 237). Although the empirical findings have been well tested and formulated in a mathematically sound manner, it is not possible to simply derive causal-theoretical statements from a descriptive thesis of this kind. This would be a fallacy of the type post hoc, ergo propter hoc: the possibility of a predictability of the practical behavior of the overall system on the basis of the description of the behavior of order parameters does not allow the conclusion that the behavior of the order parameters is actually causal for the behavior of the overall system. Moreover, Meinhard Kuhlmann (2009, pp. 310

ff., 324 ff.) considers it unresolved how the circular causality between an overall system and the order parameters that represent it, as Hermann Haken calls it, is connected to the other components of the system. This interaction can be understood in terms of causal interactions within the level of the system components (elements, sub- and part-systems) as well as between this level and the level of the overall system. The latter is then something like a downward causality, since the order parameters would have a causal effect from both the one and the other level at the same time. Discussions such as these show that there are still gaps in our deeper understanding of phase transitions. It seems clear that and why these have not always been successful so far. However, it should also have become clear that blaming employees and parts of middle management does not help, but probably does more harm than good.

Chapter 6
L 2: Systems Management

Systems Management (SM) deals with the control of systems in the system life cycle (Fig. 3.1) from the start of operation (SOP) to the start of decommissioning (palliation). In terms of cognitive science, it is about completing *tasks* (Sect. 3.2). Systems Management is also referred to as *Operations Management* (e.g. Thonemann 2015). In practice and in business theory, however, management is often associated exclusively with political leadership (G-part-system in the AGIL scheme according to Fig. 2.4) (e.g. Thommen and Achleitner 2006, p. 106; Staehle et al. 1999, p. 192). Such a narrowing is expressly not intended in the following. Instead, this term is understood in the sense of its etymological origin (from the Latin "mansionem agere" for "to run the house (for the owner")) as housekeeping according to primarily economic aspects. The term "housekeeping" can be used to refer to all public and private households, including those of private companies or non-profit organizations.

Overcoming the barrier between the actual and the target state is no longer a problem in the course of controlling systems in the sense of the word (Fig. 3.15). Thanks to good Systems Engineering, which we assume has been successful here, both the necessary means (resources) and the necessary knowledge (knowledge, know-how) are now available, because one of the main goals of the design was to reduce complexity by absorbing *contingency* (Luhmann 2018, p. 153 ff.; see Sect. 6.1). This manifests itself concretely above all in the fact that not only is creativity no longer actually required (the actuality will be explained later), but that it is even explicitly prevented.

This is most pronounced in the case of purely *technical automation*. Such automation is always designed in such a way that it *deterministically* links input and output in the sense of what Heinz von Foerster (1993, p. 247 ff.) calls the *trivial machine* (TM) and thus corresponds to the IPO model (Fig. 2.6). The function according to which this is performed is completely known and controllable. Any deviation is an error that must be either prevented in advance by appropriate design measures such as *Poka Yoke* (Niemann et al. 2021, p. 63 ff.) or sustainably eliminated during operation through a *process of*

V. Ahrens, *Systems Engineering & Management*, https://doi.org/10.1007/978-3-658-46875-0_6

continuous improvement (Kaizen) (e.g. Belz et al. 2015, p. 50 ff.). An innovative machine that runs without interruption would be a contradiction in terms.

However, there are two reasons why at least a certain degree of creativity is also required in ongoing operations. Although this is not aimed at creating something new or significantly developing something that already exists in a design sense, a certain form of creativity is also required to maintain a dynamic equilibrium.

The first reason for this lies in the *constructivist limits of human knowledge* explained in Section 2.1. All results of human design are initially based on model-like assumptions that are only partially compatible with reality. Although their mostly assumed viability (Glaserfeld 1996, p. 43) suggests that the solutions developed on this basis basically do what they were created for, Experience shows that a company that has experience in building airplanes can also make a newly developed or significantly improved airplane fly. And yet, even such a company would be well advised to test a new or significantly modified aircraft extensively under real conditions before handing it over for operation. The fact that this usually results in a number of improvements required is usually not due to a lack of development expertise, but rather to the necessarily model-based view of the real world.

These constructivist limits of human knowledge are so extensive that even the most extensive testing cannot reveal all the eventualities that may arise during operation, especially since the limits of knowledge remain, even if the regular operation brings with it a continuous confrontation with reality, much more than the design phase. This manifests itself, for example, in attempts to identify the causes of problems quickly and sufficiently comprehensively, which do not always succeed. Methods such as asking the why five times (5W method) should help to at least identify viable causal chains. Once the actual cause of a problem occurring during operation has been identified, creativity may well be required to close the gap that may become apparent between the causal model-based assumptions made during design and the real-world conditions that become apparent during operation. Not infrequently, improvisation is then also required – at least temporarily (Lt. Commander Montgomery Scott: Scotty).

The second reason why the avoidance of creativity during operation is only really possible was explained in Chapter 1 with the fundamental integration of technical systems into psycho-technical (Fig. 2.54) or socio-technical systems (Fig. 2.55) (Fig. 1.1). However, since mental systems and, by extension, social systems inherently have a certain autonomy, they can only be determined to a limited extent: *self-organization* is unavoidable.

For those who intend to shape and direct automatisms, this means both good and bad news. The good news first:

- Peter Michael Blau (1955) has empirically demonstrated that a formal organization in the literal sense *can only function if it is complemented by an informal organization*. In this respect, many deviations from prescribed processes serve to achieve the intended, even if not all of them do.
- However, deviations (creativity, initiative) can be detrimental because the absorption of contingency associated with the determination of processes and the resulting reduction of complexity causes an increase in *efficiency* and *effectiveness* (profitability). This can be compromised by deviations.

The fact that the organization of routine activities often follows the model of the trivial machine is expressed in business administration by what is known as the *machine metaphor*. Alfred Kieser and Mark Ebers (2014, p. 97) quote Jean-Jaques Bourcart (1874, p. 110) as saying: An industrial business is best compared to a clock in which one wheel engages with another and which ultimately also shows the owner what the bell has struck. The work of the manager is very similar to that of a watchmaker, who has to set up, start and regulate the gear train. In industrial production, this approach can be traced back in particular to Frederick Winslow Taylor (2001) and to the married couple Frank Bunker and Lilian Evelyn Gilbreth and the *Scientific Management* they founded (Bokranz and Landau 2012, Vol. 1, p. 91). In business administration, the fundamental work of Max Weber (1978, pp. 124 ff.) on *bureaucracy* is particularly evident. Against this background, two perspectives are possible:

- On the one hand, technical automation and the organization of business processes are more similar than some people would like. In both cases, the aim is to design and control deterministic processes and, in particular, to eliminate freedoms (creativity, initiative) to do things differently than prescribed (programmed) as far as possible in order to maximize the efficiency and effectiveness of routine activities (tasks).
- On the other hand, it does not follow that mental and social systems as parts of socio-technical systems actually lose all freedom. In this respect, technical systems are less determinable than some engineers would like. And there will also be economists who are not happy to be confronted with the widespread insight that maximizing efficiency and effectiveness cannot be everything. Human work is only sustainable if there is a good work-life balance. In the workplace, well-calculated distribution times, for example, contribute to this (Bokranz and Landau 2012, vol. 1, p. 362 ff.). In addition, reality is far too complex for all eventualities to be thought through in advance and condensed (cemented) into structures and processes. Even with a highly differentiated tax law and a tax return based on it, ambiguities remain that can lead to queries with the tax office. And even a problem-solving cycle (chapter 4) that is worked through in great detail will not be possible to sustain in practice in an ideal way.

The following comments are set in the context of the tension between the constructivist limits of human knowledge and the natural human autonomy on the one hand and a technical-organizational determinism with its advantages in terms of efficiency and effectiveness on the other.

Furthermore, the following considerations are based on the distinction between *structural and process organization* (Thommen and Achleitner 2006, p. 778 ff.) going back to Fritz Nordsieck (1931) and Erich Kosiol (1934). theoretically, Niklas Luhmann (1996, p. 70 ff.) derives this distinction from the insight that systems, in the face of complexity (both their own and that of their environment), are forced to select those aspects that they consider relevant to the maintenance of their existence. The constructivist-based necessity of *selecting* those aspects that are considered relevant for this purpose only becomes a problem, however, if an infinite amount of time is not available for this purpose. If this were the case, everything could be related to everything. Systems must therefore be able to establish *time shifts* in relation to themselves and to their environment. They have to be able to absorb the resulting risks of momentary non-conformity. If they succeed in doing so, they can react to momentary opportunities or disruptions, for example, with longer-term measures, or postpone reactions, without immediately disintegrating (Ahrens 1998, p. 21). One example of a specific measure for creating a time lag is the training of young people or the further education of experienced professionals, who are not productive during the time set aside for this, but rather learn options for action that can be used later as needed.

For such selections, systems have two different forms available: structures and processes. In a first approximation, it can be said that structures exist timelessly and, so to speak, form the statics of a system. In this way, they reconstruct the reversible aspects of their environment within the system. Processes, on the other hand, are also understood in a first approximation as being subject to time and serve to map irreversible events in the environment within the system. On closer inspection, however, it turns out that every structuring, as far as it does not occur randomly, follows a process, and that processes in turn exhibit structures. This mutual conditional relationship must always be kept in mind when the following definitions are made. According to this, structures (organizational structure) subsume the open complexity of the possibilities of linking each element with every other element into a narrower pattern of valid, usual, expectable, repeatable or otherwise preferred relations (Fig. 2.10). Processes (operational organization), on the other hand, come about through the fact that concrete selective events follow one another in time (Luhmann 1996, p. 74). On this basis, systems can develop strategies and tactics to gain time, that is, to decouple themselves from the timing of their environment.

For example, they can remain still until an environmental constellation arises that allows them to apply one of their previously developed action patterns. They can also try to be faster than their environment with their internal processes. Such superiority can be used to simulate possible environmental developments in advance and to develop courses of action on this basis. The application of computer-aided simulation is a practical example of this (Wenzel 2018). Yet another possibility is to flee from certain environmental situations, to recover from environmental events that have already occurred, or to avoid foreseeable situations. Similarities to the feeding prevention strategies observed in nature, such as running away, attacking or playing dead, are quite intentional in this context. Another strategy that can be implemented by means of a time lag between the system and the environment is the temporalization of the system's internal complexity. This involves two aspects: First, the duration of the maintenance of the relations of an element with other elements is minimized. Instead of establishing a relational pattern itself as a structure, the corresponding system only allows meta-structures that are able to establish or dissolve relations depending on internal and external requirements. Furthermore, the existence of the elements themselves is also permanently at stake. Regardless of how well it has proven itself in the past, each element is designed in such a way that it can be eliminated from the system with the same probability after each new event. The system then draws its stability not from the stability of its established elements and relations, but from the dynamics of its structure. This insight is one of the reasons for the usefulness of the view that communication acts form the elements of social systems (Parag. 2.6.2.2), because communication fulfills the condition just mentioned. Therefore, no necessity for a hire-and-fire culture or for the employment of so-called temporary workers (day laborers) can be derived from this.

Systems essentially depend on the constant disintegration of their elements. Through continuous disintegration, space and demand for new elements is created and the energies and materials released in the process can be permanently recombined. However, this results in inherently restless systems. This makes it clear that the theory on which these insights are based differs significantly from classical equilibrium theories. The latter try to show that systems can return to stable resting positions by absorbing disturbances. In contrast, this is about reflexive production, which is only possible far from equilibrium. In the model of regulation, one therefore speaks of a dynamic equilibrium and not of a (static) equilibrium.

6.1 Organizational Structure

If we follow Niklas Luhmann (1996, p. 45), complexity is the aspect that perhaps most strongly expresses the problem experiences of recent systems research. In this context, it seems remarkable that definitions of this term usually remain unclear even where it is taken as the starting point for further considerations (e.g. Boyksen and Kotlik 2013, p. 48 ff.; Haberfellner et al. 2018, p. 35, Winzer 2016, p. 7 f.). In contrast, this paper aims to always state as clearly as possible what is meant by each of these terms.

As a starting point, the following is based on Niklas Luhmann's definition (1996, p. 46), who describes systems as complex when, due to the intrinsic limitations of the elements' linking capacity, it is no longer possible for each element to be linked to every other element at all times. This will be quantified in the following using a model calculation. In this context, the linking capacity is of crucial importance, because if it were infinitely large, it would be possible for each element to link to any other element. It is only when this capacity is limited that complexity arises. Here, it is measured in units of time, assuming, for the sake of simplicity, that the capacity of all elements is the same.

Furthermore, it should be noted that each element should not only use its *capacity for linking* with other elements, but should also perform its actual tasks. To illustrate this, the following numerical example is chosen for a manufacturing company. The actual task of an employee could then, for example, be to participate in the manufacturing of products.

In the first step, the binomial coefficient (Willke 2006, p. 129), which is also abbreviated as x in the further course, serves as the basis for the calculation:

$$\binom{n}{k} = \frac{n!}{(n-k)! \cdot k!} = x \tag{6.1}$$

In this case, k indicates the number of possible combinations. Since each element is to be coordinated with every other element, $k = 2$. n corresponds to the number of elements. In this case, Eq. 6.1 can also be reformulated as follows by complete induction:

$$x = \frac{\sum_{i=1}^{n} i}{\sum_{i=1}^{n-2} i} = \frac{1 \cdot 2 \cdot \ldots \cdot (n-2) \cdot (n-1) \cdot n}{2 \cdot 1 \cdot \ldots \cdot (n-2)} = \frac{n \cdot (n-1)}{2} \tag{6.2}$$

In the second step, only the system-internal linking capacity is to be considered. To do this, Eq. 6.2 is weighted with the available capacity of the elements of a system. This is expressed by the proportion of the time in the total available time that is required for each coordination process[1]. This proportion is given the formula symbol β. For example, if $n = 10$ elements each coordinate with each other, a total of $x = 45 coordinations$ are required. If the available time per element is, for example, $t = 8h$, a total of $80h$ is available. With a proportion of $\beta = 1\%$ for each coordination process, $x(\beta = 1\%) = 3.6h$ is required for coordination. The equation is:

$$x(\beta) = \frac{n \cdot (n-1)}{2} \cdot \beta \cdot t \tag{6.3}$$

In the third step, we now also need to take into account the fact that systems have to use some of their capacity to communicate with their environment. The time required for this is designated γ and is also set at $\gamma = 1\%$ in this numerical example. Since this time component reduces the time available for completing the actual tasks to the same extent as the time component β for system-internal communication, the two time components can be added. For the sake of simplicity, the total is designated $\beta + \gamma = \alpha$ in the following.

The graphical representation in Fig. 6.1 shows the course of this coordination time $x(\alpha = 2\%)$ (solid line) and the total available time (dashed line) for this numerical example. This makes it clear that it is possible to coordinate each element with every other element up to a certain limit. At this limit, the entire available time is required for coordination. Beyond this limit, more time would be needed for coordination than is available overall. In this respect, it is no longer possible to relate every element to every other element beyond this limit: the system is complex according to Luhmann's definition (1996, p. 46). In this case, the *complexity threshold*, which is to be given the formula symbol n_g[2], is at $n_g = 101$ elements.

A variation of the time share α leads to a shift of the complexity threshold. Fig. 6.1 shows exemplarily by means of the dotted line that this threshold is reduced to $n_g = 51$ elements when α is doubled to 4%.

Mathematically, the dependency of the complexity threshold on the capacity of the elements, here expressed as the proportion of the total available time that can be used for coordination, can be expressed by the following equation:

$$n_g = \frac{2}{\alpha} + 1 \tag{6.4}$$

This dependency is shown in Fig. 6.2. It shows that the complexity threshold is quickly reached in operational practice. To calculate it precisely, it is first necessary to estimate what proportion of the total available working time is taken up by communication activities. This includes not only formal meetings and informal discussions about work-related aspects, but also, for example, reading documents (work orders, work plans, work instructions, technical drawings, etc.), filling out forms (data collection, quality records, etc.) and much

[1] Only formal communication is taken into account here. Informal communication is taken into account by the personal distribution time and is included in the standard times for productive activities.

[2] The index g is to indicate the set n of elements as a border set.

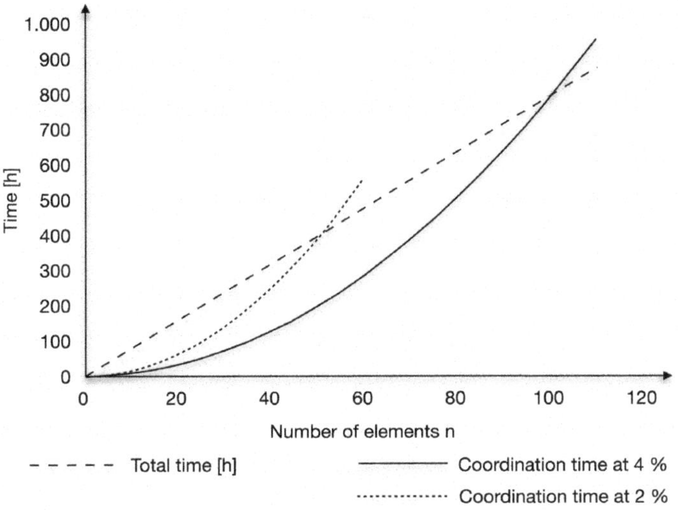

Fig. 6.1 Synchronization time (linkage capacity) and total time as a function of the number of elements (Ahrens 2015, p. 4)

more. In many companies, this quickly adds up to more than one would expect at first glance.

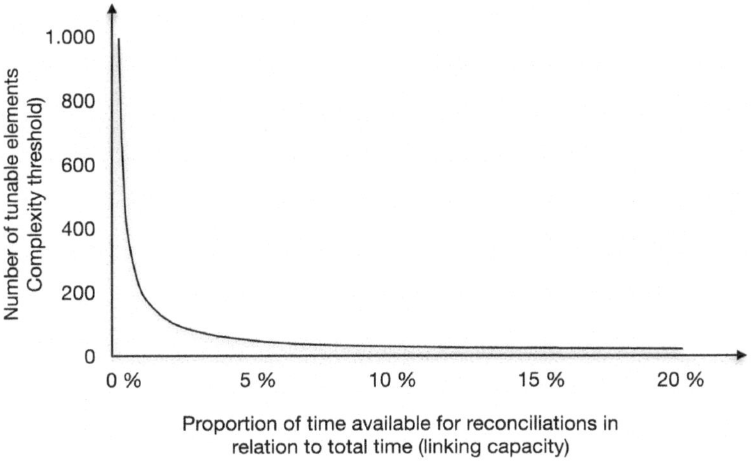

Fig. 6.2 Dependency of the capacity threshold on the linking capacity (Ahrens 2015, p. 5)

In view of this, the assumption made for the following considerations that internal system coordination and coordination with the environment take up a total of $\alpha = 4\%$ of total working hours is deliberately set low. It seems achievable in companies such as those in the metal and electrical industry, where the avoidance of waste in the sense of activities that do not directly add value is already very advanced (Liker 2016, p. 57 ff.). In this case, the complexity threshold is a company size of 51 employees. This size is interesting for the considerations presented here in that it can be assumed that most employees in such companies are actually productively active, while a level of management that is more or less exclusively limited to coordinating internal activities and would thus distort the assumption made can be largely neglected.

An analysis by the Gesamtmetall employers' association (2013) has shown that around 47% of all companies in the metal and electrical industry employ up to 49 people. In view of the complexity threshold of 51 employees, it can therefore be said that about half of all companies in this industry can be classified as complex according to the definition of Niklas Luhmann (1996, 46) and its specification by Eq. 6.4.

However, the complexity threshold calculated using Eq. 6.4 only indicates a theoretical maximum number of coordinable elements, because it is based on the assumption that employees use their entire working time for mutual coordination when the complexity threshold is reached, leaving no time to actually do the work. If we were to expect that only the portion of the total available working time that each individual employee is allowed for a coordination process should be used for coordination, then the maximum number of elements that can be coordinated is n = 3, because in this constellation, the following always applies:

$$x(\alpha) = \frac{n \cdot (n-1)}{2} \cdot \alpha \cdot t = n \cdot \alpha \cdot t = 3 \cdot \alpha \cdot t \tag{6.5}$$

So if 3 employees, each working 8 h a day, or 24 h in total, each spend 4% of their working time coordinating with each other and interacting with their environment, they each spend about 19 minutes on it, or 0.96 h in total. This in turn corresponds to 4% of the total working time. Even with one more element, i.e. $n = 4$, this ratio would shift significantly: if each individual were to continue to use $\alpha = 4\%$ of their working time for each coordination process, a total of $x(\alpha) = 1.92h$ would be required. Of the 32 h of total working time, this would already be 6%. This proportion increases linearly with the number of elements and reaches 100% at the capacity threshold of $n_g = 51$ elements in this calculation example.

6.1.1 Hierarchy

The calculations on complexity show that systems must prevent the need for each of their elements, part- and sub-systems[3] to be coordinated with each other. One of the most influential measures is hierarchical structuring, as generally outlined in Fig. 2.13 and which finds a whole range of concrete manifestations in practice, some of which are schematically shown in Fig. 6.3.

[3] Furthermore, part- and sub-systems are included for the sake of simplicity when elements are discussed.

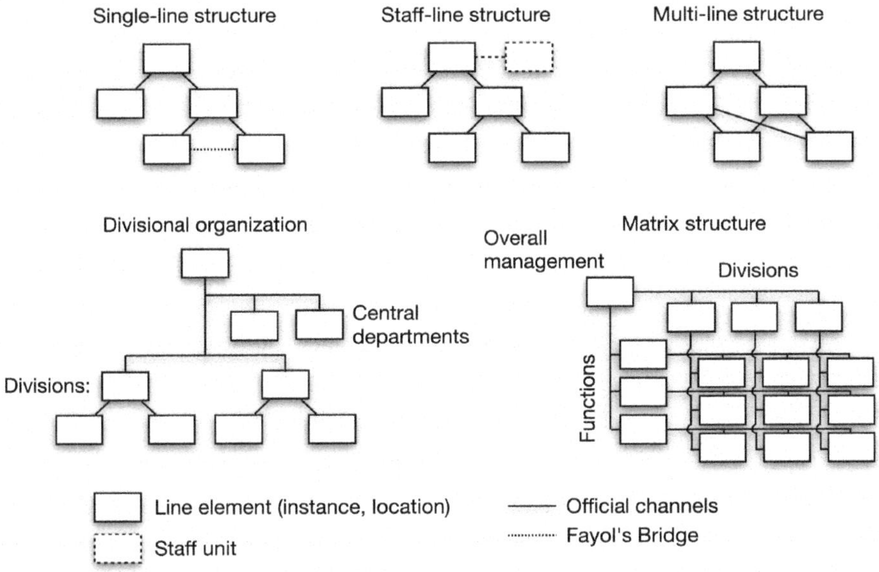

Fig. 6.3 Variants of the hierarchically structured organization plan

- The *functional organization*, which can be traced back to Frederick Winslow Taylor, links the elements (instances, positions) in a linear fashion.
  - The *single-line structure* is characterized by the fact that each element is only connected to one higher-level element, i.e. it only receives instructions from one instance (Thommen and Achleitner 2006, p. 816). Following Henry Fayol, this is also referred to as the unity of the order placement. The connections (relations) are also referred to as official channels. Since strict adherence to these communication relationships is very long and consequently not always useful for cross-connections, these are optionally permitted as so-called *Fayol bridges* to connect elements at the same hierarchy level. The advantages mentioned are tight control of communication channels, transparency and simplicity, as well as clear assignment of responsibilities. The disadvantages are the rigidity of the structure, the length of the official channels, and a heavy burden on intermediate instances.
  - One variant of the single-line structure is the *staff-line structure* (Thommen and Achleitner 2006, p. 819 f.), in which line elements with decision-making authority are supported by staff who themselves do not assume any management responsibility (Heimann and Hofinger 2022, p. 185 ff.). Rather, they provide professional support to line elements and in this respect serve the interpenetration between professional and managerial work, which will be explained in detail later. Furthermore, they contribute to the establishment of time shifts, as explained in the introduction to Chapter 5, by providing additional processing capacity. The risk that conflicts may arise as a result of the separation between decision preparation, decision-making and decision enforcement is considered disadvantageous if staff suggestions are not

sufficiently taken into account in their view when decisions are made. Furthermore, by having the opportunity to accumulate knowledge, staffs can develop into so-called éminences grise, who compete with decision-making bodies. Another objection to the work of staffs is that they work in a way that is too divorced from reality, especially when they are staffed with young employees.

- The *multi-line structure* proposed by Frederick Winslow Taylor (Thommen and Achleitner 2006, p. 817) replaces the principle of unity of the order by the principle of the shortest path. The advantages cited are the use of the advantages of specialization, short distances between elements, and high motivation through alignment with the specific abilities and skills of the people involved. The disadvantages are considered to be the risk of task overlaps, possible responsibility conflicts, and the complexity of a large number of elements.
- In a *divisional organization*, the overall system is divided into largely autonomous sub-systems (divisions). The so-called object principle is applied in that, for example, similar product or customer groups or geographically definable regions are used as a demarcation criterion. Performance-related functions such as production and sales are then assigned to the divisions. For reasons of specialization, central departments then only take on certain functions such as human resources and accounting. In this way, heterogeneous overall systems can be transferred into several homogeneous sub-systems. Concrete forms of the divisional organization include the cost center and the profit center. The risk of divisions working against each other, the incomplete exploitation of synergy effects and the necessity for internal cost allocation are considered to be disadvantages.
- The line structure is contrasted with the *matrix structure*, which basically corresponds to a multi-line structure, but is essentially characterized by the fact that it forms elements of a hierarchy level according to two or more equally ranked criteria, for example both according to functions and according to divisions or both according to projects and according to processes (Thommen and Achleitner 2006, p. 827). This is intended, among other things, to prevent one-sided representation of interests. The advantages seen are the possibility of participating in problem solving, the introduction of different perspectives, the relief of the management as a result of delegating decisions, and direct communication channels. However, the disadvantages are the potential for conflict due to overlapping responsibilities, the risk of so-called lazy compromises, and a high level of communication effort.

In the following, the success of hierarchical structures in reducing complexity will be quantified. For this purpose, it is assumed that coordination of employees and interaction with the environment is carried out exclusively by a superior[4]. The superior uses a defined part β of his time to instruct an employee; the same amount of time is spent by the instructed employee. In addition, γ time is required for the superior to interact with the environment. Assuming that the amount of necessary coordination with the environment increases linearly with the size of the system (measured here by the number of employees), the supervisor therefore uses a total of $\alpha = (n - 1) \cdot (\beta + \gamma)$ for coordination[5]. Since the

[4] The superior and the employee have the same working hours.

[5] During the time in which the superior does not communicate, he works on the completion of the actual tasks himself.

number of employees increases with the size of the system, while the control should only be carried out by one superior, it is immediately apparent that the number of tunable elements is limited by the capacity of the superior, so that in the following only this needs to be further examined.

If the supervisor did not need time for communication, he could control an infinitely large system. If he had to use $\alpha = 1\%$ of his capacity of 8 h per day to instruct one employee, he could coordinate $n_g = 101 employees$, at $\alpha = 2\%$ it would be $n_g = 51 employees$ and at $\alpha = 100\%$ it would be exactly $n_g = 2 employees$, including himself. This series of numbers is described by the following simple equation:

$$n_g = \frac{1}{\alpha} + 1 \tag{6.6}$$

If, as in the first example, we calculate that $\beta = 2\%$ of an employee's total working time is required for their instruction and that $\gamma = 2\%$ of the supervisor's working time is required for communication with the environment for each employee ($\alpha = 4\%$), then a supervisor can coordinate a maximum of $n_g = 26 employees$, including themselves, using their total working time. However, the system's total working time is not used up for coordination, but only the portion that can be calculated using the following equation:

$$x(\alpha) = 1.5 \cdot \alpha \cdot (1 - \frac{1}{n}) \cdot t \tag{6.7}$$

With $n = 26$ employees to be coordinated, only about 5.8% of the total available working time would be needed for coordination. With a working time of 8 h per employee per day, this would be 12 h per day, of which 8 h would be spent with the superior and about 10 minutes (0.16 h) with each of the 25 employees. 196 h per day, i.e. about 94.2% of the total working time, remains for the performance of the actual tasks.

Fig. 6.4 shows a comparison between control of systems by coordinating each element with every other element, and control by a hierarchically superior, central instance. While the number of hierarchically coordinable elements is limited to 26 employees due to the limited capacity of the superior, up to 51 employees can coordinate with each other in the case of decentralized control, albeit at the cost that at the maximum number there is no time left to complete the actual tasks. Both control concepts have their maximum linkage capacity with the maximum net working time yield[6] at 26 employees, whereby the same net working time yield is also achieved with 25 employees in the case of decentralized control.

The comparison between centralized and decentralized control becomes even clearer when the ratio between the respective net working times is formed in Fig. 6.5. This shows that with 2 and 3 employees, decentralized control is more efficient than hierarchy. With 4 employees, both concepts are on a par. It is only from a staff of 5 that hierarchy becomes increasingly more efficient. This does not change with a variation in the proportion of time required for coordination. However, the maximum efficiency advantage changes: the

[6] Net working time or net working time yield is understood here to mean the working time that remains after deducting the time needed to coordinate the employees to complete the actual tasks.

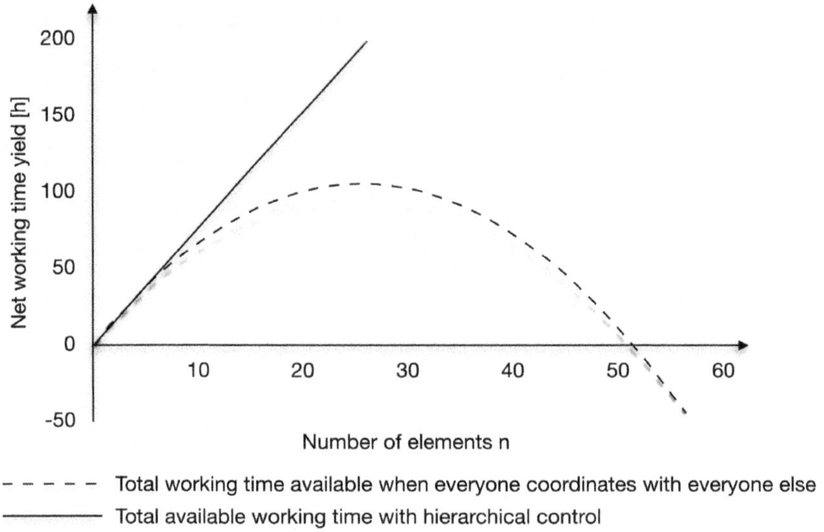

Fig. 6.4 Comparison of centralized and decentralized control based on the respective net working hours yield (Ahrens 2015, p. 7)

greater the coordination effort, the smaller the advantage of hierarchy. For example, with a coordination effort of 20%, the advantage is only 50%. This dependency is approximately linear.

Finally, the results of the previous investigations can be interpreted to mean that the realistic *complexity threshold* for decentrally controlled systems can be approximated in exactly the same way as the complexity threshold for centrally controlled systems, namely according to Eq. 6.4, because, as Fig. 6.5 shows, the net labor productivity reaches its maximum at this threshold in both cases.

The magnitude that the number of coordinable elements can assume through hierarchical structuring can be estimated by expanding the previous considerations to the model outlined below. In this model, 26 sub-systems are to be formed from 26 elements each, with hierarchical control taking place within the sub-systems as well as in the control of the sub-systems. However, it is assumed that the sub-systems are not controlled by an element, but by one of the 26 sub-systems, comparable to the fact that within the sub-systems 26 elements are controlled by one of them. Within each sub-system, there are 25 relations, making a total of 650 relations. In addition, there are the relations between the controlling part-system and the remaining 25 sub-systems, making another 25 relations. This results in a total of 675 relations.

By comparison, relating each of the 676 elements to every other element would result in 2,281,500 relations. Hierarchical structuring reduces this number to 0.3%, meaning that 99.7% of the relations that would actually be necessary are no longer needed.

While the focus so far has been on the possibility of hierarchical structuring to manage complexity, it will be shown below that, on the other hand, it is accompanied by considerable limitations. This initially affects the influence of individual elements on the system. As will

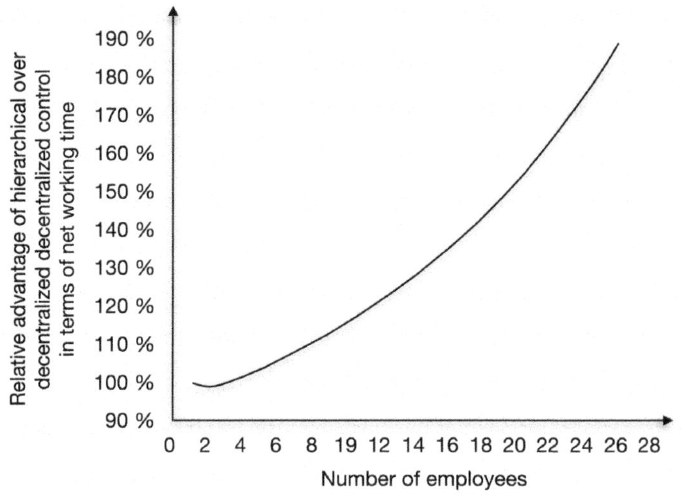

Fig. 6.5 Ratio of net working hours with centralized and decentralized control (Ahrens 2015, p. 7)

be shown, systems follow their own logic, largely independent of the influence of the elements, which is referred to as emergent in systems theory. Specifically, with regard to the relationship between the system and its environment, this means that the connection via the superior is not sufficient to influence the system from the outside. In this respect, systems behave autonomously. However, they form the typical functional structures shown in Fig. 2.4 in the form of the AGIL scheme introduced by Talcott Parsons (Münch 1992, p. 63), which enable coupling in the form of mutual expectations. This is explained in the following Section using the example of managerial accounting.

In the previous model calculation, a hierarchical structuring led to the fact that only 0.3% of all theoretically possible relations between 676 elements remained for their mutual coordination[7]. According to this model, a single employee bound by instructions spends just 2% of his working time, at 8 hours a day, less than 10 minutes, coordinating his activities with those of others. The rest of the time is spent on the actual work. Although a small amount of time is usually allowed for free communication, the standard times calculated using systems of predetermined times such as MTM (Methods Time Measurement) to an accuracy of 0 .036 s (Bokranz and Landau 2012, p. 95), are often calculated so precisely that the order of a few minutes per day is usually not significantly exceeded.

[7] This does not mean that there are no further relations between the elements. In fact, the elements are related to each other in the context of their collaboration in the completion of the actual tasks, which are often referred to as productive. The restriction of the relations here refers exclusively to the part of the coordination of the collaboration that is not predetermined by defined and thus quasi-automatic work sequences. For example, passing on a component to be assembled from one assembler to the next in a flow assembly line usually does not require any situational coordination between the assemblers, since the handover cannot otherwise be carried out sensibly without structures such as the spatial arrangement of the workplaces, work plans, etc.

It follows that systems only ever use a very small part of the linking capacities of their elements. These are replaced by a very tight corset of predetermined relations, in practice, for example, identifiable in the form of documented process descriptions (Ahrens 2012, p. 231 ff.) and spatial arrangements. Within this framework, the individual elements of larger systems only make relatively small contributions to the whole as a result of specialization based on the division of labor. In this respect, only a small part of their often much larger coordinative performance potential is called upon.

This sketch of current organizational forms, derived from Systems Theory, reveals at first glance their proximity to *Scientific Management* (Taylor 1919), to the *Industrial Engineering* that emerged from it (e.g., Bokranz and Landau 2012) and to the largely comparable concept of *Lean Production* (e.g. Dombrowski and Mielke 2015, Liker 2016). In view of approaches such as the humanization of work (Kieser 2014, p. 133 ff.), which also and above all attempt to overcome the division of labor, which is often perceived as too strong, and the associated one-sided burden, deskilling and heteronomy through concepts such as *job enlargement, job enrichment, job rotation* or *group or team work*, one gets the impression that this form of work organization is no longer up to date. But a sober look at the daily production routine shows the limited effect of such efforts (e.g. Goeudevert 1996, p. 160 ff.). This becomes particularly clear when one considers that a very large proportion of production now takes place in low-wage countries as a result of economic globalization, under conditions that are comparable in almost every respect to those that existed in the classic industrialized countries at the beginning of industrialization. But the precarization of employment relationships in high-wage countries, which affects many of the day laborers now known as temporary workers, or the increasing disregard for occupational safety rights, such as the Working Hours Act (DGB 2014, p. 27 ff.) among trainees, point in this direction. Employers (capital) and managers can only be held responsible for this if and to the extent that they disregard the law and morality. Otherwise, their pursuit of profit maximization and, consequently, cost minimization is inherent in the system of a market economy and thus justified. This means that the responsibility for ensuring a life-serving economy lies with the state. However, delving deeper into this would go beyond the scope of this article.

Instead, it is important to note once again that it is in the nature of systems to greatly restrict the degrees of freedom of their elements (Willke 2006, p. 142). As a consequence, the behavior of systems cannot be derived from the actions of their elements. Instead, it is exactly the other way around: the options for action of the elements are shaped by the system[8]. It is not the parts that produce the whole, but the whole that creates its parts (Luhmann 1996, p. 43, Willke 2006, p. 143). From a systems theory perspective, however, this does not refer to persons or even people, but to their communication; each system only ever allows a very specific type of communication (Parag. 2.6.2.2). For typical systems, this can be specified very precisely and will be explained in the following chapter using an example.

If the logic of a system, i.e. the nature and content of communication in terms of fulfilling a specific function, cannot be explained as the sum of the logics of its elements, then the

[8] This calls into question the current practice of Human Resource Management to the extent that a very extensive consideration of personality traits in the recruitment and development of personnel appears superfluous and thus uneconomical with regard to their extensive lack of influence on the company as a system.

question arises as to how the system logic comes about by other means. Systems theorists explain this with *emergence* (ibid., p. 142). A simple explanation of this phenomenon is provided by the popular saying with the now well-known insight, attributed to Aristotle (Wolf 2011, p. 161), that a whole is often more (more aptly: different) than the sum of its parts. The whole and the sum of its parts are virtually incommensurable; the transition between the two cannot be described any more than, for example, the ratio between the circumference and the radius of a circle depends on the irrational number μ and naturally cannot be completely determined.

The last hope that remains is to be able to control a system from the outside. However, the simple calculations carried out above already suggest that such an approach would quickly reach its limits. In the model last considered, it was assumed that the system's connection to its environment is established and maintained exclusively by the controlling instance. All other elements deal only with internal system matters. And the controlling instance also only uses a small part of its time for the connection to the environment, namely $\gamma = 2‰$, on a typical 8-hour day, therefore $(n-1) \cdot \gamma \cdot t = (26-1) \cdot 2‰ \cdot 8h = 4h$. That is less than 2‰ of the total available working time in the system. It is evident that a system can hardly be influenced by its environment via such a weak connection. Heinz von Foerster (1985, p. 51) also comes to a comparable conclusion with his calculations quoted in Section 2.1.

In fact, the behavior of a system as a whole is dominated by its emergent order. And this order, this logic, is generated by the system itself. If an external order were to be imposed on it that was powerful enough to superimpose or even destroy the internal order, the system would cease to be a system. In this respect, the system and its internal order are inseparably linked.

This manifests itself above all in the fact that a system delimits itself from its environment (Luhmann 1996, pp. 35 ff.). It is important to emphasize that the boundary is drawn by the system and not by its environment. In this respect, too, a system is autonomous. Of course, the boundary does not mark the end of all relations with the environment, but, to use Luhmann's words, cross-border processes are subject to different conditions of continuation. This applies less to energy and material flows than to information flows. Simply put, a system only hears what it wants to hear and makes of it what it wants. In systems theory, this phenomenon is referred to as *autonomy* and as *operative closure* (Luhmann 1996, pp. 35 ff., Willke 2006, pp. 62 f.). Information, including that which is intended to influence the system, is perceived by the system only as a stimulus or disturbance (irritation, perturbation) to behave; whether and, if so, in what form and intensity the system reacts depends exclusively on the system itself. Umberto Rosmesín Maturana (1985, quoted from Willke 2006, pp. 62 ff.) refers to this kind of relationship between system and environment as structural coupling. Karl Edward Weick (1976, p. 1 ff.) has coined the term loose coupling for this (Wolf 2011, p. 173 ff.). What this means in concrete terms will be explained in the next Section using the example of managerial accounting.

At this point, a connection can be made once again to mathematical, scientific and engineering approaches. In this sense, hierarchy could be described as a kind of *heuristic* that approximately solves a complex problem. Furthermore, it can be determined that it depends in particular on the processing capacity, which is complex and what is not. As a result of the ever-increasing processing speed of computers, as is currently expected, especially of *quantum computers* (e.g. Homeister 2018), those problems that can be solved algorithmically will be classified as less complex or no longer complex. John McCarthy (1956, p. 177 ff.) classified problems of this kind as well-structured. It is hoped that computer programs with so-called *artificial intelligence* (e.g. Russel and Norvig 2012) will help to solve problems that are poorly structured according to this classification. However, this should not obscure the fact that this will only slightly shift the boundary of what is to be classified as complex. Basically, however, complexity will continue to shape by far the greatest part of what determines human action.

6.1.2 Functional Differentiation

Despite the limited influence of individual elements on systems, orders do not arise by chance. In fact, through functional differentiation, the human condition leads to a structure (Münch 1992, p. 34 ff.), which for the sake of simplicity will be outlined below using the AGIL scheme (Fig. 2.4) by Talcott Parsons (1986, 14 ff.; 19 96, pp. 13 ff.), without taking into account the distinction into structural-functional and functional-structural approaches, which is undoubtedly important for systems theorists but which would take us too far afield here (e.g. Willke 2006, pp. 5 ff.). Thanks to its scale invariance (ibid., p. 136) or, more precisely, its self-similarity, the scheme, which is often used to model societies, can also be used to map companies[9].

Talcott Parsons identifies four sub-systems within social systems (Münch 1992, p. 30 ff., Willke 2006, p. 136 f.):

- The A-system (A: adaptation) serves to adapt the overall system to its environment; in companies, the actual value added serves this purpose.
- In the G system (G: goal attainment), political decisions are made to give the system as a whole a goal (contingency absorption); in companies, this is the function of corporate management.
- The I-system (I: integration) ensures that all elements see themselves as a community, and
- the L system (L: Latency) has the function of rationally understanding the overall system through reflection and providing arguments for specific actions. This is often also referred to as corporate culture and gives the system what system theorists mean by meaning (Luhmann 1996, 92 ff., Willke 2000, 43 ff.).

Of these four part-systems, only two, the A and G systems, will be examined in more detail in the further course of the study, since the other two systems are less important for the statements intended here.

Each part-system is distinguished from the others by the use of specific semantic codes, according to which communication is selected as system-relevant or system-irrelevant (Willke 2006, p. 67). Following Niklas Luhmann's coding (ibid., p. 213), the A-system (also

[9] The key aspects of the scheme have already been explained in Sections 2.1.2 and 2.6.2.3, among others, but should be briefly addressed again here to avoid constant back-and-forth reading. This is followed by the more in-depth considerations.

referred to below as the economic part-system) is all about money. Anything that cannot be expressed in monetary terms is irrelevant in this part-system. The specific function of this part-system is the allocation of scarce resources. In the G-system (hereinafter also referred to as the political part-system), on the other hand, it is not about money, but about power, about exercising domination.

Metaphorically speaking, we can say that very different languages are spoken within the two part-systems, which makes it difficult for them to communicate directly with each other. To put it bluntly, this means, for example, that company management is only interested in whether the company makes money to the extent that it serves to gain and maintain their own power. This can even go so far that managers have an interest in losses, namely when they can use these against competitors to gain an advantage over them. In the context of conventional views, such a statement undoubtedly seems absurd, but precisely for this reason it can be particularly helpful for understanding a systems-theoretical perspective, because it emphasizes the autonomy, self-reference and operational closure of social systems (Willke 2006, p. 61 ff.).

Compared to these dominant characteristics of the part-systems, the relationships between them and their mutual possibilities for influence are much less pronounced. In general, five possibilities for the interaction of the part-systems can be identified: dominance, adaptation, isolation, reconciliation, conflict and interpenetration (Münch 1992, p. 23). However, since only interpenetration is suitable, only it will be examined here. The starting point for this is the realization that the boundary between two systems does not lead to the complete breakdown of mutual relations. But cross-border processes [...] are [...] placed under different conditions of continuation [...] when the border is crossed (Luhmann 1996, pp. 35 f.).

6.1.2.1 Function of the Political Part-System

In terms of the economic and political part-systems, this means, for example, that money as the language of economics is only understood in the political part-system to the extent that it appears useful for acquiring and maintaining power. To illustrate this, imagine that two managers competing for a more powerful position can use economic arguments if and to the extent that these are suitable for profiling. Within the political part-system, however, such arguments are no more or no less helpful than any other argument that allows profiling, such as emphasizing personality traits, claiming patronage or dynastic relationships. Similarly, decisions in the political system are not made primarily with the economic success of the company in mind, but to gain and maintain the power of the decision-maker. Political decisions are only economically advantageous indirectly, since the purpose of a company as a whole primarily consists of generating an economic profit, so that political decisions ultimately cannot completely ignore this criterion for success. However, since the effect of each individual decision on profitability is not always immediately and precisely verifiable, there is a great deal of leeway in the political part-system for primarily power-oriented action.

In the course of the interpenetration of the economic and political part-systems, the semantic code "money" is thus translated into the semantic code "power". This may seem unusual at first glance, but from a systems theory perspective it is extremely useful, because both functions are essential for the existence of the overall system, but can only be performed if they develop according to their own laws.

The starting point for identifying the specific function of the political part-system is the situation, described by Niklas Luhmann (2000, p. 132 ff.) as contingency, that every action can be carried out in this way, but also in a different way, but not arbitrarily. The function of the political part-system is to transform this "so, but also differently" into a "so and not differently" (Luhmann 1988, p. 170). Corresponding decisions make a significant contribution to reducing complexity. They ensure certainty where this does not result from the facts themselves. It is not necessary to decide that 1 + 1 = 2. However, whether an investment in one technology or another, or even foregoing it, will be successful cannot ultimately be predicted with certainty, despite all the calculations that are usually carried out in the economic part-system (Section 4.6.4). Therefore, the final decision is not left to economists, but is made by politicians, in this specific case by executives. While the responsibility of the technicians and economists is, for example, to apply appropriate calculation methods without error, the responsibility of the managers begins where the calculations reach their limits[10]. This makes it clear that these are two different spheres of activity, which must be preserved in their difference so that they can each fulfill their specific functions.

This, of course, contradicts the widespread conviction that management decisions, in particular, are made rationally from an economic point of view and must therefore be explained causally (e.g. Göbel 2018, p. 75 ff.). Contingency, the absorption of which is an essential task of political decisions, means, however, that every decision is also possible differently and is therefore not economically unambiguous, i.e. causally conclusive from an economic point of view. In addition, this is usually observed by those affected by political decisions; the possibility that a different decision could have been made remains in consciousness even after a decision has been made (Martens and Ortmann 2006, p. 437 f.). Among other things, this leads to decisions being questioned even in retrospect and can be used for later attribution of responsibility. In this respect, the absorption of contingency is only partially successful in most cases. This is the reason why at least a pretense is made to justify decisions causally (Zimmermann 2004, pp. 315 ff.) and, in particular, to use economic arguments in companies because of their profit orientation. It can be used that usually only a few people are aware of the appearance of causal explanations. In this respect, objectification also helps to reduce contingency. It serves what Theresa Zimmermann (ibid., p. 316) calls the *causality illusion*.

To be able to decide, one must have the corresponding power. However, gaining and maintaining power takes place according to one's own and, depending on the type of rule (Weber 1972, p. 122 ff., Wolf 2011, p. 63 ff.), very different rules. These constitute the autonomy of the political part-system and are necessary within the political part-system. However, a pure autonomy would ultimately lead to Machiavellian power politics (Münch 1992, p. 17), to obsession with power and forgetfulness of power. It is the task of

---

[10] Limitations arise, for example, from uncertainty due to underlying assumptions and forecasts, from the influence of non-monetizable factors such as image/prestige, from the calculation methods used, for example when exact procedures are replaced by heuristics, etc.

interpenetration to prevent this and to put power at the service of the overall system without hindering autonomy.

6.1.2.2 Function of the Economic Part-System

The economic part-system, in which Systems Engineering & Management is also applied, has the function of ensuring adaptation to the environment of the overall system (Fig. 2.58). The environment of companies is formed by the economic part-system at the societal level, where the AGIL scheme is applied in a comparable way (e.g. Münch 1992, p. 30 ff., Willke 2006, p. 136 f.) as it is here at the organizational level. In this area, success is measured according to mathematically calculable and thus unambiguous criteria, in particular whether a contract is concluded, whether an exchange of goods/services/rights for money actually takes place and whether this results in a business profit.

Systems theory and practitioners agree that the economic part-system requires the greatest flexibility and dynamics (energy) compared to the other three part-systems (Fig. 2.58). This requires a maximum of freedom. System-internal restrictions, for example, due to the observance of bureaucratic rules, slow down action and restrict the scope of movement. This explains well that and why complaints about too many rules at both the economic and the operational level are heard primarily from the economic part-system. The function of politics, on the other hand, is to select very specific options from a wide range of possibilities with a view to certain goals, which could also be set differently (contingency), and thus exclude others. This already results in a restriction of dynamics and flexibility. The community part-system goes even further, limiting political decisions by aligning them with additional values such as social and ecological compatibility. The most far-reaching restrictions ultimately result from the cultural part-system, which, for example, demands that everything should result in a consistent pattern, that decisions should therefore be in line with the company's tradition, for example.

The economic part-system usually regards restrictions of this kind as hindering. And if these restrictions were to suddenly direct all economic activity, flexibility and dynamism would in fact be severely impeded. At the macroeconomic level, this could be observed in the centrally planned economies of socialist states (Willke 2000, pp. 195 ff.). In market-oriented economies, the privatization of state-owned enterprises such as the German Federal Railways or the German Federal Post Office is justified[11].

However, a complete unleashing of the economy would elevate economic activity to an end in itself. In fact, however, the economy serves only to direct scarce resources to where they provide the greatest benefit (*allocation*). Moreover, the economy is at the service of the overall system to ensure the prosperity of the whole of society at the macroeconomic level and to guarantee all members of a company an adequate income under acceptable working conditions and taking into account other constraints (A system) at the microeconomic level. In addition, the resources generated serve to achieve certain goals (G system), maintain a community (I system) and serve higher purposes (L system), because in the sense of *Corporate Citizenship*, a company is not limited to its economic dimension, but also

[11] This should not be understood to mean that privatizations can generally be advocated on the basis of systems theory. In fact, there may also be reasons against privatizations.

always part of society, from which not only the same comprehensive rights, but also the duties that citizens have as individuals result internally and externally (Backhaus-Maul et al. 2010).

6.1.2.3 Interpenetration Between the Political and Economic Part-System

The functionally specialized part-systems must now interact in such a way that they support each other without giving up their own autonomy or disturbing the autonomy of the other part-systems. In concrete terms, this means that a part-system that exerts greater constraints than another, for example the political part-system, must not dominate a more dynamic part-system such as the economic part-system. Conversely, a dynamic part-system such as the economic part-system must not force a more restrictive part-system such as the political part-system to adapt. Equally problematic would be the isolation of the part-systems, i.e. a lack of interaction (Münch 1992, p. 23). What remains is what Talcott Parsons calls interpenetration (Willke 2000, p. 207 f.).

Fig. 6.6 shows some of the services that the part-systems provide for each other in this way. In addition, this figure can be used to illustrate how interpenetration takes place between part-systems that are closed in operational terms and governed by their own laws. To do this, Talcott Parsons subdivided each of the four part-systems into four further part-systems (Münch 1992, p. 129, Willke 2000, p. 208), which will be referred to below as sub-part-systems to indicate the system level. For example, the A part-system consists of the AA, AG, AI and AL sub-part-systems. This makes it clear that within the economic part-system (A system) a sub-part-systems, namely the AG sub-part-system, performs the management function, while the actual value creation takes place primarily in the AA sub-part-system. In practice, this can be seen from the fact that in the AG sub-part-system of companies, for example in production, there are managers such as department heads, foremen or supervisors who, despite their management tasks, are usually not assigned to the company management because they play only a very limited political role.

The corresponding sub-part-system form so-called *interpenetration zones*, which mutually permeate each other and provide the necessary translation services. For the topic discussed here, the permeation of the GG sub-part-system with the AG sub-part-system is relevant and will be examined in detail below.

6.1.2.4 The Role of Managerial Accounting

The sub-part-systems absorb elements of different sub-part-systems that mediate between them (Münch 1992, p. 14). How this looks in practice is to be outlined with regard to the role of managerial accounting in companies, initially on the basis of a concrete example at the societal system level. The Deutsche Bundesbank[12] (BBk) is used for this purpose:

[12] For the sake of simplicity, the BBk's links with other institutions, such as the European Central Bank (ECB) or the German Federal Financial Supervisory Authority (BaFin), are not shown here. The role of the US Federal Reserve corresponds to that of the BBk).

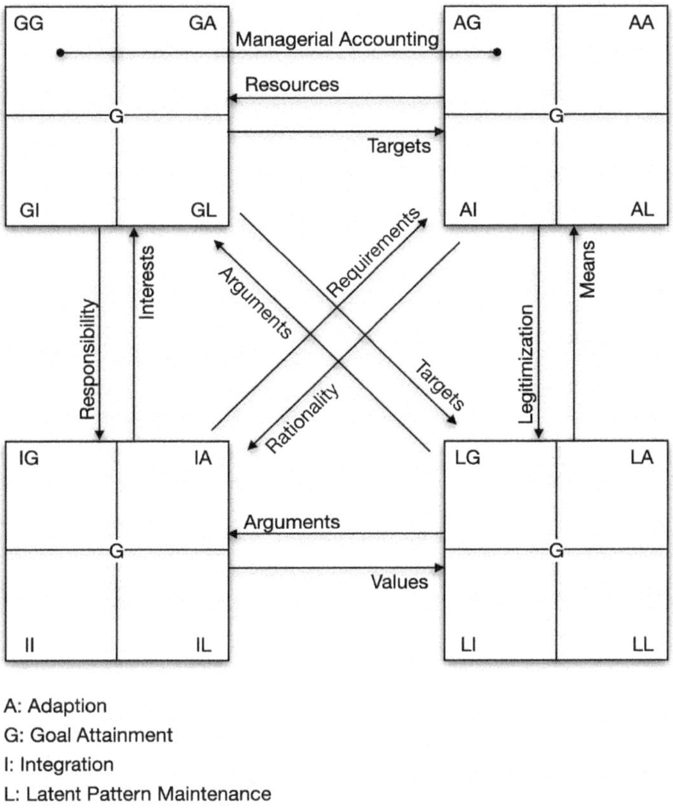

Fig. 6.6 AGIL scheme with input-output relations (Ahrens 2015, p. 15)

- Its anchoring in the political part-system can be seen from the fact that, as a federal legal entity under public law (Section 2 BbankG), it belongs to the indirect public administration. Furthermore, the board of directors is appointed by politicians (Section 7 (3) BbankG). And its primary mandate, to contribute to ensuring a stable price level (Section 3 BbankG), is also politically motivated.
- At the same time, the BBk is anchored in the economic part-system. This is evident at first glance from the fact that it acts as a bank and not as an authority. Its most important instrument is monetary policy, which, at least according to the idea, it conducts independently of political directives (Section 12 BbankG). In doing so, it influences the semantic code that guides action in the economic part-system. This influence is strengthened by the fact that it is also responsible for supervising private credit and financial services institutions, which in turn exert a decisive influence on the semantic code. However, the Bundesbank has no authority to issue instructions to market participants.

The BBk thus creates a loose link between the political and economic part-systems. Direct power is replaced by indirect influence:

- Although the BBk has an advisory function in the political arena (Section 13 BbankG), it has no voting rights (e.g. Section 9 (3) of the rules of procedure of the Stability Council).
- Banking supervision (Section 7 of the German Banking Act) also does not allow direct intervention in individual transactions of the banks, but only the definition of framework provisions and the monitoring of compliance with them.
- Specifically, the setting of the key interest rate, for example, only defines the costs at which commercial banks can obtain cash and credit; whether and, if so, to what extent they do so is left to their discretion.
- Analyses of financial and monetary stability serve only as a basis that other actors can use in their decisions, but are not obliged to do so.

In this way, the respective autonomy of the power-based political part-system and the money-based economic part-system is maintained without getting lost in it. Political decisions are influenced by economic constraints just as economic action is put at the service of politics. This gives the economy the necessary freedom to allocate scarce resources, which, thanks to political influence, are used not only for individual but also for collective goals.

6.1.2.5 Status Quo

In organizations, the link between politics and economics is often not nearly as loose as it is at the level of modern societies. Here, other interaction strategies are usually found, such as the dominance of politics over economics or the conflict between the two part-systems.

Domination is particularly recognizable on the basis of the widespread notion of a hierarchical order in which the management as a political authority is understood to be at the top (Fig. 6.3). This results in the assumption that it has direct access to all other, subordinate instances through a continuous chain of command. However, this is not compatible with systems theory models, because they recognize boundaries between functionally differentiated part-systems that require more than simple instructions to overcome (Willke 2005).

Another indication of the dominance of the political part-system over all the others is the widespread practice of political actors almost automatically receiving higher salaries than actors in other part-systems. At the societal level, this is different: in the last federal election campaign, former chancellor candidate Peer Steinbrück pointed out, politically unwise but factually correct, that even a savings bank director as head of a public-law institution can earn a significantly higher salary than the Federal Chancellor. From a systems theory perspective, however, there is nothing wrong with this, at least in principle, because according to this theory, the part-systems do not differ in terms of their greater or lesser importance or responsibility for the system as a whole, from which, among other things, the level of salaries could be derived, but only in terms of their specific function. In this respect, there is nothing wrong in principle with differentiating pay, but there is with the often-observed automatism with which this is derived from belonging to a particular part-system.

The problems resulting from the dominance of the political part-system over other part-systems include, for example, the high attractiveness of management positions, which leads many to prefer a management career to a specialist career, even if they would be much better suited to the latter and would thus be of more use to the system as a whole. In this way, companies lose a great deal of potential (see also Peter and Hull 2001).

What weighs more heavily, however, is that the economic part-system is constantly threatened in its autonomy in this way, which limits its ability to perform its function optimally. In practice, this can be seen, for example, when the sales department, as part of the economic part-system, is confronted with the wishes of customers and has to react flexibly and quickly to them, but at the same time is also subject to the ideas of the political leadership of its own company, which naturally has less of a customer focus and more of an owner focus. Flexibility and speed often fall by the wayside when dealing with the resulting conflicts.

Although it can be justified on the basis of *new institutional economics* that a central control of economic processes is characteristic of companies because they differ from markets in precisely this respect (Coase 1937, pp. 386 ff., Williamson 1975), this only justifies the necessity of management, but not its specific design. From a systems theory perspective, the central administrative allocation of resources can be limited to the economic part-system; control then takes place through the AG sub-part-system, whereby the autonomy of the A-system can be maintained without having to dispense with central administration. And, of course, political control of the economic part-system is also necessary. However, this must be done through a loose coupling, similar to the coupling at the societal level, and with regard to the role that managerial accounting plays in this, the role of the German Bundesbank can be a suitable model.

In fact, however, managerial accounting today sees itself primarily as part of the management system (G system). This can be exemplified by means of the managerial accounting conceptions of Péter Horváth (2011, p. 121) and Hans-Ulrich Küpper (2005, p. 28 ff.), which will be discussed below in the interpretation of Urs Bramsemann et al. (2004, p. 559 ff.). According to this, the task of managerial according to Péter Horváth (2011) is to coordinate the various part-system of the management system, namely the planning, control and information supply systems, and to align them with result targets. The necessity for this is justified, among other things, by the limited rationality of the actors and by the potentially opportunistic pursuit of self-interest (Horváth 2011, p. 121). And Hans-Ulrich Küpper (2005, p. 28 ff.) also sees the primary function of controlling as being to coordinate within the management system (Bramsemann et al. 2004, p. 560). This is also based, among other things, on the assumption that the actors pursue individual goals (Küpper 2005, p. 69).

From a systems theory perspective, several corrections need to be made to this before the actual discussion of the role of managerial accounting:

- Firstly, the assumption of bounded rationality seems to imply that there is only one permissible type of rationality, namely an economic rationality. However, based on the AGIL scheme, it is clear that each functionally differentiated part-system has a system-specific rationality, its own semantic code. So when political action is taken within the leadership system, it is not bounded rational and, moreover, no more oriented towards private goals than action in other part-systems. Rather, it is just as fully rational[13] as it is fully rational to act economically within the economic part-system (A-system). What is missing is merely an intermediary system, such as the German Bundesbank, which replaces the previous dominance of the political part-system with an interpenetration of the two part-systems.

- Secondly, from a systems theory perspective, it is not the actors as persons who are important, but rather their communication. If a person in the control system acts politically, this does not primarily result from the person's personal interest, their personality traits or other characteristics associated with the person, but rather from the fact that within the control system, any action other than political action is not possible. Conversely, the same person could only communicate economically within the economic part-system (A-system) because otherwise they would not be understood and could not have any effect. It is not the elements of a system that influence the system as a whole; the system produces the elements of which it consists (Luhmann 1996, p. 43, Willke 2006, p. 143). And the elements of social systems are not persons, but communication acts (Parag. 2.6.2.2). While persons can move in different systems, but are forced to use the specific semantic code in each case in order to understand, be understood and be able to have an effect, each semantic code is virtually trapped in its system and can only leave it if it translated by a mediation system into the semantic code of the target system, but in doing so loses some of its power, just as a poem written in one language often loses some of its expressiveness when translated into another language.
- Thirdly, the pursuit of self-interest is not limited to executives. Every person also pursues their own interests, which is perfectly legal (legally permissible) and legitimate (morally justifiable) within the usual limits (Parag. 2.6.2.3). And in well-functioning social systems, incentives not to lose sight of the common good despite one's own interests are not limited to economic incentives. In addition, there are also solidarity obligations generated in the I-system, as well as cognitive insights and convictions communicated in the L-system. Despite all the weaknesses that can be identified in them, modern societies are fundamentally good examples of how a well-functioning interpenetration of the part-systems can result in a well-balanced overall system (Münch 1992).

In this context, it is clear that it is fundamentally necessary to bring economic aspects to bear within the management system. And this can be the task of managerial accounting. However, this will not succeed if and to the extent that managerial accounting sees itself as part of the management system or perceives itself as being affiliated with the management system. Instead, managerial accounting, similar to the Deutsche Bundesbank, must detach itself from the management system, make itself independent of it and take on a mediating role in order to provide translation services, and to do so in both directions: from the management system to the economic part-system by translating political decisions into economic action, and from the economic part-system to the management system by influencing political decisions through economic conditions.

[13] ... as far as a system acts rationally at all – see Parag. 2.6.2.3.

6.1.2.6 Managerial Accounting as a Mediation Function

As already indicated above, part-systems take up elements of various part-systems that mediate between them (Münch 1992, p. 14). In this case, managerial accounting is placed as an element both in the GG sub-part-system of the political sub-part-system and in the AG sub-part-system of the economic sub-part-system. This is indicated by a corresponding connection in Fig. 6.6.

Analogous to Article 88 of the Basic Law for the Federal Republic of Germany (GG), the role of managerial accounting and its independence should be anchored in the corporate constitution. In addition, the primary goal should be defined at this point, which, analogous to the goal of the Deutsche Bundesbank to contribute to price stability, can be to contribute to the economic stability of the company. It may even be possible to limit the goal of managerial accounting to pure profit maximization (shareholder value) if it is ensured that other corporate goals, which result from taking into account all stakeholders (stakeholder value), are incorporated by other instances (Sect. 4.2.4.7).

The influence of political leadership on managerial accounting is ensured by the right to propose candidates for management positions in managerial accounting. However, the political leadership itself does not appoint or dismiss the managerial accounting body, but rather the shareholders' meeting or the supervisory board does so – analogous to the fact that the members of the BBk's board of directors are appointed by the Federal President (Section 7 (3) BBankG). An entry in the commercial register is not required because, unlike the management or board of directors, managerial accounting has no external impact.

Analogous to economic policy at the societal level, a division is necessary at the operational level, which puts fiscal policy in the hands of political leadership, while managerial accounting receives its own set of instruments that corresponds to monetary policy. In this context, a constructive and critical examination of the concept of "Beyond budgeting" (e.g. Hope and Fraser 2003, Horváth 2011, pp. 219 ff., Rieg 2008) as it shows clear parallels to the approach formulated here in its intentions, but significant deficits in terms of its scientific foundation (Rieg 2008, pp. 208 ff.).

The critical part of the discussion refers, for example, to the fact that beyond budgeting has so far lacked a clear model of what an alternative organization of companies should look like in concrete terms. Instead, it is recommended to imitate companies that are outlined as examples (Hope and Fraser 2003, p. 43 ff.; Rieg 2008, p. 196). The debate can be constructive, for example, to the effect that the general and thus difficult-to-operationalize demand for a delegation and decentralization of decisions (Rieg 2008, p. 201 ff.) can be substantiated and justified with the help of the approach presented here.

Ideas in the context of beyond budgeting to reduce a manager to the role of a mentor who is supposed to discipline himself not to intervene, even if he is tempted to do just that (ibid., p. 202), appear, as they have been formulated so far, to be unrealistic, because they are tantamount to a demand for self-denial. In fact, they are leaders because they see their role as leading, and if they do not lead, they are not leaders. The approach formulated here, on the other hand, envisages limiting the political leadership of the overall system to the G-part-system and economic leadership to the A-part-system, and to use controlling as a mediating instance between the two part-systems in order to transform the different rationalities (codes) without hindering, ignoring or dominating the respective autonomy of the two part-systems. This embeds the general demand for delegation and decentralization of responsibilities in a formal model and substantiates it on this basis.

And other ideas of beyond budgeting can also be both relativized and substantiated on the basis of the AGIL scheme, for example, the demand for the abolition of budgeting. The relativization consists in the fact that a complete abolition of budgeting is not demanded here because the chosen approach does not provide any arguments for it. And empirical findings, such as those cited by Rieg (2008, p. 208 ff.) against beyond budgeting, also speak against such a far-reaching step. Nevertheless, not all the reasons given by beyond budgeting against budgeting in its previous form can be completely dismissed, such as the immense effort often associated with it. In this context, the demand for the abolition of budgeting on the basis of the model used here can be substantiated to the effect that budgeting is limited to the economic part-system and placed in the sole responsibility of managerial accounting. This removes it from political power interests and makes it subject only to economic rationality. Nevertheless, the connection to politics remains in that the goals are set politically and the results have to be justified to the political part-system – just as the Deutsche Bundesbank is bound to politically set goals and has to justify the success or failure of its actions in the economic part-system to the political part-system, to the point that even Bundesbank profits are channeled to the political part-system.

In any case, the approach formulated here does not in any way intend to take control of the economic part-system completely out of the hands of the political part-system. Such neoliberalism has not proved successful at the societal level either (Ahrens 1999). Just as the political leadership at the societal level reserves the instrument of fiscal policy and general legislation for itself, the political part-system also retains comparable instruments at the organizational level to guide the economic part-bsystem. However, these instruments must be system-compliant. This means, for example, that economic actions are not directly instructed, but supported, for example, by offering subsidies. This leaves the economically motivated decision to use such offers in the economic part-system, so that its autonomy is maintained.

In addition, the political part-system has the function of setting the framework for the economic part-system, which is reflected at the societal level in legislation and its enforcement. These framework conditions must, however, be formulated as generally as possible in order to avoid any approach that ultimately seeks to directly influence individual economic aspects through this channel. Specifically, such framework conditions can, for example, define the business areas and markets in which a company operates, determine product groups and production technologies, and regulate overarching administrative processes.

6.1.2.7 Reducing Complexity Through Functional Differentiation

In the following, the model of functional differentiation based on the AGIL scheme will be examined in the light of the previous considerations regarding the influence of different organizational forms on the complexity of companies. On the one hand, there is coordination of each individual with each other, with the disadvantage of maximum complexity and the advantage of differentiated possibilities for adaptation to a complex environment in the sense of the basic law of the required variety (Ashby 1956, p. 206 ff.). On the other hand, there is hierarchical control with the advantage of a minimum of complexity and the disadvantage of very limited possibilities for adaptation to a complex environment.

Functional differentiation offers a compromise between these two extremes, which can be quantified using the model calculations introduced above. For this purpose, it is assumed that 26 sub-systems are to be coordinated in such a way that each sub-system is coordinated with every other subsystem within the system and between the sub-systems, but without all elements coordinating with all other elements. Instead of allowing only 0.3% of all possible relations, as calculated for hierarchical control, this type of coordination allows 3.85% of all relations. This means that complexity increases only slightly in relation to overall complexity, while the number of all possible relations compared to hierarchical control increases significantly, and with it the possibility of adapting to a complex environment.

The assignment of managerial accounting to the business organization as described here requires a slightly different business organization than the one that is common today. The realization that there is a need for further development of organizational forms is not new. In the context of managerial accounting, it is justified, among other things, by the concept of beyond budgeting (e.g. Hope and Fraser 2003, Horváth 2011, p. 219 ff., Rieg 2008). Empirical findings such as those of Robert Rieg (2008, p. 208 ff.), however, show that the concepts presented so far are not being applied, and assumptions such as that the reason for this is, for example, a lack of willingness on the part of companies to make far-reaching changes, appear too simple. It is more likely that the concepts available so far are not (yet) sufficiently convincing. Robert Rieg (ibid.) has outlined possible reasons for this: (1) the problem of induction, according to which conclusions drawn from particular cases (examples of successful implementations) cannot lead to scientifically proven generalizations; (2) the search for confirmation rather than the refutation attempt common in critical rationalism (Popper 1984); or (3) the failure to examine failed companies. However, it is equally unjustified to conclude from this that there is no alternative to the prevailing form of business organization. The problems, as also addressed by Beyond Budgeting, are too obvious and too profound for that. Against this background, the considerations presented here are part of a systems theory approach that attempts to take up the problems and solutions identified and developed by Beyond Budgeting, among others, but to avoid some of the mistakes: for example, a deductive approach is chosen that starts from a model that has already proven itself in the description of social systems (e.g., Münch 1992) and which, according to the ideas of its creator (Parsons 1986; Parsons 1996), is transferable to other levels of social systems on the basis of its scale invariance or, more aptly formulated, its self-similarity (Ahrens 1996; Ahrens 1999). This is also an example of the practical application of systems theory thinking.

A complete transformation of the model is still pending. However, it is already foreseeable that the prevailing form of operational organization will be challenged to a similar extent as it was by Beyond Budgeting, and that a whole range of ideas hinted at in Beyond Budgeting will be found again as a result of the intended transformation process, albeit systematized in a formal model. In this model, for example, the general and unspecific demand for delegation and decentralization (Rieg 2008, p. 201 ff.) is further developed into a functional differentiation, which, among other things, provides for a separation into a political and an economic part-system, which are loosely coupled to each other by managerial accounting. This will make it possible to determine more precisely than before who plays which role, in order to fulfill which function, and according to which rules the contributions of individuals to the whole are to be provided.

6.2 Process Organization

Like the organizational structure in the previous Section, the organizational structure is also not considered from a business perspective, but rather from a system-theoretical perspective. This ties in with the introductory considerations regarding the utilization phase in Section 3.4, so that the explanations there are only repeated here to the extent that it seems helpful to avoid constant back and forth between the pages.

While *Process Management* (Gaitanides 2012) emphasizes the necessity of designing processes consistently across interfaces, for example, within companies across departmental boundaries or in the sense of *Supply Chain Management* (SCM) (Schönsleben 2016) across company boundaries, a control loop management system aims to achieve the same thing, but emphasizes automation and the associated increase in efficiency and effectiveness on the one hand, and the necessity of maintaining dynamic equilibria (*homeostasis*), i.e. the largely automated avoidance of deviations, on the other. While Process Management is more like control from a technical point of view, regulation supplements it with a built-in feedback loop. In this very important respect, control-loop management is superior to Process Management (Heinrich and Schneider 2019) and corresponds to the second principle of the *Toyota Production System* (TPS) identified by Jeffrey K. Liker (2016, p. 135 ff.) as the pursuit of continuously flowing processes with the aim of bringing problems to light and ensuring process-immanent quality (jidoka). This, in turn, is linked to the fourth TPS principle of balanced utilization of the resources involved in a process (heijunka) (Liker 2016, p. 169 ff.).

One advantage of the systems theory perspective is that it can help to shed much of the ballast that subject areas such as business administration or Production Management bring with them. Another advantage is that it can help to overcome paradigms in order to open up opportunities beyond the beaten track.

For example, three different management systems for Quality Management (DIN EN ISO 9001:2015-11), Occupational Health and Safety (DIN ISO 45001:2018-06) and Environmental Protection (DIN EN ISO 14001:2014-11, EMAS) can be seen as *ballast*.

- Firstly, the contents of the standards are deliberately kept very similar, but as a result they are largely redundant and to that extent superfluous.
- Secondly, each of the aforementioned standards contains a lot of text, much of which is redundant and some of which is completely superfluous. The latter applies, for example, to definitions such as that of top management as a person or group of people who manage and control an organization at the highest level (DIN EN ISO 9000:2015-12, p. 25) – who would have thought it.
- Thirdly, the standards lead numerous other standards in tow, which further increase the abundance.

As will become apparent in the further course, some of this can be reduced to a comparatively few simple rules without losses, thus saving a great deal of effort (Ahrens and Hofmann-Kamensky 2000).

In this context, the *paradigms* that need to be overcome include, above all, Process Management (Ahrens 2018).

- Firstly, the embedding of Process Management (strictly speaking, control-loop management) in the system life cycle (Sect. 3.1) shows that this is only one of a total of five process models that are needed to do justice to all relevant aspects of Systems Engineering & Management (life cycle, design, operation, transient phase between design and control (Change Management), and exceptional situations).
- Secondly, processes are thought of as chains (Supply Chain Management), i.e. as a linear topology (Fig. 2.11a) (Sect. 2.4.2), while in practice they often interact so strongly with each other (e.g. DIN EN ISO 9001:2015-11, p. 10 f.), that they correspond more to a network topology (Fig. 2.12), so that one would have to speak, if at all, not of Process Management, but of Network Management.
- Thirdly, processes are conceived as controls that lack automatic feedback, so that deviations are recognized only with difficulty and often too late, for example, only when not the customer but the goods come back.

The basic functioning of control loops has already been explained in Section 3.4. In order to avoid repetition, we will first refer to this and then, using a practical example, summarize what is at stake and then focus on deepening selected aspects.

Fig. 6.7 sketches a quality control loop in concrete terms. The strategic target $w(t)$ is often based on the zero-error principle (Deckert et al. 2015, p. 80 ff.). As a controlled system, production has the function of manufacturing products of impeccable quality as an output variable $x(t)$. However, due to disturbances $z(t)$, this is not always possible. Therefore, the products x(t) are continuously measured and checked by the quality assurance system as the measuring element FM. The results are compared in processed form as the feedback variable $r(t)$ at the point of comparison with the reference variable $w(t)$. The resulting control difference $e(t) = w(t) - r(t)$ is then to be minimized by Production Management (controller) FR by means of appropriate measures. The manipulated variable $y(t)$ serves as the output variable of the controller. The measures to be taken in which situations are specified in the rule. For example, criteria are specified under which circumstances suspect products are to be subjected to further testing or when something is to be reworked or sent to the scrap.

As a rule, standard operation is relatively simple: if a customer orders 100,000 washers according to DIN 125 A 10.5 f. M 10 made of stainless steel A4 (V4A), he will receive 100,000 of these washers on the agreed delivery date and under the agreed delivery conditions. So it is not the regular operation that is crucial here, but the question of what happens when it gets out of sync. This can have numerous causes, which in turn require numerous measures. Some of these will be discussed in more detail below:

- So control loops are by their very nature not designed correctly and completely in every respect. Since they greatly reduce complexity, they regularly come into conflict with requirements that arise spontaneously and for which they are unprepared. For example, German tax law has actually regulated every conceivable case and diligent computer

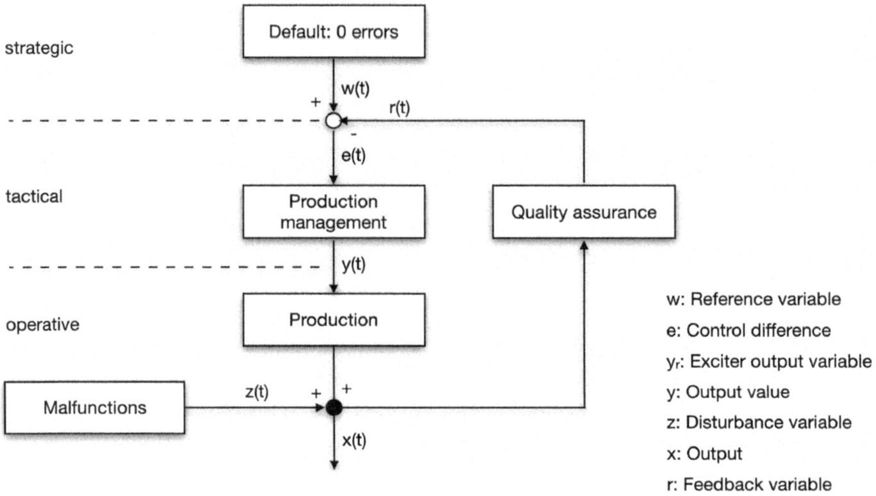

Fig. 6.7 Management control loop, for example for quality management

scientists have programmed everything into ELSTER[14], and yet someone suddenly calls the tax office and has a question. And clever engineers build perfect machines, and yet suddenly a dimension of a product is out of tolerance.

- Another significant disruptive factor in regular operations is owners. This may come as a surprise at first glance, but there are good reasons for it, which will be explained along with the countermeasures.
- Less surprising, on the other hand, is the fact that employees cause disruptions when they do not follow regulations due to a lack of knowledge and skills or because they know better but choose to ignore the rules (informal organization).
- Another common source of disruption is customers, for example when they deviate at short notice from orders that have already been placed or do not keep their part of the agreement.

With just a little imagination, this list could be extended almost indefinitely to include even the occasional freak weather conditions. This in turn makes it clear once again that control loops cannot be designed to cover all eventualities. It may be all the more impressive how well organizations such as companies or public administrations usually manage to establish and operate at least sufficiently stable control loops, even if they may not be aware that they are operating control loops. Bringing this to mind will not revolutionize the world, but it can possibly contribute to making the good even better. In the fierce international competition, this in turn can make a decisive difference.

The next step is to take a closer look at the owners, who have just been accused of occasionally disrupting the normal running of the company. As already mentioned, they may do so legitimately and will usually have good reasons for doing so. Family businesses

[14] (English: magpie, named after the bird that is considered thieving in Europe. At least the tax office shows a little humor in this way).

in particular, where owners can exert a very direct influence on day-to-day operations, are generally considered to be very successful. Nevertheless, or perhaps for that very reason, it makes sense to examine their role in more detail, for example to understand the deeper meaning of standardized management systems.

This understanding can start with *ownership* and the *property rights* derived from it (e.g. Pasetta 2005). Art. 14 (1) of the German Constitution protects property as a fundamental right and § 903 of the German Civil Code defines this as the control of a person over an object, provided that this does not conflict with the law or the rights of third parties. Although this is also subject to the proviso of the public interest (Art. 14 (2) of the German Constitution), within this framework every owner has the right to dispose of his property at his own discretion, and of course this also applies to ownership of companies. For example, an entrepreneur has the right to impose rules on his employees regarding their use of his property, but he himself does not have to abide by these rules. In short, the owner sets the law within his company, enforces the law and passes judgment. He is therefore the legislative, executive and judiciary in personal union, but is himself above the law. Strictly speaking, however, this only applies to owner-managed family businesses. In contrast, external managing directors in such companies and the boards of directors of stock corporations are themselves employees and answerable to the owners. This means that their freedom is subject to corresponding restrictions when exercising power of attorney.

With Max Weber (1972, pp. 124 ff.), these two types of legitimate rule can be classified as *traditional rule* in the first case and as *legal rule with a bureaucratic administrative staff* in the second case (Mikl-Horke 2000, pp. 93 f.). However, since the legal ownership, together with the associated bureaucratic apparatus, already has what is to be derived here, the further considerations remain with the traditional ownership for a while, which enables the owners to exercise their rights of disposal almost without restriction – one could also say: arbitrarily or at will. In this case, domination is not even necessarily tied to the qualifications of those in power. Succession in the management of family businesses is usually regulated in a dynastic way: daughters or sons take over, whether they can (and want to) or not.

If and to the extent that the owners act at their *discretion* in their company, taking into account the legally regulated restrictions (compliance with laws, protection of the rights of third parties, in this case in particular also of employees, as well as the obligation to serve the public good), this is their right and cannot be objected to. In particular, management does not have to be rational in every respect. A certain amount of *irrationality* is quite common in owner-managed family businesses, but this is allowed and the widespread success of numerous companies of this type shows that irrationality does not have to stand in the way (Captain James Tiberius Kirk). However, and this brings us back to where we started, this also affects the control loops, without which family businesses cannot manage either. While employees can be obliged to strictly follow the rules that go with it, under threat of sanctions, owners do not have to do so themselves. Even if it may not be wise and may confuse everything, they have the right to intervene at any point in a control loop at any time.

Now that freedom has been emphasized, the question of how the *conditions* of this freedom can take effect without restricting freedom in principle and beyond what is necessary will be discussed below. This is particularly linked to the question of how a sufficiently high *degree of rationality* and thus predictability can be established (Commander Spock), without unduly restricting the fundamentally justified irrationality (lust and whim, personal taste, emotion, etc.).

The first hurdle to be overcome is the *asymmetry of information*. As explained in Section 2.6, every mental and therefore also every social and socio-technical system is operationally closed. Whether Alter communicates information and what it may contain is initially entirely up to him (Parag. 2.6.2.2). And whether and, if so, how Ego ultimately understands what has been communicated depends on his hermeneutic abilities (in particular empathy). The inevitable information asymmetry that results from this is both a condition and an opportunity for numerous phenomena, all of which would go beyond the scope to discuss. What is relevant here is that asymmetrically distributed information can, under certain circumstances, lead to *moral hazards* and so-called *adverse selections*, the consequences of which are not Pareto-optimal (Erlei et al. 2016, p. 95 ff.).

One relevant form of adverse selection results from *hidden characteristics* of a product. A customer (principal) cannot assess the quality of a product before the transfer of risk (ex ante), and even after the customer has received the product (ex post), it would incur disproportionately high costs if it tried to check all quality-relevant characteristics (Akerlof 1970, pp. 488 ff.).

This leads him to subtract the monetarily quantified risk of buying inferior goods from the price. A seller (agent) knows the quality of his product better. If it is good, he will not be willing to accept the buyer's risk discount. Therefore, a contract will only be concluded if the seller himself knows about the inferior quality of his product (adverse selection). This in turn leads to market failure because ultimately only inferior goods are traded (not Pareto-optimal). George Arthur Akerlof received the Nobel Prize in Economics for insights such as these and conclusions such as those explained below.

To solve such problems, which result from the information asymmetry, there are basically two approaches available, which from the perspective of the person concerned are referred to as external and internal constraints. Both approaches allow free individuals to be subject to the conditions of freedom. External commitment reduces the information asymmetry through regulations, through disclosure obligations such as the obligation to file a tax return at regular intervals, and through control and sanctioning options (civil and criminal law). Self-commitment reduces the information asymmetry in particular through trust.

External regulation always emanates from a more powerful instance, in civilized societies, therefore, from the state. Its instruments include laws, ordinances, implementing provisions and the like. In addition, however, institutions such as employers' liability insurance associations in Germany are also authorized to administer themselves as public corporations, but to fulfill the legal mandate to prevent occupational accidents and diseases (Section 14 of the German Social Security Code (SGB VII)) and to issue binding rules for this purpose with accident prevention regulations (UVV). However, the further strategies of external control include not only demanding, but also promoting. Examples of this are state funding of consumer protection organizations or the production and broadcasting of documentary programs on public radio and television. In this way, some of the other conditions of freedom are also enforced, in particular the public welfare obligation of property, for example through progressive taxation[15], and the protection of the rights of third parties, such as the prevention of accidents in the workplace. New Institutional Economics, and in particular the principal-agent theory, refers to this as screening, since the controlling instance, the principal, goes to the trouble of obtaining reliable information from the controlled instance, the agent, who has more information than the principal in the relevant respects.

However, there are a number of reasons to subject free individuals and thus also companies as legal entities to the conditions of freedom, without forcing them to do so by a superior power (laws) or at least putting them under pressure (consumer protection).

- The first argument mentioned is the one that has the greatest significance in this context: freedom itself. In liberal societies, this should be restricted as little as possible and, against this background, a (too) strong effect is attributed to external constraints.
- In addition, the state is increasingly overwhelmed (Nassehi 2021; Willke 2014). This is due to an expansion of state problems and tasks. One example is the protection of the natural environment, which has only been increasingly and very intensively shaped and controlled by the state since the 1970s, but without having the desired effect at the necessary speed and to the required extent.
- At the same time, the possibilities for state control are steadily diminishing.
  - In particular, in the context of economic globalization, a nationally oriented policy can achieve less and less (Ahrens 1999, p. 4 f.).
  - At the same time, the influence of individuals with vast wealth, such as Bill Gates, Jeff Bezos, Elon Musk or Alexei Alexandrowitsch Mordashow, and globally active companies such as the American bank Goldman Sachs, on global events is growing ever greater, without being democratically legitimized.
  - And last but not least, world powers such as the United States of America (USA), Russia and China are now engaged in power struggles among themselves under conditions that are understood as global anarchy due to the lack of a superordinate power (Waltz 2010). These struggles entail developments that are difficult to calculate and thus difficult to shape and control for everyone, but especially for states that are not world powers themselves. The armed invasions of Iraq by the United States and its allies in March 2003 and of Ukraine by Russia in February 2022, as well as China's threat of such an invasion of Taiwan, are just three of many examples (Krastev and Holmes 2019).

In this context, *self-regulation* is added to external regulation. It is intended to supplement or even replace the requirements and controls of a higher power (principal) with *trust-building measures* of free individuals (agents). This includes, for example, *reputation*, which can be actively influenced by *marketing*, and by providing *guarantees* (as distinct from state-imposed warranty obligations) that signal to the customer that a quality defect that only becomes apparent after the transfer of risk and probably already existed at the time of the transfer of risk will not be at the customer's expense and therefore does not have to be deducted from the price as a monetarily quantified risk.

---

[15] The fact that this only applies to income, but not to wealth, reveals the limits of the practical implementation of the public-interest obligation of property.

However, *management systems* such as those for quality, Occupational Safety and Environmental Protection Management are of particular importance here, as they provide solutions for all the problems mentioned so far, such as the free disposal of owners over their companies. information asymmetry and the attempt to limit freedom as little as possible.

- With regard to the latter aspect, they are a solution in that they are part of the set of instruments for self-commitment. Only in sectors characterized by oligopolies and thus affected by a corresponding market failure are suppliers more or less forced by their powerful customers to introduce and maintain management systems. Under normal conditions, however, they are free to do so or not.
- Furthermore, management systems essentially require one thing above all: *bureaucracy*. In this way, they transfer owner-managed family businesses, which, in the sense of Max Weber (1972, p. 124), are subject to *traditional rule*, at least in those respects in which third parties (customers, employees, natural environment) are affected, into what Max Weber called a *rational form of rule*, without, however, curtailing the freedom of the owners in other respects. Management systems ensure that owners cannot act as they please in the respective functional areas covered, such as Quality Management, Occupational Health and Safety Management and Environmental Protection Management, but must also adhere to the relevant rules themselves. This is expressed, for example, by the fact that all the relevant standards address top management in great detail and make it very clear in a variety of ways what is expected of them. For example, it is not enough to carry out audits to verify the effectiveness of management systems (e.g. DIN EN ISO 9001:2015-11, p. 45 f.). In addition, top management is required to carry out its own review in the form of a management review (ibid., p. 46 f.).
- Finally, management systems contribute to solving the problem of information asymmetry by signaling to the respective target groups (stakeholders) that the company in question has structures and processes in place that are suitable for manufacturing products of impeccable quality, preventing accidents and illnesses in the workplace, and not polluting the natural environment more than necessary. This includes providing the necessary resources (personnel with the required qualifications, materials, money, etc.), delegating the appropriate responsibilities, and maintaining documentation that makes it possible to determine whether and, if so, how damage has occurred and who can be held accountable. At best, this results in the target groups trusting that their interests are being safeguarded even if they are unable to ensure this in every individual case by means of their own control (information asymmetry). This trust can be strengthened further by leaving the control to independent and thus trustworthy third parties, for example auditors from accredited certification companies, who, however, also do not test every single product, but do check compliance with the rules according to which products are manufactured in a qualitatively flawless, safe and environmentally friendly manner and according to which this is effectively monitored by internal controls (control cycles). The fact that the rules are documented in standards, which in turn document the state of the art, ensures that they are ultimately suitable for ensuring that the required standards are met. This is ultimately achieved by involving all relevant stakeholders in their formulation.

6.2.1 Bureaucracy

The practical effectiveness of organizational control loops is therefore achieved through bureaucracy. However, this does not sound particularly attractive, because this term has negative connotations not only thanks to the Vogons and their third-worst poetry (Forgács 2016, p. 15). Presumably, in a variation on a saying by Winston Churchill, it can be said that bureaucracy is indeed the worst of all forms of administration, except for all the others. In the following, however, bureaucracy is to be understood in the neutral sense of Max Weber (1972, p. 124 ff.) as a set of rules that, among other things, forms the common ground of the many seemingly different management systems. These are to be put in brackets, so to speak. As in mathematics, this also simplifies practical application. The parentheses then contain only a few terms that meet the specific requirements of the respective stakeholders to whom the management systems are aligned.

If we continue to follow Max Weber, a bureaucratic set of rules is characterized by the demand for consistent compliance by all those who belong to the corresponding social system. The rules must (1) therefore be *rational* so that they can be understood and accepted by all those involved and affected. This means that the creation and application of the rules must be *transparent*. Anything that remains hidden quickly arouses suspicion of not complying with the rules. In business ethics, this is known as the *principle of publicity*: According to Kant, the correctness of a maxim can be easily verified by asking whether one may 'let it be heard', that is, whether it has the ability to be made public. If you would rather keep a matter secret because you fear resistance otherwise, this is a strong indication of its injustice [...]: All actions related to the rights of other people whose maxims are incompatible with publicity are unjust (Göbel 2020, p. 35). This transparency also applies to management systems. The underlying *standards*, including those that regulate the review (e.g. DIN EN ISO 19011:2018-10), are publicly accessible and compliance with them is signaled by the publication of corresponding *certificates*.

Max Weber (1972, p. 125) further explains (2) that the typical legal lord has the role of a *superior*, in that he does give orders, but also obeys the impersonal order to which he orients his orders. Max Weber explicitly extends this to include superiors who are not civil servants themselves, but political leaders up to and including the owner himself (Weber cites the president as an example). Therefore, a subordinate does not obey the superior person, but his or her person and thus what this person embodies, namely the rules to which he or she is also subject[16]. It also follows that the loyalty of subordinates is limited only to the factual area of responsibility of their superior. This can be found, for example, in an *organization's manual*. At the same time, subordinates owe their performance only within the scope of their objectively *defined performance obligations*, as expressed, for example, in employment contracts and job descriptions[17]. Corresponding specifications are also required by management system standards, in that top management has to ensure "that the responsibilities and authorities for relevant roles are assigned, communicated and understood throughout the entire organization" (DIN EN ISO 9001:2015-11, p. 23).

[16] This area of responsibility is often also referred to as competence, but this concept of competence differs significantly from the concept of competence introduced in Chapter 1 and therefore avoided in this context.

[17] Exceptions to this are possible in exceptional situations (Sect. 3.7.2.2) in the course of recruitment: if the dike is in danger of bursting, even a professor will have to carry sandbags.

Max Weber (1978, pp. 125-126) cites the hierarchy of office (Fig. 6.3) as the third basic principle (3) of *bureaucratic order*. He adds to the top-down chain of command the right to hear the recipients of instructions and their right of appeal or complaint, which are to be processed appropriately according to rules that are to be defined and, if necessary, must lead to the amendment of instructions. This also promotes rationality and prevents superiors from acting as they please (lord of the manor type).

All persons involved in a regulated operation must (4) have a factually required and proven *qualification*. The standards for management systems also say essentially the same thing here: The organization shall determine [...] the necessary competence for persons; ensure that these persons are competent on the basis of appropriate education, training or experience [and] where applicable, initiate measures to acquire the necessary competence and evaluate the effectiveness of the measures taken (DIN EN ISO 9001:2018-11, p. 29).

Furthermore, work equipment is provided by the employer (5). Employees do not normally have to use private resources. In principle, administrative employees work in the *office* and not in the private household (Weber 1972). The application of this principle is particularly evident in occupational safety. For example, the employer is obliged to provide personal protective equipment (PPE) (Section 3 ArbSchG). Although employees can also work from home, the employer remains responsible for occupational safety and health and, in particular, for setting up ergonomic workplaces, including providing the necessary work equipment. The only exceptions are for so-called mobile working.

Persons cannot derive any rights (6) from the filling of a position, neither for permanent occupation nor for exploitation for purposes that go beyond the purely objective performance of the directly related tasks. The practical application of this principle is particularly evident in the prohibition of the *acceptance of benefits* (Section 332 of the German Criminal Code (StGB)). A corresponding Compliance Management system can be based on the international standard DIN ISO 37301:2021-11.

The principle of compliance with the *records* (7) requires that even if oral discussions are the rule or even a requirement, at least preliminary discussions and applications, as well as final decisions, orders and instructions, must be recorded in writing (Weber 1972, p. 126). The great importance attached to *documentation* in management systems (DIN EN ISO 9001:2018-11, p. 30; DIN EN ISO 14001:2015-11, p. 32; DIN ISO 45001:2018-06, p. 43), can be justified by the fact that Max Weber sees in it what he calls the key point of all modern administrative action (printed in bold in the original to emphasize this). In fact, documentation of all administrative acts is an essential prerequisite for ensuring that responsibility can be traced and attributed in the event of damage.

Furthermore, employees usually *work full-time* for one employer, so that they are unreservedly committed to them and do not experience *loyalty conflicts* due to other significant employment relationships or activities. In addition, they usually receive a fixed salary (civil servants: pay) in the form of money as well as pension entitlements (civil servants: pension entitlements), and these are at an appropriate level so as not to provide any incentive to accept bribes. Furthermore, the salary should be high enough to ensure a decent living, in order to avoid the employee having to earn additional income through secondary employment. This could not only lead to conflicts of loyalty, but could also prevent the employee from devoting his full attention to his main job. In addition, an employee can expect to move up the official hierarchy if he performs well, so he does not need to change employers to advance his career. This also promotes loyalty. Finally, with regard to the application of the control-loop model, it is relevant that employees are subject to strict official discipline and control (Weber 1972, p. 127). The uniformity refers to the adherence to rules and the control to the feedback loop contained in the control loop.

Purely bureaucratic, that is to say, [...] record-based administration, as Max Weber (ibid., p. 128) summarizes, is, according to all experience, the most precise, consistent, disciplined, streamlined and reliable, that is to say, predictability for the boss and for the interests, intensity and extensiveness of the performance, formally universal applicability to all tasks, technically perfectible to the highest level of performance, and in all these respects: formally the most rational form of exercising power. However, these advantages are also juxtaposed with disadvantages that should not be ignored. First and foremost is probably the concern, and sometimes the experience, that bureaucracy can be exaggerated (*bureaucratism*). The A38 pass is a well-known example of this[18]. One reason for this is the high degree of abstraction of rules, which inevitably means that they cannot be applied directly to all individual cases. This leads to the supplementation of general rules with special provisions, which over time makes the rulebook so complex that rules are needed to guide people through the tangle of rules, which in turn increases the complexity of the rulebook, and so on. In technical automation, such developments are reflected, for example, in the emergence of so-called *software robots* (RPA bots), i.e. software that is used to operate other software via its user interface.

Decentralization, i.e. a local limitation of the effect of control loops, and the *granting* of discretionary powers, are seen as an approach to avoiding bureaucracy. The latter arises in particular from the *informal organization*, which, as mentioned above, is an essential prerequisite for the functioning of the formal organization (Blau 1955). However, it is not possible to avoid the difficulty of finding the right *balance* between strict rules and discretionary leeway, between formal and informal organization. Particularly in view of the prospect of achieving efficiency and effectiveness gains by establishing both organizational and technical control loops, there is a risk that efficiency and effectiveness will ultimately be lost again if automation goes too far. This applies above all to the increasing digitalization of administrative processes. While in industrial production, where technical automation was introduced much earlier than in administration, it has long been recognized that manual work in certain areas can be more effective than a (too) high degree of automation (e.g. Schlick et al. 2018, p. 462), the automation of administrative processes seems to be (still) based on the principle of "the more the better".

What has been said so far, primarily with regard to the organization, forms the basis for the possibility of technical automation on the one hand, and on the other hand, the principles explained can be applied to it. For example, the efficiency and effectiveness of a business process will not be significantly improved by introducing support from a computer program if the business process itself is inefficiently and ineffectively organized. The programming of a software is based on the organization and only simplifies those data processing and communication processes in which the computer is superior to humans, for example in the speed and precision of calculations (Fig. 4.39). And even the movements that are to be performed by a robot arm instead of a human arm should first have been minimized according to comparable findings as those of REFA and MTM, such as those

[18] Albert Uderzo and René Goscinny (1976) Les Douze Travaux d'Astérix.

of a human. With this in mind, technical automation will not be explained in detail here. Please refer to the relevant literature (e.g. Heinrich et al. 2020).

Instead, a possible practical use of the previous explanations regarding organization will be concisely summarized. Such a use can consist of limiting the countless redundancies and the excessive accessories of the many management systems to the essentials. To do this, the following regulations can be put in parentheses and thus applied equally to all management aspects:

- *Control loops*: A control loop is set up for each stakeholder group, as shown in Fig. 6.7, to ensure that all legal and legitimate demands of the corresponding stakeholder group are met to the best of our knowledge and belief at all times (Ahrens and Kleinfeld 1999, p. 64 ff.)18.[19]
- *Stakeholder Value Management*: When there are conflicting goals in meeting the requirements of different stakeholders, balanced compromises are sought. The rules of discourse ethics (Göbel 2010, p. 42 ff.) are applied (Sect. 4.3.4.2).
- *Transparency*: The rules mentioned so far, along with evidence of consistent compliance, are disclosed to the respective stakeholders.
- *Universal validity*: All parties involved adhere to all relevant rules.
- *Provision of all necessary means*: All participants and stakeholders are provided with the means to properly fulfill their contributions to the production and use of the services (products, services, rights).
- *Documentation*: All processes concerning the requirements of stakeholders are documented in such a way that, in the event of damage, the relevant responsibility can be attributed to those responsible.
- *Decent livelihood*: All parties[20] involved in the provision of services19 are treated and remunerated in such a way that they can devote all their contractually owed energy to providing the service and do not need to ensure a decent livelihood elsewhere (significant paid secondary employment, acceptance of benefits, etc.).

These universal principles, which apply equally to all management systems, are specified with regard to the respective stakeholders addressed and in the light of the company's circumstances. For example, while the focus for occupational safety is on injury and illness-related operations, environmental protection is concerned, among other things, with keeping the air, water and soil clean, using resources (energy, materials, substances, etc.) sparingly and contributing to the organization of a circular economy. In addition, there are also synergies between specific aspects of management systems, such as the avoidance of noise between occupational safety and environmental protection.

[19] This prevents the company from being held liable under Section 823, paragraph 1 of the German Civil Code (BGB) for organizational errors, which occurs when responsibility is delegated to unsuitable employees due to a selection error, when work instructions are missing, incorrect or incomplete due to a fault in the instructions, or when controls are missing, incorrect or incomplete due to a fault in the monitoring.

[20] This also applies, of course, to subcontractors (suppliers, service providers) and their employees, as well as to their subcontractors and their employees, etc.

6.2.2 Continuation of Design Measures

While the principles of a regulation of routine activities explained so far are the result of a corresponding design, measures have already been initiated in the course of the design that must be continued during operation. These include Risk Management and Stakeholder Value Management (SVM). When routine operation begins, the stakeholders identified up to that point (Sect. 4.2.4.6 4.2.4.6) and their known requirements (Sect. 4.2.4.8 4.2.4.8 and 4.3.3) are incorporated into the work to date. However, changes are to be expected in this regard during operation, which must continue to be taken into account (DIN EN ISO 9001:2015–11, p. 18 f., 32 ff.; DIN EN ISO 14001:2015-11, p. 22; DIN ISO 45001:2018-06, p. 28). However, if significant changes occur, this requires a return to the design phase, because making changes to a running control loop will fail due to the basic function of control loops, which is to maintain dynamic equilibrium. In this respect, any attempt at change is perceived as a disturbance and eliminated. This is not a mistake. Quite the opposite: the more effectively this happens, the better the design has done its job.

Furthermore, the management of opportunities and risks must be continued, which also began in the course of the design (Sect. 4.2.4.7), because even during operation it is to be expected that originally assumed probabilities of occurrence and levels of damage (always analogous to the extent of opportunities) will be adapted to new insights and that, under certain circumstances, opportunities and risks may have to be taken into account (DIN EN ISO 9001:2015-11, p. 23 f.; DIN EN ISO 14001:2015-11, p. 26; DIN ISO 45001:2018-06, p. 34 f.). If and to the extent that significant changes have to be made to control loops that are already running, this also requires a return to the design phase. Each return to the design phase also marks the beginning of a new Change Management process (Sect. 3.5), which is particularly challenging when current control loops are to be changed. Therefore, the involvement of the relevant stakeholders continues even during the operating phase (DIN ISO 45001:2018-06, p. 31 ff.).

These considerations require a distinction to be made from the continuous improvement process (CIP) anchored in the Toyota Production System (TPS) and applied by many companies. In a control loop, it is not possible to interpret an improvement as a significant change in the structure of the loop or in the rules that apply within it, because the control loop would defend itself against this; as stated above, its core function is to absorb changes. Improvements during operation can therefore only relate to the consistency and accuracy of the application of the rules. Of course, this does not exclude small changes as long as they remain below the perception threshold. In this context, it is worth recalling the insight explained above that control loops are particularly allergic to change proposals that come from outside. It is therefore particularly important during ongoing operations that control loops develop from the inside out through participation. However, one should not have too many expectations in this regard.

Epilogue

The writing of the present work in the nominal style suggests objectivity. This is common practice, but it can only be justified because, in terms of content, I have professed constructivism, which in principle excludes any possibility of a person gaining reliable access to real reality. Therefore, it seems appropriate to me to point out the subjectivity of my arguments, at least at the end of my work.

This does not mean that I am not convinced of it or that I have not done extensive research and tried to quote correctly. But even the latter was only an attempt, which was by no means successful in every respect, to understand others and to bring what I believe I have understood into a context that seems systematic to me, with other things that I believe I have understood and with what I myself think about it and believe I have learned. In this respect, it is first and foremost a small excerpt from my very individual world of thoughts and experiences.

After all, I have been researching and teaching the content presented here for more than 17 years at a university of applied sciences that exclusively offers dual study programs. The practical relevance of theoretical content is of particular importance and many students, but above all graduates, have repeatedly confirmed the usefulness of the content in everyday business life, even after many years of practical experience. At least in this respect, my explanations seem to be confirmed intrasubjectively. And I myself have also benefited from them, both in the 10 years in which I held responsible positions in industry and in the context of research projects.

Ultimately, however, my explanations are above all a suggestion to engage with them constructively and critically in order to ultimately form one's own opinion. In the systems theory sense, they are not much more than irritations or perturbations for the benevolent reader.

V. Ahrens, *Systems Engineering & Management*, https://doi.org/10.1007/978-3-658-46875-0

References

Adam, Martin (2000) *Lebensfähigkeit sozialer Systeme: Stafford Beer's Viable-System-Modell im Vergleich*. VII, St. Gallen, Dissertation.

Adams, Douglas Noël (1979) *The Hitchhiker's Guide to the Galaxy*. Pan Books, London (UK).

Ahrens, Volker (1996a) *New organization concepts for distributed production systems*. In: Kybernetes – The International Journal of Systems and Cybernetics. Ausg. 25, No. 2. MCB University Press, Bradford. pp. 24 et seq.

Ahrens, Volker (1998) *System Models and Concepts for Distributed Production Planning and Control*. In: Scherer, Eric (ed.) Shop Floor Control – A Systems Perspective: From Deterministic Models toward Agile Operations Management. Springer, Berlin et al.

Ahrens, Volker (1999) *Globalisierung der Wirtschaft als notwendiger Schritt auf dem Weg zur Weltgesellschaft*. Contribution to the conference Globale Welt – Was tun? Lösungsperspektiven zur Globalisierungsproblematik of the Social Science Study Society in the Lutheran Church of Hannover on November 5, 1999 in Hannover.

Ahrens, Volker/Kleinfeld, Annette (1999) *Nach bestem Wissen und Gewissen: Ein unternehmensethischer Ansatz zur Integration von Managementsystemen*. In: Qualität und Zuverlässigkeit (QZ): Qualitätsmanagement in Industrie und Dienstleistung. Vol. 44, No. 1/99. Carl Hanser, Munich.

Ahrens, Volker (2001) *Business Excellence durch Stakeholder Value Management*. In: Ahrens, Volker/ Hofmann-Kamensky, Matthias (ed.) Integration von Managementsystemen: Ansätze für die Praxis. Vahlen, Munich. pp. 41–52.

Ahrens, Volker/Hofmann-Kamensky, Matthias (ed.) (2001) *Integration von Managementsystemen: Ansätze für die Praxis*. Vahlen, Munich.

Ahrens, Volker (2012) *Prozessorganisation im Kontext alternativer Organisationsformen*. In: Plate, Georg (ed.) Forschung für die Wirtschaft. Cuvillier, Göttingen, pp. 231–244.

Ahrens, Volker (2015) *Was Controller von Odysseus lernen können*. Working papers of the NORDAKADEMIE, No. 2015-02, Elmshorn.

V. Ahrens, *Systems Engineering & Management*, https://doi.org/10.1007/978-3-658-46875-0

Ahrens, Volker (2016) *Systemisches Controlling*. In: SEM Radar. Vol. 15, No. 1, Dornbirn. pp. 3–28.

Ahrens, Volker (2017) *Systemtheorie als wissenschaftliche Grundlage des Wirtschaftsingenieurwesens*. In: Technologie & Management, pp. 10–18.

Ahrens, Volker/Schnoor, Florian (2017) *SysML als Bindeglied zwischen Wirtschafts- und Ingenieurwissenschaften*. Arbeitspapiere der NORDAKADEMIE, No. 2017-02, Elmshorn.

Ahrens, Volker (2018) *Complementarity of Project and Process Management*. Working papers of the NORDAKADEMIE, No. 2018-02, Elmshorn.

Ahrens, Volker (2019) *Arbeitssystemgestaltung mit Cardboard Engineering*. Working papers of the NORDAKADEMIE, No. 2019-02, Elmshorn.

Ahrens, Volker (2020) *Abschlussarbeiten richtig gliedern in Naturwissenschaften, Technik und Wirtschaft*. 2nd ed. v/df Hochschulverlag, Zurich.

Ahrens, Volker (2021) *Karakuri: weniger ist mehr – Mechanische Automatisierung auf dem Prüfstand*. In: YOKOTEN: Magazin für Operational Excellence, Agilität und Leadership. No. 3, CETPM, Ansbach. pp. 6–8.

Ahrens, Volker (2022) *Management in Ausnahmesituationen*. In: Duales Studium: Personal in Hochschule und Betrieb gemeinsam entwickeln. No. 1, DUZ, Berlin. pp. 119–129.

Akerlof, George Arthur (1970) *The Market for „Lemons": Quality Uncertainty and the Market Mechanisms*. In: Quarterly Journal of Economics, Vol. 84(3), pp. 488–500.

Alam, Daud/Gühl, Uwe (2020) *Projektmanagement für die Praxis: Ein Leitfaden und Werkzeugkasten für erfolgreiche Projekte*. 2nd ed. Springer Vieweg, Berlin.

Allweyer, Thomas (2005) *Geschäftsprozessmanagement: Strategie, Entwurf, Implementierung, Controlling*. W3L, Herdecke/Bochum.

Altschuller, Genrich Saulowitsch (1986) *Erfinden: Wege zur Lösung technischer Probleme*. 2nd ed. VEB, Berlin.

Amabile, Teresa M. (1984) *The motivation to be creative*. In: Isaksen, S. (ed.) Frontiers of Creativity Research: Beyond the basics. Bearly Ltd., Buffalo, pp. 223–254.

Andler, Nicolai (2015) *Tools für Projektmanagement, Workshops und Consulting: Kompendium der wichtigsten Techniken und Methoden*. 6th ed. Publics Publishing, Erlangen.

Arbeitskreis Deutscher Qualifikationsrahmen (AK DQR) (2011) *Deutscher Qualifikationsrahmen für lebenslanges Lernen* (DQR). Berlin.

Argyris, John/Faust, Gunter/Haase, Maria/Friedrich, Rudolf (2017) *Die Erforschung des Chaos: Dynamische Systeme*. 3rd ed. Springer Vieweg, Berlin.

Arnold, Bozena (2013) *Werkstofftechnik für Wirtschaftsingenieure*. Springer Vieweg, Heidelberg et al.

Arrow, Kenneth Joseph (1951) *Social choice and individual values*. Yale University Press, Yale.

Arthur, W. Brian (2009) *The Nature of Technology: What it is and how it evolves*. Penguin Books, London.

Ashby, William Ross (1956) *An Introduction to Cybernetics*. 2nd ed. Chapman & Hall, London. Ast, Friedrich (1808) Grundlinien der Grammatik, Hermeneutik und Kritik. Jos Thomann, Landshut.

Atteslander, Peter (2010) *Methoden der empirischen Sozialforschung*. 13th ed. Erich Schmidt, Berlin.

Bachman, Leonard (2007) *Postindustrial Architecture, Dynamic Complexity and the Emerging Principles of Strategic Design*. Enquiry The ARCC Journal for Architectural Research, Ausg. 4, Nr. 1.

Backhaus-Maul, Holger/Biedermann, Christiane/Nährlich, Stefan/Polterauer, Judith (ed.) (2010) *Corporate Citizenship in Deutschland: Gesellschaftliches Engagement von Unternehmen – Bilanzen und Perspektiven*. 2nd ed. VS für Sozialwissenschaften, Wiesbaden.

Bähr, Johannes (2016) *Werner von Siemens 1816 – 1892*. C. H. Beck, Munich.

Baecker, Dirk (1994) *Experiment Organisation*. In: Lettre International, No. 24, pp. 22–26.

Bahill, A. Terry/Botta, Rick (2008) *Fundamental Principles of Good System Design*. In: Engineering Management Journal, Vol. 20, No. 4, pp. 9–17.

Balzert, Helmut/Schröder, Marion/Schäfer, Christian (2017) *Wissenschaftliches Arbeiten: Ethik, Inhalt & Form wiss. Arbeiten, Handwerkszeug, Quellen, Projektmanagement, Präsentation*. 2nd ed. Springer, Heidelberg.

Bamberg, Günter/Coenenberg, Adolf Gerhard/Krapp, Michael (2008) *Betriebswirtschaftliche Entscheidungslehre*. 14th ed. Vahlen, Munich.

Barabási, Albert-László/Pósfai, Márton (2016) *Network Science*. Cambridge University Press, Cambridge.

Barbuto, John E./Scholl, Richard W. (1998) *Motivation Sources Inventory: Development and Validation of New Scales to Measure an Integrative Taxonomy of Motivation*. In: Psychological Reports, Bd. 82, No. 3, pp. 1011–1022.

Barnard, Chester Irving (1938) *The functions of the executive*. Harvard University Press, Cambridge.

Barnsley, Michael F./Hurd, Lyman pp. (1996) *Bildkompression mit Fraktalen*. Vieweg, Braunschweig/Wiesbaden.

Bass, Frank M. (1967) *A New Product Growth Model For Consumer Durables*. Institute Paper No. 175. Institute for Research in the Behavioral, Economic and Management Science. Lafayette. Bateson, Gregory (1987) Geist und Natur: Eine notwendige Einheit. Suhrkamp, Frankfurt am Main.

Baur, Nina/Blasius, Jörg (ed.) (2019) *Handbuch der empirischen Sozialforschung*. 2nd ed. Springer VS, Wiesbaden.

Baylis, John/Smith, Steve/Owens, Patricia (2008) *Globalization of World Politics: An Introduction to International Relations*. Oxford University Press.

Beck, Ulrich (1986) *Risikogesellschaft: Auf dem Weg in eine andere Moderne*. Suhrkamp, Frankfurt a. M.

Becker, Hans Paul/Peppmeier, Arno (2022) *Investition und Finanzierung: Grundlagen der betrieblichen Finanzwirtschaft*. 9th ed. Springer Gabler, Wiesbaden.

Beer, Anthony Stafford (1981) *Brain of the Firm*. 2nd ed. John Wiley & Sons, Chichester.

Belardi, Nando (2018) *Supervision und Coaching: Grundlagen, Techniken, Perspektiven*. C. H. Beck, Munich.

Belz, Thore/Große-Heitmeyer, Volker/Sobiech, Hans-Jürgen (2015) *Kontinuierlicher Verbesserungsprozess*. In: Dombrowski, Uwe/Mielke, Tim (ed.) Ganzheitliche Produktionssysteme: Aktueller Stand und zukünftige Entwicklungen. Springer Vieweg, Berlin/Heidelberg. pp. 50 ff.

Bendel, Oliver (2017) *Die Industrie 4.0 aus Sicht der Ethik*. In: Reinheimer, Stefan (ed.) Industrie 4.0: Herausforderungen, Konzepte und Praxisbeispiele. Springer Vieweg, Wiesbaden. pp. 161–172.

Bender, Beate/Gericke, Kilian (ed.) (2021) *Pahl/Beitz Konstruktionslehre: Methoden und Anwendung erfolgreicher Produktentwicklung*. 9th ed. Springer Vieweg, Berlin.

Bender, Beate/Gericke, Kilian (2021) *Entwickeln der Anforderungsbasis: Requirements Engineering*. In: ders. (ed.): Pahl/Beitz Konstruktionslehre: Methoden und Anwendung erfolgreicher Produktentwicklung. 9th ed. Springer Vieweg, Berlin.

Benhabib, Seyla (1995) *Selbst im Kontext: Gender Studies*. Suhrkamp, Frankfurt am Main.

Berger, Peter L./Luckmann, Thomas (2003) *Die gesellschaftliche Konstruktion der Wirklichkeit: Eine Theorie der Wissenssoziologie (Gesellschaften)* 19th ed. Fischer, Frankfurt am Main.

Bertalanffy, Ludwig von (1951) *General Systems Theory: A new approach to unity of Science*. Ch. (ed.), Human Biology, Vol. 23, John Hopkins, Baltimore (MD).

Bertalanffy, Ludwig von, (1969) *General Systems Theory: Foundations, Development, Applications*. G. Braziller, New York.

Bester, Helmut (2017) *Theorie der Industrieökonomik*. 7th ed. Springer Gabler, Berlin/Heidelberg.

Bjørke, Øyvind (1995) *Manufacturing Systems Theory*. Tapir, Trondheim.

Bieri, Peter (2007) *Was die Sprache mit uns macht*. ZEIT Magazin Leben, No. 52, pp. 44–45.

Bion, Wilfried (2018) *Erfahrungen in Gruppen und andere Schriften*. Klett-Cotta, Stuttgart.

Blau, Peter Michael (1955) *The dynamics of bureaucracy: a study of interpersonal relations in two government agencies*. University of Chicago Press, Chicago.

Blumer, Herbert (1969) *Symbolic Interactionism: Perspective and Method*. Englewood Cliffs, New Jersey.

Böhm, Rolf (2005) *Methoden und Techniken der System-Entwicklung*. 5th ed. v/d|f Hochschulverlag, Zurich.

Bokranz, Rainer/Landau, Kurt (2012) *Handbuch Industrial Engineering: Produktivitätsmanagement mit MTM*. 2nd ed. Schäffer-Poeschel, Stuttgart.

Bolten, Jürgen (1985) *Die Hermeneutische Spirale: Überlegungen zu einer integrativen Literaturtheorie*. In: Poetica 17, Heft 3/4, pp. 362 ff.

Bortal, Karim (2016) *Task Force Management: Leitfaden für Manager*. Springer Vieweg, Berlin/Heidelberg.

Boujut, Jean-François/Blanco, Eric (2003) *Intermediary objects as a means to foster co-operation in engineering design*. Computer Supported Cooperative Work 12, pp. 205–219.

Bourcart, Jean-Jaques (1874) *Die Grundsätze der Industrie-Verwaltung: Ein praktischer Leitfaden*. Orell Füssli, Zurich.

Bourdieu, Pierre (1993) *Soziale Fragen*. Suhrkamp, Frankfurt am Main.

Bourdieu, Pierre (2003) *Die feinen Unterschiede: Kritik der gesellschaftlichen Urteilskraft*. Suhrkamp, Frankfurt am Main.

Bracht, Uwe/Geckler, Dieter/Wenzel, Sigrid (2018) *Digitale Fabrik: Methoden und Praxisbeispiele*. 2nd ed. Springer Vieweg, Berlin.

Bracke, Stefan (2022) *Technische Zuverlässigkeit: Datenanalytik, Modellierung, Risikoprognose*. Springer Vieweg, Berlin.

Bramsemann, Urs/Heineke, Carsten/Kunz, Jennifer (2004) *Verhaltensorientiertes Controlling: Konturierung und Entwicklungsstand.* DBW 64, No. 5, pp. 550–570.

Braun, Anton (2019) *Dynamische Systeme: Modellierung mit den Methoden der Laplace-Transformation*. Springer Vieweg, Wiesbaden.

Brüggemann, Holger/Bremer, Peik (2015) *Grundlagen Qualitätsmanagement: Von den Werkzeugen über Methoden zum TQM*. 2nd ed. Springer Vieweg, Wiesbaden.

Bruhn, Manfred (2022) *Marketing: Grundlagen für Studium und Praxis*. 15th ed. Springer Gabler , Wiesbaden.

Bryon, Kirstin/Khazanchi, Shalini/Nazarian, Deborah (2010) *The relationship between stressors and creativity: A meta-analysis examining competing theoretical models*. In: Journal of Applied Psychology, Bd. 95, No. 1, pp. 201–212.

Büchel, Alfred (1969) *Systems Engineering: Eine Einführung*. In: Industrielle Organisation, Ausg. 38, No. 9, pp. 373–385.

Bullinger, Hans-Jörg/Joachim Warschat (ed.) (2013) *Forschungs- und Entwicklungsmanagement: Simultaneous Engineering, Projektmanagement, Produktplanung, Rapid Product Development*. Springer, Wiesbaden.

Bungartz, Hans-Joachim/Zimmer, Stefan/Buchholz, Martin/Pflüger, Dirk (2009) *Modellierung und Simulation: Eine anwendungsorientierte Einführung*. Springer, Berlin/Heidelberg.

Burggräf, Peter/Schuh, Günther (ed.) (2021) *Fabrikplanung: Handbuch Produktion und Management 4*. 2nd ed. Springer Vieweg, Berlin.

Burggräf, Peter/Dannapfel, Matthias/Ebade Esfahani, Matthias/Pulz, Christian/Attig, Philipp (2021) F *ertigungsmittel- und Personaleinsatzplanung*. In: Burggräf, Peter/Schuh, Günther (ed.) Fabrikplanung: Handbuch Produktion und Management 4. 2nd ed. Springer Vieweg, Berlin.

Burghardt, Manfred (2013) *Einführung in Projektmanagement: Definition, Planung, Kontrolle, Abschluss*. 6th ed. Publics Publishing, Erlangen.

Carrier, Martin (2021) *Wissenschaftstheorie zur Einführung*. 5th ed. Junius, Hamburg.

Carlile, Paul R. (2002) *A Pragmatic View of Knowledge and Boundaries: Boundary Objects in New Product Development*. Organization Science 13 (4), pp. 442–455.

Coase, Ronald Harry (1937) *The Nature oft he Firm*. In: Economica, New Series, Ausgabe 4, Nummer 16, Seiten 386–405.

Cerman, Zdenek/Bartholtt, Wilhelm/Nieder, Jürgen (2005) *Erfindungen der Natur: Bionik – Was wir von Pflanzen und Tieren lernen können*. 3rd ed. Rowohlt Taschenbuch, Leipzig.

Chesbrough, Henry William (2006) *Open Innovation: The New Imperative for Creating and Profiting from Technology*. Harvard Business School Press, Cambridge.

Chi, Michelene T. H./Glaser, Robert/Farr, Marshall J. (1988) *The Nature of Expertise*. Psychology Press, New York.

Clark, Charles Hutchison (1958) Brainstorming: *The Dynamic New Way to Create Successful Ideas*. Melvin Powers Wilshire Book, Woodland Hills.

Cohen, Michael D./March, James Gardner/Olsen, Johan Peder (1972) *A Garbage Can Model of Organizational Choice*. In: Administrative Science Quarterly, Bd. 17, pp. 1–25.

Coles, John Morton (1979) *Experimental archaeology*. Acad. Press, London.

Conceição, Carolina Souza da/Silva, Ggislaine Cyrino Capistrano da/Broberg, Ole/ Duarte, F.rancisco (2012) *Intermediary objects in the workspace design process: means of experience transfer in the offshore sector*. IOS Press. pp. 127–135.

Cooper, Robert G. (2010) *Top oder Flop in der Produktentwicklung – Erfolgsstrategien: von der Idee zum Launch*. Wiley-VCH, Weinheim.

Cyert, Richard Michael/March, James Gardner (1963) *A Behavioral Theory of the Firm*. Englewood Cliffs, N. J.: Prentice Hall, New Jersey.

Czarnietzki, Reiner (2008) *Entwicklung rechnergestützter Methoden zur Belegungsplanung auf einer Kompaktwerft und deren Anwendung zur Untersuchung ausgewählter Fertigungsreihenfolgen bei der Paneel- und Sektionsfertigung*. Dissertation at the University of Rostock.

Czichos, Horst (2019) *Mechatronik: Grundlagen und Anwendungen technischer Systeme*. 4th ed. Springer Vieweg, Wiesbaden.

David, Paul A. (1985) *Clio and the Economics of QWERTY*. In: American Economic Review, Vol. 75, Nr. 2, pp. 332–337.

Deckert, Frank/Krenkel, Philipp/Mielke, Tim (2015) *Null-Fehler-Prinzip*. In: Dombrowski, Uwe/ Mielke, Tim (ed.) Ganzheitliche Produktionssysteme: Aktueller Stand und zukünftige Entwicklungen. Springer Vieweg, Berlin/Heidelberg. pp. 80 ff.

Deming, William Edwards (1950) *Elementary Principals of the Statistical Control of Quality*. Nippon Kagaku Gijutsu Remmei.

Deserti, Alessandro/Real, Marion/Schmittinger, Felicitas (ed.) (2019) *Co-creation for Responsible Research and Innovation: Experimenting with Design Methods and Tools*. Springer, Cham.

Deutsche Bundesbank (2022) *Finanzstabilitätsbericht 2022*. Selbstverlag. Frankfurt a. M.

Dewey, John (1997) *How we think*. Dover Publications, Mineola. Erstausgabe 1910.

DGB – Deutscher Gewerkschaftsbund (2014) *Ausbildungsreport 2014*. Berlin.

Diekmann, Andreas (2020) *Empirische Sozialforschung: Grundlagen, Methoden, Anwendungen*. 13th ed. Rowohlt Taschenbuch, Reinbek bei Hamburg.

Dietrich, Jürgen R./Meitinger, Thomas Heinz (2021) *Erfinderhandbuch: Innovations- und Patentmanagement für Erfinder, Ingenieure und mittelständische Unternehmen*. Springer Vieweg, Berlin.

Dijkstra, Edsger Wybe (1959) *A note on two problems in connexion with graphs*. In: Numerische Mathematik 1, pp. 269–271.

Dilthey, Wilhelm (1924) *Ideen über eine beschreibende und zergliedernde Psychologie*. In: Wilhelm Diltheys gesammelte Schriften, Bd. V: Die geistige Welt: Einleitung in die Philosophie des Lebens. Teubner, Leipzig/Berlin. pp. 139–240.

Ditfurth, Hoimar von (1988) *So lasst uns denn ein Apfelbäumchen pflanzen: Es ist soweit*. Droemersche Verlagsanstalt Th. Kaur Nachf., Munich.

Dittes, Frank-Michael (2015) *Optimierung: Wie man aus allem das Beste macht*. Springer Vieweg, Berlin/Heidelberg.

Dittes, Frank-Michael (2021) *Komplexität: Warum die Bahn nie pünktlich kommt*. 2nd ed. Springer, Berlin.

Dixius, Dieter (2013) *Simultane Projektorganisation: Ein Leitfaden für die Projektarbeit im Simultaneous Engineering*. Springer, Berlin u. a.

Doran, George T. (1981) *There's a S.M.A.R.T way to write management's goals and objectives*. In: Management Review, Ausg. 70, No. 11, pp. 35–36.

Dörner, Dietrich (1979) *Problemlösen als Informationsverarbeitung*. 2nd ed. Kohlhammer, Stuttgart/Berlin/Köln/Mainz.

Dombrowski, Uwe/Mielke, Tim (2015) *Ganzheitliche Produktionssysteme: Aktueller Stand und zukünftige Entwicklungen*. Springer, Berlin/Heidelberg.

Domschke, Wolfgang/Drexl, Andreas/Klein, Robert/Scholl, Armin (2015) *Einführung in Operations Research*. 9th ed. Springer Gabler, Berlin/Heidelberg.

Domschke, Wolfgang/Drexl, Andreas/Mayer, Gabriela/Tadumadze, Giorgi (2018) *Betriebliche Standortplanung*. In: Tempelmeier, Horst (ed.) Planung logistischer Systeme. Springer Vieweg, Berlin. pp. 1–28.

Doran, George T. (1981) *There's a S.M.A.R.T way to write management's goals and objectives*. Management Review, Ausg. 70, No. 11. pp. 35–36.

Dudgeon, Piers (2001) *Braking Out of the Box: The Biography of Edward De Bono, the man who created the concept of Lateral Thinking*. Headline Book Publishing, London.

Dürrenmatt, Friedrich (1985) *Die Physiker*: Comedy. Diogenes, Zurich.

Duncker, Karl (1974): *Zur Psychologie des produktiven Denkens*. 3rd ed. Springer, Berlin/Heidelberg/New York.

Durkheim, David Émile (1992) *Über soziale Arbeitsteilung: Studie über die Organisation höherer Gesellschaften*. Suhrkamp, Frankfurt am Main.

Ebel, Nadin (2011) *PRINCE2:2009 – Projektmanagement mit Methode – Grundlagenwissen und Zertifizierungsvorbereitung für die PRINCE:2009-Foundation-Prüfung*. Addison-Wesley, Bonn.

Ebert, Christof (2019) *Systematisches Requirements Engineering: Anforderungen ermitteln, dokumentieren, analysieren und bewerten*. 6th ed. dpunkt.verlag, Heidelberg.

Edler, Frank/Soden, Michael/Hankammer, René (2015) *Fehlerbaumanalyse in Theorie und Praxis: Grundlagen und Anwendung der Methode*. Springer Vieweg, Berlin/Heidelberg.

Ehlert, Stefanie/Seltman, Stefan (2017) *Die erste Woche Krisenprojekt aus Sicht des Krisenmanagers*. In: ProjektMagazin – Das Fachportal für Projektmanagement. pp. 4–20.

Ehrlenspiel, Klaus/Meerkamm, Harald (2017) *Integrierte Produktentwicklung: Denkabläufe, Methodeneinsatz, Zusammenarbeit*. 5th ed. Hanser, Munich.

Ehrlenspiel, Klaus/Kiewert, Alfons/Lindemann, Udo/Mörtl, Markus (2020) Kostengünstig Entwickeln und Konstruieren: Kostenmanagement bei der integrierten Produktentwicklung. 8th ed. Springer Vieweg, Berlin.

Eifler, Stefanie/Leitgöb, Heinz (2019) Experiment. In: Baur, Nina/Blasius, Jörg (ed.) *Handbuch Methoden der empirischen Sozialforschung*. 2nd ed. Springer VS, Wiesbaden. pp. 203–218.

Ekardt, Felix (2021) *Theorie der Nachhaltigkeit: Ethische, rechtliche, politische und transformative Zugänge – am Beispiel von Klimawandel, Ressourcenknappheit und Welthandel*. 3rd ed. Nomos Verlagsgesellschaft, Baden-Baden.

Eisenführ, Franz/Weber, Martin/Langer, Thomas (2010) *Rationales Entscheiden*. 5th ed. Springer, Berlin.

Erlach, Christof (2014) *Systematisches Requirements Engineering: Anforderungen ermitteln, analysieren und bewerten*. 5th ed. dpunkt.verlag, Heidelberg.

Erlei, Mathias/Leschke, Martin/Sauerland, Dirk (2016) *Neue Institutionenökonomik*. 3rd ed. Schäffer-Poeschel, Stuttgart.

Fakultäten- und Fachbereichstag Wirtschaftsingenieurwesen e. V. (FFBT)/Verband Deutscher Wirtschaftsingenieure (VWI) e. V. (ed.) (2019) *Qualifikationsrahmen Wirtschaftsingenieurwesen*. 3rd ed. Steinbeis-Edition, Stuttgart.

Fayol, Henry (1916) *Administration industrielle et générale: prévoyance – organisation – commandement – coordination – contrôle*. H. Dunod et e. Pinat, Editeurs, Paris.

Feist, Gregory John (1998) *A Meta-Analysis of Personality in Scientific and Artistic Creativity*. In: Personality ans Social Psychology Review. Bd. 2, No. 4, pp. 209–309.

Felkai, Roland/Beiderwieden, Arndt (2015) *Projektmanagement für technische Projekte: Ein Leitfaden für Studium und Beruf*. 3rd ed. Springer Vieweg, Wiesbaden.

Festinger, Leon (1978) *Theorie der kognitiven Dissonanz*. Hans Huber, Bern.

Flechtner, Hans Joachim (1972) *Grundbegriffe der Kybernetik: Eine Einführung*. 5th ed. Hirzel, Stuttgart.

Filho, Walter Leal/Azul, Anabela Marisa/Brandli, Luciana/Lange Salvia, Amanda/ Wall, Tony (ed.) (2021) *Affordable and Clean Energy (Encyclopedia of the UN Sustainable Development Goals* Springer, Cham.

Fischermanns, Guido (2013) *Praxishandbuch Prozessmanagement*. 11th ed. Schmidt, Wettenberg.

Foerster, Heinz von (1993) *Wissen und Gewissen: Versuch einer Brücke*. Schmidt, Siegfried J. (ed.). Suhrkamp, Frankfurt am Main.

Foerster, Heinz von (1985) *Sicht und Einsicht*. Vieweg & Sohn, Braunschweig/ Wiesbaden.

Foerster, Heinz von (1995) *Cybernetics of Cybernetics: The Control of Control and the Communication of Communication*. Carl Auer, Heidelberg.

Förtsch, Gabi/Meinholz, Heinz (2018) *Handbuch Betriebliches Umweltmanagement*. 3rd ed. Springer Vieweg, Wiesbaden.

Forrester, Jay Wright (2013) *Industrial Dynamics*. Martino Publishing, Mansfield Centre. Erstaus- gabe 1961.

Forgács, Peter D. (2016) *Der ausgelieferte Beamte: Über das Wesen der staatlichen Verwaltung*. Böhlau, Vienna.

Frawley, Usama M./Piatetshky-Shapiro, Gregory I./Smyth, Padhraic (1996) *From Data Mining to Knowledge Discovery in Databases*. In: AI Magazine 17, pp. 37–54.

Freeman, R. Edward (2010) *Strategic Management: A Stakeholder Approach*. Cambridge University Press, Cambridge.

Freud, Siegmund (1911) *Formulierungen über zwei Prinzipien des psychischen Geschehens*. Gesammelte Werke, Band 8. Gemeinfrei.

Freund, Jakob/Rücker, Bernd (2014) *Praxishandbuch BPMN 2.0*. 4th ed. Hanser, Munich.

Frick, Detlev/Gadatsch, Andreas/Kaufmann, Jens/Lankes, Birgit/Quix, Christoph/ Schmidt, Andreas/Schmitz, Uwe (ed.) (2021) *Data Science: Konzepte, Erfahrungen, Fallstudien und Praxis*. Springer Vieweg, Wiesbaden.

Fritzsche, Ursula (2001) *Externe Faktoren in der Dienstleistungsproduktion: Ansätze zur Lösung von Erfassungs- und Bewertungsproblemen*. Deutscher Universitätsverlag, Wiesbaden.

Früh, Werner (2017) *Inhaltsanalyse: Theorie und Praxis*. 9th ed. UKV Verlagsgesellschaft, Konstanz/Munich.

Funke, Joachim (2003) *Problemlösendes Denken*. Kohlhammer, Stuttgart.

Gadamer, Hans-Georg (2010) *Wahrheit und Methode: Grundzüge einer philosophischen Hermeneutik*. 7th ed. Mohr Siebeck, Tübingen.

Gadatsch, Andreas/Schreiber, Dirk (2021) *Management von Big Data Projekten*. In: Frick, Detlev/ Gadatsch, Andreas/Kaufmann, Jens/Lankes, Birgit/Quix, Christoph/Schmidt, Andreas/Schmitz, Uwe (ed.) Data Science: Konzepte, Erfahrungen, Fallstudien und Praxis. Springer Vieweg, Wiesbaden. pp. 41–62.

Gaitanides, Michael (2012) *Prozessorganisation: Entwicklung, Ansätze, Programme des Managements von Geschäftsprozessen*. 3rd ed. Franz Vahlen, Munich.

Galison, Peter (1997) *Die Ontologie des Feindes. Norbert Wiener und die Vision der Kybernetik*. In: Rheinberger, H.-J./Hagner, M./Wahrig-Schmidt, B. (ed.): Räume des Wissens. Repräsentierung, Codierung, Spur. Oldenbourg, Berlin. pp. 281–325.

Gansler, Jacques (1989) *Affording Defense*. MIT Press, Cambridge.

Gehlen, Arnold (2016) *Der Mensch: Seine Natur und seine Stellung in der Welt*. ed.: Rehberg, Karl-Siegbert. Vittorio Klostermann, Frankfurt am Main. Erstausgabe 1950.

Gehring, Hermann/Gabriel, Roland (2022) *Wirtschaftsinformatik*. Springer Gabler, Wiesbaden.

Geiger, Walter/Kotte, Willi (2008) *Handbuch Qualität: Grundlagen und Elemente des Qualitätsmanagements: Systeme - Perspektiven*. 5th ed. Vieweg, Wiesbaden.

Geirhos, Matthias (2017) *IT-Projektmanagement: Was wirklich funktioniert – und was nicht*. 2nd ed. Rheinwerk, Bonn.

Gerhold, Lars/Holtmannspötter, Dirk/Neuhaus, Christian/Schüll, Elmar/ Schulz- Montag, Beate/Steinmüller, Karlheinz/Zweck, Axel (ed.) (2015) *Standards und Gütekriterien der Zukunftsforschung: Ein Handbuch für Wissenschaft und Praxis*. Springer VS, Wiesbaden.

Gesamtmetall (2013) *Verteilung der Betriebe in der Metall- und Elektroindustrie in Deutschland nach Beschäftigtengrößenklassen im September 2013*. Statista. Zugriff am 05. März 2015. Verfügbar unter http://de.statista.com/statistik/daten/studie/6139/umfrage/ ueberblick-ueber-die-betriebe-in-der-metallindustrie-nach-groesse-in-2007/.

Gessler, Michael/Kaestner, Rolf (2011) *Projektphasen*. In: Deutsche Gesellschaft für Projektmanagement/Gessler, Michael (ed.) Kompetenzbasiertes Projektmanagement. 4th ed. Deutsche Gesellschaft für Projektmanagement.

Gigrenzer, Gerd (2013) *Risiko: Wie man die richtigen Entscheidungen trifft*. 3rd ed. Bertelsmann, Munich.

Göbel, Markus/Schmidt, Johannes F. K. (1998) *Inklusion/Exklusion: Karriere, Probleme und Differenzierungen eines systemtheoretischen Begriffspaars*. In: Soziale Systeme No. 4, Heft 1, pp. 87–117.

Göbel, Elisabeth (2010) *Unternehmensethik*. 2nd ed. Lucius & Lucius (UTB), Stuttgart.

Göbel, Elisabeth (2020) *Unternehmensethik: Grundlagen und praktische Umsetzung*. 6th ed. UKV, Munich.

Goeudevert, Daniel (1996) *Wie ein Vogel im Aquarium: Aus dem Leben eines Managers*. Rowohlt, Berlin.

Goll, Joachim/Hommel, Daniel (2015) *Mit Scrum zum gewünschten System*. Springer Vieweg, Wiesbaden.

Goldratt, Eliyahu M./Cox, Jeff (2013) *Das Ziel: Ein Roman über Prozessoptimierung*. Campus, Frankfurt am Main/New York.

Goleman, Daniel (1996) *EQ: Emotionale Intelligenz*. Hanser, Munich.

Gorecki, Pawel/Pautsch, Peter R. (2012) *Cardboard Engineering*. In: Productivity Management Vol. 17(2), pp. 29–31.

Gibson, James Jerome (1973) *Die Sinne und der Prozess der Wahrnehmung*. Hans Huber, Bern.

Glancey, Jonathan (2022) *Architektur: Die visuelle Geschichte*. Dorling Kindersley, Munich.

Glaserfeld, Ernst von (1996) *Radikaler Konstruktivismus: Ideen, Ergebnisse, Probleme*. Suhrkamp, Frankfurt am Main.

Gloger, Boris (2016) *Scrum: Produkte zuverlässig und schnell entwickeln*. 5th ed. Hanser, Munich.

Glosauer, Tobias (2016) *Elementar(st)e Gruppentheorie: Von den Gruppenaxiomen bis zum Homomorphiesatz*. Springer, Wiesbaden.

Goetz, Hans-Werner (2014) *Proseminar Geschichte: Mittelalter*. 4th ed. Ulmer (UTB), Stuttgart.

Grandke, Sven/Schmitt, Peter/Emmerich, Herbert (1998) *Schlüsselqualifikationen in neuen Organisationsformen: Ein Kriterienkatalog für die Praxis*. Beuth, Berlin.

Graën, Gerhard (2014) *Krisenobjekte – Anforderungen an den Sanierer und Wege aus der Krise*. In: ProjektMagazin – Das Fachportal für Projektmanagement No. 3. pp. 1–9.

Grimme, Christian/Bossek, Jakob (2018) *Einführung in die Optimierung: Konzepte, Methoden und Anwendungen*. Springer Vieweg, Wiesbaden.

Grund, Maike (2015) *Implementierung von schichtadditiven Fertigungsverfahren: Mit Fallbeispielen aus der Luftfahrtindustrie und Medizintechnik*. Springer Vieweg, Berlin/Heidelberg.

Guilford, Joy Paul/Hoepfner, Ralph (1971) *The analysis of intelligence*. McGraw-Hill, New York.

Gutenschwager, Kai/Rabe, Markus/Spieckermann, Sven/Wenzel, Sigrid (2017) *Simulation in Produktion und Logistik: Grundlagen und Anwendungen*. Springer Vieweg, Berlin.

Haberfellner, Reinhard/de Weck, Oliver L./Fricke, Ernst/Vössner, Siegfried (2018) *Systems Engineering: Grundlagen und Anwendung*. 14th ed. Orell Füssli, Zurich.

Habermas, Jürgen (1992) *Faktizität und Geltung: Beiträge zur Diskurstheorie des Rechts und des demokratischen Rechtsstaats*. Suhrkamp, Frankfurt am Main.

Hagen, Christian (2009) *Cardboard Engineering und MTM*. MTMaktuell 04, pp. 10–12.

Hahne, Markus (2019) *Systematisches Konstruieren: Praxisnah und prägnant*. Springer Vieweg, Wiesbaden.

Hahnl, Wolfgang (2015) *Praktische Methoden des Erfindens: Kreativität und Patentschutz*. Springer, Berlin/Heidelberg.

Haka, Andreas T. (2022) Die Geschichte von Verbundwerkstoffen im 19. und 20. Jahrhundert. Springer Vieweg, Wiesbaden.

Haken, Hermann (2012) *Synergetics: Introduction and Advanced Topics*. Springer, Berlin etc.

Hamann, Heiko (2019) *Schwarmintelligenz*. Springer, Berlin.

Hammer, Michael, Champy, James (2003) *Business Reengineering: Die Radikalkur für das Unter- nehmen*. 7th ed. Campus, Frankfurt am Main/New York.

Hansen, Morton T. (1999) *The Search-Transfer Problem: The Role of Weak Ties in Sharing Know- ledge across Organization Subunits*. In: Administrative Science Quarterly, No. 44, pp. 82–111. Hartmann,

Nicolai (1949) *Ethik*. 3rd ed. Walter de Gruyter, Berlin.

Harrington, David M./Chin-Newman, Christina pp. (2017) *Conscious Motivations and Adolescent Visual Artists and Creative Writers: Similarities and Differences*. In: Creativity Research Journal, Vol. 29, No. 4, pp. 442–451.

Hartmann, Peter (2019) *Mathematik für Informatiker: Ein praxisbezogenes Lehrbuch*. 7th ed. Springer Vieweg, Wiesbaden.

Haug, Christoph von (2016) *Erfolgreich im Team: Praxisnahe Anregungen für effizientes Teamcoaching und Projektarbeit*. 5th ed. Beck-Wirtschaftsberater im dtv, Munich.

Hauschildt, Jürgen/Salomo, Sören/ Schultz, Carsten/Kock, Alexander (2023) *Innovationsmanagement*. 7th ed. Vahlen, Munich.

Hegel, Georg Wilhelm Friedrich (2013) *Wissenschaft der Logik*. ed.: Holzinger, Michael. CreateSpace Independent Publishing Plattform.

Heimann, Rudi/Hofinger, Gesine (2022) *Stabsarbeit – Konzept und Formen der Umsetzung*. In: dies. (ed.) Handbuch Stabsarbeit: Führungs- und Krisenstäbe in Einsatzorganisationen, Behörden und Unternehmen. 2nd ed. Springer, Berlin. pp. 185–192.

Heinrich, Berthold/Schneider, Wolfgang (2019) *Grundlagen der Regelungstechnik: Einfache Übungen, praktische Beispiele und komplexe Aufgaben*. 5th ed. Springer Vieweg, Wiesbaden.

Heinrich, Berthold/Linke, Petra/Glöckner, Michael (2020) *Grundlagen Automatisierungstechnik: Erfassen – Steuern – Regeln*. 3rd ed. Springer Vieweg, Wiesbaden.

Hempel, Carl Gustav/Oppenheim, Paul (1948) *Studies in the Logic of Explanation in Philosophy*. In: Philosophy of Science, No. 15, pp. 135–175.

Henderson, Bruce (1968) *The Experience Curve*. In: Boston Consulting Group (ed.) Perspectives,No. 16.

Herbst, Phillip G. (1976) *Alternatives to Hierarchies*. Martinus Nijhoff Social Sciences Division, Leiden.

Hering, Ekbert/Schloske, Alexander (2022) *Quality Function Deployment (QFD): Methode zur effizienten Produktentwicklung orientiert am Kunden unter Berücksichtigung des Wettbewerbs*. Springer Vieweg, Berlin.

Hertneck, Christian/Kneuper, Ralf (2011) *Prozesse verbessern mit CMMI® for Services: Ein Praxisleitfaden mit Fallstudien*. dpunkt.verlag, Heidelberg.

Heylighen, Francis/Joslyn, Cliff (2001) *Cybernetics and Second-Order Cybernetics*. In: Meyers,

Robert A. (ed.) *Encyclopaedia of Physical Science & Technology*. 3rd ed. Academic Press, New York.

Higgs, Peter Ware (1964) *Broken symmetries, massless particles and gauge fields*. In: Physical Letters No. 12, pp. 132–133.

Hippel, Eric von (1986) *Lead Users: A Source of novel product concepts*. In: Management Science, Bd. 32, pp. 791–805.

Hirzel, Matthias/Alter, Wolfgang/Niklas, Cornelia (ed.) (2019) *Projektportfolio-Management: Strategisches und operatives Multi-Projektmanagement in der Praxis*. 4th ed. Springer Gabler, Wiesbaden.

Hofbauer, Josef/Sigmund, Karl (1984) *Evolutionstheorie und dynamische Systeme: Mathematische Aspekte der Selektion*. Paul Parey, Berlin/Hamburg.

Hoffmann, Dirk W. (2012) *Die Gödel'schen Unvollständigkeitssätze: Eine geführte Reise durch Kurt Gödels historischen Beweis*. Springer, Berlin/Heidelberg.

Hoffmann, Wilfried (2017) *Risikomanagement: Kurzanleitung*. Heft 4. 2nd ed. Springer Vieweg, Berlin.

Hofinger, Gesine (2022) *Strukturiertes Entscheiden in komplexen Lagen*. In: Hofinger, Gesine/

Heimann, Rudi (ed.) *Handbuch Stabsarbeit: Führungs- und Krisenstäbe in Einsatzorganisationen, Behörden und Unternehmen*. 2nd ed. Springer, Berlin. pp. 185–192.

Hoischen, Fritz (2020) *Technisches Zeichnen: Grundlagen, Normen, Beispiele, Darstellende Geometrie, Geometrische Produktspezifikation*. 37th ed. Cornelsen, Berlin.

Holler, Manfred J./Illing, Gerhard/Napel, Stefan (2019) *Einführung in die Spieltheorie*. 8th ed. Springer Gabler, Berlin.

Homeister, Matthias (2018) *Quantum Computing verstehen: Grundlagen – Anwendungen – Perspektiven*. 5th ed. Springer Vieweg, Wiesbaden.

Hope, Jeremy/Fraser, Robin (2003) Beyond Budgeting: Wie sich Manager aus der jährlichen Budgetierungsfalle befreien können. Schäffer-Poeschel, Stuttgart.

Hornbogen, Erhard/Eggeler, Gunther/Werner, Ewald (2019) *Werkstoffe: Aufbau und Eigenschaften von Keramik-, Metall-, Polymer- und Verbundwerkstoffen*. 12th ed. Springer Vieweg, Berlin.

Horst, Dieter/Saitz, Bernd (2000) D *ie gesetzliche Pflicht zum Risikomanagement gemäß KonTraG und ihre Auswirkungen auf integrierte Managementsysteme*. In: Ahrens, Volker/ Hofmann-Kamensky, Matthias (ed.) Integration von Managementsystemen: Ansätze für die Praxis. Vahlen, Munich. pp. 93–112.

Horváth, Péter 2011: *Controlling*. 12th ed. Vahlen, Munich.

Informationstechnikzentrum Bund (2019) *V-Modell XT Bund* (Version: 2.3). Bundesrepublik Deutschland.

Holstenkamp, Lars/Radtke, Jörg (ed.) (2018) *Handbuch Energiewende und Partizipation*. Springer VS, Wiesbaden.

Iijima, Sumio (1991) *Helical microtubules of graphitic carbon*. In: Nature. Bd. 354, pp. 56–58.

Institut für angewandte Arbeitswissenschaft e. V. (ifaa) (ed.) (2017) *Handbuch Arbeits- und Gesundheitsschutz: Praktischer Leitfaden für Klein- und Mittelunternehmen*. Springer Vieweg, Wiesbaden.

Jakoby, Walter (2021) *Projektmanagement für Ingenieure: Ein praxisnahes Lehrbuch für den systematischen Projekterfolg*. 5th ed. Springer Vieweg, Wiesbaden.

Janis, Irving Lester (1972) *Victims of groupthink*. Houghton Mifflin, Boston.

Jannaschk, Kai (2018) *Infrastruktur für ein Data Mining Design Framework: Eine Untersuchung mit Fallbeispielen*. Springer Vieweg, Wiesbaden.

Jauk, Emanuel/Benedek, Mathias/Dunst, Beate/Neubauer, Aljoscha C. (2018) *The relationship between intelligence and creativity: New support for the threshold hypothesis by means of empirical breakpoint detection*. In: Intelligence, Bd. 41, No. 4, pp. 212–221.

Jenny, Joe/Gangl, Mike/Kwolek, Rick/Melton, David/Ridenour, Nancy/Coe, Martin (2010) *Modern Methods of Systems Engineering: With an Introduction to Pattern and Model Based Methods*. CreateSpace, Scotts Valley (CA).

Jonas, Hans (1984) *Das Prinzip Verantwortung: Versuch einer Ethik für die technologische Zivilisation*. Suhrkamp, Frankfurt am Main.

Jung, Carl Gustav (2001a) *Archetypen*. Deutscher Taschenbuch, Munich.

Jung, Carl Gustav (2001b) *Typologie*. Deutscher Taschenbuch, Munich.

Kaffenberger, Rüdiger/Schulze, Sven-Olaf/Weber, Hanno (2012) *INCOSE Systems Engineering Handbuch*. Hanser, Munich.

Kahneman, Daniel/Tversky, Amos (1979) *Prospect Theory: An Analysis of Decision under Risk*. In: Econometrica, Vol. 47, No. 2, pp. 263–292.

Kahneman, Daniel (2012) *Schnelles Denken, langsames Denken*. 20th ed. Penguin, Munich.

Kairies, Peter (2017) *So analysieren Sie Ihre Konkurrenz: Konkurrenzanalyse und Benchmarking in der Praxis*. 10th ed. expert, Tübingen.

Kaldasch, Joachim (2013) *Evolutionäre Diffusion und der Produktlebenszyklus von Gebrauchsgütern*. EBC Hochschule, Berlin.

Kano, Noriaki/Seraku, F./Takahashi, F./Tsuji, pp. (1984) *Attractive Quality and Must-be Quality*. In: Journal of the Japanese Society for Quality Control. Ausg. 14, No. 2, pp. 39–48.

Kaschny, Martin/Nolden, Matthias/Schreuder, Siegfried (2015) *Innovationsmanagement im Mittel- stand: Strategien, Implementierung, Praxisbeispiele*. Springer Gabler, Wiesbaden.

Kasunik, Keith J. (2015) *Optomechanical Systems Engineering*. John Wiley & Sons, Hoboken (NJ). Keim,

Matthias (2004) *Die Prüfung des Risikomanagements im Rahmen der Abschlussprüfung*. Dissertation at the University Erlangen.

Kern, Peter/Schmauder, Martin (2005) *Einführung in den Arbeitsschutz für Studium und Betriebs- praxis*. Carl Hanser, Munich/Vienna.

Kersten, Heinrich/Klett, Gerhard (2017) *Business Continuity und IT-Notfallmanagement: Grundlagen, Methoden und Konzepte*. Springer Vieweg, Wiesbaden.

Kieser, Alfred (2014): *Human Relations-Bewegung und Organisationspsychologie*. In: Kieser, Alfred/Ebers, Mark (ed.) Organisationstheorien. 7th ed. Kohlhammer, Stuttgart. Kieser, Alfred/Ebers, Mark (ed.) (2014) Organisationstheorien. 7th ed. Kohlhammer, Stuttgart.

Kim, Yeun Joon/Zhong, Chen-Bo (2017) *Ideas rises from chaos: Information structure and creativity*. In: Organizational Behavior and Human Decision Processes. Vol. 138. Amsterdam et al. pp. 15–27.

Kirchner, Eckhard (2020) *Werkzeuge und Methoden der Produktentwicklung: Von der Idee zum erfolgreichen Produkt*. Springer Vieweg, Berlin.

Klaas, Thorsten (2002) *Logistik-Organisation: Ein konfigurationstheoretischer Ansatz zur logistikorientierten Organisationsgestaltung*. Springer-Fachmedien, Wiesbaden.

Kleist, Peter (2006) *Vier Effekte, Phänomene und Paradoxe in der Medizin*. In: Swiss Medical Forum Vol. 6, No. 46, pp. 1023–1027.

Klüver, Christina/Klüver, Jürgen/Schmidt, Jörn (2021) *Modellierung komplexer Prozesse durch naturanaloge Verfahren: Künstliche Intelligenz und Künstliches Leben*. 3rd ed. Springer Vieweg, Wiesbaden.

Knight, Frank H. (1921) *Uncertainty and Profit*. In: The Quarterly Journal of Economics, Ausg. 26, Nr. 4, Oxford University Press, Boston. pp. 682–690.

Knuppertz, Thilo/Feddern, Uwe (2011) *Prozessorientierte Unternehmensführung: Prozessmanagement ganzheitlich einführen und verankern*. Schäffer-Poeschel, Stuttgart.

Köstler, Arthur (1966) *Der göttliche Funke: Der schöpferische Akt in Kunst und Wissenschaft*. Scherz, Bern/Munich.

Kohlert, Christine Elisabeth/Cooper, Scott (2018) *Space for Creative Thinking: Design Principles for Work and Learning Environments*. 2nd ed. Callway, Munich.

Kohte, Wolfhard/Faber, Ulrich/Feldhoff, Kerstin (ed.) (2017) *Gesamtes Arbeitsschutzrecht: Arbeitsschutz, Arbeitszeit, Arbeitssicherheit, Arbeitswissenschaft*. 2nd ed. Nomos, Baden-Baden.

Koltze, Karl/Souchkov, Valeri (2017) *Systematische Innovation: TRIZ-Anwendung in der Produkt- und Prozessentwicklung*. 2nd ed. Hanser, Munich.

Konstantin, Panos (2017) *Praxisbuch Energiewirtschaft: Energieumwandlung, -transport und -beschaffung, Übertragungsnetzbau und Kernenergieausstieg*. 4th ed. Springer Vieweg, Berlin.

Kornmeier, Martin (2018) *Wissenschaftlich schreiben leicht gemacht für Bachelor, Master und Dissertation*. 8th ed. Haupt, Bern.

Kosiol, Erich (1934) *Organisation und Betriebswirtschaft*. In: DBW Vol. 27, No. 4, pp. 81–85.

Kossiakoff, Alexander/Sweet, William N./Seymour, Samuel J./Biemer, Steven M. (2011) *Systems Engineering: Principles and Practice*. John Wiley & Sons, Hoboken (NJ).

Krastev, Ivan/Holmes, Stephen (2019) *Das Licht, das erlosch: Eine Abrechnung – Der neue Kalte Krieg hat bereits begonnen*. 4th ed. Ullstein, Berlin.

Kremer, Fabian (2022) *Verhaltenssteuerung in Produktinnovationsprozessen: Gestaltung innovationsorientierter Anreizsysteme aus Controlling-Sicht*. Springer Gabler, Wiesbaden.

Kreutzer, Ralf Thomas (2022) *Praxisorientiertes Marketing: Grundlagen – Instrumente – Fallbeispiele*. 6th ed. Springer Gabler, Wiesbaden.

Krips, David (2017) *Stakeholdermanagement*. 2nd ed. Springer Vieweg, Berlin.

Krohn, Wolfgang/Küppers, Günter (1990) *Selbstreferenz und Planung*. In: Niedersen, Uwe/Pohlmann, Ludwig (ed.) Selbstorganisation: Jahrbuch für Komplexität in Natur-, Sozial- und Geisteswissenschaften. Duncker & Humbolt, Berlin.

Kuckartz, Udo (2009) *Einführung in die Computergestützte Analyse qualitativer Daten*. 3rd ed. VS für Sozialwissenschaften, Wiesbaden.

Kuhlmann, Meinhard (2009) *Theorien komplexer Systeme: Nicht-fundamental und doch unverzichtbar?* In: Bartels, Andreas/Stöckler, Manfred (ed.): Wissenschaftstheorie. Mentis, Paderborn. pp. 309–314 u. pp. 324–326.

Kuhn, Thomas Samuel (2003) *Die Struktur wissenschaftlicher Revolutionen*. 18th ed. Suhrkamp , Frankfurt am Main.

Kühnapfel, Jörg B. (2021) *Scoring und Nutzwertanalysen: Ein Leitfaden für die Praxis*. Springer Gabler, Wiesbaden.

Küpper, Hans-Ulrich (2005) *Controlling: Konzepte*. 3rd ed. Schäffer-Poeschel, Stuttgart.

Küppers, Ernst Wilhelm Udo (2019) *Eine transdisziplinäre Einführung in die Welt der Kybernetik: Grundlagen, Modelle, Theorien und Praxisbeispiele*. Springer Vieweg, Wiesbaden.

Kurzhals, Kerstin/Uude, Katrin/Sormani, Eva/Chak, Choiwai Maggie/Banze, Madlen (2022) *Das Co-Creation Toolbook: Methoden für eine erfolgreiche Kooperation zwischen Hochschule und Gesellschaft*. Springer Gabler, Wiesbaden.

Kusay-Merkle, Ursula (2021) *Agiles Projektmanagement im Berufsalltag: Für mittlere und kleine Projekte*. 2nd ed. Springer Gabler, Berlin.

Lakatos, Imre (1974) *Falsifikation und die Methodologie wissenschaftlicher Forschungsprogramme*. In: Lakatos, Imre/Musgrave, Alan (ed.) Kritik und Erkenntnisfortschritt. Vieweg, Braunschweig. pp. 89–189.

Lakoff, George/Johnson, Mark (2021) *Leben in Metaphern: Konstruktion und Gebrauch von Sprachbildern (Systemische Horizonte)*. 9th ed. Carl-Auer-Systeme, Heidelberg.

Latour, Bruno (2017) *Eine neue Soziologie für eine neue Gesellschaft*. 4th ed. Suhrkamp, Frankfurt am Main.

Laotse (1921) Tao Te King: *Das Buch des Alten vom Sinn und Leben*. Aus dem Chinesischen verdeutscht und erläutert von Richard Wilhelm. Eugen Diedrichs, Jena.

Lauer, Thomas (2019) *Change Management: Grundlagen und Erfolgsfaktoren*. 3rd ed. Springer Gabler, Berlin/Heidelberg.

Laurikainen, Kalervo V. (1985) *Wolfgang Pauli and the Copenhagen philosophy*. In: Lahti, Pekka/ Mittelstaedt, Peter (ed.) Symposium on the Foundations of Modern Physics. 60 Years of the Einstein-Podolsky-Rosen Gedankenexperiment. World Scientific, Singapur. pp. 273–287.

Lausch, Huberta (2009) *Fibonacci und die Folge(n)*. Oldenbourg Wissenschaftsverlag, Munich.

Laux, Helmut/Gillenkirch, Robert M./Schenk-Mathes, Heike Y. (2018) *Entscheidungstheorie*. 10th ed. Springer Gabler, Berlin.

Lawick-Goodall, Jane van (1971) *Tool-using in primates and other vertebrates*. In: Lehrmann, Daniel S./Hinde, Robert A./Shaw, Evelin (ed.) Advances in the study of behaviour. Vol. 3. Academic Press, New York. pp. 195-249.

Leach, Lawrence pp. (2004) *Critical Chain Project Management*. 2nd ed. Artech House, Boston/ London.

Lehmann-Waffenschmidt, Marco/Peneder, Michael (ed.) (2022) *Evolutionäre Ökonomik: Konzepte, Wegbereiter und Anwendungsfelder*. Springer, Wiesbaden.

Lenk, Hans/Ropohl, Günter (1987) *Technik und Ethik*. Reclam, Stuttgart.

Lelieveld, Johannes/Kunkel, Daniel/Lawrence, Mark G. (2012) *Global risk of radioactive fallout after major nuclear reactor accidents*. In: Atmospheric Chemistry and Physics, Vol. 12, No. 9, pp. 4245–4258.

Lewin, Kurt (1958) *Group decision and social change*. In: Newcomb, Theodore M./Hartley, Eugene L. (ed.) Readings in Social Psychology. Holt, Rinehart & Winston, New York, pp. 197–211.

Liker, Jeffrey K. (2016) *Der Toyota Weg: Erfolgsfaktor Qualitätsmanagement.* 10th ed. Finanzbuch, Munich.

Lindblom, Charles Edward (1959) *The Science of Muddling Through.* Public Administration Review 19.

Lödding, Hermann (2016) *Verfahren der Fertigungssteuerung: Grundlagen, Beschreibung und Konfiguration.* 3rd ed. Springer Vieweg, Wiesbaden.

Lomnitz, Gero (2017) *Fehlentwicklungen im Projekt rechtzeitig wahrnehmen und richtig reagieren.* In: ProjektMagazin - Das Fachportal für Projektmanagement. pp. 21–28.

Lucas, Henry C. Jr./Goh, Jie Mein (2009) *Disruptive technology: How Kodak missed the digital photography revolution.* In: The Journal of Strategic Information Systems, Vol. 18, Issue 1, pp. 46–55.

Lucht, Michael/Spangardt, Gordon (ed.) (2005) *Emissionshandel: Ökonomische Prinzipien, rechtliche Regelungen und technische Lösungen für den Klimaschutz.* Springer, Berlin/Heidelberg.

Lüer, Gerd/Spada, Hans (1990) *Denken und Problemlösen.* In: Spada, Hans (Hrsg): Lehrbuch Allgemeine Psychologie. Hans Huber, Bern. pp. 189–280.

Luhmann, Niklas (1977) *Differentiation of Society.* In: Canadian Journal of Sociology. Band 2, Nummer 1. pp. 29–53.

Luhmann, Niklas (1985) *Die Autopoiese des Bewusstseins.* In: Soziale Welt No. 36, pp. 402–446. Luhmann, Niklas (1988) Wie ist Bewusstsein an Kommunikation beteiligt? In: von Gumbrecht,

Hans Ulrich/Pfeiffer, Karl Ludwig (ed.): *Mentalität der Kommunikation.* Suhrkamp, Frankfurt am Main. pp. 884–905.

Luhmann, Niklas (1994) *Die Wirtschaft der Gesellschaft.* 8th ed. Suhrkamp, Frankfurt am Main.

Luhmann, Niklas (1994) *Liebe als Passion: Zur Codierung von Intimität.* Suhrkamp, Frankfurt am Main.

Luhmann, Niklas (1996) *Soziale Systeme: Grundriß einer allgemeinen Theorie.* Suhrkamp, Frankfurt am Main.

Luhmann, Niklas (1997a) *Die Gesellschaft der Gesellschaft.* Suhrkamp, Frankfurt am Main.

Luhmann, Niklas (1997b) *Die Moral des Risikos und das Risiko der Moral.* In: Bechmann, Gotthard (ed.) Risiko und Gesellschaft: Grundlagen und Ergebnisse interdisziplinärer Risikoforschung. 2nd ed. Westdeutscher, Opladen. pp. 327–338.

Luhmann, Niklas (2005) *Komplexität.* In: Ders. (ed.): Soziologische Aufklärung: Aufsätze zur Theorie der Gesellschaft. Vol. 2. 5th ed. VS, Frankfurt am Main. pp. 255–276.

Luhmann, Niklas (2011) *Organisation und Entscheidung.* 3rd ed. Suhrkamp, Frankfurt am Main.

Luhmann, Niklas (2018) *Zweck – Herrschaft – System: Grundbegriffe und Prämissen Max Webers.* In: Lukas, Ernst/Tacke, Veronika (ed.) Niklas Luhmann: Schriften zur Organisation, Vol. 1: Die Wirklichkeit der Organisation. Springer VS, Wiesbaden. pp. 153–184.

Lunkeit, Armin/Zimmer, Wolf (2021) *Security by Design: Security Engineering informationstechnischer Systeme.* Springer Vieweg, Berlin.

Machiavelli, Niccolò (2017) *Der Fürst.* 11th ed. Insel, Berlin.

Madauss, Bernd-J. (2017) *Projektmanagement: Theorie und Praxis aus einer Hand.* 7th ed. Springer Vieweg, Berlin.

März, Lothar/Krug, Wilfried/Rose, Oliver/Weigert, Gerald (2011) *Simulation und Optimierung in Produktion und Logistik: Praxisorientierter Leitfaden mit Fallbeispielen.* Springer, Berlin/ Heidelberg.

Mack, Oliver/Khare, Anshuman/Krämer, Andreas/Burgartz, Thomas (ed.) (2016) *Managing in a VUCA World.* Springer, Heidelberg, New York, Dordrecht, London.

Malik, Fredmund (2015) *Strategie des Managements komplexer Systeme: Ein Beitrag zur Management-Kybernetik evolutionärer Systeme.* 11th ed. Haupt, Bern.

Mandelbrot, Benoît B. (1987) *Die fraktale Geometrie der Natur.* Springer, Basel.

March, James Gardner/Simon, Herbert Alexander (1993) *Organizations.* 2n ed. Badsil Blackwell, Cambridge/Oxford.

Martens, Will/Ortmann, Günther (2014) *Organisationen in Luhmanns Systemtheorie.* In: Kieser, Alfred/Ebers, Marc (ed.) Organisationen. 7th ed. Kohlhammer-, Stuttgart. pp. 427–461.

Maturana, Humberto Romesín (1985) *Erkennen: Die Organisation und Verkörperung von Wirklichkeit: Ausgewählte Arbeiten zur biologischen Epistemologie (Wissenschaftstheorie, Wissenschaft und Philosophie).* 2n ed. Vieweg+Teubner, Wiesbaden.

Maturana, Humberto Romesín/Varela García, Francisco Javier (2018) *Der Baum der Erkenntnis: Die biologischen Wurzeln menschlichen Erkennens.* 7th ed. Fischer, Frankfurt am Main.

Matzka, Stephan (2021) *Künstliche Intelligenz in den Ingenieurwissenschaften: Maschinelles Lernen verstehen und bewerten.* Springer Vieweg, Wiesbaden.

Mayer, Horst Otto (2013) *Interview und schriftliche Befragung: Grundlagen und Methoden empirischer Sozialforschung.* 6th ed. Oldenbourg, Munich.

Mayring, Philipp (2016) *Einführung in die qualitative Sozialforschung.* 6th ed. Beltz, Weinheim/Basel.

McCarthy, John (1956) *The Inversion of Functions Defined by Turing Machines.* In: Shannon, Claude E./McCarthy, John (ed.) Automata studies. Annuals of Mathematical Studies 34 (1956), S. 177–181.

McClelland, David Clarence/Saron, Clifford/Floor, Eric (1980) *The need for power, brain norepinephrine turnover and learning.* In: Biological psychology, Bd. 10, No. 2, pp. 93–102.

McClelland, David Clarence (1987) *Human motivation.* University Press, Cambridge.

McClelland, David Clarence (1995) *Achievement motivation in relation to achivement-related recall, performance, and urine flow, a marker associated with release of vasopressin.* In: Motivation and Emotion, Bd. 19, No. 1, pp. 59–76.

McMaster, Tom/Wastell, David/Henriksen, Helle Zinner (2005) *Fooling Around: The Corporate Jester as an Effective Change Agent for Technological Innovation.* In: Baskerville, Richard/ Mathiassen, Lars/Pries-Heje, Jan/DeGross, Janice I. (ed.) Business Agility and Information Technology Diffusion. Springer, Heidelberg/New York. pp. 129–144.

Meadows, Donella/Randers, Joggen/Meadows, Dennis (2020) *Grenzen des Wachstums: Das 30- Jahre-Update – Signal zum Kurswechsel.* 6th ed. Hirzel, Stuttgart.

Meister, Martin (2011) *Soziale Koordination durch Boundary Objects am Beispiel des heterogenen Feldes der Servicerobotik*. Diss. Technische Universität Berlin.

Mertens, Peter/Barbian, Dina/Baier, Stephan (2017) *Digitalisierung und Industrie 4.0 – eine Relativierung*. Springer Vieweg, Wiesbaden.

Metzger, Wolfgang (1975) *Was ist Gestalttheorie?* In: Guss, Kurt (ed.) Gestalttheorie und Erziehung. Dr. Dietrich Steinkopff, Darmstadt.

Mikl-Horke, Gertraude (2000) *Industrie- und Arbeitssoziologie*. 5th ed. R. Oldenbourg, Munich/Vienna.

Minsky, Marvin (1988) T *he Society of Mind*. Simon & Schuster, New York.

Mintzberg, Henry (1994) *The Rise and Fall of Strategic Planning*. The Free Press, New York.

Mittelstraß, Jürgen (2003) *Transdisziplinarität: wissenschaftliche Zukunft und institutionelle Wirklichkeit*. Universitätsverlag, Konstanz.

Moen, Ronald/Norman, Clifford (2009) *Evolution of the PDCA Cycle*. In: Proceedings of the 7th ANQ Congress, Tokyo, 17.09.2009.

Mozumder, Md Ariful Islam/Sheeraz, Muhammad Mohsan/Athar, Ali/Aich, Satyabrata/ Kim, Hee-Cheol (2022) *Technology Roadmap of the Future Trend of Metaverse based on IoT, Blockchain, AI Technique, and Medical Domain Metaverse Activity*. 24th International Conference on Advanced Communication Technology (ICACT). PyeongChang Kwangwoon Do, Korea, Republic, pp. 256–261.

Müller, David (2020) *Betriebswirtschaftslehre für Ingenieure*. 3rd ed. Springer Gabler, Berlin.

Müller, Klaus-Rainer (2015) *Handbuch Unternehmenssicherheit: Umfassendes Sicherheits-, Kontiuitäts- und Risikomanagement mit System*. 3rd ed. Springer Vieweg, Wiesbaden.

Müller, Stefan/Gelbrich, Katja (2015) *Interkulturelles Marketing*. 2nd ed. Vahlen, Munich.

Müller, Markus/Hörmann, Klaus/Dittmann, Lars/Zimmer, Jörg (2016) *Automotive SPICE in der Praxis: Interpretationshilfe für Anwender und Assessoren*. 2nd ed. dpunkt.verlag, Heidelberg.

Münch, Richard (1992) Die Struktur der Moderne: *Grundmuster und differtielle Gestaltung des institutionellen Aufbaus der modernen Gesellschaften*. Suhrkamp, Frankfurt am Main.

Nachtigall, Werner (2010) *Bionik als Wissenschaft: Erkennen → Abstrahieren → Umsetzen*. Springer, Berlin/Heidelberg.

Naefe, Paul (2018) *Methodisches Konstruieren: Auf den Punkt gebracht*. 3rd ed. Springer Vieweg, Wiesbaden.

Nagel, Michael/Mieke, Christian/Teuber, Stephan (2020) *Methodenhandbuch der Betriebswirtschaft*. 2nd ed. UVK, Munich.

Nakagaki, Toshiyuki/Yamada, Hiroyasu/Tóth, Ágota (2000) *Intelligence: Maze-solving by an amoeboid organism*. In: Nature, Bd. 407, pp. 470.

Nanz, Gerd (2012) *Prozess-Alignment*. In: ZFO – Zeitschrift Führung und Organisation, Nummer 03/2012, 81. Jahrgang, Seiten 210–212.

Nassehi, Armin (2021) *Unbehagen: Theorie der überforderten Gesellschaft*. C. H. Beck, Munich.

Nerdinger, Friedemann W./Blickle, Gerhard/Schaper, Niclas (2019) Arbeits- und Organisationspsychologie. 4th ed. Springer, Berlin.

Neubauer, Michael (1999) *Krisenmanagement in Projekten – Handeln, wenn Probleme eskalieren*. Springer, Berlin/Heidelberg/New York.

Neumann-Wolff, Violetta (2022) *Systemorientierte Tiefenanalyse komplexer Organisationsprobleme: Exploration und Erprobung von vernetztem Denken und Erkundungsaufstellung im Methodenmix*. Springer Gabler, Wiesbaden.

Niemann, Jörg/Reich, Benedikt/Stöhr, Carsten (2021) *Lean Six Sigma: Methoden zur Produktionsoptimierung*. Springer Vieweg, Berlin.

Nickel, Stefan/Rebennack, Steffen/Stein, Oliver/Waldmann, Karl-Heinz (2022) *Operations Research*. 3rd ed. Springer Gabler, Berlin.

Nicolai, Alexander/Kieser, Alfred (2002) *Trotz eklatanter Erfolglosigkeit: Die Erfolgsfaktorenforschung weiter auf Erfolgskurs – Erfolgsfaktoren; Managementforschung; Methoden; Praxisrelevanz*. In: Der Betriebswirt DBW 62 (2002) 6, pp. 579–596.

Noë, Alva (2010) *Du bist nicht dein Gehirn: Eine radikale Philosophie des Bewusstseins*. Piper, Munich.

Nonaka, Ikujiro/Takeuchi, Hirotaka (2012) *Die Organisation des Wissens: Wie japanische Unternehmen eine brachliegende Ressource nutzbar machen*. 2nd ed. Frankfurt am Main/New York.

Norman, Donald Arthur (2013): *The Design of Everyday Things*. Basic Books, New York.

Nordsieck, Fritz (1931) *Grundprobleme und Grundprinzipien der Organisation des Betriebsaufbaus*. In: DBW Vol. 24, No. 6, pp. 158–162.

Norton, Michael I. (2009) *The IKEA Effekt: When Labor Leads to Love*. In: Harvard Business Review. Vol. 87, No. 2, pp. 30.

Nyamsi, Eric Aristhide (2019) *Projektmanagement mit Scrum: Tools zur Entwicklung von Software*. Springer, Wiesbaden.

Nyhuis, Peter/Wiendahl, Hans-Peter (1999) *Logistische Kennlinien: Grundlagen, Werkzeuge und Anwendungen*. Springer, Berlin/Heidelberg.

Ogden, Charles Kay/Richards, Ivor Armstrong (2013) *The Meaning of Meaning: A Study of the Influence of Language Upon Thought and of the Science of Symbolism*. Martino Fine Books, Eastford.

Olbrich, Rainer (2022) *Marketing: Eine Einführung in die marktorientierte Unternehmensführung*. 3rd ed. Springer Gabler, Berlin.

Ordóñez, Lisa D./Schweitzer, Maurice E./Galinsky, D./Bazerman, Max H. (2009) *Goals Gone Wild: The Systematic Side Effects of Overprescribing Goal Setting*. In: Academy of Management Perspectives, No. 23, pp. 6–15.

Ostrom, Elinor (1999) *Die Verfassung der Allmende: Jenseits von Staat und Markt*. Mohr Siebeck, Tübingen.

Paech, Niko (2012) *Befreiung vom Überfluss: auf dem Weg in eine Postwachstumsökonomie*. oekom, Munich.

Paetzold, Kirstin (2021) *Nutzerbedürfnisse*. In: Bender, Beate/Gericke, Kilian (ed.) Pahl/Beitz Konstruktionslehre: Methoden und Anwendung erfolgreicher Produktentwicklung. 9th ed. Springer Vieweg, Berlin.

Papageorgiou, Markos/Leibold, Marion/Buss, Martin (2015) *Optimierung: Statische, dynamische, stochastische Verfahren für die Anwendung*. 4th ed. Springer Vieweg, Berlin/Heidelberg.

Pape, Helmut (2002) *Der dramatische Reichtum der konkreten Welt: Der Ursprung des Pragmatismus im Denken von Charles pp. Pierce und William James*. Vellbrück Wissenschaft, Weilerswist.

Patzak, Gerold/Rattay, Günter (2014) *Projektmanagement: Leitfaden zum Management von Projekten, Projektportfolios und projektorientierten Unternehmen*. 6th ed. Linde, Vienna.

Parsons, Talcott (1951) *The Social System*. Routledge, London.

Parsons, Talcott (1960) *Structure and process in modern societies*. Free Press, Glencoe (IL).

Parsons, Talcott (1961) *An outline of the social system: General introduction II*. In: Ders. (ed.): Sociological theory and modern society. Free Press, New York. pp. 264–296.

Parsons, Talcott (1986) *Gesellschaften: Evolutionäre und komparative Perspektiven*. 2nd ed. Suhrkamp, Frankfurt am Main.

Parsons, Talcott (1996) *Das System moderner Gesellschaften*. 4th ed. Juventa, Munich.

Pasetta, Vesna (2005) *Modelling Foundations of Economic Property Rights Theory*. Springer, Berlin/Heidelberg/New York.

Pattee, Howard (1973) *Physical basis and origin of control*. In: Ders. (ed.) Hierarchy theory: The challenge of complex systems. Braziller, New York.

Papula, Lothar (2018) *Mathematik für Ingenieure und Naturwissenschaftler: Ein Lehr- und Arbeitsbuch für das Grundstudium*. 3 Volumes. Springer Vieweg, Wiesbaden.

Pawellek, Günther (2007) *Produktionslogistik: Planung, Steuerung, Controlling*. Hanser, Munich.

Peitgen, Hans-Otto/Jürgens, Otto/Saupe, Dietmar Hartmut (1994) *Chaos: Bausteine der Ordnung*. Klett-Kotta/Springer, Stuttgart.

Peter, Laurence Johnston/Hull, Raymond (2001) *Das Peter-Prinzip. Oder die Hierarchie der Unfähigen*. Rowohlt, Reinbek.

Pfeffer, Jeffry/Sutton, Robert I. (1999) *The Knowing-Doing-Gap: How Smart Companies Turn Knowledge into Action*. Harvard Business School Press, Boston (MA).

Pfeffer, Jeffrey (2002) *Competitive Advantage through People*. In: Henry, Jane/Mayle, David (ed.) Managing Innovation and Change. 2nd ed. Open University/Sage Publications, London/ Thousand Oaks/New Delhi.

Pichler, Roman (2008) *Scrum: Agiles Projektmanagement erfolgreich einsetzen*. dpunkt. verlag, Heidelberg.

Pinnegar, John K./Polunin, Nicholas V. C./Francour, Patrice/Badalamenti, Fabio/ Chemello, Renato/Harmelin-Vivien, Mireille L./Hereu, Bernat/Milazzo, Marco/Zabala, Mikel/ D'Anna, Giovanni/ Pipitone, Carlo (2000) *Trophic cascades in benthic marine ecosystems: lessons for fisheries and protected-area management*. In: Environmental Conservation, Bd. 27, No. 2, pp. 179–200.

Platon (2014) Der Staat. Szlezák, Thomas Alexander (ed.) *Sammlung Tusculum*. De Gruyter Akademie, Berlin.

Platz, Friedrich/Kopiez, Reinhard/Lehmann, Andreas C./Wolf, Anna (2014) *The influence of deliberate practice on musical achievement: a meta-analysis*. In: Frontiers in Psychology, Bd. 5.

Plinke, Wulff/Utzig, B. Peter (2020) *Industrielle Kostenrechnung: Eine Einführung*. 9th ed. Springer Vieweg, Berlin.

Polt, Wolfgang/Rimser, Markus (2020) *Aufstellungen mit dem Systembrett: Interventionen für Coaching, Beratung und Therapie*. Ökotopia, Münster.

Poriger, Jennifer/Brücklmeier, Markus (2022) *Stress & Kreativität*. In: Landes, Miriam/Steiner, Eberhard/Utz, Tatjana (ed.) Kreativität und Innovation in Organisationen: Impulse aus Innovationsforschung, Management, Kunst und Psychologie. Springer Gabler, Berlin.

Popper, Karl Raimund (1984) *Objektive Erkenntnis: Ein evolutionärer Entwurf*. Hoffmann und Campe, Hamburg.

Popper, Karl Raimund (1989) *Auf der Suche nach einer besseren Welt: Vorträge und Aufsätze aus dreißig Jahren*. 4th ed. Piper, Munich.

Popper, Karl Raimund (2005) *Logik der Forschung*. Collected Works, Vol. 3, hg. v. Keuth, Herbert. 11th ed. Mohr Siebeck, Tübingen.

Porter, Michael Eugene (2013) *Wettbewerbsstrategie: Methoden zur Analyse von Branchen und Konkurrenten*. 12th ed. Campus, Frankfurt am Main.

Probst, Gilbert Jean Bernard (1987) *Selbstorganisation: Ordnungsprozesse in sozialen Systemen aus ganzheitlicher Sicht*. Paul Parey, Berlin.

Prorok, Stefan (2022) *Statistische Unsicherheit in der industriellen Produktion: Grundlagen und Methoden der modernen Qualitätssicherung*. Springer Vieweg, Berlin.

Rabe, Markus/Spieckermann, Sven/Wenzel, Sigrid (2008) *Verifikation und Validierung für die Simulation in Produktion und Logistik*. Springer, Berlin/ Heidelberg.

Radjou, Navi/Prabhu, Jaideep/Ahuja, Simone (2012) *Jugaad Innovation: Think Frugal, Be Flexible, Generate Breakthrough Growth*. Josses-Bass, San Francisco.

Ratter, Beate M. W. (2006) *Komplexitätstheorie und Geographie: Ein Beitrag zur Begründung einer anderen Sicht auf Systeme*. Mitteilungen der Österreichischen Geographischen Gesellschaft, Vol. 148 (Jahresband), Vienna. pp. 109–124.

RFFA (1991) *Methodenlehre der Betriebsorganisation: Planung und Steuerung*. Part 1. Hanser, Munich.

Reibnitz, Ute von (1987) *Szenarien: Optionen für die Zukunft*. McGraw-Hill, Hamburg/New York.

Reichenbach, Hans (1938) *Experience and Prediction: An Analysis of the Foundations and the Structure of Knowledge*. Chicago.

Raiffa, Howard (1973) *Einführung in die Entscheidungstheorie*. Oldenbourg, Munich.

Reinhardt, Adolf (Adi) (2003) *Geschichten zur Simulation mit der Automobilindustrie*. In: Bayer, Johann/Collisi, Thomas/Wenzel, Sigrid (ed.) Simulation in der Automobilproduktion. Springer (VDI), Berlin/Heidelberg.

Rennert, Ines/Bundschuh, Bernhard (2013) *Signale und Systeme: Einführung in die Systemtheorie*. Carl Hanser, Munich.

Rhodes, Mel (1961) *An Analysis of Creativity*. In: The Phi Delta Kappan, Ausg. 42, No. 7, pp. 305–310. Riedl, Rupert (1987) Kultur – Spätzündung oder Evolution? Antworten auf Fragen an die Evolutions- und Erkenntnistheorie. Piper, Munich.

Rieg, Robert (2008) *Planung und Budgetierung: Was wirklich funktioniert*. 2nd ed. Gabler, Wiesbaden.

Riha, Karl (1990) *Commedia dell' arte*. 7th ed. Insel, Frankfurt am Main.

Rittel, Horst Wilhelm Jakob (1992) *Planen, Entwerfen, Design: Ausgewählte Schriften zu Theorie und Methodik (Facility Management)*. Kohlhammer, Stuttgart.

Roddeck, Werner (2019) *Bondgraphen: Modellierung und Simulation dynamischer Systeme*. Springer Vieweg, Wiesbaden.

Romeike, Frank/Stallinger, Manfred (2021) *Stochastische Szenariosimulation in der Unternehmenspraxis: Risikomodellierung, Fallstudien, Umsetzung in R*. Springer Gabler, Wiesbaden.

Ropohl, Günter (1979) *Eine Systemtheorie der Technik: Grundlegung der Allgemeinen Theorie*. Carl Hanser. Munich/Vienna/New York.

Ropohl, Günter (1987) *Neue Wege, Technik zu verantworten*. In: Lenk, Hans/ Ropohl, Günter (ed.) Technik und Ethik. Reclam, Stuttgart. pp. 149–176.

Ropohl, Günter (2009) *Allgemeine Technologie: Eine Systemtheorie der Technik*. 3rd ed. Universitätsverlag Karlsruhe. Springer Vieweg, Wiesbaden.

Roth, Gerhard (1986) *Selbstorganisation – Selbsterhaltung – Selbstreferentialität: Prinzipien der Organisation der Lebewesen und ihre Folgen für die Beziehung zwischen Organismus und Umwelt*. In: Dress, Hans/Hendrichs, Hubert/Küppers, Günter: Selbstorganisation: Die Entstehung von Ordnung in Natur und Gesellschaft. Piper, Munich/Zurich.

Roth, Gerhard (1987) *Erkenntnis und Realität: Das reale Gehirn und seine Wirklichkeit*. In: Schmidt, Siegfried J. (ed.) Der Diskus des radikalen Konstruktivismus. Suhrkamp, Frankfurt am Main.

Roth, Gerhard (1997) *Das Gehirn und seine Wirklichkeit: Kognitive Neurobiologie und ihre philosophischen Konsequenzen*. Suhrkamp, Frankfurt am Main.

Roth, Heinrich (1971) *Pädagogische Anthropologie: Entwicklung und Erziehung*. Bd. 2. Schroedel, Hannover.

Rühl, Uwe (2021) *Quick Guide: Erfolgreiches Business-Continuity-Management: Wie Sie Geschäftsunterbrechungen überleben und gestärkt in die Zukunft gehen*. Springer Gabler, Berlin.

Rump, Jutta/Schabel, Frank/Alich, David/Groh, Sibylle (2010) *Betriebliche Projektwirtschaft; Eine Vermessung*. An empirical study by the Institute for Employment and Employability (IBE) on behalf of HAYS. Mannheim.

Rupp, Chris/SOPHISTen (2021) *Requirements-Engineering-Management: Das Handbuch für Anforderungen in jeder Situation*. 7th ed. Hanser, Munich.

Russel, Stuart/Norvig, Peter (2012) *Künstliche Intelligenz: Ein moderner Ansatz*. 3rd ed. Pearson, Munich.

Sachse, Pierre (2002) *Idea materialis: Entwurfsdenken und Darstellungshandeln oder über die allmähliche Verfertigung der Gedanken beim Skizzieren und Modellieren*. Logos, Berlin.

Şanal, Ziya (2020) *Mathematik für Ingenieure: verständlich erklärt für Studium und Praxis*. 4. ed. Springer Vieweg, Wiesbaden.

Schallmo, Daniel R. A./Lang, Klaus (2020) *Design Thinking erfolgreich anwenden: So entwickeln Sie in 7 Phasen kundenorientierte Produkte und Dienstleistungen*. 2nd ed. Springer Gabler, Wiesbaden.

Schauenberg, Bernd (1978) *Zur Bedeutung der Transitivitätsvoraussetzung in der Entscheidungstheorie*. In: Zeitschrift für die gesamte Staatswissenschaft, Bd. 134, Heft 1, Mohr Siebeck, Tübingen. pp. 166–187.

Scheer, August-Wilhelm/Boczanski, Manfred/Muth, Michael/Schmitz, Willi-Gerd/ Segelbacher, Uwe (2006) *Prozessorientiertes Product Lifecycle Management*. Springer, Berlin/Heidelberg.

Scheuch, Erwin Kurt (1973) *Das Interview in der Sozialforschung*. In: König, René (ed.) Hand- buch der empirischen Sozialforschung, Bd. 3a. 3rd ed. Enke, Stuttgart. pp. 60–190.

Schiersmann, Christiane/Thiel, Heinz-Ulrich (2018) *Organisationsentwicklung: Prinzipien und Strategien von Veränderungsprozessen*. 5th ed. Springer VS, Wiesbaden.

Schimank, Uwe (2007) *Gruppen und Organisationen*. In: Joas, Hans (ed.) Lehrbuch de Soziologie. 3rd ed. Campus, Frankfurt a. M. pp. 217–240.

Schleiermacher, Friedrich Daniel Ernst (1977) *Hermeneutik und Kritik*. ed.: v. Frank, Manfred. Suhrkamp, Frankfurt am Main.

Schlick, Christopher/Bruder, Ralph/Luczak, Holger (2018) *Arbeitswissenschaft*. 4th ed. Springer Vieweg, Heidelberg.

Schlicksupp, Helmut (2004) I *deenfindung: Innovation, Kreativität und Ideenfindung*. 5th ed. Vogel, Würzburg.

Schmieder, Matthias/von Regius, Bernd/Leyendecker, Bert (2018) *Qualitätsmanagement im Einkauf: Vermeidung von Produktfehlern in der Lieferkette*. Springer Gabler, Wiesbaden.

Schneeweiß, Hans (1966) *Das Grundmodell der Entscheidungstheorie*. In: Statistische Hefte, No. 7, pp. 125–137.

Schneider, Mario (2020) *Datenanalyse für Naturwissenschaftler, Mediziner und Ingenieure*. Springer, Berlin.

Schönsleben, Paul (2016) *Integrales Logistikmanagement: Operations und Supply Chain Management innerhalb des Unternehmens und unternehmensübergreifend*. 7th ed. Springer Vieweg, Berlin/Heidelberg.

Schönwandt, Walter L./Voermanek, Katrin/Utz, Jürgen/Grunau, Jens/Hemberger, Christoph (2013) *Komplexe Probleme lösen: Ein Handbuch*. Jovis, Berlin.

Schumpeter, Joseph (1997) *Theorie der wirtschaftlichen Entwicklung*. Duncker & Humbolt, Leipzig.

Schurz, Gerhard (2014) *Einführung in die Wissenschaftstheorie*. 4th ed. Wissenschaftliche Buchgesellschaft (WBG), Darmstadt.

Schuster, Heinz Georg (1994) *Deterministisches Chaos: Eine Einführung*. VCH Verlagsgesellschaft, Weinheim.

Seifert, Katharina (1999) *Durch Umkehr zur Wende: Zehn Jahre Ökumenische Versammlung in der DDR – eine Bilanz*. Benno, Leipzig.

Seiler, Christoph (2015) *Die Diskursethik im Spannungsfeld von Systemtheorie und Differenzphilosophie: Habermas – Luhmann – Lyotard*. Springer VS, Wiesbaden.

Seiwert, Lothar (2018) *Wenn du es eilig hast, gehe langsam: wenn du es noch eiliger hast, gehe einen Umweg*. 17th ed. Campus, Frankfurt a. M./New York.

Sell, Robert/Schimweg, Ralf (2002) *Probleme lösen: In komplexen Zusammenhängen denken*. 6th ed. Springer, Berlin/Heidelberg/New York.

Selvini Palazzoli, Mara/Boscolo, Luigi/Cecchin, Gianfranco/Prata, Giuliana (2011) *Paradoxon und Gegenparadoxon: Ein neues Therapiemodell für die Familie mit schizophrener Störung*. 12th ed. Klett-Cotta, Stuttgart.

Senge, Peter Michael (2008) *Die fünfte Disziplin: Kunst und Praxis der lernenden Organisation*. Schäffer-Poeschel, Stuttgart.

Shewhart, Walter Andrew (1939) *Statistical Method from the Viewpoint of Quality Control*. Department of Agriculture, Dover.

Siebertz, Karl/von Bebber, David/Hochkirchen, Thomas (2010) *Statistische Versuchsplanung: Design of Experiments (DoE)*. Springer, Heidelberg et al.

Simon, Fritz B. (2015) *Einführung in die systemische Organisationstheorie*. 5th ed. Carl-Auer- Systeme, Heidelberg.

Simon, Herbert Alexander (1997) *Administrative behaviour*. 4th ed. Free Press, New York.

Simon, Herbert Alexander (1978) *Die Architektur der Komplexität*. In: Türk, Klaus (ed.) Handlungssysteme. Westdeutscher Verlag, Opladen. Seiten 94–120.

Simon, Walter (2009) *Managementkonzepte von A bis Z: Managementtheorien, Führungsstrategien, Führungstools*. GABAL, Offenbach am Main.

Simonis, Georg (ed.) (2013) *Konzepte und Verfahren der Technikfolgenabschätzung*. Springer VS, Wiesbaden.

Sparrer, Insa (2016) *Systemische Strukturaufstellungen: Theorie und Praxis*. 3rd ed. Carl Auer, Heidelberg.

Spencer-Brown, G. 1969 *Laws of form*. Dutton, New York.

Sprenger, Reinhard K. (2000) *Mythos Motivation: Wege aus der Sackgasse*. 13th ed. Campus, Frankfurt a. M.

Stachowiak, Herbert (1973) *Allgemeine Modelltheorie*. Springer, Vienna/New York.

Staehle, Wolfgang/Conrad, Peter/Sydow, Jörg (1999) *Management: Eine verhaltenswissenschaftliche Perspektive*. 8th ed. Franz Vahlen, Munich.

Star, Susan Leigh (2010) *This is not a Boundary Object: Reflections on the Origin of a Concept*. In: Science, Technology, & Human Values 35, pp. 601–617.

Stachura, Elisabeth (2010) *Der neurobiologische Konstruktivismus: wie lassen sich neue Erkenntnisse der Hirnforschung in den konstruktivistischen Diskurs eingliedern und welche Konsequenzen ergeben sich daraus für das menschliche Selbstverständnis?* Univ. Diss. Bremen.

Stephan, Achim (1999) Emergenz: *Von der Unvohersagbarkeit zur Selbstorganisation*. Dresden University Press, Dresden/Munich.

Stegmüller, Wolfgang (1974) *Das Problem der Induktion: Humes Herausforderung und moderne Antworten. Der sogenannte Zirkel des Verstehens*. Wissenschaftliche Buchgesellschaft, Darmstadt.

Steiner, Eberhard/Landes, Miriam (2022) *Innovation und Kreativität im Spannungsfeld von Bürokratie und Freiheit*. In: Steiner, Eberhard/Landes, Miriam/Utz, Tatjana (ed.) Kreativität und Innovation in Organisationen: Impulse aus Innovationsforschung, Management, Kunst und Psychologie. Springer Gabler, Berlin. pp. 95–110.

Stock-Homburg, Ruth (2013) *Personalmanagement: Theorien, Konzepte, Instrumente*. 3rd ed. Springer-Gabler, Wiesbaden.

Sutherland, Jeff (2015) *Die Scrum-Revolution: Management mit der bahnbrechenden Methode der erfolgreichsten Methoden*. Campus, Frankfurt am Main/New York.

Taguchi, Genichi/Chowdhury, Subir/Wu, Yuin (2004) *Taguchi's Quality Engineering Handbook*. Wiley, Hoboken.

Tattersall, Emma/Nenadic, Goran/Stevens, Robert D. (2021) *Modelling trend life cycles in scientific research using the Logistic and Gempertz equations*. In: Scientometrics: An International Journal for all Quantitative Aspects of the Science of Science, Communication in Science and Science Policy. Springer . No. 126, pp. 9113–9132.

Taylor, Frederick Winslow (2001) *The Principles of Scientific Management*. Dover Publications, New York.

Toffler, Alvin (1997) *Die dritte Welle: Perspektiven für die Gesellschaft des 21. Jahrhunderts*. Goldmann, Munich.

Thommen, Jean-Paul/Achleitner, Ann-Kristin (2006) *Allgemeine Betriebswirtschaftslehre: Umfassende Einführung aus managementorientierter Sicht*. 5th ed. Gabler, Wiesbaden.

Thonemann, Ulrich (2015) *Operations Management: Konzepte, Methoden und Anwendungen*. 3rd ed. Pearson, Hallbergmoos.

Timtschenko, Felix (2021) *Professionelles Sicherheitsmanagement für Unternehmen: Leitfaden für erfolgreiche Corporate Security*. Springer Gabler, Wiesbaden.

Tomasello, Michael (2022) *Die Ursprünge der menschlichen Kommunikation*. 5th ed. Suhrkamp, Frankfurt a. M.

Traut-Mattausch, Eva/Kerschreiter, Rudolf (2018) *Kreativitätstechniken*. In: Wastian, Monika/Braumandl, Isabell/von Rosenstiel, Lutz, West, Michael A. (ed.) Angewandte Psychologie für das Projektmanagement: Ein Praxisbuch für die erfolgreiche Projektleitung. 3rd ed. Springer, Berlin. pp. 267–288.

Tsuda, Soichiro/Zauner, Klaus-Peter/Yukio-Pegio, Gunji (2007) *Robot control with biological cells*. In: Biosystems No. 87, pp. 215–223.

Uexküll, Thure/Weisack, Wolfgang (1998) *Theorie der Humanmedizin*. 3rd ed. Urban & Schwarzenberg, Munich.

Ulrich, Hans (2001) *Die Unternehmung als produktives soziales System*. 2nd ed. Haupt, Bern u. a.

Ulrich, Helmut/Weber, Hubert (2017) *Laplace-, Fourier- und z-Transformation: Grundlagen und Anwendungen*. 10th ed. Springer Vieweg, Wiesbaden.

Vajna, Sándor (2022) *Grundlagen des Integrated Design Engineering*. In: Ders. (ed.) Integrated Design Engineering: Interdisziplinäre und ganzheitliche Produktentwicklung. 2nd ed. Springer Vieweg, Berlin, pp. 95–122.

Vartanian, Oshin/Bristol, Adam S./Kaufman, James C. (2016) *Neuroscience of Creativity*. MIT Press, Cambridge.

Vinck, Dominique/Jeantet, Alain/Laureillard, Pascal (2009) *Objects and other intermediaries in the sociotechnical process of product design: an explanatory approach*. In: Perrin, Jacques/Vinck, Dominique (ed.) The Role of Design in the Shaping of Technology. European Commission Directorate-General Science, Brüssel. pp. 297–320.

Wälder, Konrad/Wälder, Olga (2017) *Methoden zur Risikomodellierung und des Risikomanagements*. Springer Vieweg, Wiesbaden.

Wagenführ, Horst (1985) *Zukunftsforschung. In: Management-Enzyklopädie*. 2nd ed. Verlag moderne Industrie, Landsberg a. L.

Wagner, Karl Werner/Patzak, Gerold (2015) *Performance Excellence: Der Praxisleitfaden zum effektiven Prozessmanagement*. 2nd ed. Carl Hanser, Munich.

Walden, David D./Shortell, Thomas A./Roedler, Garry J./Delicado, Bernardo A./ Mornas, Odile/ Yew-Seng, Yip/Endler, David (2023) *INCOSE Systems Engineering Handbook: A Guide for Systems Life Cycle Processes and Activities*. 5th ed. Wiley, Hoboken (NJ).

Wälder, Konrad/Wälder, Olga (2017) *Methoden zur Risikomodellierung und des Risikomanagements*. Springer Vieweg, Wiesbaden.

Waltz, Kenneth Neal (2010) *Theory of International Politics*. Waveband Press, Long Grove.

Wasson, Charles pp. (2005) *Systems Engineering: Analysis, Design and Development*. John Wiley & Sons, Hoboken (NJ).

Watzlawick, Paul (ed.) (2002) *Die erfundene Wirklichkeit: Wie wissen wir, was wir zu wissen glauben? Beiträge zum Konstruktivismus*. Piper, Munich/Zurich.

Watzlawick, Paul/Weakland, John H./Frisch, Richard (2013) *Lösungen: Zur Theorie und Praxis menschlichen Wandels*. 8th ed. Hans Huber, Bern.

Wawers, Welf (2022) *Bionik: Bionisches Konstruieren verstehen und anwenden*. 2nd ed. Springer Vieweg, Wiesbaden.

Weber, Max (1904) *Die protestantische Ethik und der „Geist" des Kapitalismus*: Das Problem. In: Archiv für Sozialwissenschaft und Sozialpolitik, Nummer 20, pp. 1–54.

Weber, Max (1905) *Die protestantische Ethik und der „Geist" des Kapitalismus: Die Berufsidee des asketischen Protestantismus*. In: Archiv für Sozialwissenschaft und Sozialpolitik, Nummer 21, pp. 1–110.

Weber, Max (1972) *Wirtschaft und Gesellschaft: Grundriss einer verstehenden Soziologie*. Mohr- Siebeck, Tübingen. Nachdruck der Erstausgabe zum 50. Jahrestag des Erscheinens der Erstausgabe 1922, Tübingen.

Wegener, Robert/Fritze, Agnes/Loebbert, Michael (ed.) (2011) *Coaching Entwickeln: Forschung und Praxis im Dialog*. VS- für Sozialwissenschaften, Wiesbaden.

Weibler, Jürgen/Thielmann, Gerd (2022) *Sind Bauchentscheidungen in der Stabsarbeit möglich? Ein Plädoyer für den Wert intuitiver Lernprozesse für erfolgreiches Entscheiden*. In: Hofinger, Gesine/Heimann, Rudi (ed.): Handbuch Stabsarbeit: Führungs- und Krisenstäbe in Einsatzorganisationen, Behörden und Unternehmen. 2nd ed. Springer, Berlin.

Weick, Karl Edward (1976) *Educational Organizations as Loosely Coupled Systems*. In: Administrative Science Quarterly, Vol. 21. No. 1, pp. 1–19.

Weick, Karl Edward (2015) *Der Prozess des Organisierens*. 6th ed. Suhrkamp, Frankfurt am Main.

Weick, Karl Edward (1995) *Sensemaking in organizations*. Thousand Oaks, Sage (CA).

Weicker, Karsten (2015) *Evolutionäre Algorithmen*. 3rd ed. Springer Vieweg, Wiesbaden.

Weilkiens, Tim (2014) *Systems Engineering mit SysML/UML: Anforderungen, Analyse, Architektur*. 3rd ed. dpunkt.verlag, Heidelberg.

Weinert, Franz Emanuel (ed.) (2001) *Leistungsmessung in Schulen*. 2nd ed. Beltz, Weinheim.

Welsch, Wolfgang (2008) *Unsere postmoderne Moderne*. 7th ed. Akademie, Berlin.

Wellbrock, Wanja/Ludin, Daniela/Knezevic, Irena (2022) *Letzte Meile 4.0: Potenziale innovativer Technologien für die Auslieferung im B2C-Bereich*. Springer Gabler, Wiesbaden.

Wenzel, Sigrid/Weiß, Matthias/Collisi-Böhmer, Simone/Pitsch, Holger/Rose, Oliver (2008) *Qualitätskriterien für die Simulation in Produktion und Logistik: Planung und Durchführung von Simulationsstudien*. Springer, Berlin/Heidelberg.

Wenzel, Sigrid (2018) *Simulation logistischer Systeme*. In: Tempelmeier, Horst (ed.) Modellierung logistischer Systeme. Springer Vieweg, Berlin. pp. 1–34.

Werners, Brigitte (2008) *Grundlagen des Operations Research: Mit Aufgaben und Lösungen*. 2nd ed. Springer, Berlin/Heidelberg.

West, Michael A. (2012) *Effective Teamwork*. John Wiley & Sons, Chichester.

Westermann, Thomas (2020) *Mathematik für Ingenieure: Ein anwendungsorientiertes Lehrbuch*. 8th ed. Springer Vieweg, Berlin.

Weßels, Doris/Winters, Ruth (2010) *Top-Manager in der Verantwortung: Wege und Irrwege in Krisensituationen*. In: Projekt Magazin, No. 2/2010, pp. 1–6.

Weyrauch, Timo (2018) *Frugale Innovationen: Eine Untersuchung der Kriterien und des Vorgehens bei der Produktentwicklung*. Springer Gabler, Wiesbaden.

Whitehead, Alfred North/Russell, Bertrand (1919–1913) *Principia Mathematica*. 2nd ed. University Press, Cambridge.

Wiendahl, Hans-Peter/Ahrens, Volker (1996) *Die Chaos-Forschung hält Einzug in die Produktionswissenschaften*. In: wt-Produktion und Management, No. 86, pp. 195–201. Springer, Berlin.

Wiendahl, Hans-Peter (2014) *Betriebsorganisation für Ingenieure*. 8. ed. Hanser, Munich.

Wiener, Norbert (1961) *Cybernetics or Control and Communication in the Animal and the Machine*. 2nd ed. Cambridge, MH. MIT-Press.

Williamson, Oliver (1975) *Markets and Hierarchies: Analysis and Antitrust Implications*. Free Press, New York.

Williamson, Oliver (1985) *The economic institutions of capitalism: Firms, Markets, Relational Contracting*. Free Press, London/New York.

Willke, Helmut (1995) *Ironie des Staates: Grundlinien einer Staatstheorie polyzentrischer Gesellschaft*. 2nd ed. Suhrkamp, Frankfurt am Main.

Willke, Helmut (2004) *Einführung in das systemische Wissensmanagement*. Carl-Auer, Heidelberg.

Willke, Helmut (2005) *Systemtheorie II: Interventionstheorie*. 4th ed. Lucius & Lucius (UTB), Stuttgart.

Willke, Helmut (2006) *Systemtheorie I: Grundlagen*. 7th ed. Lucius & Lucius (UTB), Stuttgart.

Willke, Helmut (2014) *Systemtheorie III: Steuerungstheorie*. 4th ed. Lucius & Lucius (UTB), Stuttgart.

Winters, Peter R. (1960) *Forecasting Sales by Exponentially Weighted Moving Averages*. In: Management Science 6(3), pp. 324–334.

Winzer, Petra (2016) *Generic Systems Engineering: Ein methodischer Ansatz zur Komplexitätsbewältigung*. Springer Vieweg, Berlin/Heidelberg.

Wirdemann, Ralf (2017) *SCRUM mit User Stories*. 3rd ed. Hanser, Munich.

Womack, James P./Jones, Daniel T./Roose, Daniel (1992) *Die zweite Revolution in der Automobilindustrie: Konsequenzen aus der weltweiten Studie aus dem Massachusetts Institute of Technology*. 5th ed. Campus, Frankfurt a. M./New York.

Wördenweber, Burkard/Eggert, Marco/Größer, André/Wickord, Wiro (2020) *Technologie- und Innovationsmanagement im Unternehmen*. 4th ed. Springer Vieweg, Berlin.
Wolf, Joachim (2011) *Organisation, Management, Unternehmensführung: Theorien, Praxisbeispiele und Kritik*. 4th ed. Gabler, Wiesbaden.
Zacher, Serge/Reuter, Manfred (2017) *Regelungstechnik für Ingenieure: Analyse, Simulation und Entwurf von Regelkreisen*. 15th ed. Springer Vieweg, Wiesbaden.
Ylli, Klevis/Manoli, Yiamos (2022) *Energy Harvesting for Wearable Sensor Systems: Inductive Architectures for the Swing Excitation of the Leg*. Springer, Singapore.
Zimmermann, Theresa (2004) *Controlling als Steuerungsinstrument in komplexen Systemen?* In: Baecker, Dirk/Dievernich, Frank E. P./Schmidt, Thorsten (ed.) Strategien der Organisation. Ressourcen – Strukturen – Kompetenzen. Springer Fachmedien, Wiesbaden. pp. 261–330.

Standards and guidelines

ANSI/EIA-632 (2003) *Processes for Engineering a System*. American National Standards Institute (ANSI)/Electronic Industries Association (EIA), Philadelphia (PA), USA.
CAN/CSA-ISO/IEC/IEEE 24748-1:2019-06-01 *Systems and software engineering – Life cycle management, Part 1: Guide for life cycle management*. Beuth, Berlin.
DIN 863-2:1999-04 *Prüfen geometrischer Größen – Meßschrauben – Teil 2: Einbaumeßschrauben, Tiefenmeßschrauben; Begriffe, Anforderungen*, Prüfung. Beuth, Berlin.
DIN 1319-1:1995-01 *Grundlagen der Messtechnik – Teil 1: Grundbegriffe*. Beuth, Berlin.
DIN EN 1325-1:1996-11 *Value Management, Wertanalyse, Funktionsanalyse, Wörterbuch – Teil 1: Wertanalyse und Funktionsanalyse*. Beuth, Berlin.
DIN ISO 2859:2014-01 *Annahmestichprobenprüfung anhand der Anzahl fehlerhafter Einheiten oder Fehler (Attributprüfung) –Teil 1: Nach der annehmbaren Qualitätsgrenzlage (AQL) geordnete Stichprobenpläne für die Prüfung einer Serie von Losen – (ISO 2859-1:1999 + Cor. 1:2001 + Amd. 1:2011)*. Beuth, Berlin.
DIN 8580:2022-12 *Fertigungsverfahren – Begriffe, Einleitung*. Beuth, Berlin.
DIN EN ISO 9241-1:2002-02 *Ergonomische Anforderungen für Bürotätigkeiten mit Bildschirmgeräten – Teil 1: Allgemeine Einführung*. Beuth, Berlin.
DIN EN ISO 9001:2015-11 *Qualitätsmanagementsysteme: Anforderungen*. Beuth, Berlin.
DIN ISO 10007:2020-10 *Qualitätsmanagement – Leitfaden für Konfigurationsmanagement*. Beuth, Berlin.
DIN ISO 37301:2021-11 *Complience-Managementsysteme – Anforderungen mit Leitlinien zur Anwendung*. Beuth, Berlin.
DIN EN 12973:2020-05 *Value Management*. Beuth, Berlin.
DIN EN 13306:2018-02 *Instandhaltung – Begriffe der Instandhaltung*. Beuth, Berlin.
DIN EN ISO 14001:2015-11 *Umweltmanagementsysteme: Anforderungen mit Anleitung zur Anwendung*. Beuth, Berlin.
DIN EN ISO 19011:2018-10 *Leitfaden zur Auditierung von Managementsystemen*. Beuth, Berlin.
DIN ISO 21500:2016-02 *Leitlinien für das Projektmanagement*. Beuth, Berlin.

DIN EN ISO 22301:2020-06 *Sicherheit und Resilienz – Business Continuity Management System – Anforderungen*. Beuth, Berlin.

DIN EN ISO 26000:2021-04 *Leitfaden zur gesellschaftlichen Verantwortung*. Beuth, Berlin.

DIN EN ISO 26800:2011-11 *Ergonomie – Genereller Ansatz, Prinzipien, Konzepte*. Beuth, Berlin.

DIN ISO 31000:2018-10 *Risikomanagement – Leitlinien*. Beuth, Berlin.

DIN EN 31010:2010-11 *Risikomanagement – Verfahren zur Risikobeurteilung*. Beuth, Berlin.

DIN 31051:2019-06 *Grundlagen der Instandhaltung*. Beuth, Berlin.

DIN 33430:2016 *Anforderungen an berufsbezogene Eignungsdiagnostik*. Beuth, Berlin.

DIN ISO 45001:2018-11 *Managementsysteme für Sicherheit und Gesundheit bei der Arbeit: Anforderungen mit Anleitung zur Anwendung*. Beuth, Berlin.

DIN 69901-1:2009-01 *Projektmanagement: Projektmanagementsysteme – Teil 1: Grundlagen*. Beuth, Berlin.

DIN EN 61131-9:2015-02 *Speicherprogrammierbare Steuerungen: Schnittstelle für die Kommunikation mit kleinen Sensoren und Aktoren über eine Punkt-zu-Punkt-Verbindung (IEC 61131-9:2013)*. Beuth, Berlin.

DIN EN IEC 61158-1:2020-04 *Industrielle Kommunikationsnetze – Feldbusse: Überblick und Leitfaden zu den Normen der Reihen IEC 61158 und IEC 61784*. Beuth, Berlin.

DIN EN IEC 61784-1:2020:08 *Industrielle Kommunikationsnetze – Profile: Feldbusprofile*. Beuth, Berlin.

DIN EN 62026-2:2020-12 *Niederspannungsschaltgeräte – Steuerung – Geräte – Netzwerke (CDIs): Actuator Sensor Interface (AS-i)*. Beuth, Berlin.

DIN 69901-2:2009-01 *Projektmanagement: Projektmanagementsysteme – Teil 2: Prozesse, Prozessmodell*. Beuth, Berlin.

DIN 69901-3:2009-01 *Projektmanagement – Teil 3: Methoden*. Beuth, Berlin.

DIN 69901-4:2009-01 *Projektmanagement – Teil 4: Daten, Datenmodell*. Beuth, Berlin.

DIN 69901-5:2009-01 *Projektmanagement – Teil 5: Begriffe*. Beuth Verlag, Berlin.

DIN 69909-1:2013-03 *Multiprojektmanagement: Management von Projektportfolios, Programmen und Projekten – Teil 1: Grundlagen*. Beuth, Berlin.

DIN 69909-2:2013-03 *Multiprojektmanagement: Management von Projektportfolios, Programmen und Projekten – Teil 2: Prozesse, Prozessmodell*. Beuth, Berlin.

DIN 69909-4:2015-11 *Multiprojektmanagement: Management von Projektportfolios, Programmen und Projekten – Teil 4: Rollen*. Beuth, Berlin.

IEEE 1220-2005-09 *Standard for Application and Management of the Systems Engineering Process*.

DIN IEC 60050-351:2014-09 *Internationales Elektrotechnisches Wörterbuch – Teil 351: Leittechnik*. Beuth, Berlin.

ISO/IEC 9075-1:2016 *Information technology – Database languages – SQL*. International Standardization Organization (ISO), Vernier.

ISO/IEC/IEEE 15288:2023 *Systems and software engineering: Systems and life cycle process*. VDE, Berlin.

SO/IEC 15504-5:2012 *Information technology – Process assessment – Part 5: An exemplar software life cycle process assessment model*.

ISO/IEC/IEEE 20148:2018 *Systems and software engineering – Life cycle processes – Requirements Engineering*. International Standardization Organization (ISO), Vernier.

ISO/IEC 27001:2013-10 *Informationstechnik – Sicherheitsverfahren – Informationssicherheitsmanagementsysteme – Anforderungen*. Beuth, Berlin.

ISO/IEC/IEEE 42010:2011-12 *System-und-Software-Engineering – Architekturbeschreibung*. Beuth, Berlin.

VDA Band 2:2020 *Sicherung der Qualität von Lieferungen, Produktionsprozess und Produktfreigabe (PPF) 6th ed. Verband Automobilindustrie e. V. (VDA)/ Qualitätsmanagement-Center (QMC)*. Berlin.

VDA Band 3:2008 *Zuverlässigkeit bei Automobilherstellern und Lieferanten: Zuvellässigkeits-Methoden und -Hilfsmittel. Teil 2*. 3rd ed. Verband der Automobilindustrie e. V. (VDA)/ Qualitätsmanagement-Center (QMC). Berlin.

VDI 2206:2021-11 *Entwicklungsmetodik für mechatronische und cyber-physikalische Systeme*. Verein Deutscher Ingenieure (VDI), Düsseldorf.

VDI 2519 Blatt 1:2001-12 *Vorgehensweise bei der Bestellung von Lasten-/Pflichtenheft*. Verein Deutscher Ingenieure (VDI), Düsseldorf.

VDI 2800 Blatt 1:2010-08 *Wertanalyse*. Verein Deutscher Ingenieure (VDI), Düsseldorf.

VDI 2803 Blatt 1:2019-01 *Funktionsanalyse – Grundlagen und Methoden*. Verein Deutscher Ingenieure (VDI), Düsseldorf.

VDI/VDE 3550 Blatt 3:2003-02 *Computational Intelligence – Evolutionäre Algorithmen – Begriffe und Definitionen*. Verein Deutscher Ingenieure (VDI), Düsseldorf.

VDI 3633 Blatt 1:2014-12 *Simulation von Logistik-, Materialfluss- und Produktionssystemen – Grundlagen*. Verein Deutscher Ingenieure (VDI), Düsseldorf.

VDI 3633 Blatt 2:1997-12 *Simulation von Logistik-, Materialfluss- und Produktionssystemen – Lastenheft/Pflichtenheft*. Verein Deutscher Ingenieure (VDI), Düsseldorf.

VDI 3633 Blatt 6:2022-12 *Simulation von Logistik-, Materialfluss- und Produktionssystemen – Abbildung von Personen in Simulationsmodellen*. Verein Deutscher Ingenieure (VDI), Düsseldorf.

VDI 4499 Blatt 1:2008-02 *Digitale Fabrik – Grundlagen*. Verein Deutscher Ingenieure (VDI), Düsseldorf.

VDI 5600 Blatt 1:2016-10 *Fertigungsmanagementsysteme (Manufacturing Execution Systems – MES)*. Verein Deutscher Ingenieure (VDI), Düsseldorf.

VDI 6220 Blatt 1:2021-09 *Bionik – Konzeption und Strategie*. Verein Deutscher Ingenieure (VDI), Düsseldorf.

Index

V. Ahrens, *Systems Engineering & Management*, https://doi.org/10.1007/978-3-658-46875-0

The manufacturer's authorised representative in the EU is Springer Nature Customer Service Centre GmbH, Europaplatz 3, 69115 Heidelberg, Germany. If you have any concerns regarding our products, please contact ProductSafety@springernature.com

Printed and bound by CPI Group (UK) Ltd, Croydon, CR0 4YY
01/10/2026
02225958-0018